T0181062

In the last two decades, remarkable progress has been made in understanding stars. This graduate-level textbook provides a systematic, self-contained and lucid introduction to the physical processes and fundamental equations underlying all aspects of stellar astrophysics.

This timely volume provides authoritative astronomical discussions as well as rigorous mathematical derivations and illuminating explanations of the physical concepts involved. In addition to traditional topics such as stellar interiors and atmospheres, the reader is introduced to stellar winds, mass accretion, nuclear astrophysics, weak interactions, novae, supernovae, pulsars, neutron stars and black holes. A concise introduction to general relativity is also included. At the end of each chapter, exercises – more than 100 in all – and helpful hints are provided to test and develop the understanding of the student.

As the first advanced textbook on stellar astrophysics for nearly three decades, this long-awaited volume provides a thorough introduction for graduate students and an up-to-date review for researchers.

Advanced Stellar Astrophysics

Advanced Stellar Astrophysics

William K. Rose
University of Maryland College Park

CAMBRIDGE
UNIVERSITY PRESS

CAMBRIDGE UNIVERSITY PRESS
Cambridge, New York, Melbourne, Madrid, Cape Town, Singapore, São Paulo

Cambridge University Press
The Edinburgh Building, Cambridge CB2 8RU, UK

Published in the United States of America by Cambridge University Press, New York

www.cambridge.org
Information on this title: www.cambridge.org/9780521581882

© Cambridge University Press 1998

First published 1998

A catalogue record for this publication is available from the British Library

Library of Congress Cataloguing in Publication data

Rose, William K. (William Kenneth), 1935–
Stellar astrophysics/William K. Rose.
     p.   cm.
Includes bibliographical references and index.
ISBN 0 521 58188 5.
ISBN 0 521 58833 2 (pbk.)
1. Stars.   2. Astrophysics.   I. Title.
QB801.R57   1998
523.8–dc21 97–11052 CIP

ISBN 978-0-521-58188-2 hardback
ISBN 978-0-521-58833-1 paperback

Transferred to digital printing 2007

Contents

Preface

Remarkable progress in understanding stellar phenomena has occurred in recent decades. This textbook discusses in some detail those equations and physical processes that are of greatest relevance to stellar interiors and atmospheres and closely related astrophysics. Motivation for writing this book came from my own research interests and also from teaching graduate astrophysics courses, especially a course on stellar interiors at the University of Maryland. Although the text emphasizes physical principles, astronomical results and unresolved issues are also described.

Introductory material on the history of stellar astrophysics, astronomical observations, star formation and stellar evolution are given in Chapter 1, which also contains a discussion of spectroscopic binaries. Differences between single and binary star evolution have explained a number of interesting observations that are described further in later chapters.

Stellar interiors is one of the most fundamental subjects in astrophysics. Although complicated physical processes are decisive in explaining some predictions of stellar model calculations, the basic principles of stellar interiors do not require a comprehensive knowledge of them. Chapter 2 gives an introductory discussion of the physics and equations of stellar interiors. It also includes a short description of numerical methods.

Statistical physics provides the theoretical basis for much of stellar astrophysics. In Chapter 3 those aspects of statistical physics that are of greatest relevance are developed in some detail. Stellar opacities play a vital role in interpreting observations. Absorption processes are described in Chapter 4. This latter chapter includes self-contained discussions of bound–bound absorption by hydrogenic ions, bound–free absorption and free–free absorption. The basic principles of stellar atmospheres and a discussion of stellar winds are presented in Chapter 5.

Thermonuclear reactions generate most of the energy that is radiated from the photospheres of stars. They also synthesize most elements. Much of the research effort in stellar atmospheres and interiors is motivated by determining abundances and providing theoretical explanations for observations. Chapter 6 discusses low-energy nuclear reactions, some important concepts of nuclear physics and nucleosynthesis. Weak interactions are described in Chapter 7.

Although most stars are in hydrostatic equilibrium, important classes of stars such as Cepheids, RR Lyrae variables and long-period variables are unstable to radial pulsations, whereas some other stars are unstable to nonradial oscillations. Asymptotic-giant-branch stars become thermally unstable. Type II supernovae are caused by dynamical core collapse. Chapter 8 discusses the theory of stellar instabilities and gives a brief description of spherical hydrodynamics.

By definition close binary systems are binary stars in which mass transfer between the two components becomes important. Chapter 9 contains discussions of binary stars and mass accretion. Stellar rotation and the related phenomenon of meridional circulation are also included. Several topics related to stellar magnetic fields are described in Chapter 10.

The end states of stellar evolution are white dwarfs, neutron stars and black holes. White dwarfs, classical novae and supernovae are discussed in Chapter 11. The occurrence of supernovae is closely connected to the Chandrasekhar limit, which is the upper limit to the mass of a white dwarf. General relativity is essential in any treatment of neutron stars and black holes. Chapter 12 contains a concise introduction to general relativity, whereas most of Chapter 13 is concerned with neutron stars and black holes. This latter chapter emphasizes observations as well as theory. X-ray astronomy has played a particularly important role in the development of neutron-star and black hole astrophysics.

The writing of this book was substantially influenced by teaching graduate-level courses and participating on various research efforts. I am very grateful to students who took my stellar astrophysics courses and research collaborators. I also wish to thank J. Hall, S. Lehr and S. Smith for performing the difficult task of typing the manuscript.

1 Star formation and stellar evolution: an overview

1.1 A short history of stellar astrophysics

Since the Sun is a star it is probably correct to say that stellar astrophysics began with Newton's well-known explanation for the Keplerian laws of planetary motion. Although J. Goodricke observed the eclipsing binary variable Algol (β Persei) in 1782, it was not until 1803 that Sir William Herschel's observations of Castor proved that two stars revolve around each other owing to their mutual gravitational attraction.

The first measurements of stellar parallax were made by F. W. Bessel and F. G. W. Struve in 1838. F. Schlesinger revolutionized stellar distance determinations in 1903 when he introduced photographic parallaxes and thereby enabled astronomers to measure parallaxes to an accuracy of about 0.01 arc seconds. K. Schwarzschild initiated photographic photometry during the years 1904–8. Photoelectric photometry of stars began shortly after the photocell was invented in 1911.

J. Fraunhofer discovered Fraunhofer absorption lines in the solar spectrum in 1814 and subsequently observed similar lines in other stars. In 1860 Kirchhoff formulated the relationship between radiative absorption and emission of radiation which is known as Kirchhoff's law. The Doppler effect and Kirchhoff's law formed the conceptual basis of early studies of stellar atmospheres. The quantum theory of blackbody radiation was introduced by M. Planck in 1900. To a first approximation most stars radiate as blackbodies with superimposed absorption and emission lines. The modern theory of radiative transfer in stellar atmospheres was initiated in 1906 by K. Schwarzschild. The Henry Draper catalogue, which was published by A. Cannon and E. C. Pickering, is the best-known early catalogue of stars. It contained classifications for 225 000 stars. In 1888 H. A. Rowland used a diffraction grating to measure the solar spectrum and subsequently produced a table specifying the wavelengths of thousands of spectral lines.

E. Hertzsprung and H. N. Russell are credited with originating the well-known Hertzsprung–Russell diagram, in which absolute stellar magnitude is plotted along the vertical axis and spectral type or color temperature along the horizontal axis. The distinction between dwarf (main-sequence) stars and giants had been recognized as early as 1905 by Hertzsprung, but the importance of this distinction was not established until 1913, when Russell used improved parallax measurements to show that most observed stars in the solar neighborhood occupy a narrow strip in the H–R diagram known as the main sequence, whereas others are giants. It is now known that there are more white dwarfs in the vicinity of the solar system than giants.

In 1924 A. S. Eddington established the mass–luminosity relationship for main-sequence stars. His book *The Internal Constitution of the Stars*, which was published in 1926, was the first stellar interiors text. It contained the basic theory of stellar pulsations but did not identify the hydrogen and helium ionization zones as the driving mechanism for pulsations in Cepheids, RR Lyrae and long-period variables. Also in 1926 Fermi–Dirac statistics was formulated by E. Fermi and P. A. M. Dirac. In December of the same year R. H. Fowler used Fermi–Dirac statistics to show that electrons are degenerate in white-dwarf interiors. S. Chandrasekhar (1931a,b) introduced the relativistic, degenerate electron gas and proved that white dwarfs have a maximum mass, which is known as the Chandrasekhar mass limit. His book *Stellar Structure* (1939) discusses white dwarfs and early work on nuclear energy generation in stars. H. Bethe and C. H. Critchfield (1938) showed that the proton–proton reaction can proceed at appreciable rates in the solar interior. C. F. von Weizsäcker (1938) and H. Bethe (1939) introduced the CN cycle. The Sun's luminosity and that of other low-mass main-sequence stars is primarily caused by the proton–proton chain. The CN cycle is operative in massive main-sequence stars, giants and novae. M. Schwarzschild's influential textbook *Structure and Evolution of the Stars* (1957) contains early numerical solutions of the equations of stellar interiors including physically realistic red-giant models. Systematic discussions of nucleosynthesis were given by Burbidge, Burbidge, Fowler and Hoyle (B^2FH) (1957) and also in papers published by A. G. W. Cameron (1955, 1957). The terms s-process and r-process were introduced in the B^2FH review article. D. D. Clayton's 1968 textbook *Principles of Stellar Evolution and Nucleosynthesis* was the first stellar astrophysics textbook to include a thorough discussion of the thermonuclear reactions and nucleosynthesis. Other references can be found in Clayton's book.

The neutron is an unstable particle in free space; however, it is a stable particle in the atomic nucleus and also in neutron stars. L. D. Landau and F. Zwicky proposed the existence of neutron stars after J. Chadwick's discovery of the neutron in 1932. The first published calculation of neutron-star models using general relativity was by J. R. Oppenheimer and G. M. Volkoff (1939). Interest in neutron stars was practically dormant until after the discovery of nonsolar x-ray sources by R. Giaconni *et al.* (1962). The discovery of objects emitting periodic radio pulses by Hewish *et al.* (1968) and subsequent interpretation of these objects as rotating neutron stars with $\sim 10^{12}$ gauss magnetic fields generated a great deal of interest in neutron-star astrophysics. S. A. Colgate and R. H. White (1966) among others had predicted that neutron stars would be formed by supernova outbursts. The discoveries of the Crab and Vela pulsars in 1968 made the neutron-star interpretation of pulsars universally accepted by astronomers. In 1971 the first x-ray satellite, *UHURU*, discovered x-ray pulsars, which are neutron stars in close binary systems. This important discovery made it possible to estimate the masses of neutron stars. The discovery of the binary pulsar PSR 1913 + 16 and subsequent interpretation of its radio emission by R. A. Hulse and J. H. Taylor (1975) proved the existence of gravitational radiation.

On the basis of Newtonian gravity and the Newtonian corpuscular theory of light, Laplace (1795) argued that light could not escape from a sufficiently massive object of given radius. The concept of a black hole was therefore proposed long before Einstein's general theory of relativity appeared in 1915. Karl Schwarzschild (1916) derived his well-known general-relativistic solution for the metric outside a spherical mass almost immediately after Einstein's work was published. Oppenheimer and H. Snyder (1939) made the first general-relativistic collapse computation by calculating the collapse of a homogeneous, pressureless gas sphere. The so-called no-hair theorem states that the physical properties of a black hole can depend only on mass, angular momentum and charge. R. P. Kerr (1963) obtained a solution for rotating black holes. A major simplification of Kerr's work was made by R. H. Boyer and R. W. Linquist in 1967. The first identified stellar-mass black hole, Cygnus X-1, was among the approximately 20 x-ray sources with discovered optical companion stars known before the end of the 1960s.

Supernovae have played a unique role in the development of stellar astrophysics. Moreover, records of supernovae extend back to about 1300 BC. Tycho's supernova of 1572, which occurred at a distance of about 2500 pc, had a visual magnitude of approximately −4.0 and was therefore comparable in brightness to Venus at maximum light. Kepler's supernova of 1604 had a visual magnitude of about −3. The Crab nebula, which is M1 in the Messier catalogue of nebulae, was formed by a supernova observed by Chinese astronomers in 1054. Nearby supernova remnants associated with relatively recent prehistoric supernovae that must have been observable without a telescope are the Vela pulsar and surrounding nebula and the Cygnus Loop. The supernova whose stellar remnant is the Vela pulsar is believed to have occurred about 8000 BC, whereas the Cygnus Loop is about 15 000 years old. The supernova remnant and strong radio source Cassiopeia A, which is only about 300 years old, may not have been sighted because of extinction by dust. However, there has been speculation that it was observed by John Flamsteed in 1680. SN 1006 is another relatively young supernova remnant that is believed associated with a very bright star that flared up in the Spring of 1006. The well-known SN I and SN II classification of supernova light curves (see Chapter 11) is due to Minkowski (1941). Recent work has shown that SN I supernovae include two distinct classes, SN Ia and SN Ib. Supernova 1987A, which was the first supernova observable without a telescope since Kepler's supernova, will be discussed further in Chapter 11.

Although not observable without a telescope, SN 1885A (S. Andromeda), which occurred in M31 (Andromeda galaxy), is of special historical significance. Its visual spectrum was known to many astronomical observers, and in 1888 O. T. Sherman published the first supernova spectrum, the blue spectrum of SN 1885A, in the *Monthly Notices of the Royal Astronomical Society*.

1.2   The Hertzsprung–Russell (H–R) diagram

Because the distance to the Sun is known, the solar luminosity $L_\odot$ is determined from the measured solar flux ($\simeq 10^6\,\mathrm{erg\,cm^{-2}\,s^{-1}}$). The directions to neighboring stars change significantly as the Earth orbits about the Sun. The parallax of a star is half the angle subtended by the mean diameter of the Earth's orbit. Therefore, we have

$$\tan \alpha = \frac{1\,\text{AU}}{d}, \tag{1.1}$$

with α equal to the parallax in radians, d equal to the distance and $1\,\text{AU}$ equal to the astronomical unit ($1.496 \times 10^{13}\,\text{cm}$), which is the average Earth–Sun distance. A parsec (pc) is defined to be the distance at which a star's parallax equals one second of arc. Since α in Equation (1.1) is small and there are $206\,265$ seconds of arc in one radian, the stellar distance d becomes

$$d = \left(\frac{206\,265}{p''}\right) 1\,\text{AU} = \frac{1}{p''}\,\text{pc}, \tag{1.2}$$

with p'' the parallax in units of seconds of arc.

The apparent-magnitude scale is defined so that the flux ratio F_1/F_2 and corresponding apparent magnitudes m_1 and m_2 are related by the equation

$$\frac{F_1}{F_2} = 10^{0.4(m_2 - m_1)}. \tag{1.3}$$

Equation (1.3) implies that a flux ratio of 10^2 corresponds to a difference of 5 magnitudes.

The absolute magnitude M is defined so that it is equal to the apparent magnitude m if the stellar distance r is equal to 10 pc. It follows that apparent and absolute magnitude are related by the equation

$$M = m + 5 - 5\log r, \tag{1.4}$$

where r is in units of parsecs. Equation (1.4) implies that if the apparent magnitude and distance to a star are measured then the absolute magnitude is determined. Since the absolute bolometric magnitude of the Sun is 4.8, the absolute magnitude of a star of luminosity L can be expressed as

$$M - 4.8 = -2.5\log\left(\frac{L}{L_\odot}\right), \tag{1.5}$$

with $L_\odot = 3.86 \times 10^{33}\,\text{erg s}^{-1}$.

Magnitude scales are defined in different spectral bands. The visual apparent magnitude m_v (or V) is a measure of the flux received from a star through a standard visual filter. Blue (B) and ultraviolet (U) magnitude scales are also commonly used at optical wavelengths. Differences in magnitude as measured with visual (V), blue (B) and ultraviolet (U) filters determine the color temperature of a star. If a star has some visual magnitude and color temperature, then a bolometric correction (BC) is determined. The bolometric magnitude m_{bol} becomes

$$m_{bol} = m_v + BC. \tag{1.6}$$

Interstellar dust particles scatter and absorb starlight. The amount of extinction of starlight caused by dust particles depends on a star's distance and particular location. Some stars are embedded in or behind molecular clouds, which contain large amounts of dust, and consequently extinction of optical radiation is large. In the galactic plane the average amount of extinction at visual wavelengths is $\simeq 0.5$–1 magnitude per kiloparsec. Color temperatures are not always indicative of photospheric temperatures because interstellar extinction causes reddening of starlight and also because some stars are sufficiently hot that measured color temperatures are insensitive to photospheric temperatures.

Stars are classified according to their spectral types, which in order of decreasing photospheric temperature are O, B, A, F, G, K and M. The relative strengths of various absorption lines determine the particular spectral type (or class). O stars have prominent ionized helium lines (He II). Neutral helium lines are strongest in B stars, which also have prominent hydrogen Balmer-series lines ($H\alpha$, $H\beta$, $H\gamma$, ...). Balmer-series absorption lines attain their maximum strengths in A stars. He II lines disappear in the A-star range, but the relative strengths of Ca II, Fe II and other singly ionized metal lines increase. Hydrogen lines are still quite strong in F stars. However, their strengths decrease continuously from earlier- to later-type F stars, whereas the strengths of singly ionized metal lines increase. In G stars ionized metals are dominant and hydrogen lines are much weaker than in A stars. The H and K lines of Ca II, which are the most prominent absorption lines in G stars, attain maximum strengths in K stars. Neutral metallic lines (e.g. Fe I) are appreciably stronger in K stars than in G stars. Hydrogen lines are weak in K stars and CH and CN bands are prominent. At visual wavelengths the spectra of M stars are dominated by Ti O bands.

Table 1.1 gives the response function for visual (V), blue (B) and ultraviolet (U) filters. Differences in the magnitude of a star as observed through different filters give a measure of the color temperature. Figure 1.1 shows $U - B$ plotted versus $B - V$ for main-sequence stars. Notice that this relation is not similar to that of blackbody spectra. Interstellar gas between stars and Earth produce atomic and molecular absorption lines. Diffuse lines are also observed. Interstellar atomic and molecular lines have much narrower line widths than stellar lines because interstellar gas is cold and therefore the line width of a particular transition is determined by the radiative lifetime rather than the Doppler width. Table 1.2 lists some interstellar lines. The wavelengths are given in ångströms.

Figure 1.2 shows how the strengths of various spectral lines vary with spectral type. As shown in Figure 1.3, much of the continuum opacity of the K giant Aldebaran is caused by the H^- negative ion, whereas infrared continuum absorption in the atmospheres of the much cooler M giants R. Leonis and Mira is mainly the result of water-vapor bands.

The H–R diagram is a plot of stellar luminosity (or equivalently absolute magnitude) versus photospheric temperature as determined either by spectral type or by color temperature. If the luminosities and photospheric temperatures of a sample of stars in a given volume of space are plotted in the H–R diagram, then it is found that most stars occupy a narrow region that extends from relatively low luminosity and photospheric temperature to much higher luminosity and photospheric temperature. Stars within this

Table 1.1. The response function S for the UBV system as given by Matthews and Sandage (1963).

$\lambda \times 10^{-2}$ (Å)	$S(U)$	$S(B)$	$S(V)$
30	0.025		
31	0.250		
32	0.680		
33	1.137		
34	1.650		
35	2.006	0.000	
36	2.250	0.006	
37	2.337	0.080	
38	1.925	0.337	
39	0.650	1.424	
40	0.197	2.253	
41	0.070	2.806	
42	0.000	2.950	
43		3.000	
44		2.937	
45		2.780	
46		2.520	
47		2.230	
48		1.881	0.020
49		1.550	0.175
50		1.275	0.900
51		0.975	1.880
52		0.695	2.512
53		0.430	2.850
54		0.210	2.820
55		0.055	2.625
56		0.000	2.370
57			2.050
58			1.720
59			1.413
60			1.068
61			0.795
62			0.567
63			0.387
64			0.250
65			0.160
66			0.110
67			0.081
68			0.061
69			0.045
70			0.028
71			0.017
72			0.007

Table 1.2. Interstellar lines. From Jaschek and Jaschek (1987).

Atomic lines		Molecular lines		Diffuse lines
Na I	λ3302, λ3303	CH	λ3137	λ4430*
	λ5890, λ5896		λ3143	λ5780
			λ3146	λ6284
K I	λ7665		λ3878	
	λ7699		λ3886	
			λ3890	
Ca I	λ4226		λ4300*	
Ca II	λ3933			
	λ3968	CN	λ3874	
			λ3875	
Ti II	λ3073			
	λ3229	CH	λ3745	
	λ3242		λ3957	
	λ3284		λ4232*	
Fe I	λ3720			
	λ3859			

*The most important feature.

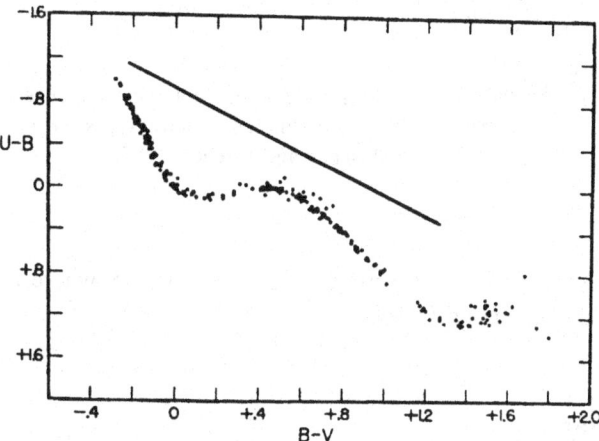

Figure 1.1. The $(U - B, B - V)$ relation for unreddened main-sequence stars. the solid line represents blackbody spectra. From Johnson and Morgan (1953).

region of the H–R diagram are known as main-sequence stars. Their luminosities are maintained by hydrogen-burning cores. Massive main-sequence stars have much higher luminosities and much shorter main-sequence lifetimes than low-mass main-sequence stars.

Figure 1.4 shows the absolute visual magnitude versus $B - V$ color temperature for nearby stars.

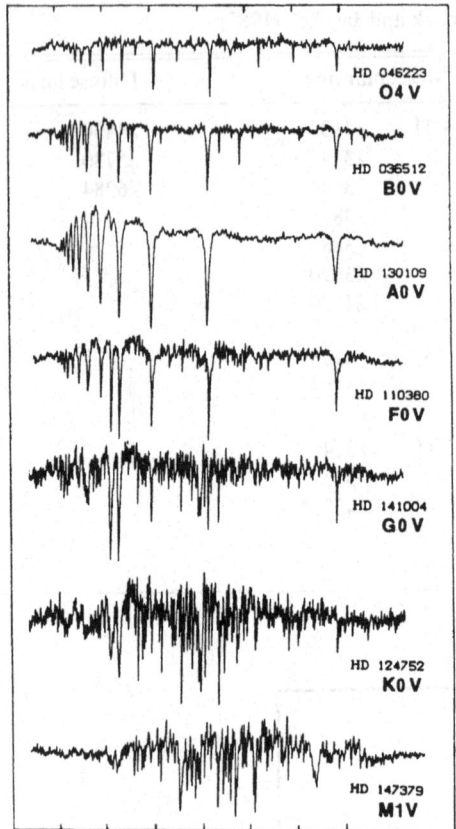

Figure 1.2. Spectra of main-sequence stars of different spectral types. From Jaschek and Jaschek (1987).

Spectral-line Doppler shifts from a binary star are shown in Figure 1.5, where it is assumed that the observer is in the plane of the orbit.

1.3   Spectroscopic binary stars

In center-of-mass coordinates the Lagrangian describing the gravitational interaction of binary stars is

$$L = \tfrac{1}{2}m(\dot{r}^2 + r^2\dot{\phi}^2) - V(r), \tag{1.7}$$

with $V(r)$ equal to the gravitational potential and $m = M_1 M_2/(M_1 + M_2)$ equal to the reduced mass derived in Equation (4.5). The Lagrangian equations of motion are

$$\frac{d}{dt}\left(\frac{\partial L}{\partial \dot{q}_i}\right) - \frac{\partial L}{\partial q_i} = 0 \quad (i = 1, 2; q_1 = r, q_2 = \phi). \tag{1.8}$$

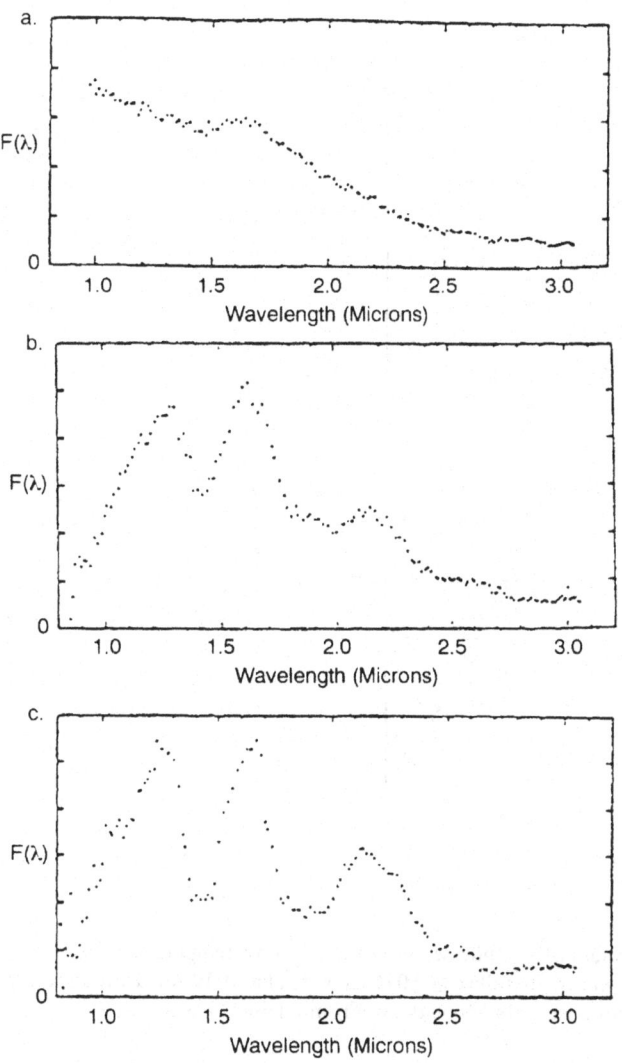

Figure 1.3. Infrared spectra of red-giant stars obtained from a balloon-borne telescope.
From Woolf, Schwarzschild and Rose (1964). a. Infrared spectrum of the K5 giant α Tauri
(Aldebaran). The intensity peak near 1.6 microns is where bound–free and free–free H⁻
absorption leaves a relatively transparent region in the continuous absorption coefficient.
b. Infrared spectrum of the M8 giant R. Leonis shows evidence for water-vapor bands at
1.4, 1.9 and 2.7 microns. c. Infrared spectrum of the M9 giant σ Ceti (Mira) shows
evidence for water-vapor bands at 1.4, 1.9 and 2.7 microns.

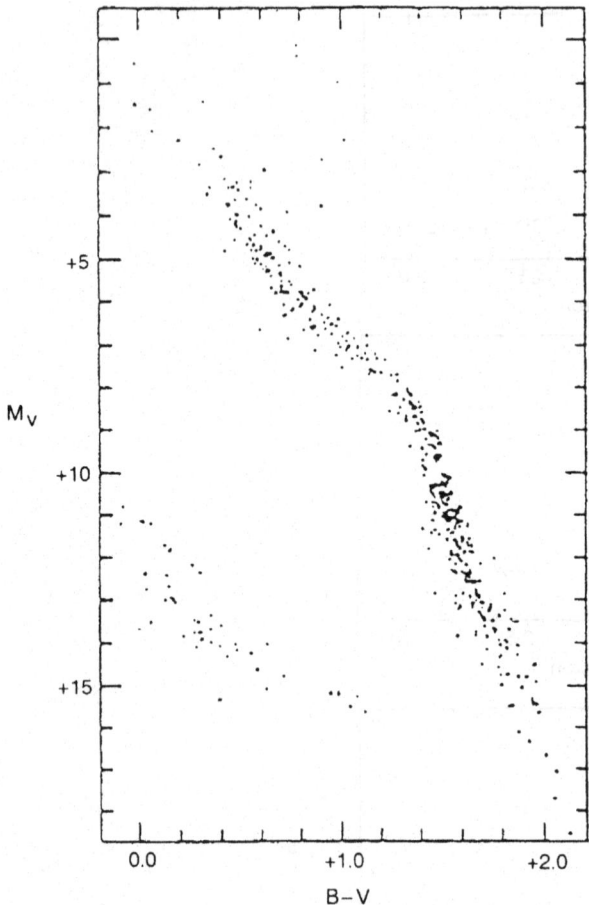

Figure 1.4. The absolute visual magnitude M_v versus $B - V$ color temperature for nearby stars (i.e. stars with trigonometric parallax $\geq$ 0.044 seconds of arc). White dwarfs are on the lower left and main-sequence stars above them. From Gleise (1978).

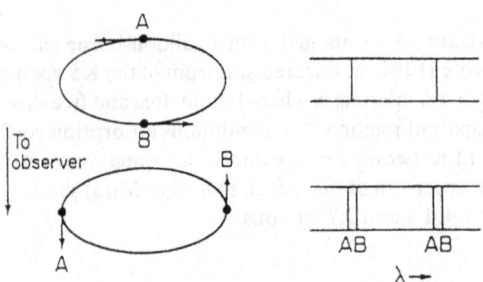

Figure 1.5. Doppler shifts of spectral lines from binary stars.

The Lagrangian L given in Equation (1.7) does not depend explicitly on ϕ, and therefore we have

$$\frac{d}{dt}\left(\frac{\partial L}{\partial \dot\phi}\right) = 0.$$
(1.9)

Equations (1.7) and (1.9) imply that

$$l = mr^2\dot\phi = \text{constant},$$
(1.10)

and therefore angular momentum is conserved. The expression $\frac{1}{2}rr\,d\phi = dA$ is the area swept out by orbital motion through angle $d\phi$. Therefore, Equation (1.10) implies that $\frac{dA}{dt} = \text{constant}.$

The total energy of binary stars in center-of-mass coordinates is

$$E = \tfrac{1}{2}m(\dot r^2 + r^2\dot\phi^2) + V(r).$$
(1.11)

Equations (1.10) and (1.11) imply that

$$\frac{dr}{dt} = \sqrt{\frac{2}{m}(E - V(r)) - \frac{l^2}{m^2 r^2}}.$$
(1.12)

From Equation (1.10) we obtain

$$d\phi = \frac{l\,dt}{mr^2}.$$
(1.13)

Using Equation (1.13) to eliminate dt from Equation (1.12) and then integrating the resultant equation gives

$$\phi = \int \frac{l\,dr}{r^2[2m(E - V(r)) - l^2/r^2]^{1/2}} + \text{constant},$$
(1.14)

with $V(r) = \dfrac{-GM_1 M_2}{r}.$

Performing the integral in Equation (1.14) leads to the equation

$$\phi = \cos^{-1}\frac{l/r - mGM_1 M_2/l}{\left[2mE + m^2\left(\dfrac{GM_1 M_2}{l}\right)^2\right]^{1/2}} + \phi_0,$$
(1.15)

with ϕ_0 equal to some constant phase angle. Equation (1.15) implies that

$$\frac{1}{r} = \frac{mGM_1 M_2}{l^2}[1 + \varepsilon\cos(\phi - \phi_0)],$$
(1.16)

with the eccentricity ε of the elliptical orbit equal to

$$\varepsilon = \left(1 + \frac{2El^2}{m(GM_1 M_2)^2}\right)^{1/2}.$$
(1.17)

From Equations (1.16) and (1.17) the semi-major axis a of the ellipse becomes

$$a = \frac{r_p + r_a}{2} = \frac{1}{2C(1 + \varepsilon)} + \frac{1}{2C(1 - \varepsilon)} = \frac{1}{C}\frac{1}{1 - \varepsilon^2}, \tag{1.18}$$

with $C = mGM_1M_2/l^2$ and r_p and r_a the periastron and apastron, which are the distances of closest approach and greatest separation, respectively, in center-of-mass coordinates. Equations (1.17) and (1.18) imply that

$$a = \frac{-GM_1M_2}{2E}. \tag{1.19}$$

From Equation (1.10) we have

$$\frac{dA}{dt} = \frac{1}{2}r^2\dot{\phi} = \frac{l}{2m}. \tag{1.20}$$

Integrating Equation (1.20) over an orbital period gives

$$\int \frac{dA}{dt}dt = A = \frac{lP}{2m}, \tag{1.21}$$

with P equal to the orbital period. The area of an ellipse is

$$A = \pi ab, \tag{1.22}$$

with the semi-minor axis $b = a\sqrt{1 - \varepsilon^2}$. From Equations (1.17) and (1.18) and the above expression for the semi-minor axis we obtain

$$b = \left(\frac{a}{C}\right)^{1/2} = a^{1/2}\left(\frac{l^2}{mGM_1M_2}\right)^{1/2}. \tag{1.23}$$

Equating the expression for A given in Equation (1.21) with that implied by Equations (1.22) and (1.23), we have

$$P^2 = \frac{4\pi^2 a^3}{G(M_1 + M_2)}. \tag{1.24}$$

Equation (1.24) is Kepler's third law of motion for elliptical orbits.

Visual binaries are binary stars sufficiently close to the solar system that separations of individual components can be measured directly. Most binary stars are too distant for positions of individual stars to be determined. However, in spectroscopic binaries, radial velocities of binary stars can be inferred from Doppler shifts of spectral lines. In some binary systems spectral-line Dopper shifts can be measured from both binary stars.

The angle of inclination i is the angle between the plane perpendicular to the line-of-sight direction and the plane of the binary orbit. Therefore, if the inclination angle i is $0°$, Doppler radial velocities are equal to zero. In general the binary orbit has an inclination angle unequal to zero and consequently measured Doppler velocities are orbital velocities projected along the line of sight, which we can choose to be the z'-axis. The line of

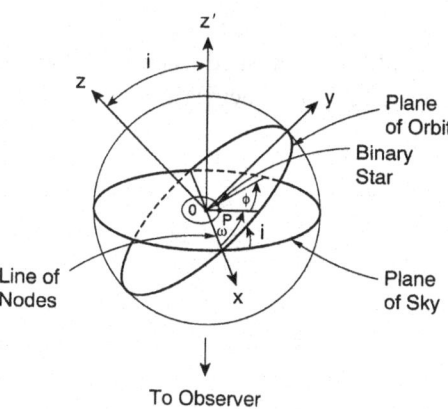

Figure 1.6. The orbit of a binary star with point O at the center of mass of the binary system and point P the periastron. The so-called line of nodes Ox is the intersection of the orbital plane with the plane perpendicular to the line of sight.

nodes is defined to be the line determined by the intersection of the plane perpendicular to the line of sight and the orbital plane. Therefore, it is perpendicular to the line-of-sight direction. We let ω equal the angle as measured in the orbital plane between the line of nodes and direction of periastron from the center of mass of the stellar orbit. The projection of the orbital radius vector onto the line-of-sight direction is (see Figure 1.6)

$$z' = r \sin(\phi + \omega) \sin i, \tag{1.25}$$

where from Equations (1.16–1.18) r satisfies

$$r = \frac{a(1 - \varepsilon^2)}{1 + \varepsilon \cos \phi}. \tag{1.26}$$

Conservation of angular momentum implies that

$$r^2 \frac{d\phi}{dt} = \frac{2\pi}{P} a^2 \sqrt{1 - \varepsilon^2}. \tag{1.27}$$

Therefore, from Equations (1.26) and (1.27) we have

$$\frac{dr}{dt} = \left(\frac{a(1 - \varepsilon^2)\varepsilon \sin \phi}{(1 + \varepsilon \cos \phi)^2} \right) \frac{2\pi a^2}{Pr^2} \sqrt{1 - \varepsilon^2}. \tag{1.28}$$

Equations (1.25), (1.27) and (1.28) imply that

$$\frac{dz'}{dt} = \left(\frac{dr}{dt} \sin(\phi + \omega) + r \cos(\phi + \omega) \frac{d\phi}{dt} \right) \sin i$$

$$= K(\cos(\phi + \omega) + \varepsilon \cos \omega), \tag{1.29}$$

with $K = \dfrac{2\pi a \sin i}{P\sqrt{1 - \varepsilon^2}}.$

The radial velocity V_R^- can be expressed as

$$V_R^- = V_c + K(\cos(\phi + \omega) + \varepsilon \cos \omega), \tag{1.30}$$

with V_c equal to the center-of-mass radial velocity. The condition that the radial velocity V_R is an extremum is $dV_R/d\phi = 0$. This condition is satisfied when $\phi + \omega = 0$ or π. It follows that the maximum and minimum orbital radial velocities are

$$V_R(\text{max}) = K(1 + \varepsilon \cos \omega) \tag{1.31}$$

and

$$V_R(\text{min}) = K(-1 + \varepsilon \cos \omega) \tag{1.32}$$

Therefore, we have

$$K = \frac{V_R(\text{max}) - V_R(\text{min})}{2} \tag{1.33}$$

and

$$\bar{V}_R = \frac{V_R(\text{max}) + V_R(\text{min})}{2} = K\varepsilon \cos \omega. \tag{1.34}$$

Equation (1.30) and the definition of ω imply that the orbital radial velocities at apastron and periastron are

$$V_R(\text{ap}) = -K \cos \omega + K\varepsilon \cos \omega \tag{1.35}$$

and

$$V_R(\text{peri}) = +K \cos \omega + K\varepsilon \cos \omega. \tag{1.36}$$

Equations (1.34–1.36) imply that

$$\bar{V}_R = \frac{V_R(\text{ap}) + V_R(\text{peri})}{2}. \tag{1.37}$$

The time of apastron and periastron differ by one-half an orbital period, and consequently the time of periastron can be determined. It follows that K, ε and $\cos \omega$ are also known, and from the definition of K given in Equation (1.29) we can calculate $a \sin i$.

Suppose that spectral-line Doppler shifts can be measured from both components of a binary system. If $x_1(x_2)$ and $y_1(y_2)$ are the periastron and apastron distances of $M_1(M_2)$ from the center of mass, then from the definition of the center of mass we have

$$M_1 x_1 = M_2 x_2,$$
$$M_1 y_1 = M_2 y_2. \tag{1.38}$$

Since the semi-major axes of M_1 and M_2 are $a_1 = (x_1 + y_1)/2$ and $a_2 = (x_2 + y_2)/2$, respectively, we obtain

$$\frac{M_1}{M_2} = \frac{a_2}{a_1}. \tag{1.39}$$

It follows directly from the conservation of angular momentum that the eccentricities ε_1

and ε_2 of the two binary stars about the center of mass are equal, and therefore, from Equation (1.29), we have

$$K_1 = \frac{2\pi a_1 \sin i}{P\sqrt{1 - \varepsilon^2}},$$

$$K_2 = \frac{2\pi a_2 \sin i}{P\sqrt{1 - \varepsilon^2}}. \tag{1.40}$$

The semi-major axis a given in Equation (1.18) is equal to $a_1 + a_2$, and therefore the sum of K_1 and K_2 given in Equation (1.40) becomes

$$K_1 + K_2 = \frac{2\pi a \sin i}{P\sqrt{1 - \varepsilon^2}}. \tag{1.41}$$

Equation (1.24) is equivalent to the equation

$$(M_1 + M_2)\sin^3 i = \frac{4\pi^2 a^3 \sin^3 i}{GP^2}. \tag{1.42}$$

Equations (1.41) and (1.42) imply that

$$(M_1 + M_2)\sin^3 i = \frac{P}{2\pi G}(1 - \varepsilon^2)^{3/2}(K_1 + K_2)^3. \tag{1.43}$$

Since $M_2 = M_1 K_1/K_2$, we obtain

$$M_1 \sin^3 i = \frac{P}{2\pi G}(1 - \varepsilon^2)^{3/2} K_2 (K_1 + K_2)^2, \tag{1.44}$$

$$M_2 \sin^3 i = \frac{P}{2\pi G}(1 - \varepsilon^2)^{3/2} K_1 (K_1 + K_2)^2. \tag{1.45}$$

Equations (1.44) and (1.45) imply that if P, ε, K_1 and K_2 are determined, then $M_1 \sin^3 i$ and $M_2 \sin^3 i$ are also known. If binary eclipses are observed, then $\sin i$ can be estimated. Because the inclination angle i is randomly oriented, it follows that if the value of $M \sin^3 i$ is obtained from measurements of a number of binary stars of similar spectral type, then we have

$$\langle M \sin^3 i \rangle = \frac{M}{2}\int_0^\pi \sin^4 i \, di = \frac{3\pi}{16} M. \tag{1.46}$$

Further discussion of binary stars is given in Chapters 9 and 13.

1.4   Star formation

Most star formation is initiated by the gravitational collapse of a large molecular cloud or region of a molecular cloud. Protostars are formed as the result of fragmentation of collapsing gas clouds. The inner region of a protostar reaches hydrostatic equilibrium first, and the velocities of the other regions of the protostar becomes supersonic with respect to the interior region. Therefore, a shock wave forms between the nearly station-ary core and infalling gas and dust. A pre-main-sequence star is formed after the completion of gravitational collapse and is in hydrostatic equilibrium. However, its central temperature is too low for hydrogen burning to occur. The luminosity of a pre-main-sequence star is maintained by gravitational contraction, and the central tempera-ture increases as a consequence of contraction. A star evolves onto the main sequence when its central temperature reaches $\simeq 10^7$ K and hydrogen burning becomes the stellar energy source.

We consider a large, massive cloud that is initially in equilibrium. The unperturbed density, temperature, pressure and gravitational potential are ρ_0, T_0, P_0 and Φ, respect-ively. The motions of the gas are governed by the equation of continuity (2.41), the equation of motion (2.26), Poisson's equation (2.50) and the equation of state of an ideal gas (2.1). Let ρ_1, P_1 and Φ_1 equal the perturbed values of the density, pressure and gravitational potential. If perturbations are assumed to occur under isothermal condi-tions, then the linearized equations describing gas motion are

$$\frac{\partial \rho_1}{\partial t} + \rho_0 \nabla \cdot v_1 = 0, \tag{1.47}$$

$$\rho_0 \frac{dv_1}{dt} = - \nabla P_1 - \rho_0 \nabla \Phi_1, \tag{1.48}$$

$$\nabla^2 \Phi_1 = 4\pi G \rho_1, \tag{1.49}$$

$$P_1 = \rho_1 \frac{kT}{\mu m_{\mathrm{H}}} = \rho_1 c_{\mathrm{s}}^2, \tag{1.50}$$

with c_{s} equal to the isothermal speed of sound. Taking the divergence of Equation (1.48) and then using Equations (1.47), (1.49) and (1.50) to eliminate v_1, P_1 and Φ_1 from the resultant equation, we obtain

$$\frac{\partial^2 \rho_1}{\partial t^2} = c_{\mathrm{s}}^2 \nabla^2 \rho_1 + 4\pi G \rho_0 \rho_1. \tag{1.51}$$

Substituting

$$\rho_1 = A e^{i(kx - st)} \tag{1.52}$$

into Equation (1.51) implies that

$$s^2 = k^2 c_{\mathrm{s}}^2 - 4\pi G \rho_0. \tag{1.53}$$

Equation (1.53) shows that plane-wave solutions to Equation (1.51) exist if s^2 satisfies the condition $s^2 > 0$. If s^2 satisfies the inequality $s^2 < 0$, then the perturbed density ρ_1 will increase exponentially as a function of time. Equation (1.53) implies that the condition for the unstable growth of perturbations (i.e. $s^2 = 0$) is

$$\lambda_J = \sqrt{\frac{\pi k T}{\mu m_H G \rho_0}}. \tag{1.54}$$

The Jeans length λ_J given by Equation (1.54) is equal to the shortest wavelength of an unstable perturbation. The corresponding Jeans mass becomes

$$M_J \simeq \rho_0 \lambda_J^3. \tag{1.55}$$

The equation of motion of a spherical gas cloud in which pressure gradients are small is

$$\frac{dv}{dt} = \frac{d^2 r}{dt^2} = -\frac{G M_r}{r^2}. \tag{1.56}$$

The velocity v can be regarded as a function of either r or M_r. Therefore, we have

$$\frac{dv}{dt} = \frac{dv}{dr}\frac{dr}{dt} = \frac{1}{2}\frac{dv^2}{dr} = -\frac{G M_r}{r^2}. \tag{1.57}$$

Integrating Equation (1.57), we obtain

$$\frac{1}{2}(v^2 - v_0^2(M_r)) = \frac{1}{2}\left[\left(\frac{dr}{dt}\right)^2 - v_0^2(M_r)\right] = G M_r\left(\frac{1}{r} - \frac{1}{r_0(M_r)}\right), \tag{1.58}$$

with $r_0(M_r)$ and $v_0(M_r)$ the initial position and velocity of M_r. We assume that the initial velocities of the various mass points M_r are equal to zero. Integrating Equation (1.58) with $v_0(M_r)$ set equal to zero gives

$$\left(\frac{2 G M_r}{r_0}\right)^{1/2} t = (r_0(M_r) - r)^{1/2} r^{1/2} + r_0(M_r)\sin^{-1}\left(\frac{r_0 - r}{r_0}\right)^{1/2}. \tag{1.59}$$

If we assume that the initial density distribution $\rho_0(M_r)$ is uniform, then Equation (1.59) becomes

$$\left(\frac{8\pi G \rho_0}{3}\right)^{1/2} t = \left(1 - \frac{r}{r_0}\right)^{1/2}\left(\frac{r}{r_0}\right)^{1/2} + \sin^{-1}\left(1 - \frac{r}{r_0}\right)^{1/2}. \tag{1.60}$$

Equation (1.60) implies that if the density distribution is uniform, then all mass points M_r reach the origin in a free-fall time equal to

$$\tau_{ff} = \left(\frac{32 G \rho_0}{3\pi}\right)^{-1/2}. \tag{1.61}$$

Numerical calculations of collapsing protostars with gas pressure included show that inner mass shells reach approximate hydrostatic equilibrium before outer mass shells. The velocities of outer mass shells become supersonic with respect to the core of the protostar, and consequently a shock wave develops at the outer core boundary.

Globules are spherical objects of stellar mass that appear to be gravitationally bound.

There is little evidence that they are undergoing collapse. However, gravitational collapse might be induced in such objects as the result of either a collision with another gas cloud or a supernova shock wave increasing the surface pressure. To a first approximation the structure of globules are isothermal gas (and dust) spheres. Although they are identified on photographic plates because of extinction by dust, globules consist mostly of molecular hydrogen and have low ($T \simeq 10\text{--}20$ K) interior temperatures.

From the ideal-gas law we have

$$P = n_{H_2}kT = \frac{kT}{\mu m_H}\rho, \tag{1.62}$$

with n_{H_2} equal to the number density of hydrogen molecules and the molecular weight $\mu = 2$ because molecular hydrogen contains two hydrogen atoms. Equations (1.62), (2.27) and (2.28) lead to the equation

$$\frac{K}{r^2}\frac{d}{dr}\left(r^2\frac{d\ln\rho}{dr}\right) = -4\pi G\rho, \tag{1.63}$$

with $K = kT/\mu m_H$.

The function ψ is defined by the equation

$$\rho = \rho_c e^{-\psi}, \tag{1.64}$$

where ρ_c is the central density. We also define the dimensionless independent variable

$$\xi = \left(\frac{K}{4\pi G\rho_c}\right)^{-1/2}r = \frac{r}{\alpha}. \tag{1.65}$$

Substituting the transformations (1.64) and (1.65) into Equation (1.63), we obtain

$$\frac{1}{\xi^2}\frac{d\left(\xi^2\frac{d\psi}{d\xi}\right)}{d\xi} = e^{-\psi}. \tag{1.66}$$

The boundary conditions required to solve Equation (1.66) are $\psi = 0$ and $d\psi/d\xi = 0$ at $\xi = 0$.

Equation (1.66) can be solved close to $\xi = 0$ by expanding ψ as a polynomial in the independent variable ξ. The boundary conditions $\psi = 0$ and $d\psi/d\xi = 0$ at $\xi = 0$ imply that the coefficients of ξ to the zeroth and first power must equal zero. Moreover, it can readily be shown that all coefficients of odd powers of ξ must also equal zero. The power-series solution of Equation (1.66) is

$$\psi = a_2\xi^2 + a_4\xi^4 + a_6\xi^6 + \ldots, \tag{1.67}$$

with $a_2 = \frac{1}{6}$, $a_4 = -\frac{1}{120}$ and $a_6 = \frac{1}{1890}$. It can be shown that the alternating power-series expansion whose first three nonzero terms are given in Equation (1.67) converges for all finite ξ.

The mass interior to some radius $r = \alpha\xi$ is

$$M(\xi) = 4\pi \int_0^r r^2 \rho \, dr = 4\pi \rho_c \alpha^3 \int_0^\xi \xi^2 e^{-\psi} d\xi$$

$$= 4\pi \rho_c \alpha^3 \int \frac{d}{d\xi} \left(\xi^2 \frac{d\psi}{d\xi} \right) d\xi = 4\pi \rho_c \alpha^3 \xi^2 \frac{d\psi}{d\xi}, \qquad (1.68)$$

where we have used Equation (1.66). The mean density interior to ξ is

$$\bar{\rho}(\xi) = \frac{4\pi \rho_c \alpha^3 \int e^{-\psi} \xi^2 d\xi}{\dfrac{4\pi}{3} \alpha^3 \xi^3} = \frac{3\rho_c}{\xi} \frac{d\psi}{d\xi}. \qquad (1.69)$$

The pressure inside an isothermal gas sphere is $P = P(\xi) = K\rho_c e^{-\psi}$, with K defined in Equation (1.63).

1.5 Stellar evolution

The Milky Way Galaxy and other spiral galaxies contain two kinematically distinct stellar populations. Population I stars are a disk population and they revolve about the Galactic center with the same sense of orbital motion. Population II stars have a spherical spatial distribution. Although they orbit the Galactic center, Population II stars have small total angular momentum because individual stars have different senses of orbital motion. Population I includes both young and old stars. However, Population II stars are old because they were formed during the same epoch as the Galaxy. The helium abundances of population I and II stars are nearly the same, whereas elements heavier than helium are appreciably more abundant in Population I stars. Extreme Population II stars have heavy-element abundances that are only about 1% as large as solar abundances. On the average, Population II stars that are relatively close to the Galactic center have higher abundances of heavy elements than those that are more distant. Both the spatial distribution and abundances of Population II stars show quite clearly that they were formed during the initial collapse of the protogalaxy. Their differing compositions reflect abundance changes that occurred during this latter epoch, which lasted for less than 10^9 years. The similar helium abundances of Population I and II stars is a consequence of the big-bang origin of the Universe.

Stars remain on the main sequence until hydrogen core burning has transformed approximately 10–15% of their mass into helium. At the present epoch, Population II main-sequence stars of $\simeq 0.8 \, M_\odot$ are evolving off the main sequence, and the age of the Galaxy can be determined from the position in the H–R diagram of the main-sequence turnoff. A comparison between the observed positions of stars evolving off the main sequence and stellar-model calculations indicates that the age of the Galaxy is $\simeq 12$–18×10^9 years. Globular clusters are nearly spherical, gravitationally bound Population II stellar systems containing $\simeq 10^5$–10^6 stars. They are particularly useful from the point of view of understanding the physical properties and evolution of Population II stars.

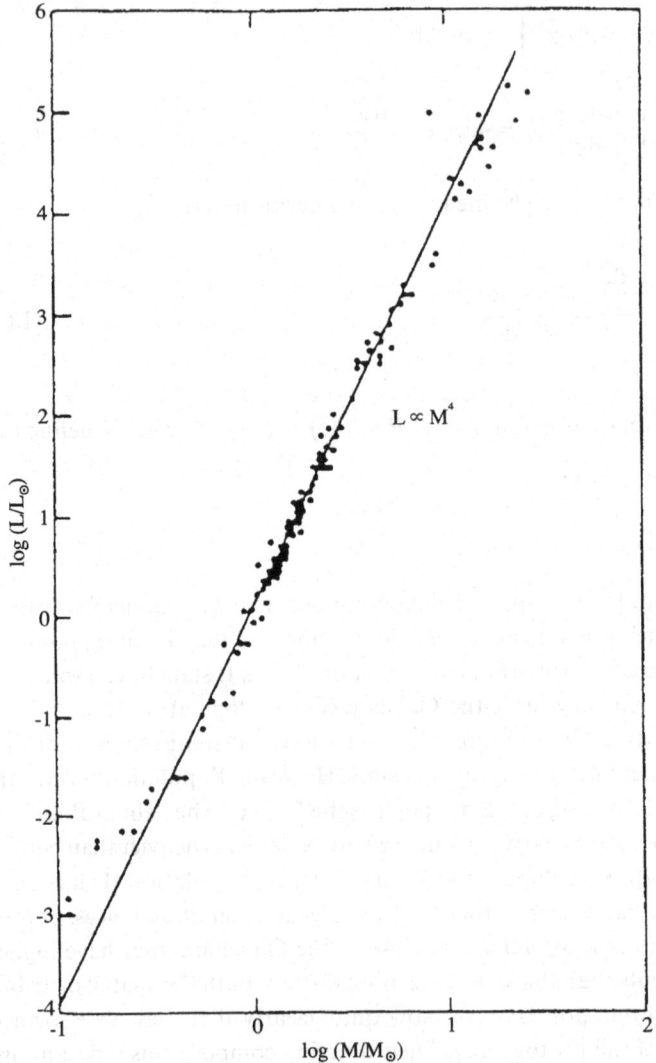

Figure 1.7. The mass–luminosity relation for stars with known masses. The figure is from Rowan-Robinson (1985) and data from Popper (1980).

Figure 1.7 shows the mass–luminosity relation for stars with known masses. As will be discussed in Chapter 2, the mass–luminosity relation should change from $L \propto M^4$ for low-mass stars to $L \propto M^3$ for massive stars, because electron scattering is the dominant opacity source.

Figure 1.8 shows the evolutionary paths in the H–R diagram of $0.5\,M_\odot$, $1\,M_\odot$ and $2\,M_\odot$ pre-main-sequence stars that are contracting toward the main sequence.

Figure 1.9 shows schematically how the Population II stars in a globular cluster are

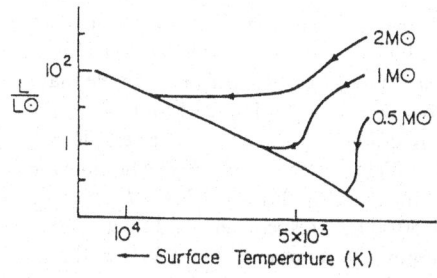

Figure 1.8. The pre-main-sequence evolutionary paths of $0.5M_\odot$, $1M_\odot$, and $2M_\odot$ stars onto the main sequence.

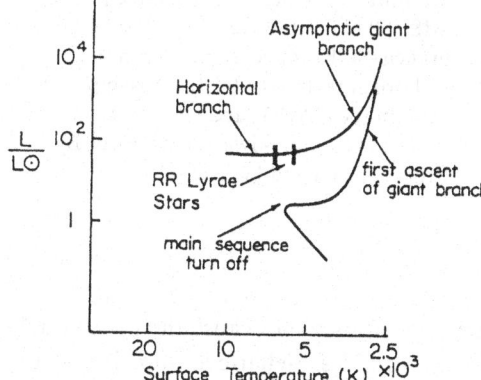

Figure 1.9. The H–R diagram of a globular cluster shown schematically.

distributed in the H–R diagram. Figure 1.10 shows the distribution of more than 10^4 stars in the color–magnitude diagram of the globular cluster M3. In Figure 1.11 the color-magnitude diagrams of several globular clusters are compared with that of the old open cluster NGC 188. The spatial distributions of globular clusters with varying iron abundances are shown in Figure 1.12. Open clusters have very different ages because they consist of Population I stars. Figure 1.13 shows the color–magnitude diagrams of a number of open clusters.

Population II stars evolve from the main sequence through a subgiant stage and then onto the red-giant branch. During their first ascent of the red-giant branch, luminosities are maintained by hydrogen-burning shells, and core contraction is concomitant with increasing luminosity. A red giant's luminosity is determined primarily by the mass of its electron-degenerate core. The core mass of a red giant increases until its central temperature reaches $\simeq 10^8$ K, which is the ignition temperature of the $3\alpha \to {}^{12}C$ (triple α) reactions. Because the pressure of an electron-degenerate gas does not depend significantly on temperature, a helium-core flash results. Although the helium-core flash is not explosive, it causes the core to undergo rapid expansion, and electron degeneracy is lifted. Population II stars evolve off the red-giant branch and onto the horizontal branch after the helium-core flash.

A horizontal-branch star has a helium-burning core and hydrogen-burning shell. Evolution on the horizontal branch proceeds slowly from lower to higher photospheric temperatures at approximately constant luminosity. RR Lyrae variables occupy a region

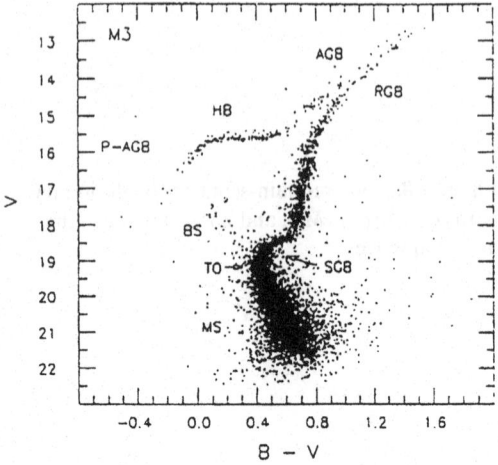

Figure 1.10. The color–magnitude diagram of the globular cluster M3. Of the 10 637 stars plotted, 9879 are from a sample that is complete down to $V = 21.5$ and the remaining 758 stars are from a sample complete down to $V = 18$. The luminosities of the two samples are $\sim 30\,000 L_\odot$ and $\sim 50\,000 L_\odot$, respectively, and their total luminosity represents about 30% of the total luminosity of M3. The various evolutionary stages are (1) main sequence (MS), (2) blue stragglers (BS), (3) subgiant branch (SGB), (4) red-giant branch (RGB), (5) horizontal branch (HB), (6) asymptotic giant branch (AGB), and (7) post-AGB evolution to white-dwarf stage (P-AGB). From Renzini and Pecci (1988).

of the horizontal branch. Hydrogen and helium ionization zones cause the pulsational instability that excites the oscillations of RR Lyrae variables. Stellar-evolutionary calculations and hydrodynamic simulations of RR Lyrae variables show that their masses are $\simeq 0.6$–$0.7\,M_\odot$ and their helium mass fractions $\simeq 25\%$. Population II stars probably lose significant amounts of mass during their first ascent of the red-giant branch because their masses at main-sequence turnoff are $\simeq 0.8\,M_\odot$.

After the completion of helium-core burning, Population II stars evolve up the red-giant branch for the second time. The luminosities of these asymptotic-branch stars are maintained by helium- and hydrogen-burning shells surrounding a degenerate carbon and oxygen core. As the core mass increases and the convective-envelope mass decreases, asymptotic-giant evolution proceeds to higher luminosity and lower photospheric temperature. A Population II star evolves off the red-giant branch and to higher photospheric temperature when the mass of the hydrogen-rich convective envelope becomes $\lesssim 0.001\,M_\odot$. Evolution to higher photospheric temperature occurs at nearly constant luminosity ($L \simeq 3$–$10 \times 10^3\,L_\odot$). The luminosity decreases after the photospheric temperature reaches $\simeq 5 \times 10^4\,K$, and the star contracts and cools until it becomes a white dwarf.

In many respects the evolution of Population II stars and low-mass ($\lesssim 1\,M_\odot$) Population I stars are similar. Population I stars have higher abundances of heavy elements than Population II stars. Therefore, for a particular stellar mass, radiation is absorbed more readily and luminosities are lower in interiors of Population I stars than in Population II stars. Although some Population I stars are nearly as old as the Galaxy, no Population I star less massive than $\simeq 0.9\,M_\odot$ has evolved off the main sequence. Low-mass main-sequence stars have convective envelopes and are therefore predicted to have stellar winds similar to the solar wind. Main-sequence stars more massive than $\simeq 1.2\,M_\odot$ burn

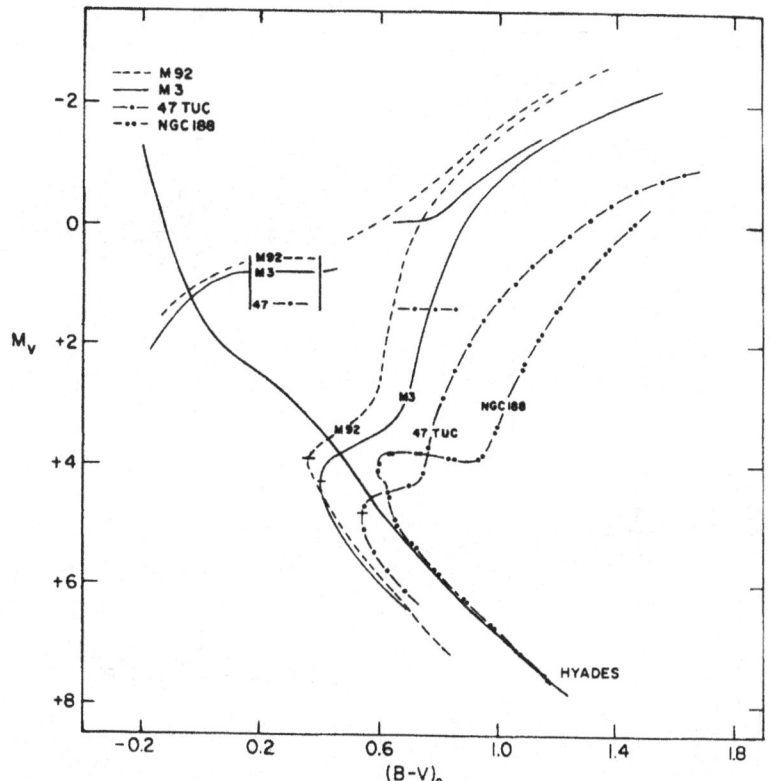

Figure 1.11. The color–magnitude diagrams for several globular clusters with different heavy-element abundances and the old ($\simeq 10^{10}$ years) open cluster NGC 188. [Fe/H] = $-$ 2 means that the iron abundance is 10^{-2} times less than solar. The normalization of absolute magnitude is obtained by setting $M_v = +0.8$ magnitudes for RR Lyrae stars in M3. From Sandage (1986).

hydrogen by means of the highly temperature-sensitive CNO cycle, and consequently they have convective cores. In less-massive main-sequence stars, hydrogen burning is completed via the proton–proton chain. Core contraction is rapid after hydrogen depletion if the CNO cycle is the principal energy source. Some clusters have a paucity of stars in the region of the H–R diagram where rapid core contraction is predicted to occur. The presence of this stellar density gap is evidence that the CNO cycle is the relevant thermonuclear energy source.

After low-mass Population I stars undergo the helium-core flash, their luminosities drop sharply to $\simeq 10^2 L_\odot$; however, unlike Population II stars, they remain on the red-giant branch as K giants during the helium-core- and hydrogen-shell-burning evolutionary stage. Old open clusters have high stellar densities in the K-giant region of the H–R diagram.

During the hydrogen- and helium-shell-burning evolutionary stage that follows helium-core exhaustion, the helium-burning shell becomes thermally unstable and helium

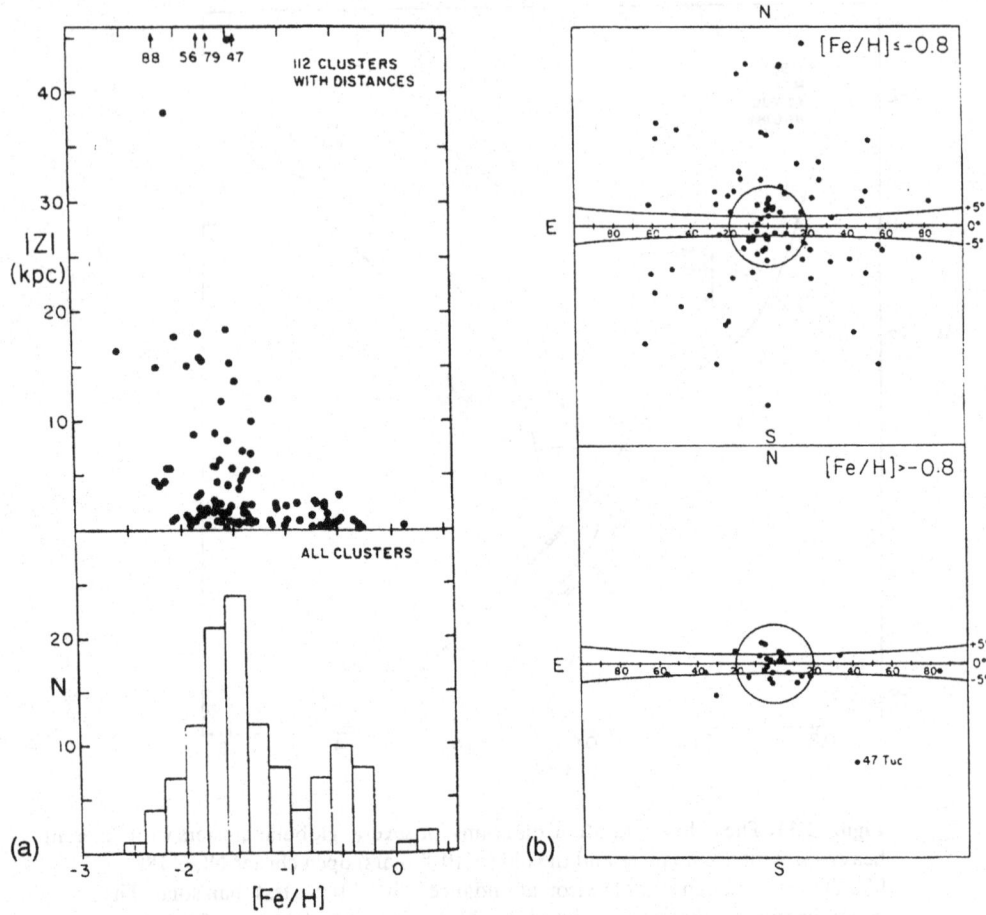

Figure 1.12. (a) Lower part of figure shows distribution of globular-cluster iron abundances and upper part gives their height above the Galactic plane. (b) Comparison of the spatial distributions of weak-lined (above) and strong-lined globular clusters. From Zinn (1985).

burning proceeds through a series of helium-shell flashes. These asymptotic-giant-branch stars become pulsationally unstable after their luminosities become $\gtrsim 2500\,L_\odot$. The pulsation periods of long-period variable stars range from $\simeq 6$ months to $\simeq 2$ years. Long-period variables experience significant mass loss. The final phase of mass loss on the red-giant branch leads to the formation of a planetary nebula. If a luminous red giant ejects a mass shell, and as a consequence the remnant star becomes nearly hydrogen deficient, then the remnant star evolves rapidly off the red-giant branch and into the region of the H–R diagram occupied by the central stars of planetary nebulae. The ejected mass shell, which is mostly molecular hydrogen soon after ejection, is ionized when the photospheric temperature of the remnant star exceeds $\simeq 30\,000$ K. The radiation from planetary nebulae is primarily in the form of emission lines such as the green lines of

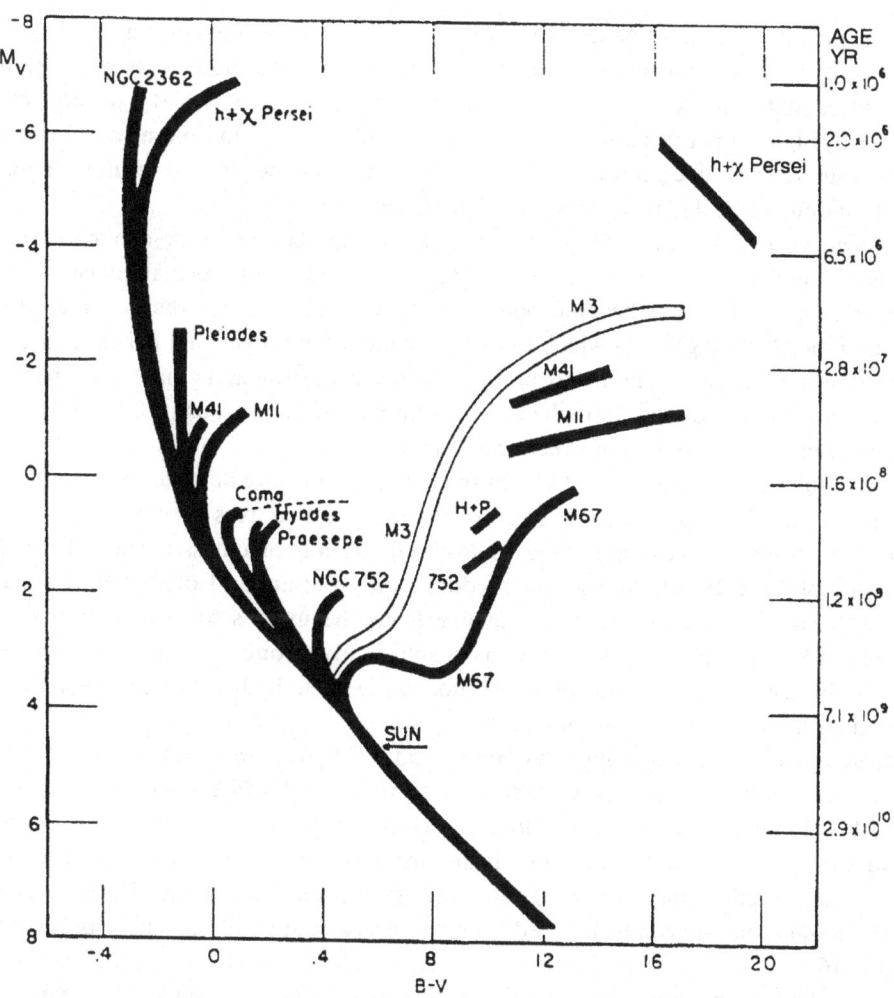

Figure 1.13. A composition H–R diagram for open clusters with various ages. NGC 2362 and the double cluster h and χ Persei are the youngest clusters ($<10^7$ years) and M67 ($\simeq 5 \times 10^9$ years) is the oldest. From Sandage (1957).

doubly ionized oxygen and hydrogen Balmer-series lines. The masses of planetary nebulae are typically 0.1–0.5 $M_\odot$. Because the central stars of planetary nebulae have exhausted their available nuclear fuel and because neutrino emission cools their degenerate cores, the luminosities of the central stars of planetary nebulae drop rapidly as they evolve into the white-dwarf end state.

Because the white-dwarf mass limit is 1.4 $M_\odot$, main-sequence stars with masses greater than 1.4 $M_\odot$ would in the absence of mass loss all produce supernova outbursts after they had exhausted their available thermonuclear energy sources. Observations of mass loss that occurs primarily on the red-giant branch and the inferred supernova rate per galaxy indicate that most main-sequence stars with masses appreciably greater than $1.4\,M_\odot$ end up as white dwarfs. Neutrino emission cools the electron-degenerate cores of red giants and consequently the cores of stars with main-sequence mass $\lesssim 8$–$9\,M_\odot$ do not reach sufficiently high temperatures for carbon burning to occur. Type Ia supernovae appear to be the result of explosive carbon burning. However, they are believed to result from mass accretion onto white dwarfs by a binary companion.

Cepheid variables, which are particularly important because they can be used as distance indicators to external galaxies, are intermediate-mass stars ($5\,M_\odot \lesssim M \lesssim 10\,M_\odot$) that become pulsationally unstable because they evolve into a region of the H–R diagram in which ionization zones of hydrogen and helium cause the longest-period oscillatory mode to be excited. Many intermediate-mass stars become carbon stars on the red-giant branch because helium-shell flashes cause ^{12}C to be dredged from their interiors into their convective envelopes.

Stars more massive than $\simeq 8$–$9\,M_\odot$ burn carbon before core densities become highly degenerate and their cores collapse after available nuclear sources have been exhausted. Type II supernovae are caused by the implosion of the core of a massive star. The stellar remnants of Type II supernovae are predominantly pulsars and other neutron stars. Although stars with mass $\lesssim 40\,M_\odot$ evolve from the main sequence onto the red-supergiant branch, more massive stars have sufficiently strong gravitational fields that their hydrogen-burning-shell sources do not cause their hydrogen-rich envelopes to expand onto the red-supergiant branch.

Stellar evolution in close binary systems is modified appreciably because mass transfer takes place between the binary stars. The formation of white dwarfs with masses $\lesssim 0.4\,M_\odot$ is one consequence of close-binary-star evolution. Nova outbursts occur because of mass accretion onto white dwarfs from binary companions. Compact x-ray sources are caused by mass accretion onto neutron stars and black holes. The existence of gravitational radiation can be inferred from the orbital motion of the binary pulsar PSR 1913 + 16.

Figure 1.14 shows how the oxygen/iron abundance ratio in stars varies with the iron/hydrogen ratio. Type II supernovae, which occur in massive stars, return large amounts of oxygen as well as some iron to the interstellar medium. Figure 1.14 indicates that Type Ia supernovae, which synthesize more iron but appreciably less oxygen than Type II supernovae, did not occur frequently until the Galactic iron abundance exceeded about 10% of the solar value.

Big-bang nucleosynthesis produces the isotopes D, ^{3}He and ^{7}Li as well as ^{4}He. Figure 1.15 gives the predicted synthesis of these isotopes for different assumed values for the average baryonic mass density of the Universe. The isotope D is destroyed when interstellar gas is recycled through stars because it burns at relatively low temperature. A reliable comparison between the D abundance at large redshifts and its present interstellar value would give an estimate of the fraction of interstellar gas that has been ejected from stars. Some ^{7}Li is synthesized in giants as well as by big-bang nucleosynthesis and cosmic-ray spallation reactions. It is formed by the nuclear reaction $^{3}He + {}^{4}He \rightarrow {}^{7}Li + p$. We note that ^{7}Li as well as D is depleted in the Sun.

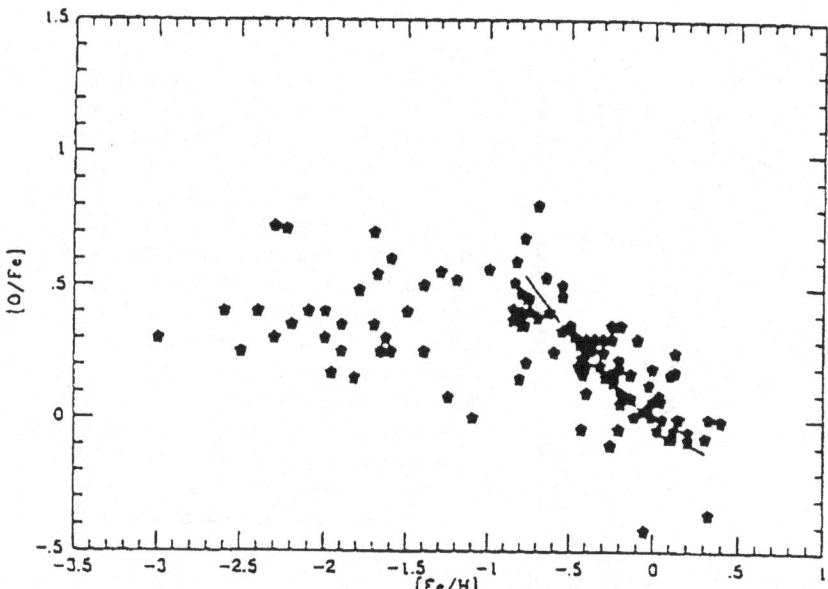

Figure 1.14. A compilation of oxygen-to-iron element-ratio measurements from the literature plotted as a function of the iron-to-hydrogen ratio. The existence of a change in the slope of the [O/Fe] versus [Fe/H] relationship indicates that the Galaxy was too young to have many type Ia supernovae when the Fe abundance was below [Fe/H] $\simeq -1$. From Gilmore, King and van der Kruit (1990).

In Figure 1.16 the distributions of H I, CO and interstellar dust are shown as a function of distance from the Galactic center. The CO molecule is a tracer of H_2, and therefore Figure 1.16 implies that the ratio atomic hydrogen/molecular hydrogen is much larger in the outer than the inner Galaxy.

1.6   Star-forming regions and very low-mass stars

Photographs of the Milky Way reveal bright regions of nebulosity and dark regions that are known to consist of cool gas and dust. The amount of interstellar extinction (i.e. scattering and absorption of starlight) depends on galactic latitude and longitude. It is caused by interstellar dust particles whose radii are typically 0.01–0.3 μm. The dust-to-gas mass ratio of about 0.01 is believed to be approximately the same in most interstellar clouds. Atomic hydrogen is found in clouds (H I regions) with a typical density of $n_H \simeq 10$–10^2 cm^{-3}, whereas dark nebulae are mostly H_2 molecules and are therefore nowadays referred to as molecular clouds. The estimated densities and temperatures of these clouds are $n_{H_2} \simeq 500$–10^4 cm^{-3} and $T \simeq 10$–50 K. Molecular clouds are often found in the vicinity of H II regions, which consist of gas photoionized by ultraviolet radiation from O- and B-type main-sequence stars.

The Orion nebular and molecular-cloud complex is the most massive nearby region of

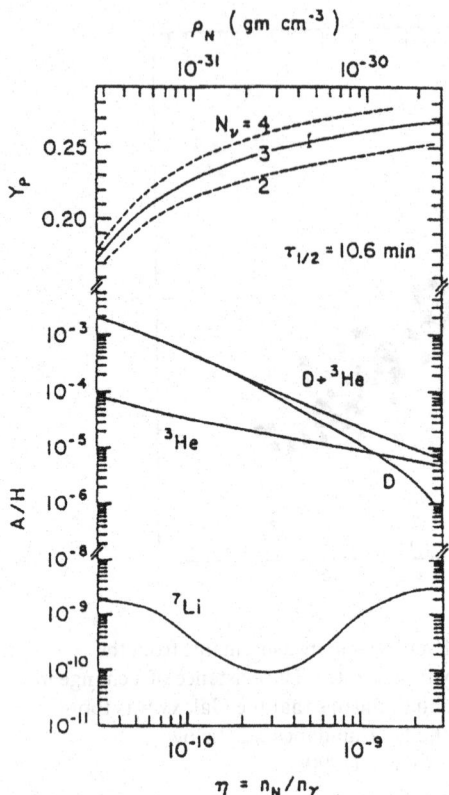

Figure 1.15. The nuclear abundances predicted by the standard big bang model versus the nucleon-to-photon ratio (lower x-axis) or the nucleon mass density (upper x-axis) (gm cm^{-3}) divided by the cube of the cosmic background temperature in units of 2.7 K. Note the expanded linear scale for Y_p, primordial ^{4}He mass fraction; the three curves are for $N_\nu = 2, 3, 4$ different neutrino types. Three is the accepted value. The neutron half-life is taken to be 10.6 min; the uncertainty marks correspond to ± 0.2 min. The results for D, ^{3}He, and ^{7}Li are plotted as ratios – by number – relative to H. Except for very small amounts of isotope production by cosmic ray interactions all isotopes not synthesized in the big bang are synthesized in stellar interiors. From Yang, Turner, Steigman, Schramm and Olive (1984). See also Boesgaard and Steigman (1985).

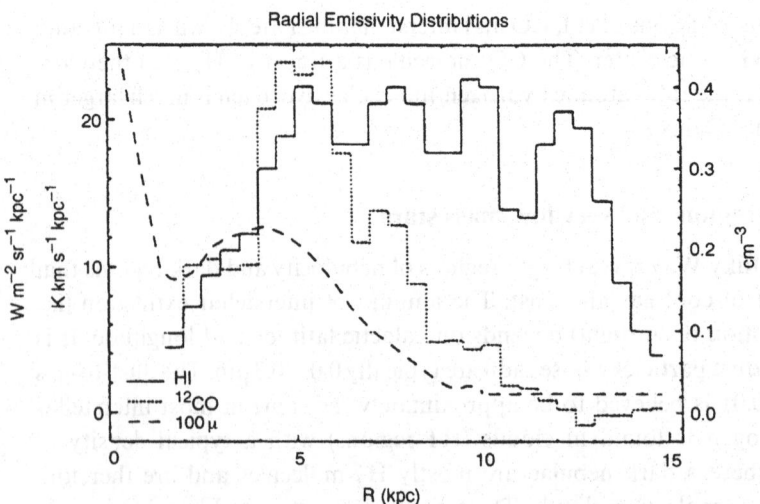

Figure 1.16. The radial distributions of HI, CO and dust with respect to the Galactic center. The solar system is at a distance of 8.5 ± 1 kpc from the center of the Milky Way. The distribution of H II regions is similar to that of giant molecular clouds (CO emission). The distribution of dust is detected from 100 μm radiation. From Burton (1988).

star formation. OH and H_2O maser sources, which are also found in mass ejected from red giants, exist in star-forming regions such as Orion. Some OH and H_2O maser sources in star-forming regions may be protostars, but others are blobs of gas in shells ejected from recently formed stars. Star formation occurs in massive and dense molecular clouds located close to the Galactic plane. Less-dense molecular clouds found well above the Galactic plane are unlikely sites of star formation.

The Coal Sack, whose distance is about 150 pc, is an example of an apparently isolated dark nebula. The highly obscured ρ Ophiuchi region is probably the closest molecular-cloud and star-forming region. Optically it has a radius of about 5 pc and contains several H II regions. There is some evidence that interstellar shocks have had an important effect on star formation in this region. Another interesting molecular cloud is Sagittarius B-2, which is located at a distance of $\simeq 200$ pc from the center of the Galaxy and has an estimated mass of $3{-}10 \times 10^6 M_\odot$. Because extinction at visual wavelengths is $\simeq 25$ magnitudes, Sag B-2 can be observed only at radio and infrared wavelengths. The presence of a stable molecular cloud close to the Galactic center places a limit on the mass of the central region of the Galaxy, including the mass of a possible black hole at the Galactic center.

Although dark nebulae and molecular clouds are the same objects, only after the discovery of H_2 and other interstellar molecules did it become possible to quantitatively determine their physical properties (e.g. density, temperature, gas kinematics and chemical state). The CO molecule is a particularly important probe of molecular clouds because it is ubiquitous and the CO/H_2 ratio ($\sim 10^{-4}$) is believed to be nearly the same in most molecular clouds. The H_2 molecule has been observed only at infrared and ultraviolet wavelengths in relatively nearby regions of interstellar space; however, the distribution of CO molecules can be observed throughout the Milky Way and also in nearby galaxies.

Star-forming regions and molecular clouds are concentrated in spiral arms. Millimeter-wavelength observations of CO have made it possible to determine the distribution of molecular clouds in spiral arms. Within the Milky Way the thickness of molecular gas is about 200 pc (full width at half maximum) within the inner region of the Galaxy and about 300 pc in the outer region. This indicates that most star formation occurs within $1{-}2 \times 10^2$ pc of the Galactic plane. The presence of relatively young stars well above the Galactic plane requires some special explanation. Stars formed close to the Galactic plane can be perturbed to orbits that extend to greater vertical distances by gravitational encounters with stars and gas clouds. Because molecular clouds are so massive ($\sim 10^5{-} 10^6 M_\odot$), they can significantly change stellar orbits. Supernova explosions in binary systems can change the spatial distributions of young stars such as pulsars and some high-velocity stars. Although pulsars are known to be formed from young, massive stars, their mean distances from the Galactic plane are appreciably greater than that of the molecular clouds from which they originated.

Development of millimeter-wavelength interferometry has made it possible to resolve molecular clouds with dimensions of ~ 30 pc (i.e. similar to the Orion molecular cloud) in Local Group galaxies Andromeda (M31) and M33. The average masses, dimensions, line widths and CO/H_2 ratios in molecular clouds within these galaxies are inferred to be similar to those in the Milky Way. Star-formation rates in spiral arms are clearly correlated with the amount of gas and dust. M51 is often referred to as a 'grand-design' galaxy because its spiral arms are particularly well defined. CO maps of M51 have shown that massive concentrations of molecular gas coincide with dust arms, and observed streaming motions are consistent with the density-wave theory for the origin of spiral structure. The density-wave theory of spiral structure asserts that gas clouds and stars are concentrated in the gravitational potential of a density wave where star formation preferentially takes place.

As the name suggests, barred spirals are spiral galaxies with stellar bars within their inner region. Stellar and gaseous bars are caused by noncircular motions. CO measurements suggest that stellar-bar gravitational potentials can cause inflows of gas from outer to central regions of galaxies and therefore possibly explain bursts of star formation in galactic nuclei (especially starburst galaxies). Such inward gas flows might also explain how the enormous radiative outputs of some active galactic nuclei are sustained. Because spiral arms emanate from the ends of stellar bars, it is likely that density waves in barred spirals are driven by the nonaxial symmetry of the stellar bar. The starburst galaxy M82, which is a member of the M81 group of galaxies, is particularly interesting from the point of view of star formation. This galaxy, which is classified as irregular, contains much more gas and dust than a typical spiral or irregular galaxy. Molecular-line measurements show an elongated structure with CO emission peaked on either side of the nucleus. The large amount of gas in M82 is probably the result of tidal capture of gas from the M81 galaxy group.

The mean initial mass function for stars more massive than approximately $1.5 \, M_\odot$ is

$$\frac{\mathrm{d}N(M)}{\mathrm{d}M} = KM^{-x}, \text{ with } x \simeq 2.5. \tag{1.70}$$

It is of obvious interest to attempt to relate the observed stellar mass function to clumpiness of gas in molecular clouds. CO observations give a clump mass distribution that corresponds to $x \simeq 1.5–2$ in Equation (1.70). This estimated clump mass distribution may only apply to stars appreciably more massive than $1.5 \, M_\odot$. Moreover, the actual stellar initial mass distribution is clearly influenced by the dynamics of collapse. This circumstance is made particularly clear by the fact that many stars form as binary or even triple stellar systems with varying orbital radii. The smallest-scale structures in molecular clouds are undoubtedly caused by turbulence. Magnetic fields can also influence protostar collapse by transporting angular momentum and affecting the directionality of gas motions.

Molecular clouds (or complexes of clouds) as massive as $10^5–10^6 \, M_\odot$ do not collapse all at once, but rather regions of a cloud collapse at somewhat different times. Much of the turbulence in interstellar space is caused by supernova explosions from young, massive stars. It is therefore plausible to assume that supernova outbursts following the collapse of a particular region of a cloud can trigger star formation in nearby molecular gas. Cloud–cloud collisions which are more frequent in spiral arms are also believed to induce star formation.

Observations of star-forming regions have shown that bipolar outflows are a common characteristic of the early evolution of low- and high-mass stars. These bipolar outflows are generally well collimated with typical outflow velocities and inferred mass-loss rates of tens of kilometers per second and 10^{-5}–$10^{-7} M_\odot$/yr, respectively. The interaction of bipolar outflows with ambient gas can cause shock-wave-excited H_2 emission. Although bipolar stellar winds are often atomic rather than molecular, they can induce motions of ambient molecular gas. In certain instances it has been possible to demonstrate that bipolar outflows emanate from an embedded infrared source.

Vega and β Pictoris are two well-known stars that are surrounded by disk material. Infrared observations have shown that Vega is surrounded by a disk of rocky materials, whereas β Pictoris is surrounded by a dusty disk. Circumstellar disks have been found around both low- and high-mass stars. T Tauri stars are of special interest because they are typically $\sim 1 M_\odot$ stars that are still in the pre-main-sequence (gravitational-contraction) stage of stellar evolution. Observed spectral energy distributions from millimeter to infrared wavelengths indicate that approximately 50% of T Tauri stars have circumstellar disks. It has been suggested that episodes of rapid mass ejection from so-called FU Ori objects results from instabilities in such disks. High-angular-resolution CO mappings of T Tauri stars have provided rather direct evidence of circumstellar disks. Evidence for disks around stars with bipolar outflows has also been obtained.

The existence of O and B stellar associations and also T associations which contain large numbers of fairly low-mass ($\simeq 1$–$3 M_\odot$) T Tauri stars demonstrates that the formation rate of massive ($M \geq 10 M_\odot$) stars is highly variable from star-forming region to star-forming region, and therefore the initial mass function given by Equation (1.70) is only applicable in some very average sense. η Carina is a star of very special interest because it is probably the most massive ($M \simeq 1$–$2 \times 10^2 M_\odot$) star in the Milky Way. Observations of η Carina were first described in the 17th century. It brightened gradually until it reached around magnitude -1 in 1843, at which time it was brighter than any star in the night sky except Sirius. After 1843 η Carina gradually faded until attaining a brightness minimum of magnitude 7.7 in 1900. Since 1900 it has been slowly brightening at optical wavelengths. Infrared observations in the 1960s showed that at 20μm η Carina is the brightest object in the sky outside the solar system. Moreover, its observed diameter depends on infrared wavelength. Larger diameters are inferred at longer wavelengths, and it is apparent that infrared emission is from dust. Estimates of luminosity range from 2–$4 \times 10^6 L_\odot$, and present power output, which is mostly at infrared wavelengths, is nearly the same as visual power output was in 1843. There is evidence for episodic mass ejection from observations of Hγ absorption dips caused by mass shells moving at different velocities. When optical emission was greatest, in 1843, η Carina was not surrounded by dust. At the present epoch η Carina is a hot ($T \sim 30\,000$ K) star embedded in dust. Bipolar jets of ionized gas are observed as well as a massive circumstellar disk. The spectrum of η Carina is dominated by emission lines from ions of He, N, Mg, Si, Fe, etc. Ultraviolet observations have identified several ionization states of N. Very massive stars have extensive convective cores, and therefore N, which is synthesized by the CNO cycle, may be mixed to the atmosphere.

H I regions contain about half of the gas in the inner Milky Way and perhaps as much as 90% in the outer Galaxy. Studies of spiral galaxies reveal spiral-arm regions of enhanced star formation that contain $\sim 10^8 \, M_\odot$ of gas. In such regions most of the gas is atomic hydrogen interspersed with molecular clouds where star formation occurs.

The faint end of the main sequence (i.e. the region of M dwarfs) is of particular interest because most stellar mass is contained in stars sufficiently low in mass ($\lesssim 0.7 \, M_\odot$) that they have not had time to evolve from the main sequence even if formed during the early history of the Milky Way. By definition a main-sequence star burns hydrogen in its core. Stars less massive than about $0.08 \, M_\odot$ never achieve central temperatures sufficiently high for the proton–proton reaction chain to generate energy. Stars too low in mass to evolve onto the main sequence are sometimes referred to as brown dwarfs. Very low-mass stars are fully convective, and therefore depletion of deuterium and lithium can occur in some brown dwarfs because both deuterium and lithium undergo thermonuclear reactions at temperatures too low for the proton–proton chain to be effective. This means that measured deuterium or lithium abundance can serve as an indicator of the mass or evolutionary state of a brown dwarf. Lithium depletion is predicted in stars more massive than $\simeq 0.06 \, M_\odot$. Such depletion has been found in at least one very low-mass star. Several brown dwarfs with masses $\simeq 0.06 \, M_\odot$ have been found.

In order to convert the observed stellar luminosity function (i.e. number of stars per unit magnitude) to a theoretical mass–luminosity relation, it is necessary to know relevant input physics accurately. Stellar opacities appear to be the main uncertainty in inferring a mass–luminosity relation from the observed stellar luminosity function. The stellar mass function flattens for masses less than about $1.5 \, M_\odot$. Estimated values of x in Equation (1.70) indicate that x is $\simeq 2$ in the mass range $1.5 \, M_\odot \geq M \geq 0.5 \, M_\odot$ and $\simeq 1$ in the mass range $0.5 \, M_\odot \geq M \geq 0.05 \, M_\odot$. A simple argument for the flattening of the initial mass function is as follows. The fragmentation process associated with star formation represents a cascade from higher- to lower-mass gas clouds. Lower-mass protostars tend to form at relatively high densities, where they are optically thick to emitted radiation and therefore less likely to fragment.

We remark that a better method of placing a limit on the mass of a black hole at the Galactic center than the tidal radius of Sag B-2 discussed above can be obtained from infrared emission-line measurements. Observations of the $12.8 \mu m$ line from [Ne II] have been interpreted as emission from a tilted, gaseous ring of radius $\simeq 2 \, pc$ around an $M \sim 1–5 \times 10^6 \, M_\odot$ black hole.

Recently (see Boss 1996 and references therein) a number of extrasolar planets and brown-dwarf binary stars have been discovered. It is anticipated that their rate of discovery will be fairly rapid over the course of the next several years. The central cores of Jupiter and Saturn, which consist of ice and rock, were probably formed by accretion within the circumsolar disk of gas and dust that existed when the Sun was formed. Accretion of gaseous hydrogen and helium occurred after central core mass became equal to about 10 Earth masses. On the other hand, stars including brown dwarfs are believed to be formed during protostar collapse. Therefore, the distinction between a star and planet is related to how it is formed. It is possible that some yet-to-be-discovered extrasolar planets will be more massive than many brown dwarfs.

Several different methods of detection have been used to discover brown dwarfs and extrasolar planets. Although fainter by a factor of 10^5 the brown-dwarf binary companion of Gliese 229 was detected directly at visual wavelengths because its large separation (44 AU) made it possible to use adaptive optics to smooth out atmospheric turbulence sufficiently that radiation from Gliese 229 could be occulted with a disk. The mass of Gliese 229B has been estimated to be approximately 20–50 times greater than the Jovian mass of $0.001\,M_\odot$.

The motion of Jupiter causes the Sun to orbit about the center of mass of the solar system with a velocity of $\simeq 10\,\mathrm{m\,s}^{-1}$. Small Doppler velocities can now be measured and reports of Jupiter-mass planets around nearby stars have been made. The Doppler method of detection favors discovery of massive planets that are close to their companion stars. Extrasolar planets with masses $0.5 M_J$ and $3.7 M_J$, respectively, have been reported to exist around 51 Pegasi and τ Boötis. The inferred distances of these planets from their central stars are only about 0.05 AU. Since it is clearly impossible for gaseous planets to form this close to their central stars, it has been suggested that these planets migrated inwards from their sites of formation. However, Gray (1997) has published additional measurements that confirm the 4.23-day period associated with 51 Peg but cast doubt on the validity of the planetary interpretation of these observations because the techniques currently used to identify very small radial-velocity variations could be associated with the intrinsic shapes of spectral lines expected when a star pulsates or has spots on its surface. A similar conclusion has been published for the planetary-system candidate τ Boötis. G. Gonzalez (*MNRAS*, 285, 403, 1997) points out that τ Boötis like 51 Peg is metal rich and a possible delta Scuti variable star.

Astrometry is a third method of discovering extrasolar planets. Two Jupiter-mass companions of the nearby star Lalande 21185 have been discovered at distances of approximately 2.5 and 10 AU. Low-mass companions have previously been reported to exist around Lalande 21185 and Barnard's star. However, these earlier results have been discredited. Hopefully the more recent measurements will be confirmed. The technique of gravitational microlensing has been used to detect very faint single and binary stars. In the future this method of detection may be used to discover planets in orbit around faint foreground stars. As discussed above, previous observations of young, solar-mass T Tauri stars have shown that approximately half of them probably have protoplanetary disks from which planets can form. However, disks of dust similar to the one around β Pictoris are not as common. We also note that two multiple-Earth-mass objects and a lunar-mass object have been inferred to orbit the pulsar PSR B1257 + 12. The supernova whose stellar remnant is PSR B1257 + 12 may have formed the accretion disk from which planetary-mass objects were created by interacting with a previously existing binary companion star.

Problems

1.   a. Give the conditions on the eccentricity ε defined in Equation (1.17) for hyperbolic, parabolic, elliptic and circular orbits.

b. Prove that $\dfrac{1}{2}\displaystyle\int r \times dr$ with $r = \hat{x}a\cos\theta + \hat{y}b\sin\theta$ integrated over an ellipse is equal to πab.

2. Assume that the star-formation rate per unit mass is equal to zero for $M < 0.05\,M_\odot$, and equal to KM^{-x} with M equal to the stellar mass, K equal to a constant and $x = 2.5$ for $M \geq 0.05\,M_\odot$. Assume that the star-formation rate has been constant for 15 billion years. In addition, assume that the main-sequence lifetime varies as $K_1 M^{-3}$ for $M \leq 10\,M_\odot$ and $K_2 M^{-2}$ for $M \geq 10\,M_\odot$. The relationship between the constants K_1 and K_2 is determined by the condition that the main-sequence lifetime is continuous at $10\,M_\odot$.

a. Determine the stellar mass M_* whose main-sequence lifetime is 15 billion years. The main-sequence lifetime of a $1\,M_\odot$ main-sequence star is about 10 billion years.

b. Let N_1, N_2, N_3 and N_4 equal the number of stars per unit volume in the mass intervals $M \leq M_*$, $M_* \leq M \leq 1.4\,M_\odot$, $1.4\,M_\odot \leq M \leq 10\,M_\odot$ and $M \geq 10\,M_\odot$ respectively. Find the predicted ratios N_2/N_1, N_3/N_1 and N_4/N_1.

3. A molecular-hydrogen cloud of uniform density has mass $M = 10^4\,M_\odot$ and temperature $T = 10\,\mathrm{K}$. Find the density of the cloud if its mass is equal to the Jeans mass M_J. What is the free-fall-collapse timescale?

4. Determine the constants a_2, a_4 and a_6 defined in Equation (1.67).

5. Consider an isothermal molecular-hydrogen gas cloud of temperature $T = 20\,\mathrm{K}$, outer radius $R = 10^{17}\,\mathrm{cm}$ and central density $\rho_c = 10^{-18}\,\mathrm{g\,cm^{-3}}$. Find the mass and mean density of the cloud. What would the temperature of the cloud have to be for its mass to equal the Jeans mass? Use the mean density to determine the Jeans mass.

6. a. Assume that the centroid of a stellar image can be measured to 0.001 seconds of arc. To what distance can stellar parallaxes be determined to an accuracy of 10%?

b. A star is at a distance of 5 kpc and there is known to be 0.5 magnitudes extinction per kpc. What is the relation between absolute and apparent magnitude?

7. a. Binary stars with a circular orbit perpendicular to the line of sight have an angular separation of 1 arc second. The orbital period is 50 years. One of the binary stars is a $1M_\odot$ main-sequence star whose apparent magnitude is $+7$. Find the mass of the companion star.

b. Suppose that the binary companion in part a is a main-sequence star of $9\,M_\odot$. Explain how measurements of angular separation can be used to determine the orientation of the orbit.

8. Let $Z = Z(t) = M_h/M_g$ with t equal to the time and M_g and M_h equal to the total interstellar mass and heavy-element mass, respectively. A generation of massive stars forms from interstellar gas and ejects a fraction of their mass as

heavy elements in the time interval δt. If δM_s is the mass of stellar remnants produced and $p \, \delta M_s$ the corresponding yield of heavy elements, then the total change in heavy-element content of interstellar gas is

$$\delta M_h = p \, \delta M_s - Z \, \delta M_s.$$

Evaluate δZ and show that

$$Z(t) = -p \ln \frac{M_g(t)}{M_g(0)}.$$

Hint: Use the mass-conservation condition $\delta M_s = -\delta M_g$.

2 Introduction to the physics of stellar interiors and the equations of stellar structure

2.1 The equation of state and the first law of thermodynamics

The gas throughout most of the interiors of stars is ionized. Even at high densities (i.e. $\rho \gtrsim 1\,\mathrm{g\,cm^{-3}}$) Coulomb interaction energy between particles in an ionized gas is much less than kinetic energy, and therefore the gas is said to be ideal (perfect). The equation of state of a fully ionized, nondegenerate gas is

$$P = nkT = \left(n_\mathrm{e} + \sum_i n_i\right)kT, \tag{2.1}$$

where n is the total number density of particles, n_e is the number density of electrons and n_i is the number density of ions of species i. The sum of the number densities of the various ions and the number density of electrons can be expressed as

$$\Sigma n_i = \sum_i \frac{X_i}{A_i}\rho N_0 \tag{2.2}$$

and

$$n_\mathrm{e} = N_0 \rho \sum_i \frac{X_i Z_i}{A_i}. \tag{2.3}$$

In Equations (2.2) and (2.3) $N_0 = 6.02 \times 10^{23}$ is Avogadro's number, ρ is the gas density, X_i is the mass fraction of ion i, A_i is the total number of neutrons and protons of ion i and Z_i is the charge of ion i. The molecular weight and electron molecular weight are

$$\frac{1}{\mu} = \frac{n}{\rho N_0} = \sum_i \frac{X_i}{A_i}(1 + Z_i) \tag{2.4}$$

and

$$\frac{1}{\mu_\mathrm{e}} = \frac{n_\mathrm{e}}{\rho N_0} = \sum_i \frac{X_i Z_i}{A_i}, \tag{2.5}$$

respectively. It follows that the total gas pressure and electron pressure are

$$P_{gas} = \frac{\rho k T}{\mu m_u} \qquad (2.6)$$

and

$$P_e = \frac{\rho k T}{\mu_e m_u} \qquad (2.7)$$

with the atomic mass unit m_u equal to 1/12 times the mass of the ^{12}C atom. In main-sequence stars approximately 98% of the mass is hydrogen and helium. Under conditions of local thermodynamic equilibrium such as exist in stellar interiors the ionization equilibria of hydrogen and helium are described by the reactions

$$
\begin{aligned}
H &\rightleftarrows p + e^-, \\
He &\rightleftarrows He^+ + e^-, \\
He^+ &\rightleftarrows He^{++} + e^-.
\end{aligned}
\qquad (2.8)
$$

The thermodynamics of partially ionized gases will be described more thoroughly in Chapters 3 and 5.

The radiation field in the interior of a star is blackbody radiation, and consequently the radiation pressure and internal energy density of the radiation field are

$$P_r = \tfrac{1}{3} a T^4 \qquad (2.9)$$

and

$$U_r = a T^4, \qquad (2.10)$$

respectively. The total pressure is

$$P = P_{gas} + P_r. \qquad (2.11)$$

The first law of thermodynamics, which is a statement of the conservation of energy, asserts that the heat added to a given amount of gas is

$$\Delta Q = \Delta U + P \Delta V, \qquad (2.12)$$

where ΔU is the change in the internal energy, ΔV the change in the volume and $P \Delta V$ the work done by the gas. For infinitesimal changes that are quasistatic and do not involve changes in ionization equilibrium or chemical composition

$$T \, dS = dQ = dU + P \, dV, \qquad (2.13)$$

with S the entropy of the gas. The internal energy U and entropy S depend only on the thermodynamic state of the gas (i.e. P, T and composition) and therefore dU and dS are perfect differentials. Equation (2.13) implies that the pressure is

$$P = -\left(\frac{\partial U}{\partial V} \right)_S. \qquad (2.14)$$

The internal energy per particle of an ionized gas equals $(3/2)kT$. If s_p is defined to be the entropy per particle, then Equation (2.13) implies that

$$T\,ds_p = c_v\,dT + P\,d\left(\frac{1}{n}\right)$$

(2.15)

with $c_v = (3/2)k$ the heat capacity at constant volume per particle and $1/n$ equal to the volume per particle. From Equations (2.1) and (2.15) it follows that

$$ds_p = \frac{3}{2}k\frac{dT}{T} - \frac{k\,dn}{n} = k\,d\ln\left(\frac{T^{3/2}}{n}\right).$$

(2.16)

Since the entropy of n particles is equal to n times the entropy of a single particle, Equation (2.16) implies that

$$s = ns_p = nk\ln\left(\frac{T^{3/2}}{n}\right) + \text{constant}.$$

(2.17)

For a given volume V the first law of thermodynamics for the radiation field is

$$T\,dS_r = d(U_r V) + P\,dV.$$

(2.18)

Equations (2.9), (2.10) and (2.18) imply that

$$dS_r = 4VaT^2\,dT + \frac{4}{3}aT^3\,dV.$$

(2.19)

Integrating Equation (2.19) gives

$$S_r = \frac{4}{3}aT^3 V.$$

(2.20)

Equations (2.17) and (2.20) imply that S, the entropy per unit volume of a mixture of ionized gas and blackbody radiation, is

$$S = nk\ln\left(\frac{T^{3/2}}{n}\right) + \frac{4}{3}aT^3 + \text{constant}.$$

(2.21)

Figure 2.1 shows in the temperature–density plane where radiation pressure dominates over gas pressure and where nonrelativistic and relativistic electron-degenerate pressures (discussed in Chapter 3) dominate.

2.2 Hydrostatic equilibrium

The gravitational force on a small mass element

$$dm = \rho\,dr\,dA$$

(2.22)

which is at a radial distance r from the center of a spherically symmetric star is

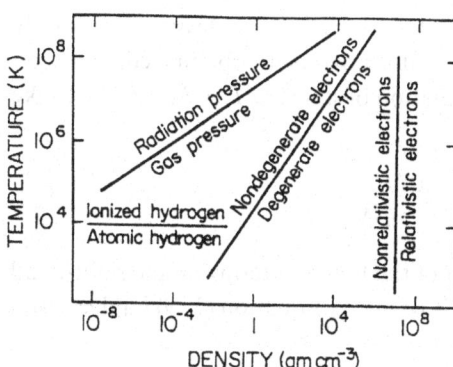

Figure 2.1. Some physical properties of a gas in thermodynamic equilibrium. Radiation pressure is dominant above the line separating radiation pressure and gas pressure. The remaining lines should be interpreted similarly.

$$F_G = -\frac{GM(r)\,dm}{r^2},$$ (2.23)

with

$$M(r) = \int_0^r \rho(r')4\pi r'^2\,dr'$$ (2.24)

The corresponding pressure force on this same mass element is

$$F_P = P\,dA - (P + dP)\,dA$$
$$= -dP\,dA.$$ (2.25)

In Equation (2.25) P is the pressure at a distance r from the center of the star and $P + dP$ equals the pressure at a radial distance $r + dr$. Equations (2.23) and (2.25) and Newton's second law of motion imply that the equation of radial motion is

$$\rho\frac{d^2r}{dt^2} = -\rho\frac{GM(r)}{r^2} - \frac{dP}{dr}.$$ (2.26)

It follows from Equation (2.26) that a spherically symmetric star will be in hydrostatic equilibrium if the equation

$$\frac{dP}{dr} = -\rho\frac{GM(r)}{r^2}$$ (2.27)

is satisfied throughout the interior.

If the pressure P is a known function of density ρ, as is the case for white dwarfs, then the structure is determined by Equations (2.27) and (2.24). The latter equation is equivalent to the differential equation

$$\frac{dM(r)}{dr} = 4\pi r^2\rho.$$ (2.28)

The pressure in the interiors of most stars is a function of temperature T as well as density

ρ, and consequently Equations (2.27) and (2.28) and the equation of state are not generally sufficient to determine a star's structure. However, it is possible to obtain a condition on the central pressure that is independent of the equation of state (i.e. $P = P(\rho, T)$). We take the derivative

$$\frac{d}{dr}\left(P + \frac{GM(r)^2}{8\pi r^4}\right) = \frac{dP}{dr} + \frac{GM(r)}{4\pi r^4}\frac{dM(r)}{dr} - \frac{GM(r)^2}{2\pi r^5}. \tag{2.29}$$

Since, from Equations (2.27) and (2.28), the sum of the first two terms on the right-hand side of Equation (2.29) is zero and the last term is negative, Equation (2.29) implies that

$$\frac{d}{dr}\left(P + \frac{GM(r)^2}{8\pi r^4}\right) < 0. \tag{2.30}$$

Since the pressure P is equal to zero at the surface and the second term within the brackets of Equation (2.30) is equal to zero at the center, Equation (2.30) implies that the central pressure P_c of a star of mass M and radius R satisfies the inequality

$$P_c > \frac{GM^2}{8\pi R^4}. \tag{2.31}$$

Equations (2.27) and (2.28) can also be used to derive the virial theorem for a spherically symmetric star. If $V(r)$ is defined by the equation

$$V(r) = \frac{4}{3}\pi r^3, \tag{2.32}$$

then it follows from the equation of hydrostatic equilibrium that

$$V(r)\, dP = -\frac{1}{3}4\pi r^2 \rho\, dr \frac{GM(r)}{r}$$

$$= -\frac{1}{3}\frac{GM(r)}{r}\, dM(r). \tag{2.33}$$

A partial integration of the left-hand side of Equation (2.33) leads to the equation

$$\int V(r)\, dP = PV\Big|_0^R - \int P\, dV. \tag{2.34}$$

Since $P = 0$ at the surface and $V(r) = 0$ at the center, Equations (2.33) and (2.34) lead to the equation

$$-\Omega = \int \frac{GM(r)}{r}\, dM(r)$$

$$= 3\int P\, dV. \tag{2.35}$$

Equation (2.35) expresses a relationship between the integral of the pressure within the interior of a star and $-\Omega$, the gravitational potential energy. For a nondegenerate ionized gas with negligible radiation pressure the virial theorem implies that the absolute value of the gravitational potential energy equals twice the thermal energy.

The equation of hydrostatic equilibrium can also be derived from a variational principle. If u is defined to be the internal energy per gram, then the total energy of a static, spherical star is

$$E = \int u \, dM_r - \int \frac{GM_r}{r} \, dM_r. \tag{2.36}$$

As will be shown below, the requirement that the total energy is an extremum when a star is in equilibrium leads to the equation of hydrostatic equilibrium.

Lagrangian displacement $\xi(r)$ of mass elements connect fluid mass elements in the unperturbed equilibrium state to corresponding elements in a perturbed state. The Lagrangian change in a variable such as the internal energy density is

$$\Delta u = u(r + \xi(r)) - u_0(r), \tag{2.37}$$

where $u_0(r)$ represents the unperturbed values of the internal energy density and $\xi(r)$ is small. The Lagrangian variation of the total energy E in Equation (2.36) is

$$\Delta E = \int \Delta u \, dM_r + \int \frac{GM_r}{r^2} \xi(r) \, dM_r, \tag{2.38}$$

with

$$\Delta u = -P\Delta\left(\frac{1}{\rho}\right) = \frac{P}{\rho^2}\Delta\rho. \tag{2.39}$$

In order to complete this derivation we must obtain a relationship between $\Delta\rho(r)$ and $\xi(r)$. The total time derivative of the density of a mass element is

$$\frac{d\rho}{dt} = \frac{\partial\rho}{\partial t} + \boldsymbol{v}\cdot\nabla\rho. \tag{2.40}$$

The equation of continuity, which is a consequence of conservation of mass, is

$$\frac{\partial\rho}{\partial t} + \nabla\cdot(\rho\boldsymbol{v}) = 0. \tag{2.41}$$

The above two equations imply that

$$\frac{d\rho}{dt} = -\rho\nabla\cdot\boldsymbol{v}. \tag{2.42}$$

Equation (2.42) implies that, for radial displacements $\xi(r)$,

$$\frac{\Delta\rho}{\rho} = -\nabla\cdot\boldsymbol{\xi}. \tag{2.43}$$

From Equations (2.38), (2.39) and (2.43) we have

$$\Delta E = \int \frac{P \Delta \rho}{\rho} 4\pi r^2 \, dr + \int \frac{GM_r}{r^2} \xi(r) \, dM_r$$

$$= - \int PV \cdot \xi 4\pi r^2 \, dr + \int \frac{GM_r}{r^2} \xi(r) \, dM_r. \tag{2.44}$$

Using the vector identity $\nabla \cdot (P\xi) = P\nabla \cdot \xi + \xi \cdot \nabla P$ and the boundary conditions $P = 0$ at the surface and $\xi = 0$ at $r = 0$, we find after a partial integration that Equation (2.44) reduces to the equation

$$\Delta E = \int \xi \cdot \left(\nabla P + \rho \frac{GM_r}{r^2} \right) 4\pi r^2 \, dr. \tag{2.45}$$

The displacement ξ is arbitrary and therefore the requirement that $\Delta E = 0$ for equilibrium implies Equation (2.27).

The above discussion of equilibrium is limited to spherical stars. For more general distributions of mass, Poisson's equation, which will be derived below, must be solved in order to determine the relationship between the density distribution and gravitational potential. Divide a star into arbitrarily small mass elements dm_i, $i = 1, \ldots, n$. The gravitational force per gram at position r_j caused by the mass elements dm_i located at positions r_i is

$$F_j = - \sum_{i=1}^{n} \frac{G(r_j - r_i) \, dm_i}{|r_j - r_i|^3}$$

$$= \nabla \sum_{i=1}^{n} \frac{G \, dm_i}{|r_j - r_i|} = - \nabla \Phi, \tag{2.46}$$

with

$$\Phi(r_j) = - \int \frac{G\rho \, dV}{|r_j - r|}.$$

Since the gravitational force at each point in a stellar interior is the gradient of the scalar function Φ, it follows that the curl of the gravitational force is zero. Suppose that a particle moves from a large distance to a position P and then out again to a large distance from the star. The condition that the curl of the gravitational force is zero and Stokes' theorem give

$$0 = \int \nabla \times F \cdot dA = \int F \cdot ds = \int_{\infty}^{P} F \cdot ds + \int_{P}^{\infty} F \cdot ds. \tag{2.47}$$

The above equation implies that the work done on a particle as it moves between any two points in a gravitational field is independent of the particle's path. Therefore a gravitational potential unique to within an arbitrary constant can be defined, and the gravitational force field is conservative.

Consider a small spherical region of radius r_0 surrounding a point P. The gravitational potential at point P is $\Phi = \Phi_o + \Phi_i$, where Φ_o and Φ_i are the contributions to the gravitational potential at point P caused by mass outside and inside the small spherical region. $F_o = -\nabla\Phi_o$ and $F_i = -\nabla\Phi_i$ are the corresponding gravitational forces. Equation (2.47) and the equality

$$\nabla^2 \frac{1}{|r - r'|} = 0 \quad (|r - r'| \neq 0) \tag{2.48}$$

imply that $\nabla \cdot F_o = -\nabla^2\Phi_o = 0$. Therefore, from Gauss's theorem we find that

$$\lim_{r_0 \to 0} \int \nabla^2\Phi \, dV = \lim_{r_0 \to 0} -\nabla \cdot F \frac{4}{3}\pi r_0^3$$

$$= \lim_{r_0 \to 0} -\int F \cdot dS = \lim_{r_0 \to 0} \left(\frac{4}{3}\pi r_0 G\rho 4\pi r_0^2\right). \tag{2.49}$$

Poisson's equation,

$$\nabla^2\Phi = 4\pi G\rho, \tag{2.50}$$

follows immediately from Equation (2.49). The gravitational energy of a star is

$$\Omega = -\frac{1}{2}\sum_j^n \left(\sum_i^n \frac{G\,dm_i}{|r_j - r_i|}\right) dm_j = \frac{1}{2}\int \Phi\rho \, d^3r, \tag{2.51}$$

with $\Phi = \Phi(r)$ equal to the gravitational potential. The factor $\frac{1}{2}$ appears in front of the above summations and integral because the summations count pairs of particles twice. Because the star is nonspherical, the Lagrangian displacements are such that $\xi = \xi(r)$. The Lagrangian variation of the energy is

$$\Delta E = \int \Delta u\rho \, d^3r + \frac{1}{2}\int \Delta\Phi\rho \, d^3r$$

$$= \int \nabla P \cdot \xi \, d^3r + \frac{1}{2}\int \Delta\Phi\rho \, d^3r. \tag{2.52}$$

The second equality follows as in the derivation of Equation (2.44).

The Eulerian displacement of a variable such as Φ is defined by the relation

$$\delta\Phi = \Phi(r) - \Phi_0(r), \tag{2.53}$$

with $\Phi_0(r)$ equal to the value of Φ at position r in the equilibrium state. It follows that the relation between the Lagrangian operator Δ and the Eulerian operator δ is given by the equation

$$\Delta = \delta + \xi \cdot \nabla. \tag{2.54}$$

Equation (2.54) implies that the Lagrangian variation $\Delta\Omega$ of the gravitational potential energy given in Equation (2.52) is

$$\Delta\Omega = \frac{1}{2}\int [\delta\Phi\rho + \rho(\xi\cdot\nabla)\Phi]\,d^3r.$$

(2.55)

Using Equations (2.47) and (2.41), and then integrating by parts, leads to

$$\delta\Phi = -G\int \frac{\delta\rho'}{|r - r'|}\,d^3r'$$

$$= G\int \frac{\nabla'\cdot(\rho'\xi')}{|r - r'|}\,d^3r' \qquad (2.56)$$

$$= -G\int \rho\xi'\cdot\nabla'\frac{1}{|r - r'|}\,d^3r'.$$

Equations (2.55) and (2.56) imply that

$$\Delta\Omega = \int \rho\nabla\Phi\cdot\xi\,d^3r, \qquad (2.57)$$

and therefore the equilibrium condition becomes

$$0 = \Delta E = \int (\nabla P + \rho\nabla\Phi)\cdot\xi\,d^3r. \qquad (2.58)$$

Since the integral is zero at each point, the equation of hydrostatic equilibrium for a static spherical or nonspherical star is

$$\nabla P + \rho\nabla\Phi = 0. \qquad (2.59)$$

Departures from spherical symmetry are caused primarily by rotation. The effects of rotation on stellar structure will be discussed in Chapter 9.

2.3   The photon gas and radiative transfer

Let r be a vector that defines the positions of points in space and θ and angle between the normal to a small area element dA and some direction defined at position r. The intensity of radiation is

$$I_\nu(r, \theta, t) = \lim_{\substack{dA,\,dt \\ d\Omega,\,d\nu\to 0}} \frac{dE}{dA\cos\theta\,d\Omega\,d\nu\,dt}\ \mathrm{erg\,cm^{-2}\,ster^{-1}\,Hz^{-1}\,s^{-1}}, \qquad (2.60)$$

with dE equal to the amount of energy that traverses the small area dA in the time interval dt, frequency interval ν, $\nu + d\nu$ and into the solid angle $d\Omega$, which makes an angle θ with respect to the normal to dA.

Consider a cylinder of cross section dA and length dl which makes an angle θ with respect to the radial direction at a distance r from the center of the star. The radiative power that enters the cylinder at radial distance r, at solid angle $d\Omega$, and in the frequency interval ν, $\nu + d\nu$ is

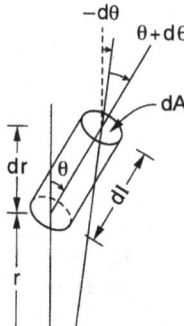

Figure 2.2. Diagram useful in deriving radiative-transfer equation. The change in I_v occurs along the displacement dl.

$$I_v(r, \theta)\, d\Omega d A d v. \tag{2.61}$$

The radiative power loss from the cylinder is

$$I_v(r + dr, \theta + d\theta)\, d\Omega d A d v \tag{2.62}$$

The rate of energy loss caused by absorption within the cylinder is

$$I_v\, d\Omega d A \kappa_v \rho d l d v, \tag{2.63}$$

with κ_v equal to the absorptive opacity and dl equal to the length of the cylinder. The corresponding energy gain caused by the emission of radiation is

$$j_v\, dA\, dl\, d\Omega\, dv. \tag{2.64}$$

In Equation (2.64) j_v is the emissivity of the gas in units of $\text{erg cm}^{-3}\, \text{s}^{-1}\, \text{Hz}^{-1}\, \text{ster}^{-1}$. The radial extent of the cylinder dr and angular displacement $d\theta$ are given by the equations

$$dr = dl \cos\theta \tag{2.65}$$

and

$$d\theta = \frac{-dl \sin\theta}{r}. \tag{2.66}$$

Figure 2.2 illustrates the derivation of the radiative-transfer equation. From the expressions given above, it follows that the equation of radiative transfer becomes

$$\frac{\partial I_v}{\partial r}\cos\theta - \frac{\partial I_v}{\partial \theta}\frac{\sin\theta}{r} + \kappa_v \rho I_v - j_v = 0. \tag{2.67}$$

If multiplied by $\cos\theta\, d\Omega = 2\pi \sin\theta \cos\theta\, d\theta$ and integrated over all solid angles, Equation (2.67) reduces to

$$c\frac{dP_v}{dr} - \frac{2\pi}{r}\int \frac{\partial I_v}{\partial \theta}\sin^2\theta \cos\theta\, d\theta + \kappa_v \rho F_v = 0, \tag{2.68}$$

with

$$P_\nu = \frac{1}{c} \int I_\nu \cos^2 \theta \, d\Omega, \qquad (2.69)$$

and the radiative flux F_ν is

$$F_\nu = \int I_\nu \cos \theta \, d\Omega. \qquad (2.70)$$

In obtaining Equation (2.68) the integral over j_ν is zero because j_ν is independent of angle. Since the radiation field is nearly isotropic in the interiors of stars, we can express $I_\nu(\theta)$ as

$$I_\nu(\theta) = B_\nu + I_1(\nu) \cos \theta, \qquad (2.71)$$

where $I_1 \ll B_\nu$, with B_ν equal to the Planck function given in Equation (3.91). It follows that Equation (2.68) reduces to the equation

$$\frac{dP_\nu}{dr} = \frac{-\kappa_\nu \rho F_\nu}{c}, \qquad (2.72)$$

with $\int P_\nu \, d\nu$ equal to the radiation pressure.

The energy density of the radiation field in the frequency interval $\nu, \nu + d\nu$ is

$$U_\nu = \frac{1}{c} \int I_\nu \, d\Omega. \qquad (2.73)$$

In stellar interiors the mean free path of a photon of frequency ν (i.e. $1/\kappa_\nu \rho$) is much less than a temperature scale height, and the radiation field is very nearly that of a blackbody (Chapter 3 contains a more complete discussion of blackbody radiation). Because the radiation field is similar to that of a blackbody, it follows that $P_\nu = (1/3)U_\nu$, and therefore integrating Equation (2.72) over all frequencies implies that

$$\frac{L_r}{4\pi r^2} = \int F_\nu \, d\nu = -\frac{c}{3\rho} \int \frac{1}{\kappa_\nu} \frac{dU_\nu}{dr} d\nu. \qquad (2.74)$$

Since

$$\frac{dU_\nu(T)}{dr} = \frac{dU_\nu}{dT} \frac{dT}{dr}, \qquad (2.75)$$

Equations (2.74) and (2.75) imply that

$$\frac{L_r}{4\pi r^2} = -\frac{c}{3\rho} \frac{dT}{dr} \int_0^\infty \frac{1}{\kappa_\nu} \frac{dU_\nu}{dT} d\nu. \qquad (2.76)$$

The Rosseland mean opacity is defined by the equation

$$\frac{1}{\kappa} = \frac{\int \frac{1}{\kappa_\nu} \frac{dU_\nu}{dT} d\nu}{\int \frac{dU_\nu}{dT} d\nu}. \qquad (2.77)$$

Using the expression for U_ν given in Equation (3.87), we obtain

$$\int U_\nu \, d\nu = aT^4$$

(2.78)

and

$$\int \frac{dU_\nu}{dT} d\nu = 4aT^3.$$

(2.79)

Equations (2.77) and (2.79) imply that Equation (2.76) can be rewritten as

$$\frac{L_r}{4\pi r^2} = -\frac{4ac}{3\rho\kappa} T^3 \frac{dT}{dr}.$$

(2.80)

Equation (2.80) is the equation of radiative transfer in stellar interiors.

2.4   Electron-scattering opacity

Stellar opacity is caused by electron scattering as well as by absorption of radiation. A free electron is accelerated by the electric field of an electromagnetic wave. According to classical electromagnetic theory the power per unit solid angle emitted by an accelerated electron is

$$\frac{dP}{d\Omega} = \frac{e^2}{4\pi c^3} a^2 \sin^2 \psi,$$

(2.81)

with a the acceleration of the electron and ψ the angle between an arbitrary position vector r and the polarization vector $\hat{\varepsilon}$ of the electromagnetic wave, which is assumed to be linearly polarized. If a Cartesian coordinate system is chosen such that the direction of the electromagnetic wave is along the z-axis and the polarization vector $\hat{\varepsilon}$ along the x-axis, then it can be shown that

$$\sin^2 \psi = 1 - \cos^2 \phi \sin^2 \Phi,$$

(2.82)

with Φ the angle between the z-axis and the arbitrary position vector r, and ϕ the angle between the x-axis and the projection of r onto the x-y plane. The average value of $\sin^2 \psi$ over all directions in the xy plane is

$$\overline{\sin^2 \psi} = 1 - \overline{\cos^2 \phi} \sin^2 \Phi = \tfrac{1}{2}(1 + \cos^2 \Phi).$$

(2.83)

Figure 2.3 shows how the angles Φ, ψ and ϕ are defined with respect to $\hat{\varepsilon}$.

An electron accelerated by an electric field

$$E = \hat{\varepsilon} E_0 \cos(kz - \omega t)$$

(2.84)

experiences an acceleration

$$a = -\hat{\varepsilon} \frac{e}{m} E_0 \cos(kz - \omega t).$$

(2.85)

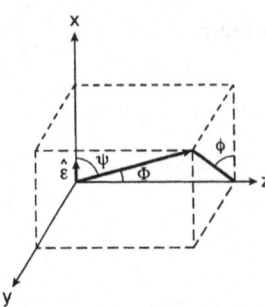

Figure 2.3. In electron scattering the direction of photon propagation is the z-axis with $\hat{\varepsilon}$ equal to the polarization of the photon. The angle Φ is the angle through which the photon is scattered and ψ is the angle between $\hat{\varepsilon}$ and direction into which photon is scattered. For an incident unpolarized wave averaging is over the angle ϕ.

The time average of a^2 is

$$\overline{a^2} = \frac{1}{2}\frac{e^2}{m^2}E_0^2. \tag{2.86}$$

It follows from the above equation that Equation (2.81) can be expressed as

$$\frac{dP}{d\Omega} = \frac{c}{8\pi}E_0^2\left(\frac{e^2}{mc^2}\right)^2 \sin^2\psi. \tag{2.87}$$

Since the power per unit area, which is given by the Poynting vector, is

$$S = \frac{c}{4\pi}\overline{E^2} = \frac{c}{8\pi}E_0^2, \tag{2.88}$$

Equations (2.83) and (2.88) imply that the differential cross section for electron scattering is

$$\frac{d\sigma}{d\Omega} = \frac{1}{S}\frac{dP}{d\Omega} = \left(\frac{e^2}{mc^2}\right)^2 \sin^2\psi = \left(\frac{e^2}{mc^2}\right)^2\frac{1}{2}(1 + \cos^2\Phi). \tag{2.89}$$

It is convenient to express (2.89) as

$$\frac{d\sigma}{d\Omega} = \sigma_T\frac{p(\cos\Phi)}{4\pi}, \tag{2.90}$$

with

$$\sigma_T = \frac{8\pi}{3}\left(\frac{e^2}{mc^2}\right)^2 \tag{2.91}$$

and

$$p(\cos\Phi) = \frac{3}{4}(1 + \cos^2\Phi). \tag{2.92}$$

The phase function $p(\cos\Phi)$ is defined so that

$$\int p(\cos\Phi)\frac{d\Omega}{4\pi} = 1. \tag{2.93}$$

Integrating Equation (2.90) over all solid angles implies that the total cross section for electron scattering, which is known as the Thomson cross section, is

$$\sigma_T = \frac{8\pi}{3}\left(\frac{e^2}{mc^2}\right)^2 = 0.667 \times 10^{-24}\,\mathrm{cm}^2. \tag{2.94}$$

Equation (2.94) implies that the opacity caused by electron scattering is

$$\kappa_s = \frac{n_e \sigma_T}{\rho}. \tag{2.95}$$

For a fully ionized gas Equation (2.95) becomes

$$\kappa_s \simeq 0.2(1 + X), \tag{2.96}$$

with X equal to the mass fraction in the form of hydrogen.

The emissivity caused by electron scattering is

$$j_\nu^s(\theta, \phi) = \frac{\kappa_s \rho}{4\pi}\int P(\theta, \phi, \theta', \phi')I_\nu(r, \theta', \phi')\,d\Omega'. \tag{2.97}$$

In the above equation (θ', ϕ') define the direction of the incoming radiation and (θ, ϕ) the direction of the scattered radiation. In the interior of a star we can, as in the previous section, let

$$I_\nu(\theta', \phi') = B_\nu + I_\nu' \cos\theta'. \tag{2.98}$$

Let $\hat{r}'$ be a unit vector in the direction of the incident radiation with Cartesian components $z' = \cos\theta'$, $y' = \sin\theta'\sin\phi'$ and $x' = \sin\theta'\cos\phi'$, and let $\hat{r}$ be a unit vector in the direction of the scattered radiation with Cartesian components $z = \cos\theta$, $y = \sin\theta\sin\phi$ and $x = \sin\theta\cos\phi$. It follows that

$$\begin{aligned}
\hat{r}\cdot\hat{r}' &= \cos\Phi = \cos\theta\cos\theta' + \sin\theta\sin\theta'(\sin\phi\sin\phi' + \cos\phi\cos\phi') \\
&= \cos\theta\cos\theta' + \sin\theta\sin\theta'\cos(\phi - \phi').
\end{aligned} \tag{2.99}$$

If we multiply Equation (2.67) by $\cos\theta$ and integrate over all solid angles for an assumed electron-scattering opacity, we obtain the equation

$$\frac{d}{dr}\int I_\nu \cos^2\theta\,d\Omega - \frac{2\pi}{r}\int\frac{\partial I_\nu}{\partial\theta}\sin^2\theta\cos\theta\,d\theta$$

$$+ \kappa_\nu^s \rho F_\nu + \frac{\kappa_\nu^s}{4\pi}\int\int P(\theta, \phi, \theta', \phi')I_\nu(\theta', \phi')\,d\Omega'd\Omega = 0. \tag{2.100}$$

The first three terms in the above equation are similar to the corresponding terms in Equation (2.68), which was derived for an assumed opacity caused by absorption processes, and consequently the second term is equal to zero. The phase function in Equation (2.100) is

$$P(\theta, \phi, \theta', \phi') = p(\cos\Phi) = \tfrac{3}{4}(1 + \cos^2\Phi). \tag{2.101}$$

Equations (2.99) and (2.101) and an elementary integration show that the fourth term in Equation (2.100) is zero. This result implies that Equation (2.100) reduces to the same equation as (2.80). It follows that electron-scattering opacity can be treated in the same manner as opacity caused by absorptive processes in stellar interiors.

2.5   Convective energy transport

The equilibrium distribution of pressure, density and entropy density throughout a layer of a star are $P_0(r)$, $\rho_0(r)$ and $s_0(r)$, respectively. The density can be expressed as $\rho = \rho(P, s)$. The condition for convective stability is derived by considering the displacement of a small but finite element of fluid. If a small element of fluid that is initially at an equilibrium position r is displaced a distance δr and the displacement occurs under adiabatic conditions (i.e. $s =$ constant), then

$$\rho'(r + \delta r) = \rho'(P_0(r + \delta r), s). \tag{2.102}$$

If the force of gravity is such that the displaced element returns to its equilibrium position, then the layer of gas is convectively stable. Otherwise the layer is convectively unstable. From Equation (2.102) the condition for convective stability becomes

$$\rho'(P_0(r + \delta r, s_0(r))) - \rho_0(P_0(r + \delta r), s_0(r + \delta r)) \begin{cases} > 0 \text{ if } \delta r > 0, \\ < 0 \text{ if } \delta r < 0. \end{cases} \tag{2.103}$$

Expanding the above expression to first order in δr implies that

$$-\left(\frac{\partial \rho}{\partial s}\right)_p \frac{ds_0}{dr} > 0. \tag{2.104}$$

The heat capacity at constant pressure is

$$c_p = T\left(\frac{ds}{dT}\right)_p. \tag{2.105}$$

It follows from implicit differentiation and the above expression for c_p that

$$\left(\frac{\partial \rho}{\partial s}\right)_p = \left(\frac{\partial \rho}{\partial T}\right)_p \Big/ \left(\frac{\partial s}{\partial T}\right)_p = \frac{T}{c_p}\left(\frac{\partial \rho}{\partial T}\right)_p. \tag{2.106}$$

For a nondegenerate gas in which radiation pressure and changes in molecular weight are small, we have

$$\left(\frac{\partial \rho}{\partial T}\right)_p = \frac{-\rho}{T} < 0. \tag{2.107}$$

Equation (2.104) and Inequality (2.107) imply that a layer of gas is convectively stable if

$$\frac{ds(r)}{dr} > 0. \tag{2.108}$$

The derivative of the entropy with respect to r becomes

$$\frac{ds}{dr} = \left(\frac{\partial s}{\partial T}\right)_p \frac{dT}{dr} + \left(\frac{\partial s}{\partial P}\right)_T \frac{dP}{dr}$$

$$= \frac{c_p}{T} \frac{dT}{dr} + \left(\frac{\partial s}{\partial \rho}\right)_T \left(\frac{\partial \rho}{\partial P}\right)_T \frac{dP}{dr}. \tag{2.109}$$

For a fully ionized, nondegenerate gas in which radiation pressure is small, the partial derivative

$$\left(\frac{\partial \rho}{\partial P}\right)_T = \frac{\rho}{P}. \tag{2.110a}$$

From the first law of thermodynamics we have

$$\left(\frac{\partial s}{\partial \rho}\right)_T = -\frac{P}{T\rho^2}. \tag{2.110b}$$

Implicit differentiation and Equations (2.110a) and (2.110b) give

$$\left(\frac{\partial s}{\partial P}\right)_T = \left(\frac{\partial s}{\partial \rho}\right)_T \left(\frac{\partial \rho}{\partial P}\right)_T = \frac{-1}{\rho T}. \tag{2.111}$$

From Equations (2.109) and (2.111) the condition for convective stability becomes

$$\frac{ds}{dr} = \frac{c_p}{T} \frac{dT}{dr} - \frac{1}{T\rho} \frac{dP}{dr} > 0. \tag{2.112}$$

It follows from the equation of hydrostatic equilibrium (i.e. Equation 2.27) that the condition for convective stability can be expressed as

$$-\frac{dT}{dr} < \frac{g}{c_p} = -\left(\frac{dT}{dr}\right)_s, \tag{2.113}$$

with $g = GM_r/r^2$. Inequality (2.113) implies that if the temperature gradient exceeds the adiabatic temperature gradient, a convective zone will occur.

When a convective zone exists, energy transport occurs by means of convection as well as radiation. In stellar interiors, convective energy transport usually exceeds radiative energy transport in a convective zone by a large factor. In order to describe the interior or atmosphere of a star in which convection occurs, it is necessary to derive an expression for the convective flux (i.e. $F_c = L_c/4\pi r^2$). For a nondegenerate, fully ionized gas adiabatic changes in a rising or falling blob of gas are described by the equation

$$P\rho^{-\Gamma_1} = \text{constant},$$
$$T^{\Gamma_1}P^{1-\Gamma_1} = \text{constant}, \tag{2.114}$$

with $\Gamma_1 = \frac{5}{3}$. The temperature excess in a rising element as compared to the surrounding gas is

$$(dT)_{ad} = \left[\left(1 - \frac{1}{\Gamma_1}\right)\frac{T}{P}\frac{dP}{dr} - \frac{dT}{dr}\right]\delta r = \Delta\nabla T \cdot \delta r, \tag{2.115}$$

and the convective-energy flux becomes

$$\frac{L_r^c}{4\pi r^2} = F_c = \Delta\nabla T\delta r c_p \rho v_c, \tag{2.116}$$

with v_c the velocity of the rising fluid element. The deficiency of the density of a rising element as compared to the surrounding gas is

$$(d\rho)_{ad} = \left[-\frac{1}{\Gamma_1}\frac{\rho}{P}\frac{dP}{dr} + \frac{d\rho}{dr}\right]\delta r = \frac{\rho}{T}\Delta\nabla T\delta r, \tag{2.117}$$

and the kinetic-energy density is

$$\frac{1}{2}\rho v_c^2 = \frac{\rho}{T}\Delta\nabla T\delta r\frac{GM_r}{r^2}\frac{1}{2}\delta r. \tag{2.118}$$

If δr is assumed equal to half the pressure scale height $\lambda = P(dr/dP)$, then Equation (2.118) implies that the average convective velocity of a rising convective blob is

$$v_c = (\Delta\nabla T)^{1/2}\left(\frac{GM_r}{Tr^2}\right)^{1/2}\frac{\lambda}{2}. \tag{2.119}$$

If the convective zone contains equal amounts of rising and falling convective blobs, then the convective flux becomes

$$\frac{L_r^c}{4\pi r^2} = F_c = c_p \rho(\Delta\nabla T)^{3/2}\left(\frac{GM_r}{Tr^2}\right)^{1/2}\frac{\lambda^2}{4}. \tag{2.120}$$

The total energy flux, which includes contributions from radiation as well as convection, is

$$\frac{L_r}{4\pi r^2} = \frac{L_r^c}{4\pi r^2} + \frac{L_r^{rad}}{4\pi r^2}, \tag{2.121}$$

with L_r^c and L_r^{rad} the convective and radiative luminosities respectively at radial distance r.

2.6   Hydrogen and helium burning

The burning of hydrogen into helium and helium into ^{12}C and ^{16}O are the energy sources for most of the radiation emitted from the photospheres of stars. The completion of hydrogen burning can occur either as a result of the proton–proton chain or the CNO cycle. In low-mass main-sequence stars, hydrogen burning takes place primarily as a result of the proton–proton thermonuclear reaction chain, which is initiated by the synthesis of the deuteron. In stellar interiors ^2D is produced either by the weak interaction

$$p + p \rightarrow {}^2D + e^+ + \nu_e \tag{2.122a}$$

or (with a much lower probability) by means of the weak interaction

$$p + e^- + p \rightarrow {}^2D + \nu_e. \tag{2.122b}$$

The cross sections for the above weak interactions are quite low.

The remaining reactions in the proton–proton chain are

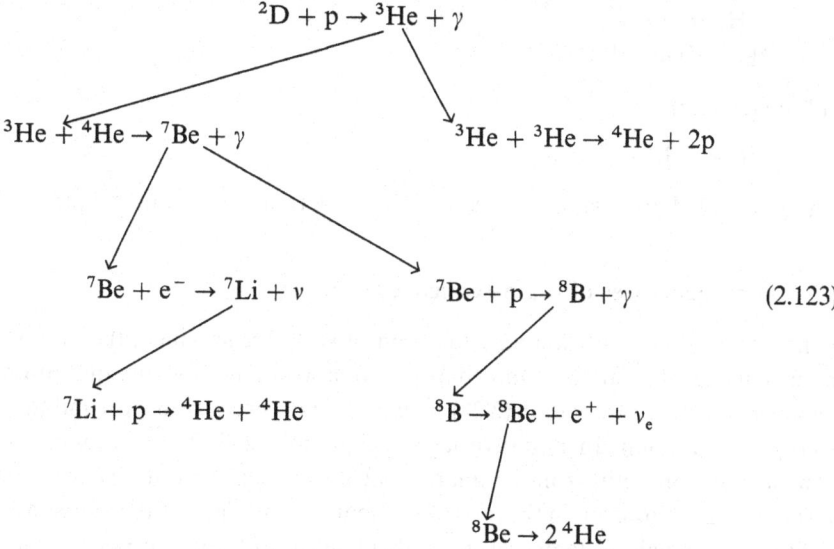

$$\tag{2.123}$$

The temperature of the gas in a star must be $\geq 10^7\,$K for energy generation by means of the proton–proton chain to become significant.

In main-sequence stars with masses $M \gtrsim 1.2\,M_\odot$, red giants and other post-main-sequence stars the core temperatures are sufficiently high (i.e. $T \gtrsim 18 \times 10^6\,$K) that hydrogen burning is complete by means of the CNO cycle. The principal way in which the CNO cycle is completed is given by the following reactions and beta decays:

$$\begin{aligned}
{}^{12}C + p &\rightarrow {}^{13}N + \gamma, \\
{}^{13}N &\rightarrow {}^{13}C + e^+ + \nu_e, \\
{}^{13}C + p &\rightarrow {}^{14}N + \gamma, \\
{}^{14}N + p &\rightarrow {}^{15}O + \gamma, \\
{}^{15}O &\rightarrow {}^{15}N + e^+ + \nu_e, \\
{}^{15}N + p &\rightarrow {}^{12}C + {}^4He.
\end{aligned} \tag{2.124a}$$

Because the last of the above reactions can produce ${}^{16}O$ but with a lower probability than ${}^{12}C + {}^4He$, hydrogen burning can also be completed by the reactions

$${}^{15}N + p \rightarrow {}^{16}O + \gamma,$$

$$^{16}O + p \rightarrow ^{17}F + \gamma,$$
$$^{17}F \rightarrow ^{17}O + e^+ + \nu_e,$$
$$^{17}O + p \rightarrow ^{14}N + ^4He.$$

(2.124b)

The relative rates of the above reactions imply that the CNO cycle transforms most of the initial ^{12}C into ^{14}N.

At temperatures $\gtrsim 10^8$ K, helium burning occurs in the interiors of stars as a consequence of the triple α reactions

$$2^4He \rightleftarrows ^8Be + \gamma,$$
$$^8Be + ^4He \rightarrow ^{12}C + \gamma$$

(2.125)

and the reaction

$$^{12}C + ^4He \rightarrow ^{16}O + \gamma.$$

(2.126)

A more detailed discussion of thermonuclear reactions is given in Chapter 6.

2.7 The structures of main-sequence stars

If the chemical composition and equation of state are specified, then the structures of main-sequence stars are determined by the equation of hydrostatic equilibrium (2.27), the equation of mass conservation (2.28), an equation of energy transport and the equation of energy conservation. In radiative layers, Equation (2.80) is the equation of radiative transport. In convective zones, Equations (2.80), (2.120), and (2.121) determine the transport of energy. However, in the interiors of main-sequence stars the temperature gradient is close to the adiabatic temperature gradient because convective energy transport is very efficient, and therefore for a fully ionized, nondegenerate gas, Equation (2.114) implies that the equation of energy transport in a convective layer is

$$\frac{P}{T}\frac{dT}{dP} = \frac{\Gamma_1 - 1}{\Gamma_1} = 0.4.$$

(2.127)

The heat added to a mass element of gas in the time interval Δt is

$$\Delta Q = \left(\frac{-dL_r}{dM_r} + \varepsilon_n - \varepsilon_\nu\right)\Delta t,$$

(2.128)

where ε_n is the total rate of thermonuclear energy generation and ε_ν the rate of neutrino emission. Because neutrinos interact with matter only via the weak interaction, their mean free paths are generally much greater than the dimension of a star, and therefore energy emitted in the form of neutrinos is rapidly removed from the interiors of stars. From Equations (2.13) and (2.128) it follows that the equation of energy conservation in stellar interiors is

$$\frac{dU}{dt} + P\frac{d\left(\frac{1}{\rho}\right)}{dt} = \varepsilon_n - \varepsilon_\nu - \frac{dL_r}{dM_r}.$$

(2.129)

For a fully ionized gas

$$U = 3/2P/\rho,$$ (2.130)

and therefore Equation (2.129) becomes

$$3/2\rho^{2/3}\frac{d(P/\rho^{5/3})}{dt} = \varepsilon_n - \varepsilon_v - \frac{dL_r}{dM_r}.$$ (2.131)

Because of the large amount of energy generated by hydrogen burning, the left-hand side of Equation (2.131) is small for main-sequence stars. The opacity κ, which appears in Equation (2.80), and the rates of thermonuclear energy generation and neutrino emission that appear in Equation (2.131) depend on the temperature and pressure (or density) of the gas. The structure of an initial main-sequence star is determined by solving Equations (2.27), (2.28), (2.80) (in convective zones Equation (2.127) replaces (2.80)) and Equation (2.129) with the time-derivative term set equal to zero. These four equations with four boundary conditions determine completely the star's structure if the chemical composition is specified. It is convenient to solve these equations for $P = P(M_r)$, $T = T(M_r)$, $L_r = L_r(M_r)$ and $r = r(M_r)$. The four boundary conditions are $r = L_r = 0$ at $M_r = 0$ and $P = 0$ and $T =$ the temperature at $M_r = M$. The temperature $T = T(M_r)$, which is close to but not equal to the effective temperature, is determined by stellar-atmosphere calculations. Once the structure of a main-sequence star has been determined, the rate of change of chemical composition in thermonuclear-burning regions can be calculated, and consequently the evolution of the star on and beyond the main sequence can be calculated by solving the above equations of stellar interiors with the time-derivative term in Equation (2.129) included in the calculations.

The structures of main-sequence stars can be determined approximately by means of dimensional analysis. If we replace Equations (2.27) and (2.28) by the dimensional relations

$$\frac{P}{R} \sim \rho\frac{GM}{R^2}$$ (2.132)

and

$$\rho \sim \frac{M}{R^3},$$ (2.133)

it follows that the central pressure is approximately

$$P \sim \frac{GM^2}{R^4}.$$ (2.134)

If radiation pressure can be neglected, then from Equation (2.6) the central temperature becomes

$$T \sim \frac{Gm_u\mu M}{kR}.$$ (2.135)

If radiative energy transport is assumed throughout the star, then Equation (2.80) implies that the luminosity is

$$L \sim \frac{ac}{3} 4\pi R^2 \frac{T^4}{R} \frac{1}{\kappa\rho} \sim \frac{ac}{3} 4\pi \left(\frac{m_u G}{k}\right)^4 \frac{\mu^4 M^3}{\kappa}. \tag{2.136}$$

If the opacity is caused primarily by electron scattering, then the dimensional relationship (2.136) implies that the luminosity of a main-sequence star varies as M^3. For low-mass main-sequence stars, absorption caused by bound–free and bound–bound transitions is important. The mass–luminosity relationship for such stars is approximately $L \propto M^4$.

Although electron scattering is the dominant source of opacity in very massive (i.e. $M \gtrsim 50\,M_\odot$) stars, the dimensional relation (2.136) does not apply, because radiation pressure is much greater than gas pressure. The luminosities of such stars are given approximately by equating the absolute value of the gravitation force on a proton to the radiative force on an electron at the surface. The condition for these two forces to have equal magnitude is

$$\frac{GMm_p}{R^2} = \sigma_T \frac{1}{3} aT^4 = \frac{4}{3}\sigma_T \frac{\sigma T^4}{c}, \tag{2.137}$$

with σ_T equal to the Thomson cross section, T equal to the effective temperature of the star and $\sigma = ac/4$. Equation (2.137) implies that

$$\frac{L}{L_\odot} = \frac{4\pi R^2 \sigma T^4}{L_\odot} = \frac{3\pi c G M m_p}{\sigma_T L_\odot} \simeq 3.3 \times 10^4 \frac{M}{M_\odot} \tag{2.138}$$

for pure hydrogen.

Very massive stars are convective throughout most of their interiors, and consequently the entropy density is also approximately a constant. Since radiation pressure greatly exceeds gas pressure in massive stars, the pressure and entropy density are given by Equations (2.9) and (2.20), respectively. Solving for T from the expression for the entropy per gram, we obtain

$$T = \left(\frac{3\rho S}{4a}\right)^{1/3}. \tag{2.139}$$

Equation (2.139) implies that the pressure can be expressed as

$$P = \tfrac{1}{3}aT^4 = K\rho^{4/3}, \tag{2.140}$$

with $K = \tfrac{1}{3}a\left(\dfrac{3S}{4a}\right)^{4/3}$. Since K is nearly constant throughout the interior, Equation (2.140) implies that the structure of a very massive star is that of a polytrope of index $n = 3$. The mass–luminosity relation in Equation (2.138) is referred to as the Eddington limit. An approximate mass for η Carina, discussed in Chapter 1, can be obtained from Equation (2.138).

2.8   **Polytropes**

If the equation of state is such that $P = P(\rho)$, then the structure of a star is determined by Equations (2.27) and (2.28). In particular, if the pressure everywhere in a stellar interior satisfies the equation

$$P = K\rho^{(n+1)/n} = K\rho^{1+(1/n)},$$
(2.141)

with K and n constants, then the stellar configuration is called a polytrope. In Equation (2.141) n is defined as the polytropic index.

Equations (2.27) and (2.28) can be readily reduced to the second-order differential equation

$$\frac{d\left(\dfrac{r^2}{\rho}\dfrac{dP}{dr}\right)}{r^2 dr} = -4\pi G\rho.$$
(2.142)

It is convenient to define the dimensionless variables

$$\rho(r) = \rho_c \theta^n(r)$$
(2.143)

and

$$r = a\xi,$$
(2.144)

with ρ_c equal to the central density and the constant a given by the equation

$$a = \left[\frac{(n+1)K\rho_c^{(1/n-1)}}{4\pi G}\right]^{1/2}.$$
(2.145)

If Equations (2.141), (2.143), (2.144) and (2.145) are substituted into Equation (2.142) then the latter equation becomes

$$\frac{1}{\xi^2}\frac{d}{d\xi}\xi^2\frac{d\theta}{d\xi} = -\theta^n.$$
(2.146)

From the definition of $\theta(r)$ and the boundary condition

$$\frac{dP}{dr} = 0 \text{ at } r = 0,$$
(2.147)

it follows immediately that $\theta(0) = 1$ and $\theta'(0) = 0$. If the polytropic index n and central density ρ_c are given, then Equation (2.146) can be integrated numerically from the center of the star to some radial distance $r = R$ where $P = 0$. The mass of a polytrope is a function of the central density ρ_c and the polytropic index. The mass of a polytrope of polytropic index n is

$$M = \int_0^{\xi(R)} \rho 4\pi r^2 \, dr = 4\pi a^3 \rho_c \int_0^{\xi(R)} \theta^n \xi^2 \, d\xi.$$
(2.148)

From Equations (2.146) and (2.148) it follows that

$$M = 4\pi a^3 \rho_c \int \frac{d\left(\xi^2 \frac{d\theta}{d\xi}\right)}{\xi^2 \, d\xi} \xi^2 \, d\xi = 4\pi a^3 \rho_c \xi^2(R) \frac{d\theta}{d\xi}\bigg|_{\xi_1 = \xi(R)} \tag{2.149}$$

From Equations (2.143), (2.144), (2.145) and (2.149) we can eliminate ρ_c and obtain

$$M = 4\pi R^{(3-n)/(1-n)} \left[\frac{(n+1)K}{4\pi G}\right]^{n/(n-1)} \xi_1^{(3-n)/(1-n)} \xi_1^2 |\theta'(\xi_1)|. \tag{2.150}$$

The ratio of the central density ρ_c divided by the mean density $\bar{\rho}$ is

$$\frac{\rho_c}{\bar{\rho}} = \frac{\rho_c \frac{4}{3}\pi R^3}{\displaystyle\int_0^R \rho 4\pi r^2 \, dr}. \tag{2.151}$$

From Equations (2.143–2.146) we obtain

$$\frac{\rho_c}{\bar{\rho}} = \frac{\xi_1^3}{3 \displaystyle\int_0^{\xi_1} \theta^n \xi^2 \, d\xi} = \frac{\xi_1}{3|\theta'(\xi_1)|}. \tag{2.152}$$

The above ratio of central density divided by mean density increases with the polytropic index. The higher the polytropic index the more centrally condensed the polytrope.

For a polytrope of index n the equation of hydrostatic equilibrium becomes

$$\frac{1}{\rho}\frac{dP}{dr} = (n+1)\frac{d(P/\rho)}{dr} = \frac{-GM_r}{r^2}. \tag{2.153}$$

Integrating Equation (2.153) gives

$$(n+1)\frac{P}{\rho} = -\Phi + \Phi_0, \tag{2.154}$$

with

$$\Phi(r) = -\int_r^\infty \frac{GM_{r'}}{r'^2} \, dr' \tag{2.155}$$

the gravitational potential at radial distance r and

$$\Phi_0 = -\frac{GM}{R} \tag{2.156}$$

the gravitational potential at the surface. The gravitational potential energy Ω of a star is

$$\Omega = \frac{1}{2}\int_0^R \Phi \, dM_r. \tag{2.157}$$

The above equations imply that

$$-\Omega = \frac{1}{2}(n+1) \int_o^R \frac{P}{\rho}\,dM_r + \frac{1}{2}\frac{GM^2}{R}$$

$$= \frac{1}{2}(n+1) \int_0^R P4\pi r^2\,dr + \frac{1}{2}\frac{GM^2}{R}. \tag{2.158}$$

The virial theorem given in Equation (2.35) and Equation (2.158) imply that Ω is given by the equation

$$-\Omega = \frac{3}{5-n}\frac{GM^2}{R}. \tag{2.159}$$

The above equation implies that $-\Omega$ becomes very large as $n \to 5$, and consequently polytropes with polytropic index $n \geq 5$ do not exist. For a fully ionized gas the virial theorem implies that the gravitational binding energy of a polytrope is $\frac{1}{2}|\Omega|$.

2.9 Numerical solutions of the equations of stellar interiors

The equations of stellar interiors expressed as functions of M_r are

$$\frac{dP}{dM_r} = -\frac{GM_r}{4\pi r^4}, \tag{2.160}$$

$$\frac{dr}{dM_r} = \frac{1}{4\pi r^2 \rho}, \tag{2.161}$$

$$\frac{L_r}{16\pi r^4} = -\frac{ac}{3\kappa}\frac{dT^4}{dM_r} \quad \text{(radiative)}, \tag{2.162a}$$

$$\frac{P}{T}\frac{dT}{dP} = 0.4 \quad \text{(convective)}, \tag{2.162b}$$

$$\frac{dU}{dt} + P\frac{d(1/\rho)}{dt} = \varepsilon_n - \varepsilon_v - \frac{dL_r}{dM_r}, \tag{2.163}$$

with the equation of state $\rho = \rho(P, T)$. If the left-hand side of Equation (2.163) is small, as is the case for main-sequence stars, then the above equations can be integrated subject to the two surface and two central-boundary conditions given in Section 2.7. A more general method, which is known as the relaxation method, is described below.

The relaxation method requires an initial estimate of the values of the physical variables that describe the structure of the star (i.e. $P = P(M_r)$, $T = (M_r)$, $L_r = L_r(M_r)$ and $r = r(M_r)$). For initial-main-sequence stars a polytrope of index $n = 3$ is a sufficiently accurate estimate. In the relaxation method, corrections to the initially estimated solutions are calculated. These corrections to the estimated values of the physical variables

P, T, L_r and r are added to the initial estimates for these solutions and the procedure is repeated iteratively until the solutions converge.

The first step in the relaxation method is to replace the differential equations (2.160–2.163) with corresponding finite-difference equations. A set of mass points $M_r(i)$, $i = 1$, $2, \ldots, m$ is defined, and initial estimated values for P_i, T_i, $L_r(i)$ and r_i are made. The differential equation (2.160) is replaced by the finite-difference equations

$$\frac{P_{i+1} - P_i}{\Delta M_r(i)} = -\frac{1}{2}\left[\frac{GM_r(i+1)}{4\pi r_{i+1}^4} + \frac{GM_r(i)}{4\pi r_i^4}\right], \quad i = 1,\ldots,m-1, \tag{2.164}$$

with $\Delta M_r(i) = M_r(i+1) - M_r(i)$. Equations (2.161–163) are in a similar manner replaced by finite-difference equations. Because it contains two time-derivative terms, Equation (2.163) is somewhat more complicated than the other equations. These time-derivative terms are backwards differenced in time so that, for example, the dU/dt terms become

$$\frac{U_{i+1}^{n+1} - U_i^n}{\Delta t} \quad i = 1,\ldots,m-1, \tag{2.165}$$

where n labels the values of the internal energy density $U = U(P,T)$ as a function of discrete time steps, and the time step $\Delta t = t_{n+1} - t_n$ is small. In Equation (2.165) the present time is labeled $n + 1$ and some recent previous time labeled n. In solving the difference equations it is assumed that the values of the physical variables P, T, L_r and r are known at time t_n. In numerically simulating the evolution of stars, the initial-main-sequence model is calculated by setting the left-hand side of Equation (2.163) equal to zero. Subsequent evolutionary stellar models are calculated with the time derivative included but with the structure of the star known at the previous time step. For stars that have evolved from the initial main sequence the values of the physical variables at the previous time step can be used as the initial estimate for the structure. In calculating the structure of a star by means of the relaxation method the finite-difference equations derived from the differential equations (2.160–2.163) are linearized. For example, Equation (2.164) becomes

$$\frac{P_{i+1} + \Delta P_{i+1} - P_i - \Delta P_i}{\Delta M_r(i)} = -\frac{1}{2}\left[\frac{GM_r(i+1)}{4\pi r_{i+1}^4} + \frac{GM_r(i)}{4\pi r_i^4}\right]$$

$$+ \frac{1}{2}\left[\frac{GM_r(i+1)}{16\pi r_{i+1}^5}\cdot\Delta r_{i+1} + \frac{GM_r(i)}{16\pi r_i^5}\cdot\Delta r_i\right]$$

$$i = 1,\ldots,m-1. \tag{2.166}$$

If we rearrange terms, Equation (2.166) can be expressed as

$$\frac{\Delta P_{i+1}}{\Delta M_r(i)} - \frac{\Delta P_i}{\Delta M_r(i)} - \frac{1}{2}\left[\frac{GM_r(i+1)}{16\pi r_{i+1}^5}\Delta r_{i+1} + \frac{GM_r(i)}{16\pi r_i^5}\Delta r_i\right]$$

$$= -\frac{1}{2}\left[\frac{GM_r(i+1)}{4\pi r_{i+1}^4} + \frac{GM_r(i)}{4\pi r_i^4}\right] - \frac{P_{i+1}}{\Delta M_r(i)} + \frac{P_i}{\Delta M_r(i)} \tag{2.167}$$

$$i = 1,\ldots,m-1.$$

All the terms on the right-hand side of Equation (2.167) are known. Moreover, the right-hand side of Equation (2.167) is not zero because the initial estimates of P_i, T_i, $L_r(i)$ and r_i are not the actual solutions for these variables. The remaining finite-difference equations derived from Equations (2.161–163) can be linearized in a similar manner, and we end up with $4m - 4$ linear equations for the $4m$ variables ΔP_i, ΔT_i, $\Delta L_r(i)$ and Δr_r. These $4m - 4$ equations and the 4 boundary conditions expressed in finite-difference form lead to $4m$ linear equations for $4m$ variables, and consequently the system of equations can be solved. The calculated solutions ΔP_i, ΔT_i, $\Delta L_r(i)$ and Δr_i $(i = 1, \ldots, m)$ are added to the estimated values of the physical variable, and we find that

$$
\begin{aligned}
P'_i &= P_i + \Delta P_i \quad (i = 1, \ldots, m), \\
T'_i &= T_i + \Delta T_i, \\
L'_r(i) &= L_r(i) + \Delta L_r(i), \\
r'_i &= r_i + \Delta r_i.
\end{aligned}
\tag{2.168}
$$

The quantities P'_i, T'_i, $L'_r(i)$ and $r'_i(i = 1, \ldots, m)$ are then used as initial estimates for the unknown solutions, and a new but similar system of linear equations is obtained. The values of ΔP_i, ΔT_i, $\Delta L_r(i)$ and Δr_i in this new system of equations can then be found and corrected estimates for P_i, T_i, $L_r(i)$ and r_i determined. This iterative procedure is repeated until solutions for the physical variables at the various mass points converge.

Problems

1. Prove that the gravitational force on a point unit mass at radial distance r is $- GM_r/r^2$ for a spherical star.

2. Find the molecular weight μ and electron molecular weight μ_e of an ionized gas with $X = 0.73$, $Y = 0.25$, $Z = 0.02$, where X, Y and Z are the mass fractions of hydrogen, helium and ^{12}C, respectively.

3. a. Prove the thermodynamic relation

$$
\left(\frac{\partial U}{\partial V} \right)_T = T \left(\frac{\partial P}{\partial T} \right)_V - P.
$$

   b. Prove that the pressure P_r and internal energy density U_r of a photon gas are related by the equation $P_r = U_r/3$.

4. Using the results of problem 3, show that $U_r = aT^4$, with a equal to a constant.

5. Two similar perfect gases with the same pressure P but at different temperatures T_1 and T_2 are initially confined to volumes V_1 and V_2. Find the change in entropy after they reach equilibrium with each other. The final gas volume is $V = V_1 + V_2$.

6. Determine how much energy is radiated from a $1\,M_\odot$ star before reaching the main sequence. Assume that a $1\,M_\odot$ star is a polytrope of index $n = 3$ and has an initial-main-sequence radius $R = R_\odot$. Estimate its pre-main-sequence lifetime.

7.   How much energy is released when one gram of hydrogen is transformed into
      ^{4}He? Estimate the main-sequence lifetime of a $1\,M_\odot$ star. Assume that 12% of its
      hydrogen is consumed on the main sequence. Estimate the temperature at the
      center of a $1\,M_\odot$ main-sequence star.

8.   The central density, pressure and temperature of a main-sequence star are ρ_c, P_c
      and T_c, respectively. The nuclear-energy generation rate ε is a known function of
      ρ and T. Find expressions for M_r, p and L_r at small radial distances from the
      center of the star. Find similar expressions for T, assuming both radiative and
      convective cores.

9.   Consider a star with radiative outer layers. Assume that $T = 0$ and $P = 0$ at
      $r = R$, and $M_r = M$ in the outer layers. Find expressions for how T depends on r
      close to $r = R$ for electron-scattering opacity (i.e. $\kappa = 0.2(1 + X)$).

10.   Consider an inhomogeneous star that is all hydrogen exterior to some mass shell
      M_r and all helium interior to M_r. Find the ratio of interior density divided by
      exterior density at the density discontinuity M_r.

11.   Prove Equation (2.82).

12.   Prove that the fourth term in Equation (2.100) is equal to zero.

13.   Derive Equations (2.107), (2.110a) and (2.110b) and show how they are modified
      if the molecular weight $\mu = \mu(T, P)$.

14.   Equation (2.164) is the finite-difference equation corresponding to the differential
      equation (2.160). Write down the difference equations that represent Equations
      (2.161–2.163), and for at least one of these equations derive an equation
      analogous to Equation (2.166).

15.   a. Let L_H and L_{He} equal the luminosities of pure-hydrogen and pure-helium
         main-sequence stars with similar mass. Calculate the ratio of their
         luminosities, assuming that the opacities of both stars are caused by electron
         scattering. Estimate the luminosity of a $0.5\,M_\odot$ helium main-sequence star.
         Assume that the mean opacity of the Sun is $\kappa = 10\,\text{cm}^2\,\text{g}^{-1}$.
      b. What does the virial theorem imply about the internal and gravitational
         energies of very massive stars $(P \gg P_{gas})$?

3 Statistical physics

3.1 The Maxwell–Boltzmann distribution

The particle distribution function $f(p)$ of a gas can be normalized so that

$$n = \int f(p)\mathrm{d}^3p,$$ (3.1)

with n equal to the number density of particles. The mean value of some physical variable such as the energy per particle becomes

$$\bar{\varepsilon} = \frac{1}{n}\int \varepsilon f(p)\mathrm{d}^3p.$$ (3.2)

Under conditions of thermodynamic equilibrium the distribution function is independent of initial conditions except for the total energy and momentum of the particles. We assume that the particle distribution function is defined in a reference frame in which the total particle momentum is zero. If a gas is in thermodynamic equilibrium, then statistical averaging of a physical variable such as the energy of a particle is equivalent to determining the time average value of the physical variable.

The entropy per particle can be defined by the equation

$$s = \frac{-k}{n}\int \ln\!f f \mathrm{d}^3p.$$ (3.3)

As will be shown below, this definition of the entropy per particle is similar to that derived in Chapter 2 to within a constant. We will derive the expression for the Maxwell–Boltzmann distribution function from the requirement that s is an extremum (maximum), subject to the conditions

$$\int f\mathrm{d}^3p = n$$ (3.4)

and

$$\int f\varepsilon\,\mathrm{d}^3p = \text{constant},$$ (3.5)

with ε equal to the energy of a particle. Using the method of Lagrangian multipliers, we add s, given by the right-hand side of Equation (3.3), a constant times the left-hand side of Equation (3.4) and another constant times the left-hand side of Equation (3.5). We then vary f inside the integral and obtain the equation

$$\int [\ln f + 1 + \alpha + \beta\varepsilon]\delta f d^3p = 0, \tag{3.6}$$

with α and β constants. The integrand in Equation (3.6) is identically zero because the small variation δf is arbitrary. Therefore, Equation (3.6) implies that

$$f = Ce^{-\beta\varepsilon}, \tag{3.7}$$

with C equal to a constant. From Equations (3.4) and (3.7) we obtain

$$C\int_0^\infty e^{-\varepsilon/kT}4\pi p^2 dp = n, \tag{3.8}$$

where the temperature of the gas has been defined by the equation

$$\beta = \frac{1}{kT}. \tag{3.9}$$

If a gas is nonrelativistic then the particle energy ε that appears in Equation (3.8) is equal to $p^2/2m$. Expressing the Maxwell–Boltzmann distribution function as a function of momentum, we find that Equations (3.7) and (3.8) lead to

$$\int f(p)4\pi p^2 dp = n, \tag{3.10}$$

with

$$f(p) = \frac{n\exp\left(\dfrac{-p^2}{2mkT}\right)}{(2\pi mkT)^{3/2}}. \tag{3.11}$$

If the particle distribution is expressed as a function of velocity, then Equations (3.10) and (3.11) become

$$\int_0^\infty f(v)4\pi v^2 dv = n, \tag{3.12}$$

with

$$f(v) = n\left(\frac{m}{2\pi kT}\right)^{3/2} e^{-mv^2/2kT}. \tag{3.13}$$

The pressure of a gas can be calculated from the expression for the distribution function given in Equation (3.13). As particles in a gas collide with a surface they exchange momentum with the surface. The force per unit area is the pressure of the gas. Let dA equal an element of surface area, and define θ equal to the angle between the normal to the surface and the direction of a particle that is incident onto the surface element. For elastic collisions, particles of speed v that are incident onto the surface with an angle θ undergo a change in momentum equal to

$$\Delta p = 2mv \cos \theta \tag{3.14}$$

in a single elastic collision. The contribution to the gas pressure caused by particles with speeds between v and $v + dv$, and incident onto the surface within the solid angle $d\Omega = 2\pi \sin \theta \, d\theta$, is

$$dP = 2mv \cos \theta \, v \cos \theta f(v)v^2 dv 2\pi \sin \theta d\theta. \tag{3.15}$$

In Equation (3.15) $f(v)$ is given by Equation (3.13) and $v \cos \theta$ is the projection of the particle's velocity along the normal to the surface. From Equation (3.15) it follows that

$$P = 4\pi \int_0^{\pi/2} \int_0^\infty mv^4 f(v) dv \cos^2 \theta \sin \theta d\theta$$

$$= \frac{4\pi nm}{3} \left(\frac{m}{2\pi kT} \right)^{3/2} \int_0^\infty e^{-mv^2/2kT} v^4 dv = nkT. \tag{3.16}$$

The thermal energy density of the gas is

$$U = 4\pi \left(\frac{m}{2\pi kT} \right)^{3/2} n \int \frac{1}{2} mv^4 e^{-mv^2/2kT} dv = \frac{3}{2} nkT. \tag{3.17}$$

Equations (3.16) and (3.17) are derivations of the expressions for gas pressure and internal energy density that were already assumed in Chapter 2.

Substituting Equation (3.13) (or 3.11) into Equation (3.3), we find that the entropy per particle is

$$s = k\ln(T^{3/2}/n) + s_0, \tag{3.18}$$

with s_0 equal to a constant. The constant s_0 that appears in Equation (3.18) can only be determined within the framework of quantum statistics, which will be discussed in the next section of this chapter. The expression for s given in Equation (3.18) diverges as $T \to 0$. At sufficiently low temperatures quantum statistics must be used to describe the physical properties of a gas.

From Equation (3.17) the heat capacity c_v per unit volume becomes

$$c_v = \left(\frac{dU}{dT} \right) = \frac{3}{2} nk. \tag{3.19}$$

Equation (3.18) implies that the entropy per unit volume can be written as

$$S = ns = nk\ln(T^{3/2}/n) + ns_0. \tag{3.20}$$

The expression for c_v given in Equation (3.19) also follows from Equation (3.20) and the relation $c_v = T(dS/dT)_v$, which is implied by the first law of thermodynamics.

3.2   Quantum statistics

The fundamental properties of matter are governed by the principles of quantum mechanics. The Maxwell–Boltzmann distribution is valid in the classical limit when individual particles are sufficiently well separated that quantum effects can be neglected. Quantum mechanics implies that the number of quantum states per unit volume that a particle can occupy in each element d^3p of momentum space is finite. As will be shown below, the number of such quantum states available to a free particle is $(2s + 1)\, d^3p/h^3$, with s equal to the spin of the particle. Those particles that have half-integer spin (i.e. 1/2, 3/2, 5/2, ...), such as electrons, protons, neutrons and some nuclei, are called fermions because they obey Fermi–Dirac statistics. Particles with spin equal to 0, 1, 2, ..., such as photons, π mesons and some nuclei, are called bosons because they obey Bose–Einstein statistics. The fermion and boson distribution functions, which determine the probability that a particular quantum state is occupied by a particle, both reduce to the Maxwell–Boltzmann distribution function in the limit of low particle density.

The Schrödinger equation

A quantum state is a particular eigenfunction (solution) of a wave equation that describes the motion of a particle (or system of particles). Associated with each eigenfunction there is an eigenvalue, which is the energy of the quantum state. Different quantum states may have the same energy. Quantum states that have the same energy are called degenerate. The Schrödinger equation describes the motions of nonrelativistic electrons and other fermions.

The energy of an electron is the sum of the kinetic and potential energy, that is

$$E = \frac{p^2}{2m} + V(x, y, z). \tag{3.21}$$

The momentum operator of a particle is

$$p = \frac{\hbar}{i}\nabla. \tag{3.22}$$

From Equations (3.21) and (3.22) we find that the Hamiltonian, which is the energy operator, is given by the expression

$$H = \frac{1}{2m}(p_x^2 + p_y^2 + p_z^2) + V(x, y, z)$$

$$= \frac{-h^2}{8\pi^2 m}\left(\frac{\partial^2}{\partial x^2} + \frac{\partial^2}{\partial y^2} + \frac{\partial^2}{\partial z^2}\right) + V(x, y, z). \tag{3.23}$$

The Schrödinger equation is

$$H\psi = \frac{-\hbar}{i}\frac{\partial\psi}{\partial t},$$

(3.24)

with the wave function ψ and its complex conjugate $\psi^{\star}$ normalized so that

$$\int \psi^{\star}\psi d^3 r = 1.$$

(3.25)

The integral given above extends over all space, and the product $\psi^{\star}\psi$ is the probability per unit volume that the electron is at a particular position in space. The Hamiltonian given in Equation (3.23) is time independent, and consequently the Schrödinger equation has stationary eigenfunctions,

$$\psi_n(x, y, z, t) = u_n(x, y, z)e^{-iE_n t/\hbar} \quad n = 1, 2, 3, \ldots$$

(3.26)

Substituting ψ_n into Equation (3.24) implies that

$$Hu_n = E_n u_n \quad n = 1, 2, 3, \ldots$$

(3.27)

with E_n equal to the eigenvalue of the eigenfunction u_n. Equations (3.23) and (3.27) imply that the Schrödinger equation can be expressed as

$$\frac{\partial^2 u}{\partial x^2} + \frac{\partial^2 u}{\partial y^2} + \frac{\partial^2 u}{\partial z^2} = -\frac{8\pi^2 m}{h^2}(E - V(x, y, z))u.$$

(3.28)

Consider an electron that is inside a cubic region of space in which $V(x, y, z) = 0$. The walls surrounding this region of space are impenetrable (i.e. V is large and positive), and consequently the eigenfunctions must vanish at the boundaries. It follows that the eigenfunctions of Equation (3.28) are

$$u = C \sin\frac{n_x \pi x}{L} \sin\frac{n_y \pi y}{L} \sin\frac{n_z \pi z}{L},$$

(3.29)

with n_x, n_y and n_z equal to positive integers and L equal to the dimension of the cubic region of space. Substituting the above expression for u into Equation (3.28) with $V(x,y,z)$ equal to zero implies that

$$E = \frac{h^2}{8mL^2}(n_x^2 + n_y^2 + n_z^2).$$

(3.30)

The required normalization of the wave function u is

$$C = \sqrt{8}L^{-3/2}.$$

(3.31)

If L is sufficiently large, then the number of quantum states with energy $\leq E$ is very nearly equal to

$$2 \cdot \frac{1}{8}\int_{-n}^{n}\int_{-n}^{n}\int_{-n}^{n} dn_x dn_y dn_z = \frac{1}{4} \cdot \frac{4\pi}{3}n^3,$$

(3.32)

with

$$n = \frac{\sqrt{8mE}}{h} L.$$

In Equation (3.32) the factor 2 appears because there are two electron spin states for each eigenfunction, and the factor $\frac{1}{8}$, because only values of n_1, n_2 and n_3 that are positive integers correspond to quantum states of the electron. Equations (3.30) and (3.32) imply that the number of quantum states per unit volume in the energy interval E, $E + dE$ is

$$g(E)dE = \frac{8\pi}{h^3} m\sqrt{2mE}\, dE = \frac{2d^3p}{h^3}. \tag{3.33}$$

In electromagnetic theory, conservation of charge is a direct consequence of Maxwell's equations and is expressed by the equation

$$\frac{\partial \rho}{\partial t} + \nabla \cdot J = 0, \tag{3.34}$$

with ρ and J equal to the charge density and current density, respectively. A mathematically similar equation that guarantees conservation of probability follows from the Schrödinger equation. Multiplying the Schrödinger equation given by Equations (3.23) and (3.24) by $\psi^\star$ implies that

$$-\psi^\star \frac{\hbar}{i} \frac{\partial \psi}{\partial t} = -\frac{\hbar^2}{2m} \psi^\star \nabla^2 \psi + V\psi^\star \psi. \tag{3.35}$$

Multiplying the complex conjugate of the Schrödinger equation by ψ gives

$$\frac{\hbar}{i} \frac{\partial \psi^\star}{\partial t} \psi = -\frac{\hbar^2}{2m} (\nabla^2 \psi^\star)\psi + V\psi^\star \psi. \tag{3.36}$$

Subtracting Equations (3.35) from Equation (3.36) leads to the equation

$$\frac{\partial P(r,t)}{\partial t} + \nabla \cdot S = 0, \tag{3.37}$$

with

$$P(r,t) = \psi^\star \psi$$

and

$$S(r,t) = \frac{\hbar}{2im} (\psi^\star \nabla \phi - (\nabla \psi^\star)\psi).$$

Equation (3.37) shows that the interpretation of $\psi^\star \psi$ as the probability density is physically meaningful because probability is conserved exactly.

The Dirac equation

The Schrödinger equation, which we have used to derive the number density of free-electron quantum states, is a nonrelativistic wave equation. In the dense interior regions of some stars (e.g. massive white dwarfs) electrons have relativistic energies, and the Dirac equation must be used to describe electron motions.

The energy and momentum of a free particle are related by the equation

$$E^2 = p^2 c^2 + (mc^2)^2. \tag{3.38}$$

The Dirac equation is derived by introducing the Hamiltonian

$$H = c(\alpha_x p_x + \alpha_y p_y + \alpha_z p_z) + \beta mc^2, \tag{3.39}$$

with α_x, α_y, α_z and β equal to 4×4 matrices. The momentum operator p in Equation (3.39) is defined in Equation (3.22). If we assume stationary solutions, the wave equation implied by the Hamiltonian given in Equation (3.39) is

$$H\psi = c(\alpha_x p_x + \alpha_y p_y + \alpha_z p_z)\psi + \beta mc^2 \psi = E\psi. \tag{3.40}$$

Equation (3.40) can be written as

$$[E - c\alpha \cdot p - \beta mc^2]\psi = 0. \tag{3.41}$$

Multiplying Equation (3.41) from the left by the operator $[E + c\alpha \cdot p + \beta mc^2]$ we find that

$$\left[E^2 - \sum_{k=1}^{3} (\alpha^k)^2 (p^k)^2 c^2 - \beta^2 (mc^2)^2 - \frac{1}{2} \sum_{k=1}^{3} \sum_{l=1}^{3} (\alpha^k \alpha^l + \alpha^l \alpha^k) p^k p^l c^2 \right.$$

$$\left. - \sum_{k=1}^{3} (\alpha^k \beta + \beta \alpha^k) mc^3 p^k \right] \psi = 0. \tag{3.42}$$

The Dirac equation follows from Equation (3.42) if we add the condition that the relationship between energy and momentum must satisfy Equation (3.38). This latter requirement implies that the matrices $\alpha^1 = \alpha_x$, $\alpha^2 = \alpha_y$, $\alpha^3 = \alpha_z$ and β satisfy the equations

$$\begin{aligned} (\alpha^k)^2 &= 1, & \alpha^k \alpha^l + \alpha^l \alpha^k &= 2\delta_{kl}, \\ \beta^2 &= 1, & \alpha^k \beta + \beta \alpha^k &= 0. \end{aligned} \tag{3.43}$$

In Appendix 2 we obtain values for the 4×4 matrices α_x, α_y, α_z and β that satisfy conditions (3.43) and show that in free space the Dirac equation has the plane-wave solutions

$$e^{i(k \cdot r - \omega t)} \tag{3.44}$$

with the electron momentum and energy equal to $p = \hbar k$ and $E = \hbar \omega$, respectively.

In order to obtain the number density of quantum states we choose an arbitrary cubic region of space with dimension L and impose periodic boundary conditions, that is

$$\begin{aligned} \psi(x + L, y, z) &= \psi(x, y, z), \\ \psi(x, y + L, z) &= \psi(x, y, z), \\ \psi(x, y, z + L) &= \psi(x, y, z). \end{aligned} \tag{3.45}$$

The above boundary conditions imply that plane-wave solutions of the Dirac equation are of the form

$$e^{i\left[\frac{2\pi}{L}(n_1 x + n_2 y + n_3 z) - \frac{E(n_1, n_2, n_3)t}{\hbar}\right]}, \tag{3.46}$$

with n_1, n_2 and n_3 equal to integers. Plane-wave solutions exist for both positive and negative integers because electrons in either spin state can propagate in both positive and negative directions. From Equations (3.42) and (3.43) we find that the energies and momenta of the plane-wave solutions given in Equation (3.46) satisfy the equation

$$E^2(n_1, n_2, n_3) = h^2 \left(\frac{2\pi c}{L}\right)^2 n^2 + (mc^2)^2, \tag{3.47}$$

with $n^2 = n_1^2 + n_2^2 + n_3^2$. Because L can be chosen arbitrarily large, the number of quantum states with energies less than E given by Equation (3.47) becomes

$$2 \int_{-n}^{n} \int_{-n}^{n} \int_{-n}^{n} dn_x dn_y dn_z = \frac{8\pi}{3} n^3. \tag{3.48}$$

Since energy E and momentum p are related by Equation (3.38) and n is large, we find from Equations (3.47) and (3.48) that the number of quantum states with values of n in the interval $n, n + dn$ is

$$8\pi n^2 dn = \frac{8\pi p^2 dp L^3}{h^3}. \tag{3.49}$$

Equation (3.49) implies that the number of quantum states per unit volume in the small region of momentum space $d^3 p$ is

$$2\frac{d^3 p}{h^3}. \tag{3.50}$$

Photon quantum states

A particle with spin angular momentum s equal to 1 and nonzero rest mass would have a degeneracy of $2s + 1 = 3$. The spin angular momentum of a photon is equal to 1. However, because photons have zero rest mass, they travel at the speed of light, and the spin of a photon must be along its direction of motion. A photon can exist in either a right-handed circularly polarized state or a left-handed circularly polarized state. Photons, like electrons, have plane-wave solutions in free space. Because there are two photon polarization states, there exists a one-to-one correspondence between plane-wave solutions of photons and free electrons. Therefore, the number density of photon quantum states in a small volume element of momentum space is the same as for electrons. Since the momentum of a photon is $p = hv/c$, the number density of photon quantum states in the frequency range $v, v + dv$ becomes

$$\frac{2d^3 p}{h^3} = \frac{8\pi v^2 dv}{c^3}. \tag{3.51}$$

Particle distribution functions

The basic requirement of Fermi statistics is that no more than one particle can occupy a particular quantum state. We divide energy space into different particle energy levels $\varepsilon_k (k = 1, 2, 3, \ldots)$. If g_k is defined to be the number density of quantum states with energy ε_k, and n_k is equal to the occupation number of particle energy level ε_k, then for fermions we have

$$n_k \leq g_k \quad k = 1, 2, 3, \ldots \tag{3.52}$$

If the n_k particles of energy ε_k were distinguishable, then they could be assigned to g_k quantum states in

$$g_k(g_k - 1)(g_k - 2) \ldots (g_k - n_k + 1) = \frac{g_k!}{(g_k - n_k)!} \tag{3.53}$$

different ways. However, a basic principle of quantum mechanics is that particles (fermions and bosons) with the same energy ε_k are indistinguishable. Therefore, the number of different ways in which n_k fermions can be assigned to g_k quantum states becomes

$$\frac{g_k!}{n_k!(g_k - n_k)!}. \tag{3.54}$$

Equation (3.54) implies that the total number of different ways fermions can be assigned with occupation numbers n_k $(k = 1, 2, 3, \ldots)$ to particle energy levels ε_k (which have degeneracies g_k) is the product

$$W = \prod_k \frac{g_k!}{n_k!(g_k - n_k)!}. \tag{3.55}$$

The index k in Equation (3.55) extends over all particle energy levels ε_k, and W is equal to the total number of quantum states.

Because we are concerned with systems that contain large numbers of particles we can assume that n_k and g_k are large numbers. The factorial of a large number n is

$$n! \simeq \sqrt{2\pi n}\left(\frac{n}{e}\right)^n. \tag{3.56}$$

Equation (3.56), which is known as the Sterling approximation, implies that

$$\ln n! = n\ln n - n - \frac{1}{2}\ln(2\pi n). \tag{3.57}$$

The last term on the right-hand side of Equation (3.57) is small compared to the other terms, and therefore from Equations (3.55) and (3.57) we find that

$$\ln W = \sum_k [(n_k - g_k)\ln(g_k - n_k) - n_k\ln n_k + g_k\ln g_k]. \tag{3.58}$$

The fermion equilibrium distribution function follows from the requirement that $\ln W$

is a maximum, subject to the conditions that the total number of particles n and total energy E remain unchanged. The above conditions can be expressed as

$$\delta n = \sum_k \delta n_k = 0,$$

$$\delta E = \sum_k \varepsilon_k \delta n_k = 0. \tag{3.59}$$

Using the method of Lagrangian multipliers, we find that $\ln W$ in Equation (3.58) is an extremum (maximum) subject to the above two conditions if

$$\sum_k \left(\ln \frac{n_k}{(g_k - n_k)} + \alpha + \beta \varepsilon_k \right) \delta n_k = 0, \tag{3.60}$$

with α and β equal to constants. Because the variations δn_k represent arbitrary small changes in n_k, the terms within the parenthesis in Equation (3.60) are all equal to zero, and consequently Equation (3.60) leads to the expression

$$n_k = \frac{g_k}{e^{\alpha + \beta \varepsilon_k} + 1}. \tag{3.61}$$

More than one photon (or other boson) can exist in a particular quantum state, and therefore condition (3.52) is not relevant for photons (bosons). Let n_k bosons of energy ε_k be assigned to the g_k quantum states, which have energy ε_k. A total of $g_k - 1$ partitions are required to separate g_k quantum states. The total number of different ways in which n_k different particles can be separated by means of $g_k - 1$ partitions is

$$(n_k + g_k - 1)!. \tag{3.62}$$

It is immediately clear that particle distributions with different partitions interchanged are physically equivalent. Moreover, the principles of quantum mechanics imply that the n_k particles with energy ε_k are indistinguishable. It follows that the number of physically different ways in which n_k particles can be assigned to g_k quantum states is

$$\frac{(n_k + g_k - 1)!}{n_k!(g_k - 1)!}. \tag{3.63}$$

The above expression implies that the total number of eigenstates is

$$W = \prod_k \frac{(n_k + g_k - 1)!}{n_k!(g_k - 1)!}. \tag{3.64}$$

With the Sterling approximation, Equation (3.64) leads to the equation

$$\ln W = \sum_k [(n_k + g_k)\ln(n_k + g_k) - n_k \ln n_k - g_k \ln g_k]. \tag{3.65}$$

If we require that $\ln W$ is an extremum (maximum) subject to conditions (3.59), we obtain

$$\sum_k \left[\ln \frac{n_k}{(g_k - n_k)} + \alpha + \beta \varepsilon_k \right] \delta n_k = 0, \tag{3.66}$$

with α and β equal to constants. Since δn_k represent arbitrary small changes in n_k, the boson equilibrium distribution function implied by Equation (3.66) becomes

$$n_k = \frac{g_k}{e^{\alpha + \beta \varepsilon_k} - 1}. \tag{3.67}$$

Photons have zero rest mass and therefore can be formed with arbitrarily low energy. It follows that the first of conditions (3.59) is not relevant for photons. Omission of this latter condition is equivalent to setting α equal to zero in Equation (3.66), and therefore the photon distribution function becomes

$$n_k = \frac{g_k}{\exp(\beta \varepsilon_k) - 1}. \tag{3.68}$$

Although we have already derived the distribution function for a gas in the low-density (i.e. classical) limit, it is useful from the point of view of understanding the difference between classical and quantum statistics to derive the classical particle distribution function by a method similar to that used to derive the fermion and boson equilibrium distribution functions. According to classical physics it is possible to measure the positions and momenta of particles to arbitrarily high accuracy. This consequence of classical physics implies that, at least in principle, the exact locations of particles can be determined, and therefore different particles can be distinguished.

As in the above derivation of the quantum distribution functions, we divide energy space into energy levels ε_k ($k = 1, 2, 3, \ldots$). The number of states and occupation numbers of energy levels ε_k are denoted by g_k and n_k, respectively. The number of different ways in which a total of n distinguishable particles can be assigned to energy levels ε_k ($k = 1, 2, 3, \ldots$) is

$$W = \prod_k \frac{n!}{n_k!} g_k^{n_k}, \tag{3.69}$$

with

$$n = \sum_k n_k.$$

Because n and n_k are large numbers, Equation (3.69) leads to the equation

$$\ln W = n \ln n + \sum_k [n_k \ln g_k - n_k \ln n_k]. \tag{3.70}$$

If we require that $\ln W$ in Equation (3.70) is an extremum (maximum) subject to conditions (3.59), we obtain

$$n_k = \frac{g_k}{e^{\alpha + \beta \varepsilon_k}}. \tag{3.71}$$

Atoms, ions with one or more electrons, molecules and nuclei exist in discrete energy levels. From the point of view of statistical physics such discrete energy levels are distinguishable, and therefore under equilibrium conditions the relative number of atoms (or ions with electron(s), molecules or nuclei) in two different energy levels is given by the Boltzmann relation

$$\frac{n_l}{n_k} = \frac{g_l}{g_k} e^{-(\varepsilon_l - \varepsilon_k)/kT}, \tag{3.72}$$

with g_k and g_l the degeneracies of energy levels ε_k and ε_l respectively.

The fermion, boson and classical distribution functions can be summarized as

$$n_k = \frac{g_k}{e^{\alpha + \beta \varepsilon_k} \pm 1(0)}. \tag{3.73}$$

In Equation (3.73) the ratio n_k/g_k is the probability that a particular quantum state with energy ε_k is occupied. The temperature T and chemical potential μ are defined by the equations

$$\beta = \frac{1}{kT} \tag{3.74}$$

and

$$\alpha = \frac{\mu}{kT}, \tag{3.75}$$

respectively. Equations (3.61), (3.74) and (3.75) imply that the probability for a particular electron quantum state to be occupied is

$$f = \frac{1}{e^{(\varepsilon - \mu)/kT} + 1}. \tag{3.76}$$

From Equations (3.50) and (3.76) we find that the number density of occupied quantum states with momenta in the interval $p, p + dp$ is

$$n_e(p)dp = f \frac{8\pi p^2 dp}{h^3}. \tag{3.77}$$

Equation (3.77) implies that the number density of electrons is

$$n_e = \frac{8\pi}{h^3} \int \frac{p^2 dp}{e^{(\varepsilon - \mu)/kT} + 1}. \tag{3.78}$$

An electron gas is degenerate if $\mu \gg kT$.

Physical arguments similar to those used to obtain the pressure of a nondegenerate, ideal gas, given in Equation (3.16), imply that the pressure of an electron gas (degenerate or nondegenerate) is

$$P_e = \int \int (2p\cos\theta)(v\cos\theta)n_e(p)\,dp\frac{\sin\theta d\theta}{2}$$

$$= \frac{8\pi}{3h^3}\int_0^\infty \frac{p^3 v\,dp}{e^{(\varepsilon-\mu)/kT}+1}, \tag{3.79}$$

with $\varepsilon^2 = p^2 c^2 + (mc^2)^2$. The internal energy density of an electron gas is

$$U_e = \frac{8\pi}{h^3}\int_0^\infty \frac{\varepsilon^1 p^2\,dp}{e^{(\varepsilon-\mu)/kT}+1}, \tag{3.80}$$

where ε^1 is equal to $p^2/2m$ if the gas is nonrelativistic and $\varepsilon^1 = \varepsilon - mc^2$ for a relativistic gas because the internal energy density does not include rest-mass energy.

If an electron gas is fully degenerate (i.e. μ is $\gg kT$), then the function f given in Equation (3.76) is close to unity for values of electron energy $\varepsilon < \mu$ and close to zero for electron energies $\varepsilon > \mu$. In the limit of complete degeneracy the electron number density n_e becomes

$$n_e = \frac{8\pi}{3h^3}p_F^3, \tag{3.81}$$

with $\mu^2 \equiv \varepsilon_F^2 = p_F^2 c^2 + (mc^2)^2$. For a nonrelativistic, completely degenerate electron gas, Equations (3.79) and (3.80) imply that the electron pressure P_e and internal energy density U_e are

$$P_e = \frac{8\pi}{15h^3 m}p_F^5 \tag{3.82}$$

and

$$U_e = 3/2P_e, \tag{3.83}$$

respectively. From Equations (3.79) and (3.80) the corresponding expressions for an ultrarelativistic, completely degenerate electron gas become

$$P_e = \frac{2\pi c}{3h^3}p_F^4 \tag{3.84}$$

and

$$U_e = 3P_e. \tag{3.85}$$

Equations (3.51), (3.67) and (3.74) imply that the equilibrium number density of photons in the frequency interval v, $v + dv$ is

$$n_v dv = \frac{8\pi}{c^3}\frac{v^2 dv}{e^{hv/kT}-1}. \tag{3.86}$$

From Equation (3.86) we find that the blackbody photon energy density in the frequency interval v, $v + dv$ is

$$U_v dv = \frac{8\pi}{c^3} \frac{hv^3 dv}{e^{hv/kT} - 1}.$$

(3.87)

If we integrate Equation (3.87) over all frequencies, the energy density of the blackbody radiation field becomes

$$U_r = \int_0^\infty U_v dv = aT^4,$$

(3.88)

with $a = 8\pi^5 k^4/15c^3 h^3 = 7.56 \times 10^{-15}$ erg cm^{-3} deg^{-4}. From Equation (3.86) and physical arguments similar to those used to derive Equations (3.16) and (3.79) we find that the radiation pressure of a blackbody radiation field is

$$P_r = \frac{1}{3} aT^4.$$

(3.89)

If the intensity of radiation is I_v, then the energy density of the radiation field in the frequency interval $v, v + dv$ is

$$U_v = \frac{1}{c} \int I_v d\Omega.$$

(3.90)

A blackbody radiation field is isotropic, and therefore the mean intensity of radiation and intensity of radiation I_v are identical. Therefore, from Equations (3.87) and (3.90) we have

$$I_v = B_v = \frac{2}{c^2} \frac{hv^3}{e^{hv/kT} - 1},$$

(3.91)

where $B_v = B_v(T)$ is known as the Planck function. In the high-frequency limit (i.e. $hv \gg kT$) Equation (3.91) reduces to the equation

$$B_v = \frac{2hv^3}{c^2} e^{-hv/kT}.$$

(3.92)

In the low-frequency limit (i.e. $hv \ll kT$) Equation (3.91) becomes

$$B_v = \frac{2}{c^2} v^2 kT = \frac{2kT}{\lambda^2}.$$

(3.93)

Planck's constant does not appear in Equation (3.93), which is referred to as the Rayleigh–Jeans approximation. It can readily be shown by differentiating B_v with respect to frequency v that B_v has a maximum value when

$$hv_{max} = 2.82kT.$$

(3.94)

If the radiation field at the surface of a star is that of a blackbody, then the flux F integrated over all frequencies is

$$F = 2\pi \int_0^{\pi/2} B_v \cos\theta \sin\theta d\theta dv = \pi \int B_v dv = \sigma T^4,$$

(3.95)

with $\sigma = ac/4 = 5.67 \times 10^{-5} \, \mathrm{erg\, cm^{-2}\, s^{-1}\, K^{-4}}$. Since the flux F in Equation (3.95) is the power per unit area through the surface of the star, the luminosity L becomes

$$L = 4\pi R^2 \sigma T^4, \tag{3.96}$$

with R equal to the radius of the star.

3.3   Thermodynamic functions of atomic and fermion gases

Nondegenerate, ionized gases

In thermodynamics the free energy F is defined by the equation

$$F = U - TS, \tag{3.97}$$

with U equal to the internal energy and S equal to the entropy. For a low-density gas of N particles Equations (3.17), (3.20) and (3.97) imply that

$$F = \frac{3}{2}NkT - NkT\ln(T^{3/2}/n) - NTs_0. \tag{3.98}$$

The Gibbs free energy G is defined by the expression

$$G(p, T, N) = U - TS + PV. \tag{3.99}$$

From Equation (3.99) and the first law of thermodynamics we obtain

$$dG(P, T, N) = VdP - SdT + \mu dN, \tag{3.100}$$

with $\mu = (\partial G/\partial N)_{P,T}$ the chemical potential. Since changes in dG under conditions of constant T and P are proportional to dN, we conclude that

$$G = \mu N. \tag{3.101}$$

Since $N = nV$ and $P = nkT$, Equations (3.97), (3.98), (3.99) and (3.101) lead to the equation

$$\mu = 5/2kT - kT\ln(T^{3/2}/n) - Ts_0. \tag{3.102}$$

From Equation (3.20) and the first law of thermodynamics the heat capacity c_v per particle becomes

$$c_v = \left(\frac{dU}{dT}\right)_v = T\left(\frac{dS}{dT}\right)_v = 3/2k. \tag{3.103}$$

From Equation (3.20) and the relation $P = nkT$ we find that the heat capacity c_p per particle is

$$c_p = T\left(\frac{dS}{dT}\right)_P = 5/2k. \tag{3.104}$$

In a low-density gas the classical and quantum-mechanical expressions for the number densities of electrons (fermions) in the momentum interval $p, p + dp$ must be the same for all values of p. From Equations (3.10) and (3.11) $n_e(p)$ becomes

$$n_e(p) = \frac{4\pi n p^2}{(2\pi m k T)^{3/2}} e^{-p^2/2mkT}.$$

(3.105)

The corresponding equation for $n_e(p)$ from the quantum-mechanical expression evaluated at low densities (i.e. μ large and negative) follows from Equations (3.76) and (3.77). We obtain

$$n_e(p) = e^{\mu/kT} \frac{8\pi p^2}{h^3} e^{-p^2/2mkT}.$$

(3.106)

Equating Equations (3.105) and (3.106) leads to the equation

$$e^{\mu/kT} \frac{8\pi}{h^3} = \frac{4\pi n}{(2\pi m k T)^{3/2}}.$$

(3.107)

Solving for the chemical potential μ in Equation (3.107), we obtain

$$\mu = - kT\ln(T^{3/2}/n) - \frac{3}{2} kT\ln\left(\frac{2\pi m k}{h^2}\right) - kT\ln 2.$$

(3.108)

The expressions for μ given in Equations (3.102) and (3.108) do not appear to be the same. However, we can choose the constant of integration in the derivation of the entropy per particle so that the chemical potential μ in Equation (3.102) defined by the relation $G = \mu N$ and the chemical potential μ given in Equation (3.108) are the same. The expressions for μ given in Equations (3.102) and (3.108) are equal if the constant s_0 in Equation (3.102) is set equal to

$$s_0 = \frac{5}{2} k + \frac{3}{2} k\ln\left(\frac{2\pi m k}{h^2}\right) + k\ln 2.$$

(3.109)

If the value for the integration constant s_0 given in Equation (3.109) is substituted into Equation (3.20), the entropy per electron becomes

$$s = k\ln(T^{3/2}/n) + \frac{5}{2} k + \frac{3}{2} k\ln\left(\frac{2\pi m k}{h^2}\right) + k\ln 2.$$

(3.110)

Atomic gases

In an atomic gas, such as a gas consisting of hydrogen atoms, the atoms can exist in various discrete energy levels. The internal energy per particle of an atomic gas is

$$u = \frac{3}{2} kT + \frac{\sum_n g_n \varepsilon_n e^{-\varepsilon_n/kT}}{\sum_n g_n e^{-\varepsilon_n/kT}},$$

(3.111)

with g_n ($n = 1, 2, 3, \ldots$) equal to the degeneracies of the various energy levels. Equation (3.111) can be written as

$$u = \frac{3}{2}kT + kT^2\frac{\partial \ln Z}{\partial T} \tag{3.112}$$

with the partition function $Z = \sum_n g_n e^{-\varepsilon_n/kT}$.

Because the energy levels ε_n and degeneracies g_n cannot depend on thermodynamic variables such as T, the heat capacity c_v per particle becomes

$$c_v = \frac{3}{2}k + 2kT\frac{\partial \ln Z}{\partial T} + kT^2\frac{\partial^2 \ln Z}{\partial T^2}. \tag{3.113}$$

From the first law of thermodynamics we have

$$s = \int \frac{c_v dT}{T} + \int \frac{P}{T}d\left(\frac{1}{n}\right), \tag{3.114}$$

with s equal to the entropy per particle. Equations (3.110), (3.113) and (3.114) imply that

$$s = k\ln(T^{3/2}/n) + \frac{5}{2}k + \frac{3}{2}k\ln\left(\frac{2\pi mk}{h^2}\right) + k\ln g$$

$$+ 2k\int \frac{\partial \ln Z}{\partial T}dT + k\int T\frac{\partial^2 \ln Z}{\partial T^2}dT, \tag{3.115}$$

where g is the degeneracy of the atom. From the equality

$$\frac{\partial}{\partial T}\left(kT\frac{\partial \ln Z}{\partial T}\right) = kT\frac{\partial^2 \ln Z}{\partial T^2} + k\frac{\partial \ln Z}{\partial T}, \tag{3.116}$$

it follows that Equation (3.115) can be expressed as

$$s = k\ln(T^{3/2}/n) + \frac{5}{2}k + \frac{3}{2}k\ln\left(\frac{2\pi mk}{h^2}\right) + k\ln g$$

$$+ k\ln Z + kT\frac{\partial \ln Z}{\partial T}$$

$$= k\ln(T^{3/2}/n) + \frac{5}{2}k + \frac{3}{2}k\ln\left(\frac{2\pi mk}{h^2}\right) + k\ln g$$

$$+ k\ln\sum_n g_n e^{-\varepsilon_n/kT} + \frac{1}{T}\frac{\sum_n g_n \varepsilon_n e^{-\varepsilon_n/kT}}{\sum_n g_n e^{-\varepsilon_n/kT}}. \tag{3.117}$$

Weak degeneracy

Equations (3.78), (3.79) and (3.80) give expressions for the number density n_e, pressure P_e and internal energy density U_e, respectively, that are valid for both degenerate and nondegenerate electron gases. For a weakly degenerate gas, μ/kT is large and negative. Under such physical conditions the probability for a particular quantum state to be occupied that is given in Equation (3.76) can be approximated as

$$f = \frac{1}{e^{(\varepsilon - \mu)/kT}(1 + e^{-(\varepsilon - \mu)/kT})}$$

$$\simeq e^{-(\varepsilon - \mu)/kT}(1 - e^{-(\varepsilon - \mu)/kT}). \tag{3.118}$$

Substituting the above expression for f into Equation (3.78), and introducing the variable $x = p/\sqrt{2mkT}$, with $\varepsilon \ll mc^2$, we obtain

$$n_e = \frac{8\pi(2mkT)^{3/2}}{h^3}\left[e^{\mu/kT}\int_0^\infty e^{-x^2}x^2 dx - e^{2\mu/kT}\int_0^\infty e^{-2x^2}x^2 dx\right]$$

$$= \frac{2(2\pi mkT)^{3/2}}{h^3}e^{\mu/kT}\left(1 - \frac{e^{\mu/kT}}{2^{3/2}}\right)$$

$$\simeq \frac{2(2\pi mkT)^{3/2}}{h^3}e^{\mu/kT}\left(1 - \frac{e^{\mu_0/kT}}{2^{3/2}}\right), \tag{3.119}$$

with μ_0 equal to the chemical potential of a nondegenerate electron gas, given in Equation (3.108).

Solving Equation (3.119) for the chemical potential μ and using the expression for μ_0 given in Equation (3.108), we have

$$\mu = -kT\ln\left(\frac{2\pi mk}{h^2}\right)^{3/2} - kT\ln(2T^{3/2}/n_e) - \frac{n_e}{2^{1/2}}\left(\frac{h^2}{2\pi mkT}\right)^{3/2}. \tag{3.120}$$

Substituting the approximate expression for f from Equation (3.118) into Equation (3.80) and performing integrations similar to those in Equation (3.119), we obtain

$$U_e = 3kT\left(\frac{2\pi mkT}{h^2}\right)^{3/2}e^{\mu/kT}\left[1 - \frac{e^{\mu/kT}}{2^{5/2}}\right]. \tag{3.121}$$

Dividing Equation (3.121) by (3.119), we find that

$$\frac{U_e}{n_e} = \frac{3}{2}kT\frac{\left[1 - \dfrac{e^{\mu/kT}}{2^{5/2}}\right]}{\left[1 - \dfrac{e^{\mu/kT}}{2^{3/2}}\right]}. \tag{3.122}$$

Since μ/kT is large and negative, we can expand the denominator of equation (3.122) by means of the binomial expansion and obtain

$$U_e \simeq \frac{3}{2} n_e kT \left[1 - \frac{e^{\mu/kT}}{2^{5/2}} + \frac{e^{\mu/kT}}{2^{3/2}} \right] \simeq \frac{3}{2} n_e kT \left(1 + \frac{e^{\mu_0/kT}}{2^{5/2}} \right). \tag{3.123}$$

Substituting the expression for μ_0, the chemical potential of a nondegenerate gas, into Equation (3.123), we have

$$U_e = \frac{3}{2} n_e kT \left[1 + \frac{n_e}{2^{7/2}} \left(\frac{h^2}{2\pi m kT} \right)^{3/2} \right]. \tag{3.124}$$

Equation (3.124) shows that weak degeneracy leads to an increase in the internal energy of a gas. The pressure P_e of a nonrelativistic electron degenerate gas is $(2/3) U_e$, regardless of the degree of degeneracy.

With Equation (3.124) the heat capacity c_v per unit volume becomes

$$c_v = \frac{3}{2} n_e k \left[1 - \frac{n_e}{2^{9/2}} \left(\frac{h^2}{2\pi m kT} \right)^{3/2} \right]. \tag{3.125}$$

Equation (3.125) shows that the heat capacity c_v decreases as a result of weak degeneracy.

Next we assume that an electron gas is weakly degenerate but ultrarelativistic (i.e. $\varepsilon \gg mc^2$). Under equilibrium conditions a weakly degenerate, ultrarelativistic gas contains almost as many positrons as electrons. Introducing the variable $x = pc/kT$ with $\varepsilon = pc$ in Equation (3.78) and using Equation (3.118), we obtain

$$n_e = 8\pi \left(\frac{kT}{hc} \right)^3 \left[e^{\mu/kT} \int_0^\infty x^2 e^{-x} dx - e^{2\mu/kT} \int_0^\infty x^2 e^{-2x} dx \right]$$

$$= 16\pi \left(\frac{kT}{hc} \right)^3 e^{\mu/kT} \left[1 - \frac{e^{\mu/kT}}{2^3} \right]. \tag{3.126}$$

If similar substitutions are made in Equation (3.80), the internal energy density U_e becomes

$$U_e = 8\pi kT \left(\frac{kT}{hc} \right)^3 \left[e^{\mu/kT} \int_0^\infty x^3 e^{-x} dx - e^{2\mu/kT} \int_0^\infty x^3 e^{-2x} dx \right]$$

$$= 48\pi kT \left(\frac{kT}{hc} \right)^3 e^{\mu/kT} \left(1 - \frac{e^{\mu/kT}}{2^4} \right). \tag{3.127}$$

From Equation (3.78) the chemical potential of an ultrarelativistic, nondegenerate gas is related to the electron number density n_e by the equation

$$n_e = \frac{8\pi}{(hc)^3} e^{\mu_0/kT} \int_0^\infty e^{-\varepsilon/kT} \varepsilon^2 d\varepsilon$$

$$= 16\pi \left(\frac{kT}{hc} \right)^3 e^{\mu_0/kT}. \tag{3.128}$$

Equation (3.128) implies that the chemical potential of an ultrarelativistic, nondegenerate gas is

$$\mu_0 = kT\ln(n_e/16\pi) - 3kT\ln(kT/hc). \tag{3.129}$$

Dividing Equation (3.127) by Equation (3.126) and assuming that $\mu \simeq \mu_0$, we have

$$\frac{U_e}{n_e} = 3kT\frac{\left(1 - \dfrac{e^{\mu/kT}}{2^4}\right)}{\left(1 - \dfrac{e^{\mu/kT}}{2^3}\right)} \simeq 3kT\left(1 + \frac{e^{\mu/kT}}{16}\right), \tag{3.130}$$

where we have expanded the denominator by the binomial expansion. Substituting the expression for μ_0 from Equation (3.129) as μ in Equation (3.130), we obtain

$$U_e \simeq 3n_e kT\left[1 + \frac{n_e}{256\pi}\left(\frac{hc}{kT}\right)^3\right]. \tag{3.131}$$

The pressure P_e of an ultrarelativistic electron gas is $\frac{1}{3}U_e$.

Equation (3.131) implies that the heat capacity of an ultrarelativistic, weakly degenerate electron gas is

$$c_v = \left(\frac{dU_e}{dT}\right)_v \simeq 3n_e k\left[1 - \frac{n_e}{128\pi}\left(\frac{hc}{kT}\right)^3\right]. \tag{3.132}$$

If we substitute $\mu = \mu_0$, with μ_0 given by Equation (3.129), into the final exponential term in Equation (3.126), we have

$$\ln n_e \simeq \ln\left[16\pi\left(\frac{kT}{hc}\right)^3\right] + \mu/kT + \ln\left(1 - \frac{e^{\mu_0/kT}}{2^3}\right),$$

$$\simeq \ln\left[16\pi\left(\frac{kT}{hc}\right)^3\right] + \mu/kT - \frac{e^{\mu_0/kT}}{2^3}. \tag{3.133}$$

Therefore, from Equations (3.129) and (3.133) the chemical potential μ of an ultrarelativistic, weakly degenerate electron gas becomes

$$\mu \simeq -kT\ln\left[\frac{16\pi}{n_e}\left(\frac{kT}{hc}\right)^3\right] + \frac{e^{\mu_0/kT}}{2^3}$$

$$= -kT\ln\left[\frac{16\pi}{n_e}\left(\frac{kT}{hc}\right)^3\right] + \frac{n_e}{2^7\pi}\left(\frac{hc}{kT}\right)^3. \tag{3.134}$$

Strong degeneracy

In order to determine the thermodynamic functions of a strongly degenerate ($\mu \gg kT$) electron gas we must evaluate integrals of the form

$$I = \int_{mc^2}^{\infty} f(\varepsilon)g(\varepsilon)d\varepsilon, \tag{3.135}$$

with $f(\varepsilon)$, the probability that a particular quantum state is occupied, given by the equation

$$f(\varepsilon) = \frac{1}{e^{(\varepsilon-\mu)/kT} + 1}. \tag{3.136}$$

For strong degeneracy the derivative $df/d\varepsilon = f'(\varepsilon)$ has a sharp maximum at $\varepsilon = \mu$ and is small at values of ε that are either significantly less than or greater than $\varepsilon = \mu$. The functions $g(\varepsilon)$ vary much more slowly as a function of ε than $f(\varepsilon)$. Integrating Equation (3.135) by parts gives

$$I = \int_{mc^2}^{\infty} f(\varepsilon)g(\varepsilon)d\varepsilon = f(\infty)G(\infty) - f(mc^2)G(mc^2) - \int_{mc^2}^{\infty} f'(\varepsilon)G(\varepsilon)d\varepsilon, \tag{3.137}$$

with

$$G(\varepsilon) = \int_{mc^2}^{\varepsilon} g(\varepsilon')d\varepsilon'.$$

Since $f(\infty)$ and $G(mc^2)$ are equal to zero, Equation (3.137) becomes

$$I = -\int_{mc^2}^{\infty} f'(\varepsilon)G(\varepsilon)d\varepsilon. \tag{3.138}$$

Making the variable transformation $x = (\varepsilon - \mu)/kT$ in Equation (3.138) and then expanding the function $G(x)$ in a Taylor series expansion about $x = 0$, we obtain

$$G(x) = \sum_{n=0}^{\infty} \frac{x^n}{n!} G^n(0), \tag{3.139}$$

with

$$G^0(0) \equiv \int_{mc^2}^{\mu} g(\varepsilon)d\varepsilon \tag{3.140}$$

and

$$G^n(0) = (kT)^n \left(\frac{d^{n-1}g(\varepsilon)}{d\varepsilon^{n-1}}\right)_{\varepsilon=\mu} \equiv (kT)^n g^{n-1}(\mu) \quad (n = 1, 2, 3, \ldots). \tag{3.141}$$

Substituting Equations (3.139–3.141) into Equation (3.138) leads to the equation

$$I = -G(0) \int_{mc^2}^{\infty} f'(\varepsilon)d\varepsilon - \sum_{n=1}^{\infty} \frac{(kT)^n}{n!} g^{n-1}(\mu) \int_{x=-(\mu-mc^2)/kT}^{\infty} x^n f'(x)dx. \tag{3.142}$$

The integral that appears in the first term of Equation (3.142) is

$$\int_{mc^2}^{\infty} f'(\varepsilon)d\varepsilon \simeq 1. \tag{3.143}$$

The above expression is approximately unity because $f'(\varepsilon)$ has the mathematical properties of a δ function in a strongly degenerate gas.

From Equation (3.136) it can readily be shown that the derivative $f'(x)$ is

$$f'(x) = -\frac{1}{kT}\frac{e^x}{(e^x + 1)^2} = -\frac{1}{kT}\frac{1}{(e^x + 1)(e^{-x} + 1)}. \tag{3.144}$$

The second of the above equalities shows clearly that $f'(x)$ is symmetric in x (i.e. $f'(x) = f'(-x)$). Because $\mu - mc^2$ is $\gg kT$, the limits of integration that appear in the second term of Equation (3.142) can be considered to extend from $-\infty$ to $+\infty$. It follows that only even values of n will contribute to the second term on the right-hand side of Equation (3.142). Therefore, only odd derivatives of the function g will appear in the final expression for the integral I.

Because $x = \mu/kT$ is positive, the exponential terms in Equation (3.144) can be expanded by means of a binomial expansion. We obtain

$$\frac{1}{(e^x + 1)(e^{-x} + 1)} = \frac{e^{-x}}{(1 + e^{-x})^2} = -\sum_{m=1}^{\infty}(-1)^m m e^{-mx}. \tag{3.145}$$

Equation (3.145) implies that the integrals appearing inside the summation over n given in Equation (3.142) are equal to

$$\left(\frac{2}{kT}\right)\sum_{m=1}^{\infty}(-1)^m m \int_0^{\infty} x^n e^{-mx}\,dx = \frac{2n!}{kT}\sum_{m=1}^{\infty}\frac{(-1)^m}{m^n} \quad (n \text{ even}). \tag{3.146}$$

From Equations (3.140) and (3.144–3.146) we find that the integral I, which was initially defined in Equation (3.135), becomes

$$I = \int_{mc^2}^{\mu} g(\varepsilon)\,d\varepsilon - 2\sum_{n=1}^{\infty}(kT)^{2n}\left(\frac{d^{2n-1}g(\varepsilon)}{d\varepsilon^{2n-1}}\right)_{\varepsilon=\mu}\sum_{m=1}^{\infty}\frac{(-1)^m}{m^{2n}}. \tag{3.147}$$

Retaining only the first two terms in the summation in Equation (3.147) and using the well-known mathematical expressions

$$-\sum_{m=1}^{\infty}\frac{(-1)^m}{m^2} = \frac{\pi^2}{12}\quad\text{and}\quad -\sum_{m=1}^{\infty}\frac{(-1)^m}{m^4} = \frac{7\pi^4}{720} \tag{3.148}$$

leads to the equation

$$I = \int_{mc^2}^{\infty}\frac{g(\varepsilon)\,d\varepsilon}{e^{(\varepsilon-\mu)/kT} + 1}$$

$$= \int_{mc^2}^{\mu} g(\varepsilon)\,d\varepsilon + \frac{\pi^2}{6}(kT)^2\left(\frac{dg}{d\varepsilon}\right)_{\varepsilon=\mu} + \frac{7\pi^4}{360}(kT)^4\left(\frac{d^3g}{d\varepsilon^3}\right)_{\varepsilon=\mu} + \cdots \tag{3.149}$$

The thermodynamic functions of a strongly degenerate electron gas can now be calculated by substituting into Equation (3.149) the appropriate function $g(\varepsilon)$ and the derivatives of $g(\varepsilon)$.

Writing the expression for the electron number density n_e given in Equation (3.78) as a function of ε with $\varepsilon^2 = p^2c^2 + (mc^2)^2$, we obtain

$$n_e = \frac{8\pi}{h^3c^3} \int_{mc^2}^{\infty} \frac{\varepsilon\sqrt{\varepsilon^2 - (mc^2)^2}\,d\varepsilon}{e^{(\varepsilon-\mu)/kT} + 1}. \tag{3.150}$$

Equation (3.150) implies that

$$g(\varepsilon) = \frac{8\pi}{h^3c^3} \varepsilon\sqrt{\varepsilon^2 - (mc^2)^2}. \tag{3.151}$$

Differentiating the above expression for $g(\varepsilon)$ gives

$$\left.\frac{dg(\varepsilon)}{d\varepsilon}\right|_{\varepsilon=\mu} = \frac{8\pi}{h^3c^3} \frac{2\mu^2 - (mc^2)^2}{\sqrt{\mu^2 - (mc^2)^2}}. \tag{3.152}$$

Therefore, using Equations (3.149–3.152) and retaining only the first two terms on the right-hand side of Equation (3.149) leads to the equation

$$n_e = \frac{8\pi}{h^3c^3} \left[\int_{mc^2}^{\mu} \varepsilon\sqrt{\varepsilon^2 - (mc^2)^2}\,d\varepsilon + \frac{\pi^2}{6}(kT)^2 \left(\frac{2\mu^2 - (mc^2)^2}{\sqrt{\mu^2 - (mc^2)^2}} \right) \right]$$

$$= \frac{8\pi}{h^3c^3} \left[\left.\frac{(\varepsilon^2 - (mc^2)^2)^{3/2}}{3}\right|_{mc}^{\mu} + \frac{\pi^2}{6}(kT)^2 \left(\frac{2\mu^2 - (mc^2)^2}{\sqrt{\mu^2 - (mc^2)^2}} \right) \right]$$

$$= \frac{8\pi}{h^3c^3} \left[\frac{(\mu^2 - (mc^2)^2)^{3/2}}{3} + \frac{\pi^2}{6}(kT)^2 \left(\frac{2\mu^2 - (mc^2)^2}{\sqrt{\mu^2 - (mc^2)^2}} \right) \right]. \tag{3.153}$$

The above rather complicated expression can be simplified in the limits of nonrelativistic and ultrarelativistic degeneracy.

For a nonrelativistic electron gas we can redefine the chemical potential μ as

$$\mu^1 = \mu - mc^2, \tag{3.154}$$

with μ^1 a new symbol. Squaring Equation (3.154) and neglecting $(\mu^1)^2$, we have

$$\mu^2 \simeq (mc^2)^2 + 2\mu^1 mc^2. \tag{3.155}$$

Substituting the expression for μ^2 given in Equation (3.155) into Equation (3.153) gives

$$n_e = \frac{32\pi(m\mu^1)^{3/2}}{3\sqrt{2}h^3} \left[1 + \frac{\pi^2}{8} \left(\frac{kT}{\mu^1} \right)^2 \right]. \tag{3.156}$$

Since μ^1/mc^2 is $\ll 1$, Equation (3.156) leads to the equation

$$\frac{1}{\mu^1} = \frac{1}{\mu_0} \left[1 + \frac{\pi^2}{8} \left(\frac{kT}{\mu^1} \right)^2 \right]^{2/3}$$

$$\simeq \frac{1}{\mu_0} \left[1 + \frac{\pi^2}{8} \left(\frac{kT}{\mu_0} \right)^2 \right]^{2/3}, \tag{3.157}$$

with

$$\mu_0 = \left(\frac{h^2}{8m}\right)\left(\frac{3n_e}{\pi}\right)^{2/3}.$$

Since μ_0, which is the chemical potential at zero temperature, is assumed to be large compared to kT, we have

$$\mu^1 = \mu_0\left[1 - \frac{\pi^2}{12}\left(\frac{kT}{\mu_0}\right)^2\right].$$ (3.158)

For an ultrarelativistic gas, μ is $\gg mc^2$, and therefore Equation (3.153) becomes

$$\frac{1}{\mu^3} = \frac{1}{\mu_0^3}\left[1 + \pi^2\left(\frac{kT}{\mu}\right)^2\right] \simeq \frac{1}{\mu_0^3}\left[1 + \pi^2\left(\frac{kT}{\mu_0}\right)^2\right],$$ (3.159)

with $\mu_0^3 = 3n_e h^3 c^2/8\pi$. Since kT is $\ll \mu_0$, Equation (3.159) implies that the chemical potential of a degenerate, ultrarelativistic gas is

$$\mu \simeq \mu_0\left[1 - \frac{\pi^2}{3}\left(\frac{kT}{\mu_0}\right)^2\right].$$ (3.160)

If the expression for the pressure of an electron-degenerate gas given in Equation (3.79) is written as a function of particle energy ε and momentum p, with $\varepsilon^2 = p^2c^2 + (mc^2)^2$, then we have

$$P_e = \frac{8\pi c^2}{3h^3}\int_0^\infty \frac{(p^4/\varepsilon)dp}{e^{(\varepsilon-\mu)/kT} + 1} = \frac{8\pi}{3h^3c^3}\int_0^\infty \frac{(\varepsilon^2 - (mc^2)^2)^{3/2}}{e^{(\varepsilon-\mu)/kT} + 1}\,d\varepsilon.$$ (3.161)

Equation (3.161) implies that the function $g(\varepsilon)$ in Equation (3.149) is

$$g(\varepsilon) = \frac{8\pi}{3h^3c^3}(\varepsilon^2 - (mc^2)^2)^{3/2}.$$ (3.162)

Differentiating $g(\varepsilon)$ with respect to energy, we obtain

$$\left(\frac{dg(\varepsilon)}{d\varepsilon}\right)_{\varepsilon=\mu} = \frac{8\pi\mu}{h^3c^3}(\mu^2 - (mc^2)^2)^{1/2}.$$ (3.163)

If we express the integral in Equation (3.149) as a function of momentum and retain only the first derivative term, it follows that Equations (3.161–3.163) lead to the equation

$$P_e = \frac{8\pi}{3h^3c^3}\left[(mc^2)^4\int_0^{x_0} \frac{x^4dx}{(1+x^2)^{1/2}} + \frac{\pi^2}{2}(kT)^2\mu[\mu^2 - (mc^2)^2]^{1/2}\right],$$ (3.164)

with $x_0 = p_Fc/mc^2$ and $\mu^2 = p_F^2c^2 + (mc^2)^2$.

From the equality

$$d[x^3(1+x^2)^{1/2}] = 3(1+x^2)^{1/2}x^2dx + \frac{x^4dx}{(1+x^2)^{1/2}}$$ (3.165)

and an integration by parts, the integral in Equation (3.164) becomes

$$
I_0 = \int_0^{x_0} \frac{x^4 dx}{(1 + x^2)^{1/2}} = x_0^3(1 + x_0^2)^{1/2} - 3 \int_0^{x_0} (1 + x^2)^{1/2} x^2 dx
$$

$$
= x_0^3(1 + x_0^2)^{1/2} - \frac{3}{4}x_0(x_0^2 + 1)^{3/2} + \frac{3}{8}x_0(x_0^2 + 1)^{1/2} + \frac{3}{8}\ln(x_0 + (x_0^2 + 1)^{1/2})
$$

$$
= \frac{3}{8}x_0(1 + x_0^2)^{1/2}\left(\frac{2x_0^2}{3} - 1\right) + \frac{3}{8}\ln(x_0 + (x_0^2 + 1)^{1/2}). \tag{3.166}
$$

For a nonrelativistic electron gas, x_0 is $\ll 1$. Expanding the $(1 + x_0^2)^{1/2}$ term in Equation (3.166) by a binomial expansion and using the Taylor expansion

$$
\ln[x + (1 + x^2)^{1/2}] = \sin h^{-1} x \simeq x - \frac{x^3}{2 \cdot 3} + \frac{1 \cdot 3 x^5}{2 \cdot 4 \cdot 5}, \tag{3.167}
$$

we find that the integral (3.166) becomes

$$
I_0 \approx \frac{1}{5}x_0^5, \tag{3.168}
$$

with $x_0 = p_F c/mc^2 = (2\mu_0/mc^2)^{1/2}$.

For an ultrarelativistic electron gas in which x_0 is $\gg 1$ the integral I_0 is approximately

$$
I_0 \simeq \frac{1}{4}x_0^4, \tag{3.169}
$$

with $x_0 = p_F c/mc^2$.

Using the approximation for the integral I_0 in Equation (3.168), Equation (3.158) and the approximation $\mu = mc^2 + \mu_0 = mc^2 + p_F^2/2m$, the expression for the pressure of a nonrelativistic electron gas in Equation (3.164) becomes

$$
P_e \simeq \frac{2}{5}n_e\mu^1\left(\frac{\mu^1}{\mu_0}\right)^{3/2}\left[1 + \frac{5\pi^2}{8}\left(\frac{kT}{\mu_0}\right)^2\right]
$$

$$
= \frac{2}{5}n_e\mu_0\left[1 + \frac{5\pi^2}{12}\left(\frac{kT}{\mu_0}\right)^2\right]. \tag{3.170}
$$

From Equations (3.79) and (3.80) it can readily be shown that the internal energy density of a nonrelativistic electron gas is $U_e = (3/2)P_e$. Dividing the expression for U_e implied by Equation (3.170) by the expression for n_e given in Equation (3.156) and using Equation (3.158), we have

$$
\frac{U_e}{n_e} = \frac{3}{5}\mu^1 \frac{\left[1 - \frac{\pi^2}{12}\left(\frac{kT}{\mu_0}\right)^2\right]\left[1 + \frac{5\pi^2}{8}\left(\frac{kT}{\mu_0}\right)^2\right]}{\left[1 + \frac{\pi^2}{8}\left(\frac{kT}{\mu_0}\right)^2\right]}. \tag{3.171}
$$

Expanding the temperature-dependent terms to order $(kT/\mu_0)^2$ in Equation (3.171) gives

$$U_e = \frac{3}{5} n_e \mu^1 \left[1 + \frac{5\pi^2}{8}\left(\frac{kT}{\mu_0}\right)^2\right]$$

$$\simeq \frac{3}{5} n_e \mu_0 \left[1 + \frac{5\pi^2}{12}\left(\frac{kT}{\mu_0}\right)^2\right]. \tag{3.172}$$

From Equation (3.172) the heat capacity c_v becomes

$$c_v = \left(\frac{dU_e}{dT}\right)_v = \frac{\pi^2}{2}\frac{n_e k^2 T}{\mu_0}. \tag{3.173}$$

Equation (3.173) shows that the heat capacity c_v of a nonrelativistic electron-degenerate gas becomes small as T decreases.

For an ultrarelativistic electron-degenerate gas the expression for the electron number density given in Equation (3.153) is approximately

$$n_e \simeq \frac{8\pi}{3h^3c^3}\mu^3\left[1 + \pi^2\left(\frac{kT}{\mu_0}\right)^2\right]$$

$$\simeq \frac{8\pi}{3h^3c^3}\mu_0^3\left[1 - \pi^2\left(\frac{kT}{\mu_0}\right)^2\right]\left[1 + \pi^2\left(\frac{kT}{\mu_0}\right)^2\right]$$

$$\simeq \frac{8\pi}{3h^3c^3}\mu_0^3. \tag{3.174}$$

From Equations (3.164), (3.166) and (3.169) and the relation $U_e = 3P_e$, which is valid for an ultrarelativistic electron gas, we have

$$P_e = \frac{2\pi\mu_0^4}{3h^3c^3}\left[1 + 2\pi^2\left(\frac{kT}{\mu_0}\right)^2\right] = U_e/3. \tag{3.175}$$

Dividing U_e, we obtained from Equation (3.175), by n_e, given in Equation (3.174), we find that

$$U_e = \frac{3}{4} n_e \mu_0\left[1 + 2\pi^2\left(\frac{kT}{\mu_0}\right)^2\right] \tag{3.176}$$

and

$$c_v = \left(\frac{dU_e}{dT}\right)_v = 3\pi^2 n_e\frac{k^2 T}{\mu_0}. \tag{3.177}$$

3.4 Ionization equilibrium

Consider a nondegenerate, partially ionized hydrogen gas contained in an arbitrary volume V. The number of electrons is equal to the number of protons. The electron degeneracy of each of the energy levels ε_n of atomic hydrogen is $2n^2$. Since the spin I of the proton is $1/2$, the total degeneracy of each energy level becomes $g_n = (2I + 1) 2n^2 = 4n^2$. For an arbitrary volume V the number of quantum states with the hydrogen atom in the momentum interval $p, p + dp$ and in the energy level ε_n is

$$G_n(p)dp = g_n \frac{4\pi p^2 dp}{h^3} V. \tag{3.178}$$

Equation (3.178) follows from Equation (3.78). The latter equation also implies that the number of quantum states with electrons in the momentum interval $p_1, p_1 + dp_1$ and protons in the momentum interval $p_2, p_2 + dp_2$ is

$$G_e(p_1)dp_1 G_p(p_2)dp_2 = \frac{8\pi p_1^2 dp_1}{h^3} \frac{8\pi p_2^2 dp_2}{h^3} V^2. \tag{3.179}$$

Equations (3.72), (3.178) and (3.179) imply that under equilibrium conditions we have

$$\frac{N_e(p_1)dp_1 N_p(p_2)dp_2}{N_n(p)dp} = \frac{V 8\pi p_1^2 e^{-p_1^2/2mkT} dp_1 8\pi p_2^2 e^{-p_2^2/2m_p kT} dp_2}{h^3 g_n e^{-\varepsilon_n/kT} 4\pi p^2 e^{-p^2/2m_H kT} dp}, \tag{3.180}$$

with $N_e(p_1)\,dp_1$ equal to the number of electrons in the momentum interval $p_1, p_1 + dp_1$, $N_p(p_2)\,dp_2$ equal to the number of protons in the momentum interval $p_2, p_2 + dp_2$ and $N_n(p)\,dp$ equal to the number of hydrogen atoms in energy level n and in the momentum interval $p, p + dp$. In Equation (3.180) the Boltzmann relation has been used to assign relative populations of quantum states. Integrating over electron and proton momenta in Equation (3.180), and then inverting the resultant equation and integrating over the momenta of the hydrogen atom, we obtain

$$\frac{N_n V}{N_e N_p} = \frac{h^3 g_n e^{-\varepsilon_n/T}}{4(2\pi mkT)^{3/2}} \left(\frac{m_H}{m_p}\right)^{3/2}. \tag{3.181}$$

Since $m_H \simeq m_p$, Equation (3.181) expressed in terms of particle number densities becomes

$$\frac{n_n}{n_e n_p} = \frac{g_n h^3 e^{-\varepsilon_n/kT}}{4(2\pi mkT)^{3/2}}, \tag{3.182}$$

with ε_n negative because the energy levels of atomic hydrogen have negative energies. Since the total number density of hydrogen atoms is $n_H = \Sigma_{n=1} n_n$ and $n_e = n_p$, Equation (3.182) leads to the equation

$$\frac{n_H}{n_e^2} = \frac{h^3 \Sigma g_n e^{-\varepsilon_n/kT}}{4(2\pi mkT)^{3/2}}. \tag{3.183}$$

Equation (3.183) is known as the Saha equation.

Hydrogen and helium are the most abundant elements in stars. The ionization equilibrium of a mixture of hydrogen and helium is described by the equations

$$\frac{n_{\rm H}}{n_{\rm p}n_{\rm e}} = \frac{h^3 \Sigma g_n e^{-\varepsilon_n/kT}}{4(2\pi mkT)^{3/2}}, \tag{3.184}$$

and

$$\frac{n_{\rm He}}{n_{\rm He^+}n_{\rm e}} = \frac{h^3(\Sigma g_n^1 e^{-\varepsilon_n^1/kT})e^{\chi_{\rm He}/kT}}{2(2\pi mkT)^{3/2}(\Sigma g_n e^{-4\varepsilon_n/kT})} \tag{3.185}$$

and

$$\frac{n_{\rm He^+}}{n_{\rm He^{++}}n_{\rm e}} = \frac{h^3 \Sigma \frac{g_n}{2} e^{-4\varepsilon_n/kT}}{2(2\pi mkT)^{3/2}}. \tag{3.186}$$

In Equation (3.185) $\chi_{\rm He}$ is the ionization potential of neutral helium and g_n^1 is the degeneracy of energy level ε_n^1 of neutral helium. Singly ionized helium (He$^+$) is a hydrogenic ion, and therefore its energy levels have energies $Z^2\varepsilon_n = 4\varepsilon_n$ with ε_n the energy of an energy level of atomic hydrogen. The spin of an α particle is 0, and therefore the denominators of Equation (3.185) and (3.186) contain a factor of 2 rather than the factor of 4 that appears in Equation (3.184).

3.5   Molecules

The velocity V of the center of mass of a molecule is

$$V = \frac{\displaystyle\sum_{i=1}^{N} m_n v_n}{\displaystyle\sum_{i=1}^{N} m_n} \equiv \frac{\displaystyle\sum_{i=1}^{N} m_n v_n}{M}, \tag{3.187}$$

with m_n and v_n equal to the mass and velocity of a particular particle. The Hamiltonian of a molecule can be expressed as

$$H = \frac{-\hbar^2}{2M}\nabla^2 - \frac{\hbar^2}{2m}\sum_i \nabla_i^2 - \sum_n \frac{\hbar^2}{2M_n}\nabla_n^2 - \sum_n \sum_i \frac{Z_n e^2}{r_{ni}}$$

$$+ \frac{1}{2}\sum_m \sum_n \frac{Z_m Z_n e^2}{r_{mn}} + \frac{1}{2}\sum_i \sum_j \frac{e^2}{r_{ij}}. \tag{3.188}$$

The first term on the right-hand side of Equation (3.188) represents the translational kinetic energy of the molecule. The second and third terms represent the kinetic energies of the electrons and nuclei measured with respect to the center-of-mass frame of reference. The remaining terms are potential-energy terms that result from nuclei–nuclei, nuclei–electron and electron–electron Coulomb interactions. The translational-kinetic-energy term is evaluated in the laboratory frame of reference. In the remaining kinetic-energy terms, particle positions are measured with respect to the center of mass of the molecule.

The Schrödinger equation for the molecule is

$$H\psi_m(\bar{R}, r, R) = \frac{-\hbar}{i} \frac{\partial \psi_m}{\partial t}(\bar{R}, r, R),$$

(3.189)

with H equal to the Hamiltonian given in Equation (3.188). In Equation (3.189) $\bar{R} = (\bar{x}, \bar{y}, \bar{z})$ is the position of the center of mass, and the coordinates of the electrons and nuclei are represented by r and R, respectively. The molecular wave function ψ_m given in Equation (3.189) can be written as

$$\psi_m(\bar{R}, r, R) = \psi_{tr}(\bar{R})\psi(r, R).$$

(3.190)

Substituting the above wave function into Equation (3.189) causes the equation to separate, and the wave function ψ_{tr} satisfies the equation

$$\frac{-\hbar^2}{2M} \nabla^2 \psi_{tr} = \frac{-\hbar}{i} \frac{\partial \psi_{tr}}{\partial t}.$$

(3.191)

It can readily be shown that Equation (3.191) has plane-wave solutions. Substituting

$$\psi_{tr} = e^{i(k \cdot r - \omega t)},$$

(3.192)

with $k = p/\hbar$ and $\omega = \varepsilon_{tr}/\hbar$, into Equation (3.191), we find that

$$\varepsilon_{tr} = \frac{\hbar^2 k^2}{2M} = p^2/2M.$$

(3.193)

Equation (3.193) implies that ε_{tr} is an eigenvalue and ψ_{tr}, given in Equation (3.192), an eigenfunction of Equation (3.191).

   Because the mass of an electron is much less than the masses of nuclei, electron motions inside a molecule are much more rapid and the electron energy levels can be determined with the coordinates of the nuclei held stationary. The above physical argument allows us to express the wave function $\psi(r, R)$ as

$$\psi(r, R) = \psi_e(r; R)\chi_N(R).$$

(3.194)

In Equation (3.194) r and R are equal to the positions of electrons and nuclei, respectively, $\psi_e(r; R)$ is the electron wave function with the nuclei at positions R, and $\chi_N(R)$ is the wave function that describes the motions of the nuclei. Substituting Equation (3.194) into Equation (3.188), with the translational-kinetic-energy term omitted, causes the Schrödinger equation to be separated into the two equations

$$H_e \psi_e(r; R) = \left[\frac{-\hbar^2}{2m} \sum_i \nabla_i^2 - \sum_n \sum_i \frac{Z_n e^2}{|R_n - r_i|} + \frac{1}{2} \sum_i \sum_{\substack{j \\ i \neq j}} \frac{e^2}{r_{ij}} \right] \psi_e(r; R)$$

$$= E_e(R)\psi_e(r; R)$$

(3.195)

and

$$H_n \psi(r, R) = \left[-\sum_n \frac{\hbar^2}{2M_n} \nabla_n^2 + E_e(R) + \frac{1}{2} \sum_m \sum_{\substack{n \\ m \neq n}} \frac{Z_m Z_n e^2}{|R_m - R_n|} \right] \psi(r, R)$$

$$= E\psi(r, R). \tag{3.196}$$

In Equations (3.195) and (3.196), R (i.e. $R_1, R_2, \ldots, R_N$) denotes the positions of the N nuclei with respect to the center of mass. In Equation (3.195) $E_e(R)$ is an eigenvalue of the equation with the nuclei at positions R. Solutions of Equation (3.195) with the nuclei at positions R give both the wave functions $\psi_e(r; R)$ and the eigenvalues $E_e(R)$. Equation (3.195) has solutions that correspond to different electronic states of the molecule as well as different positions of the nuclei. In the discussion below, we will assume that the molecule is in the lowest-energy electronic state.

Equation (3.196) can be written as

$$\left[-\sum_n \frac{\hbar^2}{2M_n} \nabla_n^2 + U(R) \right] \chi_N(R)\psi_e(r; R) = E\chi_N(R)\psi_e(r; R), \tag{3.197}$$

with $U(R) = E_e(R) + \frac{1}{2} \sum_m \sum_{\substack{n \\ m \neq n}} \frac{Z_m Z_n e^2}{|R_m - R_n|}$. Equation (3.197) can be re-expressed as

$$\psi_e(r; R) \left[-\sum_{n=1}^N \frac{\hbar^2}{2M_n} \nabla_n^2 + U(R) - E \right] \chi_N(R)$$

$$= \sum_{n=1}^N \frac{\hbar^2}{2M_n} [\chi_N(R)\nabla_n^2 \psi_e(r; R) + 2\nabla_n \chi_N(R)\nabla_n \psi_e(r; R)]. \tag{3.198}$$

To a first approximation we can assume that the right-hand side of Equation (3.198) is equal to zero because the motions of nuclei are small, and therefore the $\nabla_n \psi_e$ terms are small because the motions of electrons are much greater than those of nuclei. Setting the right-hand side of Equation (3.198) equal to zero leads to the simpler equation

$$\left[-\sum_{n=1}^N \frac{\hbar^2}{2M_n} \nabla_n^2 + U(R) \right] \chi_N(R) = E\chi_N(r). \tag{3.199}$$

If the molecule is a diatomic molecule (i.e. a molecule with two nuclei), then Equation (3.199) can be expressed as

$$\left[\frac{-\hbar^2}{2M_r} \nabla^2 + U(r) \right] \chi(r, \theta, \phi) = E\chi(r, \theta, \phi), \tag{3.200}$$

with $M_r = \dfrac{M_1 M_2}{M_1 + M_2}$. In Equation (3.200) r is the radial distance between the two nuclei.

The Laplacian ∇^2 in Equation (3.200) is given by the equation

$$\nabla^2 = \frac{1}{r^2} \frac{\partial r^2}{\partial r} \frac{\partial}{\partial r} + \frac{1}{r^2 \sin \theta} \frac{\partial}{\partial \theta} \left(\sin \theta \frac{\partial}{\partial \theta} \right) + \frac{1}{r^2 \sin^2 \theta} \frac{\partial^2}{\partial \phi^2}. \tag{3.201}$$

The square of the angular momentum is

$$L^2 = (rxp) \cdot (rxp) = r^2 p^2 - (r \cdot p)^2. \tag{3.202}$$

Equation (3.202) implies that

$$p^2 = p_r^2 + \frac{L^2}{r^2}, \tag{3.203}$$

with $p^2 = -\hbar^2 \nabla^2$. Equations (3.201–3.203) imply that

$$p_r^2 = -\hbar^2 \frac{1}{r^2} \frac{\partial}{\partial r} \left(r^2 \frac{\partial}{\partial r} \right) \tag{3.204}$$

and

$$L^2 = -\hbar^2 \left[\frac{1}{\sin\theta} \frac{\partial}{\partial\theta} \left(\sin\theta \frac{\partial}{\partial\theta} \right) + \frac{1}{\sin^2\theta} \frac{\partial^2}{\partial\phi^2} \right]. \tag{3.205}$$

As will be shown in Section 4.1, the spherical harmonics $Y_K^m(\theta, \phi) = C_K^m P_K^m(\theta) e^{im\phi}$, with K and m integers, $P_K^m(\theta)$ the associated Legendre functions and C_K^m normalization constants, are eigenfunctions of the equation

$$L^2 Y_K^m = K(K+1)\hbar^2 Y_K^m \quad (m \leq K), \tag{3.206}$$

with $K = 0, 1, 2, 3, \ldots$ Equations (3.201) and (3.206) imply that Equation (3.200) becomes

$$\left[\frac{-\hbar^2}{2M_r} \frac{1}{r^2} \frac{\partial}{\partial r} \left(r^2 \frac{\partial}{\partial r} \right) + \frac{L^2}{2M_r r^2} + U(r) \right] \chi(r, \theta, \phi) = E\chi(r, \theta, \phi). \tag{3.207}$$

If we let

$$\chi(r, \theta, \phi) = \frac{Y(r)}{r} Y_K^m(\theta, \phi), \tag{3.208}$$

then Equation (3.207) reduces to the equation

$$\frac{-\hbar^2}{2M_r} \frac{d^2 Y(r)}{dr^2} + W(r) Y(r) = E Y(r), \tag{3.209}$$

with

$$W(r) = \frac{\hbar^2 K(K+1)}{2M_r r^2} + U(r).$$

The function $W(r)$ has a minimum value $W(r_0)$ for some equilibrium separation r_0 between the nuclei. Expanding $W(r)$ in a Taylor expansion about r_0, we obtain

$$W(r) = W_0 + \frac{1}{2} K_0 (r - r_0)^2, \tag{3.210}$$

with

$$W_0 = \varepsilon_0 + \frac{\hbar^2 K(K+1)}{2M_r r_0^2}.$$

Equation (3.209) can now be written as

$$\frac{-\hbar^2}{2M_r}\frac{d^2 Y(r)}{dr^2} + \left[W_0 + \frac{1}{2}K_0(r - r_0)^2\right]Y(r) = E Y(r). \tag{3.211}$$

The above equation, which is similar to that of a simple harmonic oscillator, has the eigenvalues

$$E = \varepsilon_0 + \frac{\hbar^2 K(K + 1)}{2M_r r_0^2} + \hbar\left(\frac{K_0}{M_r}\right)^{1/2}(v + 1/2), \tag{3.212}$$

with $K = 0, 1, \ldots$ equal to the rotational quantum numbers and $v = 0, 1, 2, \ldots$ equal to the vibrational quantum numbers. Equation (3.212) implies that if a molecule is assumed to be in its lowest electronic energy level, the energy levels of a molecule can be expressed as

$$\varepsilon_{v,K} = \varepsilon_0 + \hbar\omega(v + 1/2) + \frac{\hbar^2}{2I}K(K + 1), \tag{3.213}$$

with ε_0 negative, the moment of inertia $I = M_r r_0^2$, and $M_r = (M_1 M_2)/(M_1 + M_2)$. Therefore, the energy of the molecule becomes

$$\varepsilon = \varepsilon_{tr} + \varepsilon_{v,K}, \tag{3.214}$$

with ε_{tr} equal to the translational kinetic energy. An energy-level diagram of the H_2 molecule is given in Figure 3.1.

Having determined the energy levels of a diatomic molecule, we are now in a position to determine the thermodynamic functions. If the electrons are in the lowest energy level, the partition function associated with the discrete quantum states of a diatomic molecule is

$$Z = g_s e^{-\varepsilon_0/kT} Z_{rot} Z_{vib}, \tag{3.215}$$

with

$$Z_{rot} = \sum_{K=0}^{\infty} (2K + 1)e^{-\frac{\hbar^2 K(K+1)}{2kTI}}, \tag{3.216}$$

$$Z_{vib} = \sum_{v=0}^{\infty} e^{-(\hbar\omega/kT)(v + 1/2)}, \tag{3.217}$$

and g_s equal to the degeneracy of the lowest electronic level. In most diatomic molecules the spin of the lowest electronic energy level is 0, and therefore $g_s = 2s + 1 = 1$.

The vibrational partition function Z_{vib} can be readily evaluated because it contains a geometric series. We obtain

$$Z_{vib} = e^{-(\hbar\omega/2kT)}\sum_{v=0}^{\infty} e^{-(\hbar\omega/kT)v} = \frac{e^{-\hbar\omega/2kT}}{1 - e^{-\hbar\omega/kT}}. \tag{3.218}$$

The average vibrational energy per molecule is

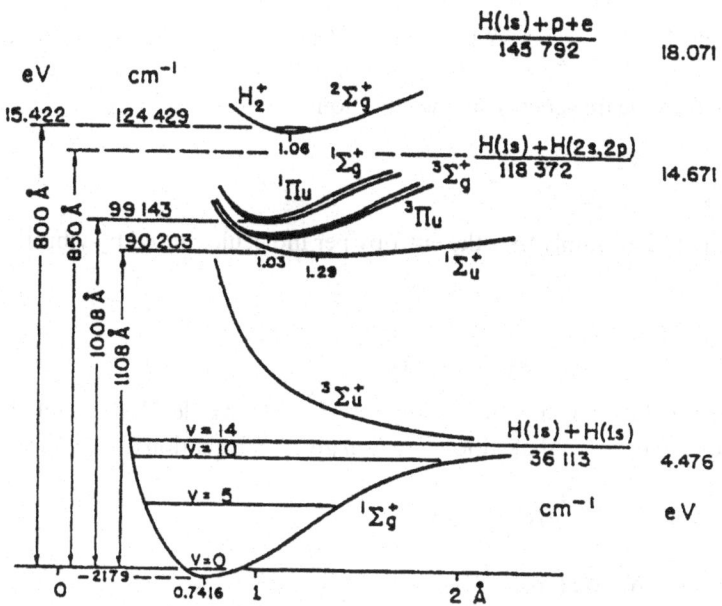

Figure 3.1. Schematic energy level diagram of H_2. Energies are measured with respect to the $v = 0$ level of the ground state. From Field, Somerville and Dressler (1966).

$$\varepsilon_{vib} = \frac{\Sigma n\hbar\omega e^{-(\hbar\omega/kT)v}}{\Sigma e^{-(\hbar\omega/kT)v}} = -kT^2 \frac{\partial \ln}{\partial T} \Sigma e^{n(\hbar\omega/kT)v}. \tag{3.219}$$

Summing the geometric series in Equations (3.219) and then differentiating the sum of the series gives

$$\varepsilon_{vib} = \frac{\hbar\omega}{e^{\hbar\omega/kT} - 1}. \tag{3.220}$$

At high temperatures (i.e. $kT \gg \hbar\omega$)ε_{vib} becomes kT. From Equation (3.220) the heat capacity per molecule becomes

$$c_{vib} = k\left(\frac{\hbar\omega}{kT}\right)^2 \frac{e^{\hbar\omega/kT}}{(e^{\hbar\omega/kT} - 1)^2}. \tag{3.221}$$

For high temperatures (i.e. $kT \gg \hbar\omega$) Equation (3.221) implies that

$$c_{vib} = k. \tag{3.222}$$

At low temperatures the vibrational heat capacity c_{vib} becomes

$$c_{vib} = k\left(\frac{\hbar\omega}{kT}\right)^2 e^{-(\hbar\omega/kT)}. \tag{3.223}$$

The vibrational free energy F_{vib} per molecule is equal to

$$F_{\text{vib}} = -kT \ln Z_{\text{vib}} = -\frac{\hbar\omega}{2} + kT \ln(1 - e^{-\hbar\omega/kT}). \tag{3.224}$$

The entropy is related to the free energy by the equation

$$S = -\left(\frac{\partial F}{\partial T}\right)_V. \tag{3.225}$$

Equations (3.224) and (3.225) imply that the entropy per molecule caused by vibrational energy levels is

$$S_{\text{vib}} = -k \ln(1 - e^{-\hbar\omega/kT}) + \frac{\hbar\omega}{T(e^{\hbar\omega/kT} - 1)}. \tag{3.226}$$

If the temperature is sufficiently high that many rotational energy levels are excited (i.e. if kT is $\gg \hbar^2/2I$), then the rotational partition function given in Equation (3.216) becomes

$$Z_{\text{rot}} = \int (2K + 1)e^{-\frac{\hbar^2 K(K+1)}{2kTI}} \, dK. \tag{3.227}$$

Substituting $K(K + 1) = N^2$, we have

$$(2K + 1)dK = 2N \, dN. \tag{3.228}$$

Equation (3.228) implies that Equation 3.227 can be expressed as

$$Z_{\text{rot}} = 2\int_0^\infty N e^{-\hbar^2 N^2/2IkT} dN = \frac{2IkT}{\hbar^2}. \tag{3.229}$$

Using the above expression for Z_{rot}, the rotational energy ε_{rot}, free energy F_{rot} and entropy S_{rot} become

$$\varepsilon_{\text{rot}} = -kT^2 \frac{\partial \ln Z_{\text{rot}}}{\partial T} = kT, \tag{3.230}$$

$$F_{\text{rot}} = -kT \ln Z_{\text{rot}} = -kT \ln(2IkT) + 2kT \ln \hbar \tag{3.231}$$

and

$$S_{\text{rot}} = -\left(\frac{\partial F_{\text{rot}}}{\partial T}\right)_v = k \ln(2IkT) + k(1 - 2\ln \hbar). \tag{3.232}$$

From Equation (3.232) the heat capacity c_{rot} becomes

$$c_{\text{rot}} = T\left(\frac{dS}{dT}\right)_v = k. \tag{3.233}$$

At low temperatures (i.e. $kT \ll \hbar^2/2I$) we have

$$Z_{\text{rot}} = 1 + 3e^{-\hbar^2/IkT}. \tag{3.234}$$

The rotational energy ε_{rot} per molecule is

$$\varepsilon_{\text{rot}} \simeq (3\hbar^2/I)e^{-\hbar^2/IkT}. \tag{3.235}$$

The free energy per molecule F_{rot} is

$$F_{rot} = -kT\ln(1 + 3e^{-\hbar^2/IkT}) \simeq -3kTe^{-\hbar^2/IkT} \tag{3.236}$$

and the entropy per molecule S_{rot} is

$$S_{rot} = -\left(\frac{\partial F_{rot}}{\partial T}\right)_V = \left(3k + \frac{3\hbar^2}{IT}\right)e^{-\hbar^2/IkT}. \tag{3.237}$$

The heat capacity per molecule c_{rot} becomes

$$c_{rot} = T\left(\frac{dS_{rot}}{dT}\right)_V = \frac{3\hbar^4}{I^2kT^2}e^{-\hbar^2/IkT}. \tag{3.238}$$

Since the translational component of the heat capacity is $(3/2)k$, the total heat capacity at high temperatures becomes

$$c_v = \frac{3}{2}k + c_{vib} + c_{rot} = \frac{7}{2}k. \tag{3.239}$$

At low temperatures both c_{vib} and c_{rot} approach zero, and therefore c_v approaches $(3/2)k$. At intermediate temperatures c_{vib} and c_{rot} must be calculated from the partition functions given above. The rotational heat capacity c_{rot} is greater than the vibrational heat capacity c_{vib} at low temperatures because rotational energy levels can be excited at lower temperatures than vibrational energy levels.

Consider a gas containing N diatomic molecules. From Equations (3.99) and (3.101) we have

$$\mu N = U + PV - TS = U + NkT - TS, \tag{3.240}$$

with μ equal to the chemical potential of the molecule. If the molecule exists at temperatures sufficiently high that many rotational energy levels are excited, then from Equations (3.220) and (3.230) we find that

$$U = \frac{3}{2}NkT + N\varepsilon_0 + N\frac{\hbar\omega}{e^{\hbar\omega/kT} - 1} + NkT. \tag{3.241}$$

Equations (3.110), (3.226) and (3.232) imply that the entropy of the gas is

$$S = \frac{5}{2}Nk + Nk\ln(T^{3/2}/n) + \frac{3}{2}Nk\ln\left(\frac{2\pi mk}{\hbar^2}\right) - Nk\ln(1 - e^{-\hbar\omega/kT})$$
$$+ \frac{N\hbar\omega}{T(e^{\hbar\omega/kT} - 1)} + Nk\ln(2IkT) + Nk(1 - 2\ln\hbar) + k\ln g_I, \tag{3.242}$$

with g_I equal to the degeneracy caused by the spin of the two nuclei. From Equations (3.240–3.242) the chemical potential μ becomes

$$\mu = kT + \varepsilon_0 + \frac{\hbar\omega}{e^{\hbar\omega/kT} - 1} - kT\ln(T^{3/2}/n) - \frac{3}{2}kT\ln\left(\frac{2\pi mk}{\hbar^2}\right)$$
$$+ kT\ln(1 - e^{-\hbar\omega/kT}) - \frac{\hbar\omega}{e^{\hbar\omega/kT} - 1} - kT\ln(2IkT) + kT(-1 + 2\ln\hbar)$$

$$- kT \ln g_I$$

$$= \varepsilon_0 - kT \ln(T^{3/2}/n) - \frac{3}{2}kT \ln\left(\frac{2\pi mk}{\hbar^2}\right)$$

$$+ kT \ln(1 - e^{\hbar\omega/kT}) - kT \ln(2IkT) + 2kT \ln \hbar - kT \ln g_I. \tag{3.243}$$

The above equation is correct for heteropolar molecules such as CO, but for homopolar molecules such as H_2 where the two nuclei are identical the third from last term in Equation (3.243) must be changed to $- kT \ln(IkT)$, because rotation of the symmetry axis through $180°$ leaves the molecule unchanged.

3.6   Reaction equilibrium

The second law of thermodynamics states that if two systems, which are labeled 1 and 2 in expression (3.244), given below, exchange an amount of heat dQ, then the total entropy change is

$$dS_1 + dS_2 \geq 0. \tag{3.244}$$

If heat exchange occurs by means of an irreversible process, then the inequality in Expression (3.244) holds. The exchange of energy that occurs when radiation flows from a hotter to a cooler gas is an example of an irreversible process. Contracting stars illustrate the second law of thermodynamics in a rather interesting manner. Pre-main-sequence stars remain in hydrostatic equilibrium as they contract. Therefore, the virial theorem, given in Equation (2.35), is satisfied. Since the equation of state of the gas inside a pre-main-sequence star is that of an ideal nondegenerate gas (i.e. $P = nkT$), it can readily be shown from the virial theorem that as a result of contraction $T \propto \rho^{1/3}$. Therefore from Equation (3.110) it follows that the entropy of the star decreases during contraction. However, the entropy of the emitted radiation is such that the total amount of entropy in the Universe increases. The second law of thermodynamics implies that

$$TdS \geq dQ = dU + PdV. \tag{3.245}$$

During an irreversible process we have $dS > dQ/T$.

The Gibbs function G is defined by the equation

$$G = U + PV - TS. \tag{3.246}$$

If an irreversible process occurs under conditions of constant P and T, then Expression (3.245) and Equation (3.246) imply that

$$dG < 0. \tag{3.247}$$

The Gibbs function for a gas consisting of different constituents such as different molecules or different nuclei can be written as

$$G = G(P, T, n_1, n_2, \ldots, n_n), \tag{3.248}$$

with $n_1, n_2, \ldots, n_n$ equal to the number densities of the various constituents. Small changes in G satisfy the equation

$$dG = -SdT + VdP + \sum_{i=1}^{n} \mu_i dn_i, \qquad (3.249)$$

with $\mu_i = dG/dn_i$. If the various constituents of a gas react with each other as, for example, molecules involved in a chemical reaction do, then changes in dn_i $(i = 1, \ldots, n)$ are such that $dn_i = n_i d\lambda$, with λ a common proportionality factor. It follows that

$$dG = G d\lambda. \qquad (3.250)$$

If changes in G occur under conditions of constant P and T, then Equations (3.249) and (3.250) imply that

$$(dG)_{P,T} = G d\lambda = \sum_{i=1}^{n} \mu_i dn_i. \qquad (3.251)$$

Since $dn_i = n_i d\lambda$, Equation (3.251) becomes

$$G = \sum_{i=1}^{n} \mu_i n_i. \qquad (3.252)$$

If constituents A_1, A_2, A_3 and A_4 are involved in the reversible reaction

$$v_1 A_1 + v_2 A_2 \rightleftarrows v_3 A_3 + v_4 A_4, \qquad (3.253)$$

with v_i $(i = 1, 2, 3, 4)$ equal to positive integers, then changes in the number densities of the various constituents are such that

$$\frac{dn_1}{-v_1} = \frac{dn_2}{-v_2} = \frac{dn_3}{v_3} = \frac{dn_4}{v_4}. \qquad (3.254)$$

Equations (3.254) imply that Equation (3.251) can be written as

$$(dG)_{P,T} = \sum_{i=1}^{4} \mu_i dn_i = dn_1 \left(\mu_1 + \frac{v_2}{v_1} \mu_2 - \frac{v_3}{v_1} \mu_3 - \frac{v_4}{v_1} \mu_4 \right). \qquad (3.255)$$

Equilibrium exists if $dG = 0$, and therefore Equations (3.245), (3.246), (3.249) and (3.255) imply that at equilibrium we have

$$v_1 \mu_1 + v_2 \mu_2 = v_3 \mu_3 + v_4 \mu_4. \qquad (3.256)$$

Equation (3.256) can be derived using a somewhat different approach. Under equilibrium conditions the first law of thermodynamics is

$$Tds = d\left(\frac{u}{n}\right) + Pd\left(\frac{1}{n}\right) \qquad (3.257)$$

with s equal to the entropy per particle, n equal to the total number density of particles and $u = u(n, s)$ the internal energy density. If the gas consists of a number of different types of particles i with concentrations $c_i = n_i/n$ then we can write

$$u = u(n, s, c_i). \tag{3.258}$$

Equation (3.258) implies that

$$d\left(\frac{u}{n}\right) = -Pd\left(\frac{1}{n}\right) + Tds + \Sigma\mu_i dc_i \tag{3.259}$$

with

$$P = \frac{-\partial\left(\dfrac{u}{n}\right)}{\partial\left(\dfrac{1}{n}\right)} = n^2\frac{\partial\left(\dfrac{u}{n}\right)}{\partial n}$$

$$T = \frac{\partial\left(\dfrac{u}{n}\right)}{\partial s}$$

$$\mu_i = \frac{\partial\left(\dfrac{u}{n}\right)}{\partial c_i} = \frac{\partial u}{\partial n_i}.$$

Since Equation (3.257) is satisfied at equilibrium Equation (3.259) reduces to the equation

$$\sum_i \mu_i dc_i = 0. \tag{3.260}$$

Equations (3.256) and (3.260) are equivalent.

In Section 3.4 we derived the equation of ionization equilibrium for a partially ionized gas (see Equation (3.183). The derivation of Equation (3.183), which is known as the Saha equation, was based on deriving expressions for the number densities of quantum states and assigning relative probabilities for the occupation of particular quantum states based on their relative energies. It is instructive to show that the equation of ionization equilibrium can also be derived from Equation (3.256).

In the limit of low density the number density of electrons n_e given in Equation (3.78) becomes

$$n_e = \frac{8\pi}{h^3}e^{\mu_e/kT}\int_0^\infty e^{-p^2/2mkT}p^2 dp$$

$$= 2e^{\mu_e/kT}\left(\frac{2\pi mkT}{h^2}\right)^{3/2}. \tag{3.261}$$

A similar equation relating proton number density n_p and chemical potential u_p follows immediately from Equation (3.261). From Equation (3.256) and a similar equation for n_p the electron and proton chemical potentials become

$$\mu_e = -kT\ln(T^{3/2}/n_e) - \frac{3}{2}kT\ln\left(\frac{2\pi mk}{h^2}\right) - kT\ln 2 \tag{3.262}$$

and

$$\mu_p = -kT\ln(T^{3/2}/n_p) - \frac{3}{2}kT\ln\left(\frac{2\pi m_p k}{h^2}\right) - kT\ln 2. \tag{3.263}$$

Next we will evaluate the chemical potential of the hydrogen atom μ_H. The internal energy per atom and entropy per atom are readily obtained from Equations (3.112) and (3.117) respectively. These latter equations and Equation (3.240) imply that the chemical potential μ_H of atomic hydrogen is

$$\mu_H = -kT\ln(T^{3/2}/n_H) - \frac{3}{2}kT\ln\left(\frac{2\pi m_H k}{h^2}\right) - kT\ln \Sigma g_n e^{-\varepsilon_n/kT}. \tag{3.264}$$

From Equation (3.256) it follows that equilibrium occurs when

$$\mu_e + \mu_p = \mu_H. \tag{3.265}$$

Using the expressions for μ_e, μ_p and μ_H given in Equations (3.262)–(3.265) respectively and setting $m_p = m_H$ we obtain

$$\ln\frac{n_H}{n_e n_p} + \ln\left[4\left(\frac{2\pi mkT}{h^2}\right)^{3/2}\right] = \ln \Sigma g_n e^{-\varepsilon_n/kT}. \tag{3.266}$$

Equation (3.266) leads directly to the equation

$$\frac{n_H}{n_e n_p} = \frac{\Sigma g_n e^{-\varepsilon_n/kT}}{4\left(\frac{2\pi mkT}{h^2}\right)^{3/2}}, \tag{3.267}$$

which is identical to Equation (3.183).

The equilibrium between hydrogen atoms and molecules is described by the reactions

$$H + H \rightleftharpoons H_2. \tag{3.268}$$

We assume that hydrogen molecules are in their lowest electronic energy level (i.e. $S = 0$), and the temperature of the gas sufficiently high that the chemical potential of the hydrogen molecule is given in Equation (3.243). The equation of equilibrium between hydrogen atoms and hydrogen molecules is

$$2\mu_H = \mu_{H_2} + 2\varepsilon_1 \tag{3.269}$$

where μ_H is given in Equation (3.264), μ_{H_2} is given in Equation (3.243), ε_1 is the energy of the lowest energy level of atomic hydrogen. The term $2\varepsilon_1$ is required in Equation (3.269) because atomic and molecular hydrogen chemical potentials must have the same zero energy. Substituting the expressions for μ_H and μ_{H_2} into Equation (3.269) we obtain

$$- 2kT\ln(T^{3/2}/n_H) - 3kT\ln\left(\frac{2\pi m_H k}{h^2}\right) - 2kT\ln\Sigma g_n e^{-\varepsilon_n/kT}$$

$$= \varepsilon_0 + 2\varepsilon_1 - kT\ln(T^{3/2}/n_{H_2}) - \frac{3}{2}kT\ln\left(\frac{2\pi m_{H_2} k}{h^2}\right)$$

$$+ kT\ln(1 - e^{-\hbar\omega/kT}) - kT\ln(IkT) + 2kT\ln\hbar - kT\ln g_I \qquad (3.270)$$

with the nuclear spin degeneracy g_I equal to $(2I + 1)(2I + 1) = 4$. Solving Equation (3.270) we have

$$\frac{n_{H_2}}{n_H^2} = \frac{h^3 m_{H_2}^{3/2} e^{-(\varepsilon_0 + 2\varepsilon_1)/kT}}{(2\pi m_H^2 kT)^{3/2}\left(\Sigma\left(\frac{g_n}{2}\right)e^{-\varepsilon_n/kT}\right)^2 (1 - e^{-\hbar\omega/kT})/\left(\frac{h^2}{IkT}\right)} \qquad (3.271)$$

with $g_n/2$ equal to the electron degeneracy of the nth energy level of the hydrogen atom (i.e. $g_n = 4n^2$).

If the temperature in the interior of a star exceeds about 10^9 K, electron–positron pairs will be produced in significant numbers, and the reaction and inverse reaction,

$$e^- + e^+ \leftrightarrow \gamma + \gamma, \qquad (3.272)$$

will go into equilibrium. We denote the number density of electrons in the absence of e^-–e^+ pair production as n_e^0. If the gas density inside a star is 10^4–10^5 g cm^{-3} so that n_e^0 is $\simeq 10^{28\text{-}29}$ cm^{-3} then the gas will be nondegenerate for temperatures $\gtrsim 10^9$ K. Such physical conditions can arise in the cores of evolved, massive stars.

As already discussed in Section 3.3, the chemical potential of the photon gas is zero because photons have zero rest mass. Therefore Equations (3.272) and (3.256) imply

$$\mu_- = - \mu_+ \qquad (3.273)$$

with μ_- and μ_+ equal to the chemical potentials of the electrons and positrons respectively. If it is assumed that the gas is nonrelativistic then the chemical potentials μ_- and μ_+ can be written

$$\mu_- = mc^2 + \mu^1$$

and

$$\mu_+ = - \mu_- = - mc^2 - \mu^1. \qquad (3.274)$$

From Equations (3.78) and (3.274) the number density of electrons becomes

$$n_e^- = n_e^0 + n_e^+ = \frac{8\pi}{h^3}e^{\mu^1/kT}\int_0^\infty e^{-\varepsilon/kT}p^2 dp = 2\left(\frac{2\pi mkT}{h^2}\right)^{3/2}e^{\mu^1/kT}. \qquad (3.275)$$

The corresponding expression for the number density of positrons is

$$n_e^+ = \frac{8\pi}{h^3} e^{\frac{-2mc^2 - \mu^1}{kT}} \int_0^\infty e^{-\varepsilon/kT} p^2 dp = 2\left(\frac{2\pi mkT}{h^2}\right)^{3/2} e^{\frac{-2mc^2 - \mu^1}{kT}} \tag{3.276}$$

Equations (3.275) and (3.276) lead directly to the equations

$$\ln n_e^0 + \ln\left(1 + \frac{n_e^+}{n_e^0}\right) = \ln\left[2\left(\frac{2\pi mkT}{h^2}\right)^{3/2}\right] + \frac{\mu^1}{kT} \tag{3.277}$$

and

$$\ln n_e^+ = \ln\left[2\left(\frac{2\pi mkT}{h^2}\right)^{3/2}\right] - \frac{2mc^2}{kT} - \frac{\mu^1}{kT}. \tag{3.278}$$

Adding Equations (3.277) and (3.278) we obtain

$$\ln n_e^0 n_e^+ \left(1 + \frac{n_e^+}{n_e^0}\right) - \ln\left[4\left(\frac{2\pi mkT}{h^2}\right)^3\right] = \frac{-2mc^2}{kT} \tag{3.279}$$

Taking the exponential of both sides of equation (3.279) gives

$$n_e^+(n_e^0 + n_e^+) = 4\left(\frac{2\pi mkT}{h^2}\right)^3 e^{-2mc^2/kT}. \tag{3.280}$$

Solving the above quadratic equation we have

$$n_e^+ = \frac{-n_e^0}{2} + \frac{n_e^-}{2}\left(1 + \frac{16}{n_e^{02}}\left(\frac{2\pi mkT}{h^2}\right)^3 e^{-2mc^2/kT}\right)^{1/2}$$

$$\simeq \frac{4}{n_e^0}\left(\frac{2\pi mkT}{h^2}\right)^3 e^{-2mc^2/kT}. \tag{3.281}$$

The number density of electrons becomes $n_e^- = n_e^0 + n_e^+$.

In the above equation we used the binomial expansion because the second factor within the square root term is $\ll 1$ for the physical conditions considered. Because we have assumed that the gas is nonrelativistic the above expressions for n_e^+ and n_e^- are accurate only if $kT \ll 2mc^2$.

3.7 Imperfect gases

An ideal (or perfect) gas is one in which the gas pressure is caused entirely by the kinetic energies of the particles. In real gases particles have potential energies because they interact with other particles in the gas. The interactions between particles cause the pressure of a gas to differ from that of a perfect gas. Consider a gas consisting of ions with charge Z and electrons. The Coulomb potential energy between two neighboring ions is $\simeq (Ze)^2/r_i$ with the distance between neighboring ions approximately equal to $(n_i)^{-1/3}$. An ionized gas is nearly perfect if the Coulomb energy per ion is very small compared to kT, and therefore n_i satisfies the inequality

$$n_i \ll \left(\frac{kT}{Z^2 e^2} \right)^3. \tag{3.282}$$

The electrons in an ionized gas will tend to be concentrated in the vicinity of ions because of attractive Coulomb forces. Moreover, ions will also tend to distribute themselves nonuniformly with respect to other ions because of the repulsive Coulomb interaction. Although ions will sometimes come close to each other in an ionized gas their statistical distribution is such that the probability of an ion being found in the vicinity of another ion is less than the probability of finding an ion in an equal but arbitrarily close volume element of space. It is reasonable to assume that in a low-density ionized gas the ion number density $n_i(r)$ and the electron number density $n_e(r)$ satisfy the Boltzmann relations

$$n_i(r) = n_i \exp(-Ze\phi(r)/kT) \tag{3.283}$$

and

$$n_e(r) = n_e \exp(e\phi(r)/kT) \tag{3.284}$$

with r equal to the distance from some particular ion, Z the charge of the ion, $\phi(r)$ the electrostatic potential, n_i the mean ion number density and n_e the mean electron number density.

The electrostatic potential $\phi(r)$, and the charged particle number densities $n_i(r)$ and $n_e(r)$ are related by the Poisson equation

$$\nabla^2 \phi(r) = -4\pi e[Zn_i(r) - n_e(r)], \quad r \neq 0. \tag{3.285}$$

Because the Coulomb energies $Ze\phi$ and $-e\phi$ are $\ll kT$, Equations (3.283) and (3.284) become

$$n_i(r) \simeq n_i \left(1 - \frac{Ze\phi(r)}{kT} \right) \tag{3.286a}$$

$$n_e(r) \simeq n_e \left(1 + \frac{e\phi(r)}{kT} \right). \tag{3.286b}$$

Using Equations (3.286a) and (3.286b) we can write Equation (3.285) as

$$\nabla^2 \phi(r) = -4\pi \left[Zn_i e - n_e e - \frac{(Z^2 n_i + n_e)e^2 \phi(r)}{kT} \right], \quad r \neq 0. \tag{3.287}$$

From charge neutrality we obtain

$$Zn_i - n_e = 0. \tag{3.288}$$

Equations (3.287) and (3.288) imply that

$$\nabla^2 \phi(r) - \kappa^2 \phi(r) = 0, \quad r \neq 0 \tag{3.289}$$

with

$$\kappa^2 \approx \frac{1}{\lambda_D^2} = \frac{4\pi(Z+1)e^2 n_e}{kT}.$$

In Equation (3.289) λ_D has the dimensions of length and is known as the Debye length. By using the expression for charge neutrality given in Equation (3.288) we have made the implicit assumption that the average number of ions within a distance $\kappa = 1/\lambda_D$ from an ion is $\gg 1$. Since the potential $\phi(r)$ in Equation (3.289) depends only on r the Laplacian ∇^2 becomes

$$\nabla^2 = \frac{1}{r^2}\frac{\partial r^2}{\partial r}\frac{\partial}{\partial r} = \frac{\partial^2}{\partial r^2} + \frac{2}{r}\frac{\partial}{\partial r}. \tag{3.290}$$

Substituting $\phi(r) = Zee^{-\kappa r}/r$ we obtain

$$\frac{1}{Ze}\frac{\partial^2 \phi(r)}{\partial r^2} = -2\kappa e^{-\kappa r}\frac{\partial\left(\frac{1}{r}\right)}{\partial r} + e^{-\kappa r}\frac{\partial^2\left(\frac{1}{r}\right)}{\partial r^2} + \frac{\kappa^2}{r}e^{-\kappa r} \tag{3.291}$$

and

$$\frac{2}{Zer}\frac{\partial\phi(r)}{\partial r} = \frac{2e^{-\kappa r}}{r}\frac{\partial\left(\frac{1}{r}\right)}{\partial r} - \frac{2\kappa e^{-\kappa r}}{r^2}. \tag{3.292}$$

Adding Equations (3.291) and (3.292) gives

$$\left(\frac{1}{Ze}\right)\nabla^2\phi(r) = \frac{2}{r}e^{-\kappa r}\frac{\partial\left(\frac{1}{r}\right)}{\partial r} + e^{-\kappa r}\frac{\partial^2\left(\frac{1}{r}\right)}{\partial r^2} + \frac{\kappa^2}{r}e^{-\kappa r} \tag{3.293}$$

Equations (3.290) and (3.293) imply that $\phi(r) = Ze(e^{-\kappa r}/r)$ is a solution of Equation (3.289) for $r \neq 0$ since the function $1/r$ satisfies Laplace's equation

$$\nabla^2\frac{1}{r} = 0 \tag{3.294}$$

if r is $\neq 0$.

At distances close to a particular ion the potential $Ze\phi(r) = Ze(e^{-\kappa r}/r)$ becomes

$$\phi(r) = \frac{Zee^{-\kappa r}}{r} \simeq \frac{Ze}{r} - eZ\kappa \tag{3.295}$$

with

$$\kappa = \frac{1}{\lambda_D} = \left(\frac{4\pi(Z+1)e^2 n_e}{kT}\right)^{1/2}.$$

Equation (3.295) implies that the Coulomb potential close to an ion Z caused by the

Coulomb interaction with all other particles in the gas is $\simeq -eZ/\lambda_D$. The Coulomb energy per ion becomes

$$u_c \simeq -\frac{Z^2 e^2}{\lambda_D} = -Z^2 e^2 \left(\frac{4\pi(Z+1)e^2 n_e}{kT}\right)^{1/2}$$

$$= -Z^{5/2}(Z+1)^{1/2} e^3 \left(\frac{4\pi}{kT}\right)^{1/2} n_i^{1/2}$$

$$= -Z^{5/2}(Z+1)^{1/2} \left(\frac{4\pi}{k}\right)^{1/2} e^3 \left(\frac{n_i}{T^{3/2}}\right)^{1/3} n_i^{1/6}. \qquad (3.296)$$

Since $n_i/T^{3/2}$ remains unchanged if variations occur under adiabatic conditions the contribution to the gas pressure from Coulomb interactions becomes

$$P_c = -\left(\frac{du_c}{d(1/n_i)}\right)_s = n_i^2 \left(\frac{du_c}{dn_i}\right)_s = \frac{n_i u_c}{6} = \frac{U_c}{6} \qquad (3.297)$$

with u_c given by Equation (3.296) and U_c equal to the Coulomb interaction energy per unit volume. The total pressure of the nondegenerate gas becomes

$$P = nkT + P_c \qquad (3.298)$$

with $n = n_e + n_i$ and P_c negative because the Coulomb potential energy is negative (i.e. the Coulomb interaction is attractive).

For a high-density and relatively low-temperature gas such as the electron-degenerate gases in the interiors of white dwarfs we can calculate the electrostatic corrections assuming that ions are at fixed positions in space. An electron cloud whose radius r_i is determined from the relations

$$\frac{1}{n_i} = \frac{Z}{n_e} = \frac{4}{3}\pi r_i^3 \qquad (3.299)$$

is assumed to exist around each ion. If the Z electrons within the radius r_i have a uniform density then the attractive electrostatic energy with the ion is

$$E_{ie} = -\int_0^{r_i} \frac{Ze^2}{r} n_e 4\pi r^2 dr = -\frac{3}{2}\frac{(Ze)^2}{r_i} \qquad (3.300)$$

with $n_e = 3Z/(4\pi r_i^3)$ equal to the number density of electrons.

The electrostatic energy caused by the repulsive Coulomb interaction between electrons, which are assumed to be uniformly distributed about an ion within a sphere of radius r_i, is

$$E_{ee} = \int_0^\infty \frac{q}{r} dq \qquad (3.301)$$

with $q = Ze(r^3/r_i^3)$ and $dq = 3Ze(r^2/r_i^3)dr$. From the above expressions the Coulomb potential energy caused by Coulomb interaction between electrons becomes

$$E_{ee} = 3\frac{(Ze)^2}{r_i^6} \int_0^{r_i} r^4 dr = \frac{3}{5}\frac{(Ze)^2}{r_i}. \tag{3.302}$$

From Equations (3.300) and (3.302) the total Coulomb energy per ion becomes

$$E_c = E_{ie} + E_{ee} = -\frac{9}{10}\frac{(Ze)^2}{r_i} = -\frac{9}{10}\left(\frac{4\pi n_i}{3}\right)^{1/3}(Ze)^2. \tag{3.303}$$

Therefore, the Coulomb contribution to the pressure is

$$P_c = -\frac{dE_c}{d\left(\frac{1}{n_i}\right)} = n_i^2\frac{dE_c}{dn_i} = -\left(\frac{3}{10}\right)\left(\frac{4\pi}{3}\right)^{1/3}(Ze)^2 n_i^{4/3} \tag{3.304}$$

with $1/n_i$ equal to the volume per ion. Expressing Equation (3.304) in terms of the electron number density n_e we have

$$P_c = -\left(\frac{3}{10}\right)\left(\frac{4\pi}{3}\right)^{1/3} Z^{2/3} e^2 n_e^{4/3}. \tag{3.305}$$

The total gas pressure becomes

$$P = P_e + n_i kT + P_c \tag{3.306}$$

with P_e the pressure of the electron-degenerate gas.

3.8   The Boltzmann equation and transport coefficients in a gas

Let $f(r, v, t)$ be a particle distribution function with r and v the position vector and velocity respectively, and t equal to the time. The distribution function $f(r, v, t)$ is normalized so that

$$\int\int f(r, v, t) d^3 r\, d^3 p = n \tag{3.307}$$

with n equal to the number density of particles. If a force F acts for a short time interval Δt we have

$$f\left(r + v\Delta t, v + \frac{F}{m}\Delta t, t + \Delta t\right) d^3 r' d^3 v' = f(r, v, t) d^3 r\, d^3 v + \left(\frac{\partial f}{\partial t}\right)_c \Delta t\, d^3 r\, d^3 v \tag{3.308}$$

where $\left(\dfrac{\partial f}{\partial t}\right)_c$ is a term that allows for collisions between particles. If the external force F depends only on the position vector r then

$$d^3 r' d^3 v' = d^3 r\, d^3 v. \tag{3.309}$$

Equation (3.309) is equivalent to stating that the Jacobian of the transformation between primed and unprimed coordinates is unity. Equation (3.309) implies that Equation (3.308) can be rewritten as

$$f(r + v\Delta t, v + F/m\Delta t, t + \Delta t) = f(r, v, t) + \left(\frac{\partial f}{\partial t}\right)_c. \tag{3.310}$$

If the left-hand side of Equation (3.310) is expanded to first-order terms then we have

$$\left(\frac{\partial}{\partial t} + v \cdot \nabla + \frac{F}{m} \cdot \nabla_v\right) f(r, v, t) = \left(\frac{\partial f}{\partial t}\right)_c \tag{3.311}$$

with $\nabla_v = \left(\dfrac{\partial}{\partial v_x}, \dfrac{\partial}{\partial v_y}, \dfrac{\partial}{\partial v_z}\right)$.

Assume that the distribution function f can be expressed as

$$f = f_0 + f' \tag{3.312}$$

with f_0 equal to the distribution function under conditions of LTE and f' small as compared to f_0. Since the gas may have a bulk velocity $u(r)$, the distribution function f_0 is equal to

$$f_0 = n(r)\left(\frac{m}{2\pi k T(r)}\right)^{3/2} \exp[-m(v - u(r))^2/2kT(r)] \tag{3.313}$$

where the number density of particles $n(r)$, temperature $T(r)$ and bulk velocity $u(r)$ are functions of positions r. Since collisions will cause the distribution function f to relax toward the equilibrium distribution f_0 we have

$$\left(\frac{\partial f}{\partial t}\right)_c = -v_c(f - f_0) = -v_c f' \tag{3.314}$$

with v_c equal to the characteristic collision frequency. If we let

$$f = f_0 + f' e^{-t/\tau} \tag{3.315}$$

with $\tau = 1/v_c$ then the distribution function f, which is $f_0 + f'$ at $t = 0$ will relax toward the equilibrium distribution function f_0 in a timescale $\tau = 1/v_c$.

Electrical conductivity of a nondegenerate, ionized gas
Ionized gases have finite electrical conductivities, and therefore the presence of an electric field will cause a current to flow through the gas. We neglect the motions of ions and assume that n and T in Equation (3.313) are independent of position. In addition, we assume that the bulk velocity u given in Equation (3.313) is small as compared to the thermal velocities of electrons. From Equations (3.311) and (3.314) we obtain

$$\frac{eE_x}{m}\frac{\partial f_0}{\partial v_x} = -v_c f' \tag{3.316}$$

with f_0 given in Equation (3.313) but with $n(r)$ and $T(r)$ independent of r, and the electric field E_x assumed to be directed along the negative x-direction since the charge of the electron is negative. Using Ohm's law we have

$$\sigma E_x = J_x = e \int v_x f \, dv_x dv_y dv_z = e \int v_x f' \, dv_x dv_y dv_z. \tag{3.317}$$

The second of the above equalities follows because the current density is equal to zero under conditions of thermodynamic equilibrium.

Multiplying Equation (3.316) by ev_x and integrating over velocity space leads to the equation

$$- v_c J_x = \frac{e^2 E_x}{m} \int \frac{\partial f_0}{\partial v_x} v_x dv_x dv_y dv_z. \tag{3.318}$$

Differentiating the distribution function f_0 with respect to v_x and assuming that u is much less than the electron thermal velocity gives

$$\frac{\partial f_0}{\partial v_x} = \frac{-m}{kT} f_0(v_x - u) \simeq \frac{-m f_0 v_x}{kT}. \tag{3.319}$$

From Equations (3.318) and (3.319) we obtain

$$J_x = \frac{e^2 E_x}{v_c kT} \int f_0 v_x^2 dv_x dv_y dv_z = \frac{ne^2 E_x}{v_c m}. \tag{3.320}$$

Equations (3.317) and (3.320) imply that the electrical conductivity is

$$\sigma = \frac{ne^2}{mv_c}. \tag{3.321}$$

In order to determine a numerical value for the collision frequency v_c we must first calculate the Rutherford scattering cross section for the scattering of an electron in the Coulomb field of an ion. In polar coordinates the energy of an electron in the electrostatic field of an ion of charge Ze is

$$E = \frac{1}{2} m \dot{r}^2 + \frac{1}{2} mr^2 \dot{\Phi}^2 - \frac{Ze^2}{r}. \tag{3.322}$$

The conservation of angular momentum of the electron is given by the equation

$$l = mr^2 \dot{\Phi} = mbv \tag{3.323}$$

with b equal to the impact parameter and v equal to the velocity of the electron at large distances from the ion. Equations (3.322) and (3.323) lead to the equation

$$\frac{dr}{dt} = \left(\frac{2}{m} \left(E + \frac{Ze^2}{r} \right) - \frac{l^2}{m^2 r^2} \right)^{1/2}. \tag{3.324}$$

Expressing Equation (3.323) as

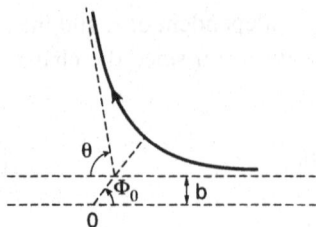

Figure 3.2. Relates the angle Φ_0 given by Equation (3.326) to the total scattering angle θ.

$$d\Phi = \frac{ldt}{mr^2}$$
(3.325)

and substituting for dt in Equation (3.324) gives

$$\Phi_0 = \int_{r_{min}}^{\infty} \frac{ldr}{r^2 \left(2m\left(E + \frac{Ze}{r}\right)^2 - \frac{l^2}{r^2}\right)^{1/2}}$$
(3.326)

with r_{min} equal to the distance of closest approach between the ion and the hyperbolic orbit of the electron. The total angle θ through which the electron is scattered is related to Φ_0 given in Equation (3.326) by the equation (see Figure 3.2)

$$\theta = |2\Phi_0 - \pi|.$$
(3.327)

Integrating Equation (3.326) gives

$$\Phi_0 = \cos^{-1} \frac{Ze^2/mv^2b}{\sqrt{1 + (Ze^2/mv^2b)^2}}.$$
(3.328)

Solving Equation (3.323) for b and using Equation (3.327) gives

$$b = \frac{Ze}{mv^2} \tan \Phi_0 = \frac{Ze}{mv^2} \cot\left(\frac{\theta}{2}\right).$$
(3.329)

Since the impact parameter b is a function of the scattering angle θ we have

$$d\sigma = -\pi b \frac{db}{d\theta} d\theta.$$
(3.330)

The minus sign occurs in Equation (3.330) because θ decreases as the impact parameter b increases. From Equations (3.329) and (3.330) it follows that

$$d\sigma = \pi \left(\frac{Ze^2}{mv^2}\right)^2 \frac{\cos \frac{1}{2}\theta}{\sin^3 \frac{1}{2}\theta} d\theta.$$
(3.331)

Dividing both sides of the above equation by the solid angle $d\Omega = 2\pi \sin \theta \, d\theta$ and using the trigonometric identity $\sin \theta = 2\sin(\theta/2)\cos(\theta/2)$ implies

$$\frac{d\sigma}{d\Omega} = \left(\frac{Ze^2}{2mv^2}\right)^2 \frac{1}{\sin^4 \frac{1}{2}\theta}.$$ (3.332)

Equation (3.332) is the Rutherford scattering differential cross section. The inverse of the collision frequency v_c that appears in Equation (3.316) is the time required for an electron to be deflected through an angle $\geq \pi/2$. If such scattering were to occur as the result of a single scattering with an ion of charge Ze then v_c would be equal to

$$(v_c)_{\frac{\pi}{2}} = nv\sigma(\theta \geq \pi/2)$$ (3.333)

with

$$\begin{aligned}
\sigma(\theta_0 \geq \pi/2) &= \left(\frac{Ze^2}{2mv^2}\right)^2 \int_{\pi/2}^{\pi} \frac{2\pi \sin\theta d\theta}{\sin^4(\theta/2)} \\
&= 8\pi \left(\frac{Ze^2}{2mv^2}\right)^2 \int_{\pi/2}^{\pi} \frac{\cos(\theta/2)}{\sin^3(\theta/2)} d\theta \\
&= (\sqrt{2}-1)\pi \left(\frac{Ze^2}{mv^2}\right)^2.
\end{aligned}$$ (3.334)

Equations (3.333) and (3.334) give a lower limit to the collision frequency v_c because electrons can be deflected through an angle $\geq \pi/2$ as a result of many (or several) small-angle scatterings. For small-angle scattering and change in the velocity of the electron Δv and the scattering angle θ are related by the equation

$$\Delta v = v\theta$$ (3.335)

and the Rutherford differential scattering cross section given in Equation (3.332) becomes

$$\frac{d\sigma(\theta)}{d\Omega} = 4\left(\frac{Ze^2}{mv^2}\right)^2 \left(\frac{1}{\theta}\right)^4.$$ (3.336)

The time rate of change of the mean square change in velocity caused by small-angle scatterings is approximately

$$\begin{aligned}
\frac{d|\Delta v|^2}{dt} &= nv \int |\Delta v|^2 \sigma(\theta) d\Omega \\
&= 8\pi n \left(\frac{Ze^2}{m}\right)^2 \frac{1}{v} \int_{\theta_{min}}^{\pi/2} \frac{d\theta}{\theta}
\end{aligned}$$ (3.337)

and the minimum scattering angle θ_{min} is approximately

$$\theta_{min} = \frac{2Ze^2}{mv^2 \lambda_D}$$ (3.338)

because the Debye length λ_D given in Equation (3.289) measures the distance over which charge neutrality occurs.

Since a deflection of $\simeq \pi/2$ occurs when $\sqrt{(\Delta v)^2} \simeq v$ we have

$$\frac{d\,|\Delta v|^2}{dt} \simeq v^2 v_c. \tag{3.339}$$

From Equations (3.337–3.339) we obtain

$$(v_c)_s \simeq 8\pi n v \left(\frac{Ze^2}{mv^2}\right)^2 \ln\left(\frac{1}{2}\frac{mv^2\lambda_D}{Ze^2}\right). \tag{3.340}$$

The approximate expression for the collision frequency $(v_c)_s$ caused by small angle scattering in Equation (3.340) is greater than the collision frequency caused by a single large angle scattering given by Equations (3.332) and (3.333–3.334) by a factor of

$$\frac{(v_c)_s}{(v_c)_{\pi/2}} \simeq \frac{8}{(\sqrt{2}-1)} \ln\left(\frac{1}{2}\frac{mv^2}{Ze^2}\lambda_D\right). \tag{3.341}$$

Since the mean value of $\frac{1}{2}mv^2$ is $\frac{3}{2}kT$ the mean value of the above ratio becomes

$$\frac{8}{(\sqrt{2}-1)} \ln\left(\frac{3}{2}\frac{kT}{Ze^2}\lambda_D\right). \tag{3.342}$$

The Debye length λ_D given in Equation (3.289) determines the minimum scattering angle θ_{min} in Equation (3.338) if $(\lambda_D/r_i) \gg 1$ with $r_i = (3/4\pi n_i)^{1/3}$. At high density (and/or low temperature) the minimum scattering angle is

$$\theta_{min} = \frac{2Ze^2}{mv^2 r_i} = 2\left(\frac{4\pi}{3}\right)^{1/3}\frac{Z^{2/3}e^2}{mv^2}n_e^{1/3} \tag{3.343}$$

and therefore Equation (3.340) becomes

$$(v_c)_s \simeq 8\pi n v \left(\frac{Ze^2}{mv^2}\right)^2 \ln\left(\frac{1}{2}\frac{mv^2}{Ze^2}r_i\right) \tag{3.344}$$

with $r_i = (3/4\pi n_i)^{1/3}$.

Since the electrical conductivity of a nondegenerate gas follows from Equations (3.321) and (3.340) (or (3.344)) we can estimate the characteristic timescale for a magnetic field to diffuse from a star. Ohm's law (i.e. $J = \sigma E$) and the Maxwell equation

$$\nabla \times E = -\frac{1}{c}\frac{\partial B}{\partial t} \tag{3.345}$$

imply

$$\frac{1}{\sigma}\nabla \times J = -\frac{1}{c}\frac{\partial B}{\partial t}. \tag{3.346}$$

Taking the curl of the Maxwell equation

$$\nabla \times \boldsymbol{B} = \frac{4\pi}{c}\boldsymbol{J} + \frac{1}{c}\frac{\partial \boldsymbol{E}}{\partial t} \tag{3.347}$$

and neglecting the $\dfrac{1}{c}\dfrac{\partial \boldsymbol{E}}{\partial t}$ term leads to the equation

$$\nabla \times \boldsymbol{J} = \frac{c}{4\pi}\nabla \times (\nabla \times \boldsymbol{B}) = \frac{c}{4\pi}[-\nabla^2\boldsymbol{B} + \nabla(\nabla \cdot \boldsymbol{B})]. \tag{3.348}$$

Since $\nabla \cdot \boldsymbol{B} = 0$ Equations (3.346) and (3.348) become

$$\nabla^2\boldsymbol{B} = \frac{4\pi\sigma}{c^2}\frac{\partial \boldsymbol{B}}{\partial t}. \tag{3.349}$$

Dimensional analysis of Equation (3.349) implies that the magnetic diffusion timescale τ_{Diff} is

$$\tau_{\text{Diff}} \sim \frac{4\pi\sigma}{c^2}R^2 \tag{3.350}$$

with R equal to the radius of the star.

Thermal conductivity of a nondegenerate, ionized gas

If the x-axis is chosen along the direction of the temperature gradient within the gas then the heat flux becomes

$$q_x = \frac{1}{2}m\int v^2 v_x f dv_x dv_y dv_z = \frac{1}{2}m\int v^2 v_x f' dv_x dv_y dv_z \tag{3.351}$$

with f' equal to the perturbed distribution function. Both n and T are assumed to vary along the x-axis, and therefore the unperturbed distribution function f_0 is

$$f_0 = n(x)\left(\frac{m}{2\pi kT(x)}\right)^{3/2}\exp[-mv^2/2kT(x)]. \tag{3.352}$$

The relevant form of the Boltzmann equation is

$$v_x\frac{\partial f_0}{\partial x} = \left(\frac{\partial f}{\partial t}\right)_c = -v_c f'. \tag{3.353}$$

Because f' is small we have set $f = f_0$ on the left-hand side of Equation (3.353). The derivative $\partial f_0/\partial x$ in Equation (3.353) can be expressed as

$$\frac{\partial f_0}{\partial x} = \frac{\partial f_0}{\partial T}\frac{dT}{dx} + \frac{\partial f_0}{\partial n}\frac{dn}{dx}. \tag{3.354}$$

If it is assumed that the pressure P is independent of x then we have

$$\frac{1}{n}\frac{dn}{dx} = -\frac{1}{T}\frac{dT}{dx}.$$

(3.355)

On the other hand, if a polytropic relation is assumed (i.e. $P = \kappa\rho^{1+1/n'}$ with n' equal to the polytropic index) then it can be readily shown that

$$\frac{1}{n}\frac{dn}{dx} = n'\frac{1}{T}\frac{dT}{dx}.$$

(3.356)

Pressure gradients must balance gravitation in stellar atmospheres and interiors, and therefore the above polytropic relation between temperature T and density n is the most relevant relation between the temperature and density gradients. The polytropic index is likely to be $\simeq 3$–1.5 in stars. Using Equation (3.355) we find that Equation (3.354) becomes

$$\frac{\partial f_0}{\partial x} = f_0\left[\frac{mv^2}{2kT^2} - \frac{5}{2T}\right]\frac{dT}{dx}.$$

(3.357)

If Equation (3.355) gives the relationship between density and temperature gradients then Equation (3.354) becomes

$$\frac{\partial f_0}{\partial x} = f_0\left[\frac{mv^2}{2kT^2} + \frac{1}{T}(n' - 3/2)\right]\frac{dT}{dx}.$$

(3.358)

The heat flux q_x is given by the equation

$$q_x = \frac{1}{2}\int mv^2 v_x f' dv_y dv_z.$$

(3.359)

Multiplying Equation (3.353) by $\frac{1}{2}mv^2 v_x$ with $\dfrac{\partial f_0}{\partial x}$ given by Equation (3.357) we obtain

$$q_x = \frac{5md T}{4v_c T dx}\int v^2 v_x^2 f_0 dv_x dv_y dv_z - \frac{m^2}{v_c 4kT^2}\frac{dT}{dx}\int v^4 v_x^2 f_0 dv_x dv_y dv_z$$

$$= -\frac{5}{2}\frac{n}{m}\frac{k^2 T}{v_c}\frac{dT}{dx}.$$

(3.360)

Multiplying Equation (3.358) by $\frac{1}{2}mv^2 v_x$, using Equation (3.359), and integrating over velocity space gives

$$q_x = -\left(\frac{n'}{2} - \frac{3}{4}\right)\frac{m}{v_c T}\frac{dT}{dx}\int v^2 v_x^2 f_0 dv_x dv_y dv_z - \frac{m^2}{4v_c kT^2}\frac{dT}{dx}\int v^4 v_x^2 f_0 dv_x dv_y dv_z$$

$$= -\frac{5}{2}\left[5(n' - 3/2) + \frac{7}{2}\right]\frac{nk^2 T}{mv_c}\frac{dT}{dx}.$$

(3.361)

The thermal conductivity is defined by the equation

$$q_x = -\kappa_c \frac{dT}{dx} \tag{3.362}$$

with dT/dx negative because heat flows in the direction of a negative temperature gradient. Therefore, the thermal conductivity κ_c is equal to

$$\kappa_c = \frac{5}{2} \frac{nk^2 T}{mv_c} \tag{3.363}$$

if Equation (3.360) is applicable, and

$$\kappa_c = \frac{5}{2} \left[5(n' - 3/2) + \frac{7}{2} \right] \frac{nk^2 T}{mv_c} \tag{3.364}$$

if Equation (3.361) is applicable.

Viscosity of a nondegenerate, ionized gas

Suppose that an ionized gas has a local mean velocity that is in the x-direction and a velocity gradient that is in the z-direction. The coefficient of viscosity η is defined by the equation

$$\eta \frac{du(z)}{dz} = - \int mv_x v_z f dv_x dv_y dv_z = - m \int v_x v_z f' dv_x dv_y dv_z. \tag{3.365}$$

Equation (3.365) implies that η is the shear stress produced by a unit velocity gradient. The distribution function f_0 is

$$f_0 = n \left(\frac{m}{2\pi kT} \right)^{3/2} \exp\left(-\frac{m}{2kT} [(v_x - u(z))^2 + v_y^2 + v_z^2] \right). \tag{3.366}$$

Substituting the above distribution function into the Boltzmann equation gives

$$\frac{mv_z}{kT} \frac{du}{dz} (v_x - u) f_0 \simeq \frac{mv_z}{kT} v_x \frac{du}{dz} f_0 = - v_c f' \tag{3.367}$$

since u is much smaller than a typical particle velocity. If Equation (3.367) is multiplied by $mv_x v_z$ and integrated over velocity space then a comparison between the resultant equation and Equation (3.365) implies that

$$\eta = nkT/v_c \tag{3.368}$$

with v_c given in Equation (3.340) or (3.344).

Thermal conduction in a degenerate electron gas

Thermal conduction by electrons is a principal source of energy transport in the interiors of white dwarfs and neutron stars and in the degenerate cores of red giants. In an electron-degenerate gas, as in an ionized, nondegenerate gas, the electron-distribution function deviates from that of thermodynamic equilibrium because there is a temperature gradient. The electron-distribution function in a degenerate gas is determined by the Boltzmann equation

$$\left(\frac{\partial}{\partial t} + \boldsymbol{v} \cdot \nabla - \frac{eE}{m} \cdot \nabla_v\right) f(r, \boldsymbol{v}, t) = \left(\frac{\partial f}{\partial t}\right)_c \tag{3.369}$$

with E equal to the electron field and the collisional term $\left(\frac{\partial f}{\partial t}\right)_c$ caused by Coulomb collisions between electrons and ions. Since we have steady-state conditions the distribution function does not depend explicitly on time, and therefore the $\frac{\partial}{\partial t}$ term in the above equation is equal to zero. Moreover, we need only consider changes along one direction, which we choose to be the x-direction. The unperturbed distribution function $f_0(v)$ is that of an electron-degenerate gas given in Equation (3.76). Because departures from thermodynamic equilibrium are small we can express the distribution function as

$$f(\boldsymbol{v}) = f_0(v) + f'(\boldsymbol{v}) \tag{3.370}$$

with $f'(\boldsymbol{v})$ small as compared to $f_0(v)$. Equation (3.369) can now be expressed as

$$- v_x \frac{\partial f}{\partial x} - \frac{eE_x}{m} \frac{\partial f}{\partial v_x} = \left(\frac{\partial f}{\partial t}\right)_c. \tag{3.371}$$

In Equation (3.371) v_x is negative because the Coulomb force $- eE_x$ is negative and E_x is along the x-direction. The net energy flow, which causes the electron-distribution function to depart from $f = f_0$, is caused by a temperature gradient, and we can assume that changes in the distribution function f along the x-direction are proportional to the temperature gradient. Therefore, the first term in Equation (3.371) becomes

$$- v_x \frac{\partial f}{\partial x} = - v_x \frac{\partial f}{\partial T} \frac{dT}{dx}. \tag{3.372}$$

Because the temperature gradient dT/dx can also be regarded as small we can substitute df_0/dT for df/dT in Equation (3.372). For a nonrelativistic electron-degenerate gas the electron-distribution function f_0 given in Equation (3.76) can be expressed as

$$f_0 = \frac{1}{\exp[(\varepsilon - \mu(x))/kT(x)] + 1} \tag{3.373}$$

with $\varepsilon = \frac{1}{2}mv^2$. Differentiating the above expression for f_0 with respect to x implies that

$$\frac{df_0}{dx} = \frac{\partial f_0}{\partial \mu}\left(\frac{d\mu}{dx} + \frac{(\varepsilon - \mu)}{T} \frac{dT}{dx}\right). \tag{3.374}$$

Because $\mu = \mu(T(x))$ and $-\dfrac{\partial f_0}{\partial \mu} = \dfrac{\partial f_0}{\partial \varepsilon}$ Equation (3.374) becomes

$$\frac{\partial f_0}{\mathrm{d}x} = -\frac{\partial f_0}{\partial \varepsilon}\left(\frac{\mathrm{d}\mu}{\mathrm{d}T} + \frac{\varepsilon - \mu}{T}\right)\frac{\mathrm{d}T}{\mathrm{d}x}. \tag{3.375}$$

Equation (3.375) and the relation $\varepsilon = \frac{1}{2}mv^2$ imply that Equation (3.371) becomes

$$\left(\frac{\partial f}{\partial t}\right)_c = v_x\frac{\partial f_0}{\partial \varepsilon}\left[eE_x - \left(\frac{\mathrm{d}\mu}{\mathrm{d}T} + \frac{(\varepsilon - \mu)}{T}\right)\frac{\mathrm{d}T}{\mathrm{d}x}\right]. \tag{3.376}$$

Equation (3.78) and the substitutions $\varepsilon = p^2/2m = \frac{1}{2}mv^2$ imply that the electron current density is

$$J_x = -\frac{2m^3e}{h^3}\int v_x f(v)\mathrm{d}^3v. \tag{3.377}$$

It is clear that there is no net flow of charge because such a flow would produce very large electrostatic forces. We can, therefore, assume that

$$J_x = 0. \tag{3.378}$$

The heat flux is given by the expression

$$q_x = \frac{2m^3}{h_3}\int \frac{1}{2}mv^2 v_x f(v)\mathrm{d}^3v. \tag{3.379}$$

We will calculate the heat flux q_x subject to the condition that $J_x = 0$. Evaluation of the heat flux q_x requires that we determine the distribution function f.

Substituting $f = f_0 + f'$ for the distribution function allows us to express the collisional term $\left(\dfrac{\partial f}{\partial t}\right)_c$ in the Boltzmann equation as

$$\left(\frac{\partial f}{\partial t}\right)_c = -\frac{f'}{\tau(v)} \tag{3.380}$$

where the relaxation time $\tau(v)$, which is the inverse of the collision frequency v_c, depends on the magnitude but not the direction of the particle velocity v. As previously discussed (see Equation (3.315)), τ is the timescale for the perturbed distribution function f' to relax to the thermodynamic equilibrium distribution function if the temperature gradient were to disappear suddenly. The usefulness of the approximation given in Equation (3.380) arises because the $(\partial f/\partial t)_c$ term in the Boltzmann equation will be expressed below as an integral over solid angles. If the approximation contained in Equation (3.380) were not made then a rather complicated integral equation would have to be solved in order to determine $(\partial f/\partial t)_c$.

At nonrelativistic energies electron scattering by ions is elastic, and therefore electron kinetic energies are conserved when electrons are scattered by ions. Electrons moving in a particular range of velocities defined by $(v, \mathrm{d}v, \mathrm{d}\Omega)$ will be scattered by ions into all other directions, and therefore the number of such electrons will decrease. On the other hand, some electrons with speed such that $|v'| = |v|$ and $dv' = dv$ but with different directions of motion will be scattered into the velocity interval $(v, dv, d\Omega)$, and therefore the number of electrons in this latter velocity interval will be increased.

Let $d\sigma(\theta, v)/d\Omega$ be the differential Rutherford scattering cross section that is given in Equation (3.332). The angle θ is the angle between the velocity v of an incoming electron and the velocity v' of the scattered electron ($|v| = |v'|$). If the number density of ions is n_i then the rate of increase of electrons in the velocity interval $(v, dv, d\Omega)$ caused by scattering of all electrons in the velocity interval (v', dv') is

$$\frac{dn_+(v)}{dt} v^2 dv d\Omega = n_i \frac{2m^3}{h^3} d\Omega dv \int_{\Omega'} v' f(v') v'^2 \frac{d\sigma}{d\Omega'}(\theta, v')(1 - f(v)) d\Omega'. \tag{3.381a}$$

The corresponding rate of loss of electrons from the velocity interval $(v, dv, d\Omega)$ is

$$\frac{dn_-(v)}{dt} v^2 dv d\Omega = n_i \frac{2m^3}{h^3} v f(v) v^2 d\Omega \int_{\Omega'} \frac{d\sigma}{d\Omega'}(\theta, v)(1 - f(v')) d\Omega'. \tag{3.381b}$$

The terms $1 - f(v)$ and $1 - f(v')$ occur in the above equations because the Pauli exclusion principle implies that scattering into a quantum state cannot occur if that state is already occupied. The net rate of change of electrons in a particular small volume of velocity space caused by collisions is

$$\left[\frac{dn_+(v)}{dt} - \frac{dn_-(v)}{dt}\right] v^2 dv d\Omega = \frac{2m^3}{h^3} \left(\frac{df(v)}{dt}\right)_c v^2 dv d\Omega. \tag{3.382}$$

Equations (3.381a–3.382) and the relation $|v| = |v'|$ imply

$$\left(\frac{\partial f}{\partial t}\right)_c = n_i v \int_{\Omega'} \frac{d\sigma}{d\Omega'}(\theta, v)[f(v')(1 - f(v)) - f(v)(1 - f(v'))] d\Omega'. \tag{3.383}$$

Expressing the distribution function $f(v) = f_0(v) + f'(v)$ with $f'(v)$ small we find that Equation (3.383) reduces to the equation

$$\left(\frac{\partial f}{\partial t}\right)_c = n_i v \int_{\Omega'} \frac{d\sigma(\theta, v)}{d\Omega'}[f'(v') - f'(v)] d\Omega'. \tag{3.384}$$

A different expression for $\left(\frac{\partial f}{\partial t}\right)_c$ than that given in Equation (3.384) is implied by Equation (3.376), which can be expressed as

$$\left(\frac{\partial f}{\partial t}\right)_c = v_x \frac{\partial f_0}{\partial \varepsilon}\left[eE_x - \left(\frac{d\mu}{dT} + \frac{(\varepsilon - \mu)}{T}\right)\frac{dT}{dx}\right] \equiv v_x F(v). \tag{3.385}$$

The function $F(v)$ depends on the magnitude v but not the direction of the velocity v. Approximating the collisional term $(\partial f/\partial t)_c$ as

$$\frac{\partial f'(\boldsymbol{v})}{\partial t} = \left(\frac{\partial f}{\partial t}\right)_c = -\frac{f'(\boldsymbol{v})}{\tau(v)} \tag{3.386}$$

and substituting the expression for $f'(\boldsymbol{v})$ given by Equation (3.386) into Equation (3.384) leads to the equation

$$\left(\frac{\partial f}{\partial t}\right)_c = n_i v \tau(v) \int \frac{d\sigma(\theta, v)}{d\Omega'}\left[\frac{\partial f'(\boldsymbol{v})}{\partial t} - \frac{\partial f'(\boldsymbol{v}')}{\partial t}\right] d\Omega'. \tag{3.387}$$

Since the function $F(v)$ in Equation (3.385) does not depend on direction Equations (3.385) and (3.387) imply

$$\left(\frac{\partial f}{\partial t}\right)_c = v_x F(v) = n_i v \tau(v) F(v) \int \frac{d\sigma}{d\Omega}(v_x - v_x') d\Omega'. \tag{3.388}$$

If Φ is defined as the angle between the x-axis and $\boldsymbol{v}$, and θ is the angle between $\boldsymbol{v}$ and $\boldsymbol{v}'$ then we have

$$v_x = v \cos \Phi \tag{3.389}$$

and

$$v_x' = v \cos \Phi \cos \theta + v \sin \Phi \sin \theta \cos \phi. \tag{3.390}$$

The differential cross section $d\sigma/d\Omega$ depends on θ only, and therefore using Equations (3.389) and (3.390) the integral in Equation (3.388) becomes

$$\int_0^{2\pi} \int_0^\pi \frac{d\sigma}{d\Omega}(\theta, v)(v \cos \Phi - v \cos \Phi \cos \theta - v \sin \Phi \sin \theta \cos \phi) \sin \theta d\theta d\phi. \tag{3.391}$$

Performing the integral over ϕ we find that Equation (3.388) reduces to

$$v_x = n_i v_x \tau(v) \int \frac{d\sigma(\theta, v)}{d\Omega}(1 - \cos \theta) 2\pi \sin \theta d\theta. \tag{3.392}$$

Equation (3.392) leads directly to the equation

$$\frac{1}{\tau(v)} = 2\pi v n_i \int \frac{d\sigma}{d\Omega}(1 - \cos \theta) \sin \theta d\theta. \tag{3.393}$$

The Rutherford differential cross section $d\sigma/d\Omega$ that appears in the above equations is given by Equation (3.332) with the minimum scattering angle θ_0 given by Equation (3.343) because the density of the electron-degenerate gas is high. Substituting the expression for $d\sigma/d\Omega$ from Equation (3.332) into Equation (3.393) and letting θ_0 be the lower limit of the integration we obtain

$$\frac{1}{\tau(v)} = 2\pi n_i v \left(\frac{Ze^2}{mv^2}\right)^2 \int_{\theta_0}^\pi \frac{\sin \theta d\theta}{(1 - \cos \theta)}$$

$$= 2\pi n_i v \left(\frac{Ze^2}{mv^2}\right)^2 \ln \frac{2}{(1 - \cos \theta_0)}. \tag{3.394}$$

Using the relation $f'(v) = (\partial f/\partial t)_c \, \tau(v)$ the expressions for the current density J_x and heat flux q_x contained in Equations (3.377) and (3.379) respectively become

$$J_x = 0 = \frac{2m^3 e}{h^3} \int v_x^2 \tau(v) \frac{\partial f}{\partial \varepsilon}\left(eE_x - \left(\frac{d\mu}{dT} + \frac{(\varepsilon - \mu)}{T}\right)\frac{dT}{dx}\right) d^3 v \tag{3.395}$$

and

$$q_x = \frac{-2m^3}{h^3} \int v_x^2 \, \varepsilon\tau(v) \frac{df_0}{d\varepsilon}\left(eE_x - \left(\frac{d\mu}{dT} + \frac{(\varepsilon - \mu)}{T}\right)\frac{dT}{dx}\right) d^3 v. \tag{3.396}$$

Since E_x and the thermodynamic functions μ and T do not depend explicitly on electron velocities they can be removed from the integrals, and Equations (3.395) and (3.396) can be written as

$$\left[eE_x - \left(\frac{d\mu}{dT} - \frac{\mu}{T}\right)\frac{dT}{dx}\right]\int \frac{df_0}{d\varepsilon} v_x^2 \tau(v) d^3 v = \frac{1}{T}\frac{dT}{dx}\int v_x^2 \varepsilon\tau(v)\frac{df_0}{d\varepsilon} d^3 v \tag{3.397}$$

and

$$\frac{h^3}{2m^3} q_x = \left[-eE_x + \left(\frac{d\mu}{dT} - \frac{\mu}{T}\right)\frac{dT}{dx}\right]\int v_x^2 \tau(v)\varepsilon\frac{df_0}{d\varepsilon} d^3 v$$

$$+ \frac{1}{T}\frac{dT}{dx}\int v_x^2 \varepsilon^2 \tau(v)\frac{df_0}{d\varepsilon} d^3 v. \tag{3.398}$$

From the expression for $\tau(v)$ given in Equation (3.394) and from the relations

$$v_x = v\cos\theta,$$
$$d^3 v = 2\pi \sin\theta d\theta v^2 dv \tag{3.399}$$

and

$$\varepsilon = \frac{1}{2} mv^2.$$

Equations (3.397) and (3.398) can be readily reduced to the equations

$$eE_x - \left(\frac{d\mu}{dT} - \frac{\mu}{T}\right)\frac{dT}{dx} = \frac{\dfrac{1}{T}\dfrac{dT}{dx}\displaystyle\int_0^\infty \frac{\varepsilon^4}{G(\theta_0)}\frac{df_0}{d\varepsilon} d\varepsilon}{\displaystyle\int_0^\infty \frac{\varepsilon^3}{G(\theta_0)}\frac{df_0}{d\varepsilon} d\varepsilon} \tag{3.400}$$

and

$$q_x = -\frac{16m}{3h^3 Z^2 e^4 n_i}\left[\left(eE_x - \left(\frac{d\mu}{dT} - \frac{\mu}{T}\right)\frac{dT}{dx}\right)\int_0^\infty \frac{\varepsilon^4}{G(\theta_0)}\frac{df_0}{d\varepsilon} d\varepsilon \right.$$

$$\left. - \frac{1}{T}\frac{dT}{dx}\int_0^\infty \frac{\varepsilon^5}{G(\theta_0)}\frac{df_0}{d\varepsilon} d\varepsilon\right] \tag{3.401}$$

with $G(\theta_0) = \frac{1}{2}\ln\left(\frac{2}{1 - \cos\theta_0}\right)$. The minimum scattering angle θ_0 is given in Equation (3.343).

The thermal conductivity κ_c is defined by the equation

$$q_x = -\kappa_c \frac{dT}{dx}.$$ (3.402)

Eliminating eE_x from Equations (3.400) and (3.401), and using Equation (3.402) implies that

$$\kappa_c = \frac{16m}{3h^3 Z^2 e^4 T n_i}\left[\frac{\left(\int_0^\infty \frac{\varepsilon^4}{G(\theta_0)}\frac{df_0}{d\varepsilon}d\varepsilon\right)^2}{\int_0^\infty \frac{\varepsilon^3}{G(\theta_0)}\frac{df_0}{d\varepsilon}d\varepsilon} - \int_0^\infty \frac{\varepsilon^5}{G(\theta_0)}\frac{df_0}{d\varepsilon}d\varepsilon\right].$$ (3.403)

Since heat fluxes are additive, the total radiative and conductive heat flux in a degenerate gas becomes

$$\frac{L_r}{4\pi r^2} = -\frac{4ac}{3\kappa\rho}T^3\frac{dT}{dr} - \kappa_c\frac{dT}{dr}.$$ (3.404)

Problems

1. Using the expression for the Maxwell–Boltzmann distribution given in Equation (3.13) find the
    a. mean speed $\langle v \rangle$
    b. mean square velocity $\langle v^2 \rangle$
    c. mean square fluctuation $\langle \Delta v^2 \rangle = \langle (v - \bar{v})^2 \rangle$.
2. Show that the expression for the entropy given in Equation (3.18) satisfies Equation (3.3) by direct substitution.
3. Give complete derivations of Equations (3.61) and (3.68).
4. a. Show that the energy levels of the hydrogen atom can be found by requiring that the electron energy

    $$E = \frac{p^2}{2m} - \frac{e^2}{r}$$

    is a minimum subject to the conditions $pr = n\hbar$ ($n = 1, 2, 3, \ldots$).
    b. Consider an electron in a cubic volume with impenetrable walls and dimension $L = (4\pi/3)^{1/3}a_0$ with a_0 equal to the Bohr radius of the hydrogen atom. What is the difference in energy between the two lowest energy levels? Explain why the zero-point energy of the lowest energy level is physically significant.

5.   a. Consider a nonrelativistic completely degenerate electron gas contained in some volume. Find the frequency of electron collisions $(cm^{-2}s^{-1})$ with the wall if the electron number density is n.

b. Determine a similar electron collision frequency for a completely degenerate, ultrarelativistic electron gas.

6.   Obtain a formula for the photon number density in a blackbody radiation field of temperature T.

7.   Consider a nonrelativistic degenerate boson gas of particles of finite mass m. Write down the expression for the particle number density n and explain why the chemical potential μ must be negative. For an assumed number density n determine the temperature T_c at which $\mu \to 0$. The temperature T_c is referred to as the Bose condensation temperature and for $T < T_c$ particles condense into the lowest energy level $\varepsilon = 0$. Show that for $T < T_c$ the number densities of bosons with particle energy $\varepsilon > 0$ and $\varepsilon = 0$ are

$$n(\varepsilon > 0) = n\left(\frac{T}{T_c}\right)^{3/2}$$

and

$$n(\varepsilon = 0) = n\left(1 - \left(\frac{T}{T_c}\right)^{3/2}\right)$$

respectively. In deriving the above expressions for $n(\varepsilon > 0)$ and $n(\varepsilon = 0)$ assume that when the inequality $T < T_c$ is satisfied bosons with $\varepsilon > 0$ are given by the equation for n that is applicable when $T \geq T_c$.

8.   a. Show that the free energy F defined in Equation (3.97) is related to pressure P, entropy S and internal energy U by the equations

$$P = -\left(\frac{\partial F}{\partial V}\right)_T, \quad S = -\left(\frac{\partial F}{\partial T}\right)_V, \quad U = -T^2\left(\frac{\partial F/T}{\partial T}\right).$$

b. Show that under conditions of constant T and V changes in F satisfy the inequality $\Delta F \leq 0$.

c. The partition function of an atom is $Z = \sum_n e^{-\varepsilon/kT}$. Show that $F = -kT\ln Z$.

9.   Boltzmann statistics are applicable when the condition $e^{\mu/kT} \ll 1$ is satisfied. Show that this latter condition is equivalent to the inequality

$$n\left(\frac{h^2}{mkT}\right)^{3/2} \ll 1.$$

10.   A gas at temperature T is assumed to consist of atoms which have only two energy levels ε_1 and ε_2 with degeneracies g_1 and g_2 respectively. Find an expression for the heat capacity c_v per atom.

11. Prove that the specific heat c_v per particle of a nondegenerate, ultrarelativistic gas is $3k$.

12. For the H_2 molecule we have $\hbar^2/2kI \simeq 85\,\text{K}$ and $\hbar\omega/k \simeq 6100\,\text{K}$. For $n_{H_2} = 10^4\,\text{cm}^{-3}$ find $c_v = 3/2\,k + c_{\text{vib}} + c_{\text{rot}}$ at temperatures $T = 10^3\,\text{K}$, $10^2\,\text{K}$, $40\,\text{K}$ and $10\,\text{K}$.

4 Absorption processes

4.1 Time-independent solutions of the Schrödinger equation

As measured in the laboratory frame of reference, the kinetic energy of two particles of mass m_1 and m_2 is

$$\frac{1}{2}m_1 v_1^2 + \frac{1}{2}m_2 v_2^2. \tag{4.1}$$

The velocity of the center of mass is

$$V = \frac{m_1 v_1 + m_2 v_2}{m_1 + m_2}, \tag{4.2}$$

and the relative velocity is equal to

$$v = v_2 - v_1. \tag{4.3}$$

Solving for v_1 and v_2 in Equations (4.2) and (4.3), we obtain

$$v_1 = V - \frac{m_2 v}{m_1 + m_2},$$

$$v_2 = V + \frac{m_1 v}{m_1 + m_2}. \tag{4.4}$$

Equations (4.4) imply that the kinetic energy of the two particles is

$$\frac{1}{2}m_1\left(V - \frac{m_2 v}{m_1 + m_2}\right)^2 + \frac{1}{2}m_2\left(V + \frac{m_1 v}{m_1 + m_2}\right)^2$$

$$= \frac{1}{2}(m_1 + m_2)V^2 + \frac{1}{2}m_r v^2$$

$$= \frac{P^2}{2(m_1 + m_2)} + \frac{p^2}{2m_r}, \tag{4.5}$$

with $m_r = (m_1 m_2)/(m_1 + m_2)$.

In the hydrogen atom or an ion containing only one electron, the potential energy depends on the radial distance between the electron and proton (ion). Therefore, in the center-of-mass frame of reference, the Hamiltonian operator becomes

$$H = -\frac{\hbar^2}{2m_r}\nabla^2 - \frac{Ze^2}{r},$$
(4.6)

with $p = \frac{\hbar}{i}\nabla$. In the discussion below we will assume that the nucleus is a proton or hydrogenic ion (i.e. an ion containing only one electron) and make the approximation $m_r = m$, the mass of the electron.

The stationary eigenfunctions and eigenvalues of a hydrogen atom or hydrogenic ion are obtained from the equation

$$-\frac{\hbar^2}{2m}\left[\frac{1}{r^2}\frac{\partial}{\partial r}\left(r^2\frac{\partial\psi}{\partial r}\right) + \frac{1}{r^2\sin\theta}\frac{\partial}{\partial\theta}\sin\theta\frac{\partial\psi}{\partial\theta} + \frac{1}{r^2\sin^2\theta}\frac{\partial^2\psi}{\partial\phi^2}\right] - \frac{Ze^2\psi}{r} = E\psi.$$
(4.7)

The above expression for the Laplacian in spherical coordinates has already been used in Equation (3.201).

Because the potential-energy term is spherically symmetric, Equation (4.7) separates into three equations. We assume that a stationary solution is of the form

$$\psi_{nlm}(r, \theta, \phi) = R_{nl}(r)Y_l^m(\theta, \phi).$$
(4.8)

Substituting the above expression for ψ_{nlm} into Equation (4.7) leads to the two equations

$$\frac{1}{r^2}\frac{d}{dr}\left(r^2\frac{dR_{nl}}{dr}\right) + \left[\frac{2m}{\hbar^2}\left[E + \frac{Ze^2}{r}\right] - \frac{\lambda}{r^2}\right]R_{nl} = 0$$
(4.9)

and

$$\frac{1}{\sin\theta}\frac{\partial}{\partial\theta}\sin\theta\frac{\partial Y_l^m}{\partial\theta} + \frac{1}{\sin^2\theta}\frac{\partial^2 Y_l^m}{\partial\phi^2} + \lambda Y_l^m = 0,$$
(4.10)

with $\lambda = l(l+1)$, as will be shown below. Substituting $Y_l^m(\theta, \phi) = \Theta(\theta)\Phi(\phi)$ into Equation (4.10) leads to the two equations

$$\frac{d^2\Phi}{d\phi^2} + \nu\Phi = 0$$
(4.11)

and

$$\frac{1}{\sin\theta}\frac{d}{d\theta}\left(\sin\theta\frac{d\Theta}{d\theta}\right) + \left(\lambda - \frac{\nu}{\sin^2\theta}\right)\Theta = 0.$$
(4.12)

The physically relevant solutions of Equation (4.11) are of the form $e^{im\phi}$, with m equal to zero or to a negative or positive integer. The requirement that m is an integer follows because both $\Phi(\phi)$ and $d\Phi/d\phi$ must be continuous from $\phi = 0$ to $\phi = 2\pi$. Substituting $\Phi(\phi) = e^{im\phi}$ into Equation (4.11) implies that $\nu = m^2$ and therefore Equation (4.12) becomes

$$\frac{1}{\sin\theta}\frac{d}{d\theta}\sin\theta\frac{d\Theta}{d\theta} + \left(\lambda - \frac{m^2}{\sin^2\theta}\right)\Theta = 0, \tag{4.13}$$

with m equal to an integer. If we make the change of variable $x = \cos\theta$, then Equation (4.13) can be written as

$$(1 - x^2)y'' - 2xy' + \left[\lambda - \frac{m^2}{1 - x^2}\right]y = 0. \tag{4.14}$$

If $m = 0$, then Equation (4.14) reduces to Legendre's differential equation. The physically relevant solutions of Legendre's differential equation are the Legendre polynomials, which can be obtained from the formula

$$P_l(x) = \frac{1}{2^l l!}\frac{d^l(x^2 - 1)^l}{dx^l}, \tag{4.15}$$

with $l = 0, 1, 2, 3, \ldots$ Substituting the Legendre polynomials into Equation (4.14) with $m = 0$ implies that $\lambda = l(l + 1)$. The physically relevant solutions of Equation (4.14) with $m = 0$ or a positive integer are obtained from the formula

$$P_l^m(x) = (-1)^m(1 - x^2)^{m/2}\frac{d^m P_l(x)}{dx^m}, \tag{4.16}$$

where the associated Legendre functions $P_l^m(x) = P_l^m(\cos\theta) = 0$ for $m > l$. The orthogonality of the Legendre functions is given by the equations

$$\int_{-1}^{1} P_l^m(x)P_{l'}^m(x)dx = 0, \quad l \neq l' \tag{4.17}$$

and the normalization by the equation

$$\int_{-1}^{1} [P_l^m(x)]^2 dx = \frac{2}{2l + 1}\frac{(l + m)!}{(l - m)!}. \tag{4.18}$$

For $m < 0$, the associated Legendre functions are equal to $(-1)^m P_l^{|m|}(\cos\theta)$, and therefore the normalization is the same as given in Equation (4.18) but with $m = |m|$.

The spherical harmonics, which are given by the equation

$$Y_l^m(\theta, \phi) = \left[\frac{2l + 1}{2}\frac{(l - |m|)!}{(l + |m|)!}\right]^{1/2}P_l^m(\cos\theta)\frac{e^{im\phi}}{\sqrt{2\pi}}, \tag{4.19}$$

with $-l \leq m \leq l$, are solutions of Equation (4.10), with $\lambda = l(l + 1)$, and they satisfy the orthonormal conditions

$$\int_0^{2\pi}\int_0^{\pi} Y_l^{m\star}(\theta, \phi)Y_{l'}^{m'}(\theta, \phi)\sin\theta\,d\theta\,d\phi = \delta_{ll'}\delta_{mm'}. \tag{4.20}$$

As can be readily verified by substitution into Equation (4.9), the three lowest-energy radial solutions of Equation (4.9) with $\lambda = l(l + 1)$ are

$$R_{10}(r) = \left(\frac{Z}{a_0}\right)^{3/2} 2e^{-\frac{Zr}{a_0}},$$

$$R_{20}(r) = \left(\frac{Z}{2a_0}\right)^{3/2}\left(2 - \frac{Zr}{a_0}\right)e^{-\frac{Zr}{2a_0}}$$

and

$$R_{21}(r) = \left(\frac{Z}{2a_0}\right)^{3/2}\frac{Zr}{a_0\sqrt{3}}e^{-\frac{Zr}{2a_0}}, \tag{4.21}$$

respectively, with the Bohr radius $a_0 = \hbar^2/me^2$.

The corresponding energy levels are

$$E_n = -\frac{Z^2e^2}{2a_0n^2}. \tag{4.22}$$

The radial solutions $R_{nl}(r)$ satisfy the orthonormal conditions

$$\int R_{nl}(r)R_{n'l'}(r)r^2\mathrm{d}r = \delta_{nn'}\delta_{ll'}. \tag{4.23}$$

4.2   Line widths caused by spontaneous emission

If an electron is in an excited state of an atom, it will decay as a result of spontaneous emission to a state of lower energy. If at time $t = 0$ an electron is in state k whose mean radiative lifetime is τ, then the time-dependent wave function of the electron can be expressed as

$$\psi_k(t) = u_k e^{-t/2\tau}e^{-i\omega_k t} \quad (t > 0), \tag{4.24}$$

where u_k is the time-independent wave function associated with a particular eigenvalue. The probability that the electron has remained in eigenstate k at some time $t > 0$ is

$$\int \psi_k^* \psi_k \mathrm{d}^3r = e^{-t/\tau}. \tag{4.25}$$

The natural line width of an atomic transition is the result of the finite lifetime of the state to spontaneous emission. Determination of the Fourier transform of the factor $e^{-t/2\tau}e^{-i\omega_k t}$ in Equation (4.24) will enable us to calculate the probability that an electron in quantum state k has a particular energy. Since $e^{-t/2\tau}e^{-i\omega_k t}$ is equal to zero for $t < 0$, its Fourier transform is

$$\phi(\omega) = \int_0^\infty e^{-t/2\tau} e^{i(\omega - \omega_k)t} dt$$

$$= \frac{1}{i(\omega - \omega_k) - \dfrac{1}{2\tau}} = \frac{-i(\omega - \omega_k) - \dfrac{1}{2\tau}}{(\omega - \omega_k)^2 + \left(\dfrac{1}{2\tau}\right)^2}. \tag{4.26}$$

We introduce the normalization constant C and require that

$$1 = C^2 \int_0^\infty \phi^\star(\omega)\phi(\omega)d\omega. \tag{4.27}$$

Making the change of variable $x = 2\tau(\omega - \omega_k)$ and using Equations (4.26) and (4.27), we obtain

$$1 = 2\tau C^2 \int_{-2\tau\omega_k}^\infty \frac{dx}{x^2 + 1} \simeq 2\tau C^2 \int_{-\infty}^\infty \frac{dx}{x^2 + 1} = 2\tau\pi C^2. \tag{4.28}$$

Equation (4.28) implies that $C^2 = 1/2\pi\tau$, and consequently Equation (4.27) becomes

$$1 = \frac{1}{2\pi\tau} \int_0^\infty \frac{d\omega}{(\omega - \omega_k)^2 + \dfrac{1}{4\tau^2}} = \frac{\gamma}{2\pi} \int_0^\infty \frac{d\omega}{(\omega - \omega_k)^2 + \dfrac{\gamma^2}{4}}, \tag{4.29}$$

with $\gamma = 1/\tau$ equal to the full width of the line at half maximum (FWHM). The natural-line profile given in Equation (4.29) is known as a Lorentzian line shape. Equation (4.29) determines the probability that a photon will be emitted with a frequency that is within some particular frequency interval close to the center of the emission-line profile. If the intensity of radiation I_ν varies slowly with frequency (i.e. I_ν changes very little over a frequency interval $\simeq \gamma$ from the center of a line), then the absorption of radiation will also have the Lorentzian line shape given in Equation (4.29).

4.3   Einstein coefficients of absorption and emission

Atoms and ions that contain one or more electrons absorb radiation at particular frequencies and undergo transitions from lower to upper energy levels. Radiative transitions from upper to lower energy levels occur as a result of both spontaneous emission and stimulated emission, which takes place when an incident photon induces a transition from an excited state to a state of lower energy. Under conditions of thermodynamic equilibrium the number of transitions from a lower energy level to an upper energy level caused by the absorption of radiation is equal to the number of radiative transitions from the upper energy level to the lower energy level caused by stimulated and spontaneous emission.

Consider two energy levels E_k and E_l such that $E_k < E_l$. The degeneracies of energy levels E_k and E_l are g_k and g_l, respectively. Under conditions of thermodynamic equilibrium the intensity of the radiation field is given by the Planck function B_v. The Einstein coefficient for absorption, B_{kl}, is defined so that the probability per unit time that an atom initially in energy level E_k will absorb a photon of frequency $v = (E_l - E_k)/h$ and undergo a transition to energy level E_l is

$$B_{kl}B_v,$$ (4.30)

with

$$B_v = \frac{2hv^3}{c^2}\frac{1}{e^{hv/kT} - 1}.$$ (4.31)

The probability per unit time that an atom initially in energy level E_l will undergo a radiative transition from energy level E_l to energy level E_k is

$$A_{lk} + B_{lk}B_v,$$ (4.32)

where A_{lk} is the Einstein coefficient for spontaneous emission and B_{lk} is the Einstein coefficient for stimulated emission.

If N_k and N_l are the number of atoms in energy levels k and l, respectively, then the requirement that the number of upward transitions per unit time must equal the number of downward transitions per unit time under conditions of thermodynamic equilibrium implies that

$$N_kB_{kl}B_v = N_l(A_{lk} + B_{lk}B_v).$$ (4.33)

The ratio N_l/N_k is given by the Boltzmann relation,

$$\frac{N_l}{N_k} = \frac{g_l}{g_k}e^{-hv/kT},$$ (4.34)

with g_k and g_l equal to the degeneracies of energy levels E_k and E_l, respectively. Therefore, Equations (4.31), (4.33) and (4.34) lead to the equation

$$B_v = \frac{2hv^3}{c^2}\frac{1}{e^{hv/kT} - 1} = \frac{g_lA_{lk}}{[g_kB_{kl}e^{hv/kT} - g_lB_{lk}]}.$$ (4.35)

Equation (4.35) implies that the Einstein coefficients are related by the equations

$$A_{lk} = \frac{2hv^3}{c^2}B_{lk}$$ (4.36)

and

$$B_{lk} = \frac{g_k}{g_l}B_{kl}.$$ (4.37)

Equations (4.36) and (4.37) imply that if one of the three Einstein coefficients is calculated, then the remaining two Einstein coefficients are also determined. The Einstein coefficients are sometimes defined with U_v, the internal energy density per unit frequency of blackbody radiation, substituted in Equations (4.31) and (4.32) instead of the Planck function B_v.

Einstein coefficients for bound–free transitions can also be defined. Let N_H equal the number of hydrogen atoms in the lowest energy level and N_p equal the number of protons and $N_e(p)$ the number of electrons in the momentum interval $p, p + dp$. Radiative bound–free transitions and the inverse process, radiative recombination, are described by the reactions

$$H + hv \leftrightarrow p + e^-. \tag{4.38}$$

Under conditions of thermodynamic equilibrium the number of photoionizations to some particular energy substate of the continuum will be equal to the number of corresponding radiative recombinations.

Let $A_{r,1}$ equal the Einstein coefficient for spontaneous emission from some continuum energy E_r to the 1S level of atomic hydrogen. $B_{1,r}$ and $B_{r,1}$ are equal to the corresponding Einstein coefficients for absorption and stimulated emission respectively. The kinetic energy of the free electron and energy of the absorbed (or emitted) photon are related by the equation

$$hv = \chi + \frac{p^2}{2m}, \tag{4.39}$$

with $\chi = 13.6\,\text{eV}$ equal to the ionization potential of hydrogen.

Setting the number of spontaneous and stimulated emissions (i.e. radiative recombinations) equal to the number of photoionizations implies that

$$N_p N_e(p)(A_{r,1} + B_v B_{r,1}) = N_H B_v B_{1,r}. \tag{4.40}$$

Equations (4.40) can be rewritten as

$$\frac{N_p N_e(p)}{N_H} = \frac{B_v B_{1,r}}{A_{r,1} + B_v B_{r,1}} = \frac{B_{1,r}/B_{r,1}}{1 + (A_{r,1}/B_v B_{r,1})}, \tag{4.41}$$

with the Planck function B_v given in Equation (4.35). From Equation (3.180) we obtain

$$\frac{N_p N_e(p)}{N_H} = \frac{g_p g_e}{g_H} \frac{4\pi p^2}{h^3} e^{-p^2/2mkT} e^{-\chi/kT}$$

$$= \frac{4\pi p^2}{h^3} e^{-p^2/2mkT} e^{-\chi/kT}, \tag{4.42}$$

since $g_p = 2$, $g_e = 2$ and $g_H = 4$. From Equations (3.10) and (3.11) we have

$$N_e(p) = \frac{N_e 4\pi p^2}{(2\pi mkT)^{3/2}} e^{-p^2/2mkT}. \tag{4.43}$$

If we set

$$\frac{B_{1,r}}{B_{r,1}} = \frac{4\pi p^2}{h^3} \tag{4.44}$$

and

$$\frac{A_{r,1}}{B_{r,1}} = \frac{2hv^3}{c^2},$$ (4.45)

then Equations (4.41–4.43) lead to the equation

$$\frac{N_p N_e}{N_H} = \frac{(2\pi mkT)^{3/2}}{h^3} e^{-\chi/kT}.$$ (4.46)

Equation (4.46) is identical to the Saha equation given in Equation (3.182), and therefore we conclude that Equations (4.44) and (4.45) give the two relations between the bound–free Einstein coefficients.

4.4 Time-dependent perturbation theory

In order to calculate the probability that a particular radiative transition will occur in the presence of an electromagnetic field we must solve the time-dependent Schrödinger equation,

$$H\psi = (H_0 + H')\psi = \frac{-\hbar}{i}\frac{\partial \psi}{\partial t}.$$ (4.47)

In Equation (4.47) the Hamiltonian H is the sum of the unperturbed Hamiltonian H_0, which is given in Equation (4.6), and a time-dependent term H', which is the result of the interaction with the electromagnetic field.

Expanding the wave function ψ in Equation (4.47) in terms of the stationary eigenfunctions of the unperturbed Hamiltonian H_0, we have

$$\psi = \Sigma a_n(t) u_n e^{-i\frac{E_n t}{\hbar}},$$ (4.48)

where the functions u_n are the eigenfunctions of the unperturbed Hamiltonian H_0 and the coefficients $a_n(t)$ are time dependent. Substituting the expression for ψ given in Equation (4.48) into Equation (4.47), we obtain

$$\Sigma i\hbar \dot{a}_n u_n e^{-i\frac{E_n t}{\hbar}} + \Sigma a_n E_n u_n e^{-i\frac{E_n t}{\hbar}}$$

$$= \Sigma a_n(t)(H_0 + H')u_n e^{-i\frac{E_n t}{\hbar}}.$$ (4.49)

If the above equation is multiplied through on the left by $u_m^\star$ and integrated over all space, it reduces to the equation

$$i\hbar \dot{a}_m e^{-i\frac{E_m t}{\hbar}} = \sum_n a_n e^{-i\frac{E_n t}{\hbar}} \int u_m^\star H' u_n d^3 r.$$ (4.50)

In obtaining Equation (4.50) we have used the equation $H_0 u_n = E_n u_n$, which is satisfied because the functions u_n are orthonormal eigenfunctions of the unperturbed Hamiltonian

H_0. Equation (4.50) can be rewritten as

$$\dot{a}_m = \frac{-i}{\hbar} \sum_n H'_{mn} a_n e^{i\omega_{mn}t},\tag{4.51}$$

with

$$H'_{mn} = \int u^*_m H' u_n \mathrm{d}^3 r$$

and

$$\omega_{mn} = \frac{E_m - E_n}{\hbar}.$$

We assume that at some initial time $t = -\infty$ the atom is in some quantum state k with energy E_k. It follows that $a_k(-\infty) = 1$ and $a_n(-\infty) = 0$ for all $n \neq k$. We assume that a pulse of radiation causes transitions to occur from quantum state k to other energy levels. However, the probability amplitudes $a_n(n \neq k)$ associated with energy levels other than E_k remain small compared to 1. It follows that Equation (4.51) becomes

$$\dot{a}_m = \frac{-i}{\hbar} H'_{mk} e^{i\omega_{mk}t}.\tag{4.52}$$

Solving Equation (4.52), we have

$$a_m(+\infty) = \frac{-i}{\hbar} \int_{-\infty}^{+\infty} H'_{mk} e^{i\omega_{mk}t} \mathrm{d}t.\tag{4.53}$$

In the presence of an electromagnetic field the Hamiltonian of a hydrogen atom or hydrogenic ion is

$$H = \frac{\left(p + \dfrac{e}{c}A\right)^2}{2m} - \frac{Ze^2}{r},\tag{4.54}$$

with A equal to the vector potential. Expanding the $\left(p + \dfrac{e}{c}A\right)^2$ term in Equation (4.54) leads to the equation

$$H = \frac{p^2}{2m} + \frac{e}{2mc}(p \cdot A + A \cdot p) + \frac{e^2}{2mc^2}A^2 - \frac{Ze^2}{r},\tag{4.55}$$

with $p = \dfrac{\hbar}{i}\nabla$. The A^2 term in Equation (4.55), which accounts for Thomson scattering, is small and will be neglected. If we choose the Coulomb gauge (i.e. $\nabla \cdot A = 0$), then the Hamiltonian in Equation (4.55) becomes

$$H = \frac{p^2}{2m} - \frac{Ze^2}{r} + \frac{e}{mc}p \cdot A, \tag{4.56}$$

where the final term is time dependent.

It is convenient to introduce ket notation for eigenfunctions and matrix elements. In ket notation a matrix element

$$H_{mk} = \int u_m^\star H u_k d^3r \tag{4.57}$$

is written as

$$H_{mk} = \langle m|H|k \rangle. \tag{4.58}$$

Since the eigenfunctions u_n $(n = 1, 2, \ldots)$ form a complete set of eigenfunctions, any function ψ can be expanded as

$$\psi = \sum_n \langle n|\psi \rangle |n\rangle, \tag{4.59}$$

where $|n\rangle$ denotes the eigenfunction u_n and the summation is over all eigenfunctions.

Consider the commutation relation

$$H_0 r - r H_0, \tag{4.60}$$

with

$$H_0 = \frac{-\hbar^2}{2m}\left(\frac{\partial^2}{\partial x^2} + \frac{\partial^2}{\partial y^2} + \frac{\partial^2}{\partial z^2}\right) - \frac{Ze^2}{r}.$$

The x-component of the above commutation relation is

$$H_0 x - x H_0 = \frac{-\hbar}{2m}\left(\frac{\partial^2 x}{\partial x^2} - \frac{x\partial^2}{\partial x^2}\right) = -\frac{\hbar^2}{m}\frac{\partial}{\partial x} = \frac{\hbar}{im}P_x. \tag{4.61}$$

The y- and z-components of commutation relation (4.60) follow immediately from Equation (4.61).

From Equations (4.56) and (4.61) we obtain

$$\frac{e}{mc}p \cdot A = \frac{ieA}{\hbar c} \cdot (H_0 r - r H_0). \tag{4.62}$$

Equations (4.56) and (4.62) imply that the matrix element H'_{mk} in Equation (4.53) becomes

$$H'_{mk} = \frac{ie}{\hbar c}A \cdot \langle m|H_0 r - r H_0|k\rangle$$

$$= \frac{ie}{\hbar c}A \cdot [\langle m|H_0 r|k\rangle - E_k\langle m|r|k\rangle]. \tag{4.63}$$

Because the eigenfunctions $|n\rangle$ form a complete set of functions, $x|k\rangle$, $y|k\rangle$, and $z|k\rangle$ can each be expanded as

$$\sum_n \langle n|\psi_i\rangle|n\rangle \quad (i = 1, 2, 3), \tag{4.64}$$

with $\psi_1 = xu_k$, $\psi_2 = yu_k$ and $\psi_3 = zu_k$. Therefore, each component of the first term inside the parenthesis of Equation (4.63) can be expressed as

$$\sum_n \langle m|H_0|n\rangle\langle n|\psi_i\rangle = E_m\langle m|\psi_i\rangle. \tag{4.65}$$

Equation (4.65) implies that Equation (4.63) reduces to the equation

$$H'_{mk} = \frac{ie}{\hbar c}(E_m - E_k)\langle m|A \cdot r|k\rangle. \tag{4.66}$$

Using the Fourier integral theorem, we obtain

$$A(r, t) = \int_{-\infty}^{+\infty} A(\omega)e^{-i\omega\left(t - \frac{\hat{n} \cdot r}{c}\right)}d\omega. \tag{4.67}$$

Equations (4.53), (4.66) and (4.67) imply that

$$a_m(\infty) = \frac{e\omega_{mk}}{\hbar c}\int_{-\infty}^{+\infty}\int_{-\infty}^{+\infty} \langle m|e^{\frac{i\omega\hat{n} \cdot r}{c}} \hat{\varepsilon} \cdot r|k\rangle A(\omega)e^{i(\omega_{mk} - \omega)t}d\omega dt, \tag{4.68}$$

with $\hat{\varepsilon}$ equal to a unit vector in the direction of A. Because the wavelength of the emitted radiation is much greater than the size of an atom, we have

$$e^{\frac{i\omega}{c}\hat{n} \cdot r} \simeq 1 + i\frac{\omega}{c}\hat{n} \cdot r. \tag{4.69}$$

Neglecting the last term in Equation (4.69) implies that Equation (4.68) reduces to the equation

$$a_m(\infty) = \frac{e}{\hbar c}\omega_{mk}\langle m|\hat{\varepsilon} \cdot r|k\rangle \int_{-\infty}^{+\infty}\int_{-\infty}^{+\infty} A(\omega)e^{i(\omega_{mk} - \omega)t}d\omega dt. \tag{4.70}$$

The approximation made in reducing Equation (4.69) to Equation (4.70) is known as the dipole approximation. Using the equality

$$\int_{-\infty}^{+\infty} e^{i(\omega_{mk} - \omega)t}dt = 2\pi\delta(\omega_{mk} - \omega) \tag{4.71}$$

in Equation (4.70) implies that

$$a_m(\infty) = \frac{2\pi e\omega_{mk}}{\hbar c}\langle m|\hat{\varepsilon} \cdot r|k\rangle A(\omega_{mk}). \tag{4.72}$$

Therefore, the transition probability $|a_m(\infty)|^2$ is equal to

$$|a_m(\infty)|^2 = \frac{4\pi^2 e^2}{\hbar^2 c^2} \omega_{mk}^2 |A(\omega_{mk})|^2 |\langle m|\hat{\boldsymbol{\varepsilon}} \cdot \boldsymbol{r}|k\rangle|^2. \tag{4.73}$$

Equation (4.73) gives the probability that an electromagnetic pulse with vector potential $A(\boldsymbol{r}, t)$ causes the atom to undergo a transition from energy level k to energy level m. The absorption line will have the same Lorentzian line shape as that given in Equation (4.29) so long as the spectrum of radiation does not vary appreciably over the natural line width. Such physical conditions are always present in stellar interiors because the radiation field is that of a blackbody.

Using the expression for a Lorentzian line profile given in Equation (4.29), it follows that at a frequency v the probability per unit v for a transition to occur between energy level k and m is

$$|a_m(\infty)|^2 \frac{\gamma/4\pi^2}{(v - v_{mk})^2 + \left(\dfrac{\gamma}{4\pi}\right)^2}, \tag{4.74}$$

and the energy absorbed is $\hbar\omega_{mk}$ times the above expression. The cross section $\sigma(v)$ is equal to the energy absorbed per unit v divided by the energy per unit area per unit v in the pulse. As will be shown below, the energy spectrum of the pulse is

$$\int_0^\infty E(\omega)\mathrm{d}\omega = \frac{1}{c}\int_0^\infty \omega^2 |A(\omega)|^2 \mathrm{d}\omega, \tag{4.75}$$

and therefore the energy per unit area per unit v at v_{mk} is

$$E(v_{mk}) = 2\pi E(\omega_{mk}) = \frac{2\pi\omega_{mk}^2}{c}|A(\omega_{mk})|^2. \tag{4.76}$$

Equation (4.73), Expression (4.74) multiplied by $\hbar\omega_{mk}$ and Equation (4.76) imply that the cross section $\sigma(v)$ is

$$\sigma(v) = \frac{2\pi e^2 \omega_{mk}}{\hbar c}|\langle m|\hat{\boldsymbol{\varepsilon}} \cdot \boldsymbol{r}|k\rangle|^2 \frac{\gamma/4\pi^2}{(v - v_{mk})^2 + \left(\dfrac{\gamma}{4\pi}\right)^2}. \tag{4.77}$$

Integrating over the line, we have

$$\sigma_{km} = \int \sigma(v)\mathrm{d}v = 2\pi\alpha\omega_{mk}|\langle m|\hat{\boldsymbol{\varepsilon}} \cdot \boldsymbol{r}|k\rangle|^2, \tag{4.78}$$

with the fine-structure constant $\alpha = e^2/\hbar c = \frac{1}{137.04}$. If the radiation field is unpolarized, as is always true in stellar interiors and usually the case in stellar atmospheres, then

$$\overline{|\langle m|\hat{\boldsymbol{\varepsilon}} \cdot \boldsymbol{r}|k\rangle|^2} = \frac{1}{3}|\langle m|\boldsymbol{r}|k\rangle|^2. \tag{4.79}$$

Therefore, for unpolarized radiation the total (i.e. frequency-integrated) cross section σ_{km}

becomes

$$\sigma_{km} = \frac{2\pi}{3} \alpha \omega_{mk} |\langle m|r|k\rangle|^2. \tag{4.80}$$

The total cross sections given in Equations (4.78) and (4.80) have the dimensions of $cm^2 s^{-1}$, whereas $\sigma(\nu)$, the cross section at a frequency ν, given in Equation (4.77), has the dimensions of cm^2. As is clear from the above discussion, the total cross section for absorption and emission of radiation between two particular quantum states is the same. The equality of such cross sections is an example of the principle of detailed balance.

The power per unit area in an electromagnetic wave is given by the Poynting vector

$$S = \frac{c}{4\pi} E \times H, \tag{4.81}$$

with $E = -\frac{1}{c}\frac{\partial A}{\partial t}$ and $H = \nabla \times A$.

From the above expressions the Poynting vector S becomes

$$S = \frac{c}{4\pi} E \times H = \frac{c}{4\pi}\left(-\frac{1}{c}\frac{\partial A}{\partial t}\right) \times (\nabla \times A). \tag{4.82}$$

Expanding the vector potential, as in Equation (4.67), implies that

$$S = -\frac{c}{4\pi}\left\{\frac{1}{c}\int_{-\infty}^{+\infty} -i\omega A(\omega)\hat{\varepsilon}e^{-i\omega\left(t-\frac{\hat{n}\cdot r}{c}\right)}d\omega\right\}$$
$$\times \left\{\frac{1}{c}\int_{-\infty}^{+\infty} i\omega' A(\omega')(\hat{n} \times \hat{\varepsilon})e^{-i\omega'\left(t-\frac{\hat{n}\cdot r}{c}\right)}d\omega'\right\}. \tag{4.83}$$

Because $\hat{\varepsilon} \cdot n = 0$ and $\hat{\varepsilon} \times (\hat{n} \times \hat{\varepsilon}) = \hat{n}$, Equation (4.83) reduces to the equation

$$S = \frac{\hat{n}}{4\pi c}\int_{-\infty}^{+\infty}\int_{-\infty}^{+\infty} \omega\omega' A(\omega)A(\omega')e^{-i(\omega+\omega')\left(t-\frac{\hat{n}\cdot r}{c}\right)}d\omega d\omega'. \tag{4.84}$$

The energy in the electromagnetic pulse becomes

$$E = \frac{1}{4\pi c}\int_{-\infty}^{+\infty}\int_{-\infty}^{+\infty}\int_{-\infty}^{+\infty} \omega\omega' A(\omega)A(\omega')e^{-i(\omega+\omega')\left(t-\frac{\hat{n}\cdot r}{c}\right)}d\omega d\omega' dt. \tag{4.85}$$

Using Equation (4.71) to perform the integration over t in Equation (4.85), we obtain

$$E = \frac{1}{2c}\int_{-\infty}^{+\infty}\int_{-\infty}^{+\infty} \omega\omega' A(\omega)A(\omega')\delta(\omega + \omega')d\omega d\omega'$$
$$= \frac{1}{2c}\int_{-\infty}^{+\infty} \omega^2 A(\omega)A(-\omega)d\omega. \tag{4.86}$$

The conditions that the vector potential A is real implies that $A^\star(\omega) = A(-\omega)$, and therefore Equation (3.86) becomes

$$E = \frac{1}{c} \int_0^\infty \omega^2 |A(\omega)|^2 d\omega = \int_0^\infty E(\omega) d\omega. \tag{4.87}$$

Equation (4.87) shows that the expression for $E(\omega_{mk})$ given in Equation (4.76) and used to derive the cross section σ_{km} in Equation (4.80) is correct.

Consider two energy levels k and l which have degeneracies g_k and g_l, respectively. We define the total cross section σ_{kl} to be the total cross section for a transition from a particular quantum state in energy level k to all quantum states in energy level l. If there are N_k atoms in energy level k, then the energy absorbed per unit time between energy level k and energy level l caused by incident radiation in solid angle $\Delta\Omega$ is

$$I_\nu \Delta\Omega \sigma_{kl} N_k, \tag{4.88}$$

with $\nu = (E_l - E_k)/h$. Under conditions of thermodynamic equilibrium, I_ν is equal to the Planck function B_ν. A blackbody radiation field is isotropic, and therefore integrating over all solid angles, we see that the absorbed energy per unit time becomes

$$4\pi B_\nu \sigma_{kl} N_k. \tag{4.89}$$

Setting the above expression equal to the left-hand side of Equation (4.33) times the energy $h\nu$ of an absorbed photon gives

$$4\pi B_\nu \sigma_{kl} N_k = h\nu N_k B_{kl} B_\nu. \tag{4.90}$$

Therefore, the Einstein coefficient for absorption, B_{kl}, is equal to

$$B_{kl} = 4\pi\sigma_{kl}/h\nu, \tag{4.91}$$

and from Equation (4.37) the Einstein coefficient for stimulated emission is

$$B_{lk} = \frac{g_k}{g_l} B_{kl} = \frac{g_k}{g_l} \frac{4\pi\sigma_{kl}}{h\nu}. \tag{4.92}$$

Equations (4.36) and (4.92) imply that the Einstein coefficient for spontaneous emission A_{lk} is

$$A_{lk} = \frac{8\pi\nu^2}{c^2} \sigma_{kl} \frac{g_k}{g_l} = \frac{8\pi\nu^2}{c^2} \sigma_{lk}, \tag{4.93}$$

where the second equality in Equation (4.93) follows from the principle of detailed balance.

Having determined the Einstein coefficient for spontaneous emission, A_{lk}, from the total cross section σ_{kl}, given in Equation (4.80), we can now calculate $\sigma(\nu)$, given in Equation (4.77), if the radiative lifetime is caused entirely by transitions from energy level l to energy level k. If radiative transitions to other energy levels also occur, then the radiative lifetime τ is

$$\frac{1}{\tau} = \sum_n A_{ln}, \tag{4.94}$$

where the summation is over all energy levels n into which spontaneous emission occurs from energy level l.

The lowest energy level has no natural line width because radiative decay cannot occur. The natural line width of the Lyman α line, which occurs between the 2P and 1S energy levels of atomic hydrogen, is caused entirely by the width of the 2P energy level. However, for most spectral lines both upper and lower energy levels have finite radiative lifetimes, and therefore the natural line profiles are caused by the widths of both upper and lower energy levels.

Consider radiation between two energy levels with finite radiative lifetimes. The Hα transitions 3S $\rightarrow$ 2P, 3P $\rightarrow$ 2S, and 3D $\rightarrow$ 2P are examples of such transitions. Let γ_1 and γ_2 equal the Lorentzian widths of the lower and upper energy levels, with E_1 and E_2 equal to the energies at the centers of the levels. Radiation is absorbed at some energy $E = E_1 + hx$ of the lower level and the atom excited to some energy substate $E' = E_2 + hx'$.

From the principle of detailed balance it follows that we can define Einstein coefficients between energy substates of each of the two energy levels such that

$$b_{12}(x, x')B_v = a_{21}(x, x') + b_{21}(x, x')B_v. \tag{4.95}$$

Each Einstein coefficient is proportional to

$$\frac{\gamma_1}{x^2 + \gamma_1^2} \frac{\gamma_2}{x'^2 + \gamma_2^2}, \tag{4.96}$$

and the constants of proportionality are determined from Equations (4.36) and (4.37) and the requirement

$$\iint b_{12}(x, x')\mathrm{d}x\mathrm{d}x' = B_{12}. \tag{4.97}$$

Conservation of energy implies that the frequency of an absorbed photon and energy-level substates E and E' are related by the equation

$$hv = E_2 + hx' - (E_1 + hx). \tag{4.98}$$

We let

$$b_{12}(x, x') = B_{12}\frac{1}{\pi^2}\frac{\gamma_1\gamma_2}{(x^2 + \gamma_1^2)(x'^2 + \gamma_2^2)}. \tag{4.99}$$

To obtain the probability for a photon in the frequency interval $v, v + \mathrm{d}v$ to be absorbed, we integrate $b_{12}(x, x')$ over x with the requirement that $x - x' = v_0 - v = \text{constant} = x_0$. Therefore, we must calculate the integral

$$\int_{-\infty}^{+\infty} b_{12}(x, x - x_0)\mathrm{d}x = B_{12}\frac{\gamma_1\gamma_2}{\pi^2}\int_{-\infty}^{+\infty} \frac{\mathrm{d}x}{(x^2 + \gamma_1^2)[(x - x_0)^2 + \gamma_2^2]}. \tag{4.100}$$

The integral in Equation (4.100) can be calculated by means of contour integration. The complex function

$$f(z) = \frac{1}{(z + i\gamma_1)(z - i\gamma_1)(z - x_0 + i\gamma_2)(z - x_0 - i\gamma_2)} \tag{4.101}$$

reduces to the integrand of the integral in Equation (4.100) for z real and has the four poles $z = \pm i\gamma_1$ and $z = x_0 \pm i\gamma_2$ in the complex plane. Consider a closed path of integration C which includes an integration along the real axis and a semicircle Γ of radius R which encloses the poles $z = i\gamma_1$ and $z = x_0 + i\gamma_2$ From Cauchy's residue theorem we have

$$\int_C f(x)dz = 2\pi i \times \text{sum of residues at poles inside } C. \tag{4.102}$$

As $R \to \infty$, the integral $\int_\Gamma f(x)dz \to 0$. Therefore,

$$\int_{-\infty}^{+\infty} f(x)dx = \lim_{R\to\infty} \int_C f(x)dz. \tag{4.103}$$

The residues R_1 and R_2 at the poles $z = i\gamma_1$ and $z = x_0 + i\gamma_2$, respectively, are

$$R_1 = \frac{1}{2i\gamma_1[i\gamma_1 - (x_0 - i\gamma_2)][i\gamma_2 - (x_0 + i\gamma_2)]},$$

$$R_2 = \frac{1}{2i\gamma_2[x_0 + i(\gamma_1 + \gamma_2)][x_0 - i(\gamma_1 - \gamma_2)]}. \tag{4.104}$$

Summing R_1 and R_2, we have

$$R_1 + R_2 = \frac{-i}{2\gamma_1\gamma_2} \frac{\gamma_1 + \gamma_2}{x_0^2 + (\gamma_1 + \gamma_2)^2}. \tag{4.105}$$

Therefore, from Equations (4.102), (4.103) and (4.105) we have

$$\int_{-\infty}^{+\infty} f(x)dx = \frac{\pi}{\gamma_1\gamma_2} \frac{\gamma_1 + \gamma_2}{x_0^2 + (\gamma_1 + \gamma_2)^2}. \tag{4.106}$$

Equation (4.106) implies that Equation (4.100) becomes

$$\int_{-\infty}^{+\infty} b_{12}(x, x - x_0)dx = B_{12} \frac{1}{\pi} \frac{\gamma_1 + \gamma_2}{(\nu - \nu_0)^2 + (\gamma_1 + \gamma_2)^2}. \tag{4.107}$$

Equation (4.97) is satisfied if Equation (4.107) is integrated over ν. Equation (4.107) shows that the natural line profile of a radiative transition between two energy levels with finite radiative lifetimes is Lorentzian, and the width γ of the line profile is $\gamma = \gamma_1 + \gamma_2$ with $1/\gamma_1$ and $1/\gamma_2$ equal to the radiative lifetimes of the two energy levels.

4.5   Calculation of absorption cross sections and radiative lifetimes

The square of the matrix element given in Equation (4.80) can be expressed as

$$|\langle m|r|k\rangle|^2 = |\langle m|x|k\rangle|^2 + |\langle m|y|k\rangle|^2 + |\langle m|z|k\rangle|^2. \tag{4.108}$$

The first two terms on the right-hand side of Equation (4.108) can be written as

$$|\langle m|x|k\rangle|^2 + |\langle m|y|k\rangle|^2 = \frac{1}{2}|\langle m|x+iy|k\rangle|^2 + \frac{1}{2}|\langle m|x-iy|k\rangle|^2. \tag{4.109}$$

Since in spherical coordinates we have

$$x = r\sin\theta\cos\phi$$

and

$$y = r\sin\theta\sin\phi, \tag{4.110}$$

it follows that

$$x + iy = r\sin\theta e^{i\phi}$$

and

$$x - iy = r\sin\theta e^{-i\phi}. \tag{4.111}$$

Therefore, from Equations (4.109), (4.110) and (4.111) we obtain

$$|\langle m|r|k\rangle|^2 = \frac{1}{2}|\langle m|r\sin\theta e^{i\phi}|k\rangle|^2 + \frac{1}{2}|\langle m|r\sin\theta e^{-i\phi}|k\rangle|^2$$

$$+ |\langle m|r\cos\theta|k\rangle|^2. \tag{4.112}$$

Since the eigenfunctions are of the form $\psi_{nlm} = R_{nl}(r)Y_l^m(\theta, \phi)$, the matrix elements that appear on the right-hand side of Equation (4.112) become

$$\langle m|r\sin\theta e^{i\phi}|k\rangle = \int R_{nl}(r)R_{n'l'}(r)r^3 dr \int Y_l^{m\star} \sin\theta e^{i\phi} Y_{l'}^{m'} d\Omega \tag{4.113}$$

$$\langle m|r\sin\theta e^{-i\phi}|k\rangle = \int R_{nl}(r)R_{n'l'}(r)r^3 dr \int Y_l^{m\star} \sin\theta e^{i\phi} Y_{l'}^{m'} d\Omega \tag{4.114}$$

and

$$\langle m|r\cos\theta|k\rangle = \int R_{nl}(r)R_{n'l'}(r)r^3 dr \int Y_l^{m\star} \cos\theta Y_{l'}^{m'} d\Omega. \tag{4.115}$$

The ϕ dependence of the spherical harmonics Y_l^m is contained in the factor $e^{im\phi}/(2\pi)^{1/2}$, and therefore the integrals over ϕ in Equations (4.113–4.115) become

$$\frac{1}{2\pi}\int_0^{2\pi} e^{i(-m+m'+1)\phi} d\phi = \delta_{m-m',1}, \tag{4.116}$$

$$\frac{1}{2\pi}\int_0^{2\pi} e^{i(-m+m'-1)\phi}d\phi = \delta_{m'-m,1} \tag{4.117}$$

and

$$\frac{1}{2\pi}\int_0^{2\pi} e^{i(-m+m')\phi}d\phi = \delta_{m'-m,0}, \tag{4.118}$$

respectively. Equations (4.116–4.118) imply that for dipole radiation the selection rule for the azimuthal quantum number m is

$$m' - m = 0, \pm 1. \tag{4.119}$$

The spherical harmonics Y_l^m given in Equation (4.19) contain the factor $P_l^m(\cos\theta)$. By integrating over the angle θ it can be shown that the matrix elements given in Equations (4.113–4.115) are zero unless $l - l' = \pm 1$. Therefore, electric-dipole transitions can occur only if the angular momentum l changes by ± 1. In particular, transitions between the 2P energy level and 1S energy level of atomic hydrogen are allowed by the dipole selection rule ($\Delta l = \pm 1$), whereas transitions between the 2S energy level and 1S energy level are not allowed because $\Delta l = 0$. We will calculate the absorption cross section for Lyman α radiation. Absorption cross sections for transitions between other energy levels of atomic hydrogen and between different energy levels of other hydrogenic ions can be calculated in a similar manner but with different wave functions.

First we will evaluate the integrals over the angle θ contained in the matrix elements (4.113–4.115) for transitions between the 1S and 2P levels. Since $l = 0$ and $m = 0$ for the 1S energy level, we have $N_l^m P_l^m(\cos\theta) = P_0(\cos\theta)/\sqrt{2} = 1/\sqrt{2}$, with $1/\sqrt{2}$ the normalization constant. The 2P energy level has a degeneracy of $2(2l+1) = 6$ but spin flip does not occur. The three 2P final-state eigenfunctions contain the terms

$$N_1^0 P_1(\cos\theta) = \sqrt{3/2}\cos\theta,$$
$$N_1^1 P_1^1(\cos\theta) = -\sqrt{3/4}\sin\theta,$$
$$N_1^{-1} P_1^{-1}(\cos\theta) = \sqrt{3/4}\sin\theta, \tag{4.120}$$

with N_1^0, N_1^1 and N_1^{-1} equal to normalization constants.

Using the above expressions for $N_L^M P_L^M(\cos\theta)$, the nonzero integrals over θ and ϕ in Equations (4.113–4.115) become

$$\frac{N_1^0}{\sqrt{2}}\int_0^\pi P_1^0(\cos\theta)\cos\theta P_0(\cos\theta)\sin\theta d\theta = \int_0^\pi \sqrt{\frac{3}{2}}\cos\theta\frac{\cos\theta}{\sqrt{2}}\sin\theta d\theta = 1/\sqrt{3}, \tag{4.121}$$

$$\frac{N_1^1}{\sqrt{2}}\int_0^\pi P_1^1(\cos\theta)\sin\theta P_0(\cos\theta)\sin\theta d\theta = -\int_0^\pi \sqrt{\frac{3}{4}}\frac{\sin^3\theta}{\sqrt{2}}d\theta = \sqrt{\frac{2}{3}}, \tag{4.122}$$

and

$$\frac{N_1^{-1}}{\sqrt{2}}\int_0^\pi P_1^{-1}(\cos\theta)\sin\theta P_0(\cos\theta)\sin\theta d\theta = \sqrt{\frac{2}{3}}. \tag{4.123}$$

Equations (4.112), (4.116–4.118) and (4.121–4.123) imply that the square of the matrix element $\langle 2P|r|1S \rangle$ is

$$\langle 2P|r|1S \rangle^2 = \left(\frac{1}{2}\frac{2}{3} + \frac{1}{2}\frac{2}{3} + \frac{1}{3}\right)\left(\int_0^\infty R_{21}(r)R_{10}(r)r^3 dr\right)^2$$

$$= \left(\int_0^\infty R_{21}(r)R_{10}(r)r^3 dr\right)^2. \tag{4.124}$$

Substituting the radial eigenfunctions R_{21} and R_{10} given in Expression (4.21) into Equation (4.124), we obtain

$$\int_0^\infty R_{21}(r)R_{10}(r)r^3 dr = \frac{a_0^{-4}}{\sqrt{6}}\int_0^\infty e^{-r/2a_0}e^{-r/a_0}r^4 dr$$

$$= \left(\frac{2}{3}\right)^5 \frac{a_0}{\sqrt{6}}\int_0^\infty e^{-x}x^4 dx = \frac{24}{\sqrt{6}}\left(\frac{2}{3}\right)^5 a_0. \tag{4.125}$$

From Equation (4.125) we have

$$\left(\int_0^\infty R_{21}(r)R_{10}(r)r^3 dr\right)^2 = \left(\frac{32}{27}\right)^3 a_0^2, \tag{4.126}$$

with the Bohr radius $a_0 = 0.53 \times 10^{-8}$ cm, and therefore from Equations (4.80), (4.124) and (4.126) the total cross section for Lyman α absorption becomes

$$\sigma_t(2P, 1S) = 21.9 v\alpha a_0^2 = 1.1 \times 10^{-2} \text{cm}^2 \text{s}^{-1}, \tag{4.127}$$

where $\alpha = e^2/\hbar c = 1/137.04$, $v = 2.467 \times 10^{15}$ Hz and $a_0 = 0.53 \times 10^{-8}$ cm. From Equations (4.91) and (4.127) the Einstein absorption coefficient B_{12} becomes

$$B_{12} = \frac{4\pi\sigma_t(2P, 1S)}{hv} = 0.85 \times 10^{10} \text{cm}^2 \text{erg}^{-1}\text{s}^{-1}. \tag{4.128}$$

Therefore, from Equations (4.93) and (4.127) we obtain

$$A_{21} = \gamma_{21} = \frac{8\pi v^2}{c^2}\sigma_t\frac{g_{1S}}{g_{2P}} = 0.63 \times 10^9 \text{s}^{-1}, \tag{4.129}$$

with γ_{21} equal to the line width. From Equation (4.129) the 2P radiative lifetime becomes

$$\tau = 1.6 \times 10^{-9} \text{s}. \tag{4.130}$$

From Equations (4.77–4.80) we have

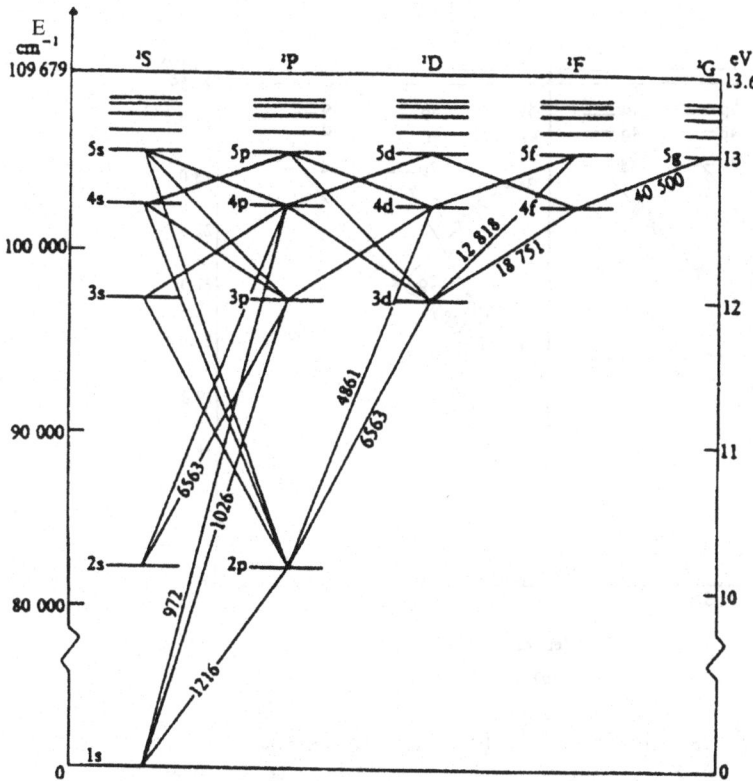

Figure 4.1. Energy levels of the hydrogen atom and wavelengths of spectral lines in ångstroms. Transitions obeying the $\Delta l = \pm 1$ selection rule are shown with solid lines. From Bransden and Joachain (1983).

$$\sigma(v) = 2\pi\sigma(\omega) = \sigma_t \frac{\gamma}{4\pi^2} \frac{1}{(v - v_0)^2 + \left(\dfrac{\gamma}{4\pi}\right)^2}. \tag{4.131}$$

Equations (4.129) and (4.131) imply that the line-center absorption cross section is

$$\sigma(v_0) = \sigma_t \frac{4}{\gamma_{21}} = 7 \times 10^{-11}\,\mathrm{cm}^2. \tag{4.132}$$

Energy levels of the hydrogen, helium and calcium atoms are given in Figures 4.1 and 4.2.

The classical oscillator

The motions of a classical oscillator are described by the equation

$$\ddot{x} + \gamma\dot{x} + \omega_0^2 x = -\frac{e}{m}E, \tag{4.133}$$

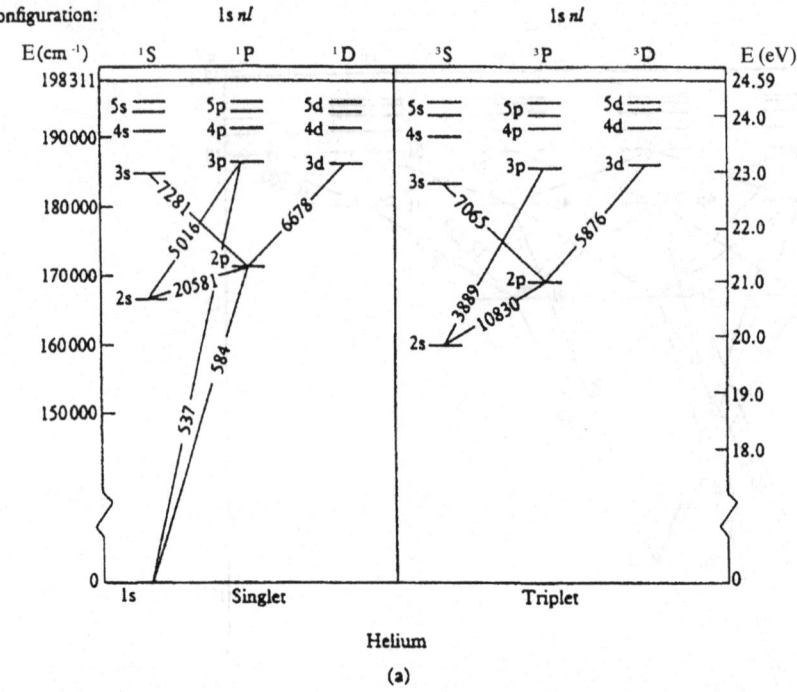

Helium

(a)

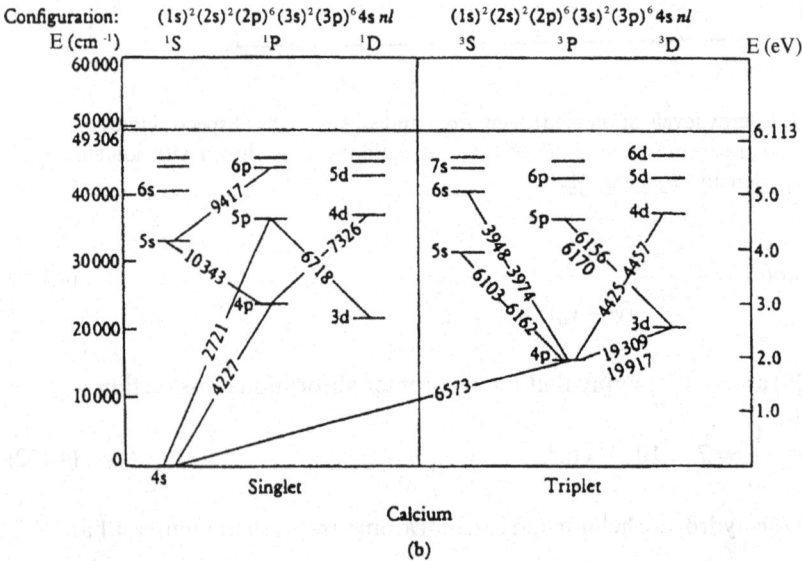

Calcium

(b)

Figure 4.2. Energy levels of helium and calcium atoms with atomic fine structure. Wavelengths are in ångstroms. From Bransden and Joachain (1983).

where γ is a damping constant, m the mass of an electron and E an external electric field. We replace the above equation with the equation

$$\ddot{z} + \gamma\dot{z} + \omega_0^2 z = -\frac{e}{m}E, \tag{4.134}$$

where z and E are complex, $\mathrm{Re}(z) = x$ and $\mathrm{Re}(E)$ equals the physical electric field. The general solution of an inhomogeneous, linear, second-order differential equation such as Equation (4.134) can be expressed as the sum of the general solution of the corresponding homogeneous equation (i.e. the differential equation obtained by setting the right-hand side of the equation equal to zero) and a particular solution of the inhomogeneous equation. We assume that steady-state conditions exist, and therefore the general solution of the homogeneous equation is unimportant.

If we let $E = E_0 e^{i\omega t}$, then the steady-state solution of Equation (4.134) becomes

$$z = \frac{-eE/m}{\omega_0^2 - \omega^2 + i\omega\gamma}. \tag{4.135}$$

For values of ω such that $|\omega - \omega_0| \ll \omega_0$ we have

$$\omega_0^2 - \omega^2 \simeq 2\omega(\omega_0 - \omega). \tag{4.136}$$

Equation (4.136) implies that the above solution of Equation (4.135) can be re-expressed as

$$z = \frac{-eE/2m\omega}{\omega_0 - \omega + i\dfrac{\gamma}{2}}. \tag{4.137}$$

The complex current density caused by the motions of bound electrons is

$$J = -n_0 e\dot{z} = \frac{n_0 e^2/2m\omega}{\omega_0 - \omega + i\dfrac{\gamma}{2}}\frac{\partial E}{\partial t}, \tag{4.138}$$

with n_0 equal to the number density of bound electrons. Taking the curl of the Maxwell Equation (3.340) implies that

$$\nabla \times (\nabla \times E) = \nabla(\nabla \cdot E) - \nabla^2 E = -\frac{1}{c}\frac{\partial}{\partial t}\nabla \times H. \tag{4.139}$$

Since

$$\nabla \cdot E = -4\pi n_0 e, \tag{4.140}$$

the gradient term in Equation (4.139) is equal to zero if n is assumed independent of position. Equations (3.342), (4.139) and (4.140) lead to the equation

$$\nabla^2 E = \frac{1}{c}\frac{\partial}{\partial t}\nabla \times H = \frac{1}{c^2}\frac{\partial}{\partial t}\left[\frac{\partial E}{\partial t} + 4\pi J\right]. \tag{4.141}$$

Using the expression for J given in Equation (4.138), we find that Equation (4.141) reduces to the equation

$$\nabla^2 E = \frac{1}{c^2}\left[1 + \frac{2\pi n_0 e^2/m\omega}{\omega_0 - \omega + i\gamma/2}\right]\frac{\partial^2 E}{\partial t^2}. \tag{4.142}$$

If the electromagnetic wave is observed to propagate in the z-direction, then the complex electric field in Equation (4.142) can be expressed as

$$E = E_0 e^{-\frac{\alpha}{2}z + i(kz - \omega t)}, \tag{4.143}$$

with α equal to the absorption coefficient. The factor $\alpha/2$ appears in Equation (4.143) because the Poynting vector $c(E \times H)/4\pi$ is proportional to the square of the electric field and $H = E \times \hat{n}$, with $\hat{n}$ equal to a unit vector along the direction of the electromagnetic wave.

Substituting the expression for E in Equation (4.143) into Equation (4.142) leads to the equation

$$\left(\frac{\alpha}{2} + ik\right)^2 = -\frac{\omega^2}{c^2}\left[1 + \frac{2\pi n_0 e^2/m\omega}{\omega_0 - \omega + i\gamma/2}\right]. \tag{4.144}$$

Setting the real and imaginary parts of the above equation equal separately, we obtain

$$\left(\frac{\alpha}{2}\right)^2 - k^2 = \left(\frac{\alpha}{2}\right)^2 - n^2\left(\frac{\omega}{c}\right)^2 = -\frac{\omega^2}{c^2}\left[1 + \frac{2\pi n_0 e^2(\omega_0 - \omega)/m\omega}{(\omega_0 - \omega)^2 + (\gamma/2)^2}\right] \tag{4.145}$$

and

$$k\alpha = n\frac{\omega}{c}\alpha = \frac{\pi n_0 e^2 \gamma/m\omega c^2}{(\omega_0 - \omega)^2 + (\gamma/2)^2}, \tag{4.146}$$

where $k = n\omega/c$ and n is the index of refraction. If the number density of oscillators n_0 is sufficiently low that n is $\simeq 1$, then Equation (4.146) becomes

$$\alpha = \frac{\pi e^2}{mc}n_0\frac{\gamma}{(\omega_0 - \omega)^2 + (\gamma/2)^2}. \tag{4.147}$$

From Equation (4.147) it follows that the absorption cross section $\sigma(v)$ is

$$\sigma(v) = \frac{\alpha}{n_0} = \frac{\pi e^2}{mc}\frac{1}{\pi}\frac{\gamma/4\pi}{(v - v_0)^2 + (\gamma/4\pi)^2}, \tag{4.148}$$

and therefore the total cross section becomes

$$\int_0^\infty \sigma(v)dv = \frac{\pi e^2}{mc} = 0.027\,\text{cm}^2\,\text{s}^{-1}. \tag{4.149}$$

The total cross sections for atomic transitions are often expressed in units of $\pi e^2/mc$, i.e.

$$\sigma_{t} = \frac{\pi e^2}{mc} f, \tag{4.150}$$

where σ_{t} is the total cross section calculated quantum mechanically and f is known as the oscillator strength (or f-value) of the transition.

4.6 Bound–free absorption

We assume that at a time $t = 0$ a hydrogen atom or a hydrogenic ion is in its lowest energy level. The Hamiltonian is given by Equation (4.56). We will calculate the cross section for the 1S electron to be photoionized into a continuum quantum state with kinetic energy $p^2/2m$. The energy of the photoionizing photon is

$$\hbar\omega = \chi_z + \frac{p^2}{2m} = \chi_z + \frac{\hbar^2 k^2}{2m}, \tag{4.151}$$

with χ_z equal to 13.6 eV for hydrogen and 13.6 Z^2 eV for hydrogenic ions of charge Z. The photoionized electron will satisfy a plane-wave solution. The normalized plane-wave solutions in a large cubic region of volume L^3 are

$$\frac{e^{i(k \cdot r - \omega t)}}{L^{3/2}}. \tag{4.152}$$

The final expression for the photoionization cross section will not depend on the dimension L because the cross section is proportional to the number of available quantum states and the square of a matrix element that contains the wave function given in Expression (4.152).

The time-dependent wave function $\psi(t)$ can be expanded as

$$\psi(t) = a_{1S}(t)u_{1S}e^{-i\frac{E_1 t}{\hbar}} + \frac{1}{L^{3/2}}\sum_m a_m(t)e^{i(k_m \cdot r - \omega t)}, \tag{4.153}$$

where m labels the plane-wave solution of the photoionized electron. Performing similar calculations to those already used to obtain Equation (4.53), we find that the probability amplitude a_m for an electron to have plane-wave solution m as the result of photoionization by a pulse of radiation is

$$a_m(+\infty) = \frac{-i}{\hbar}\int \langle m|\frac{e}{mc}A \cdot p|k\rangle e^{i\omega_{mk}t}dt, \tag{4.154}$$

where the perturbed Hamiltonian is the same as that given in Equation (4.56). After expanding the magnetic potential A in Equation (4.154) by means of the Fourier integral theorem, as in Equation (4.67), the probability amplitude a_m becomes

$$a_m(\infty) = \frac{-ie}{\hbar mc}\int_{-\infty}^{+\infty}\int_{-\infty}^{+\infty} \langle m|e^{i\frac{\omega}{c}n \cdot r}\hat{\varepsilon} \cdot p|k\rangle A(\omega)e^{-i(\omega - \omega_{mk})t}d\omega dt. \tag{4.155}$$

Performing the integration over t and using Equation (4.71), we obtain

$$a_m(\infty) = \frac{-2\pi i e A(\omega_{mk})}{\hbar m c} \langle m | e^{i\frac{\omega}{c}\hat{n}\cdot r} \hat{\varepsilon}\cdot p | k \rangle. \tag{4.156}$$

Equation (4.156) implies that the transition probability is

$$|a_m(\infty)|^2 = \frac{4\pi^2}{m^2\hbar c} \alpha | A(\omega) |^2 | \langle m | e^{i\frac{\omega}{c}\hat{n}\cdot r} \hat{\varepsilon}\cdot p | k \rangle |^2, \tag{4.157}$$

with $\alpha = e^2/\hbar c$ and $\omega = \omega_{mk}$. Because the photoionized electron can occupy only one of the two spin states, the number of available quantum states in a volume $p^2 dp d\Omega$ of momentum space is

$$\Delta n = \frac{p^2 dp d\Omega}{h^3} L^3, \tag{4.158}$$

with p equal to the magnitude of the photoionized electron's momentum and $d\Omega$ a small solid-angle element along the direction of the electron's motion.

Using Equation (4.151) implies that we can re-express Δn in Equation (4.158) as

$$\Delta n = \frac{mk\Delta\omega d\Omega}{(2\pi)^2 h} L^3. \tag{4.159}$$

From Equations (4.157) and (4.159) it follows that the energy absorbed by all quantum states Δn is

$$\frac{\hbar\omega}{2\pi} \frac{\alpha k L^3}{\hbar^2 m c} | A(\omega) |^2 | \langle m | e^{i\frac{\omega}{c}\hat{n}\cdot r} \hat{\varepsilon}\cdot p | k \rangle |^2 \Delta\omega d\Omega. \tag{4.160}$$

Equation (4.76) implies that the energy per unit area in the frequency interval $\omega, \omega + \Delta\omega$ of an electromagnetic pulse propagating in a particular direction with polarization $\hat{\varepsilon}$ is

$$\frac{\omega^2}{c} | A(\omega) |^2 \Delta\omega. \tag{4.161}$$

Dividing Expression (4.160) by Expression (4.161), we find that the differential cross section for photoionization is

$$\frac{d\sigma(\omega)}{d\Omega} = \frac{\alpha k L^3}{m h \omega} | \langle m | e^{i\frac{\omega}{c}\hat{n}\cdot r} p \cdot \hat{\varepsilon} | k \rangle |^2. \tag{4.162}$$

The factor L^3 that appears in the numerator of Equation (4.162) is canceled by the normalization factor $L^{-3/2}$ of the wave function $\langle m |$. Therefore, L is omitted from the derivation given below.

The wave function for a 1S-state electron is $\psi_{1S} = R_{10} Y_0^0$, with R_{10} given in Equation (4.21) and $Y_0^0 = 1/\sqrt{4\pi}$. Substituting the wave functions ψ_{1S} for $|k\rangle$ and $\exp(-ik \cdot r)$ for $\langle m |$ into Equation (4.162) leads to the equation

$$\frac{d\sigma(\omega)}{d\Omega} = \frac{\alpha k}{\pi m h\omega}\left(\frac{Z}{a_0}\right)^3 \left|\int e^{-ik\cdot r + i\frac{\omega}{c}\hat{n}\cdot r}\hat{\varepsilon}\cdot\frac{\hbar}{i}\nabla e^{-\frac{Zr}{a_0}}dV\right|^2. \tag{4.163}$$

The matrix element in Equation (4.163) is Hermitian and $\hat{\varepsilon}\cdot\hat{n} = 0$ because the vector potential is perpendicular to the direction of the electromagnetic wave. Therefore, we obtain

$$M_{mk} = \int e^{-ik\cdot r + i\frac{\omega}{c}\hat{n}\cdot r}\hat{\varepsilon}\cdot\frac{\hbar}{i}\nabla e^{-Zr/a_0}dV$$

$$= \hat{\varepsilon}\cdot\int\left(\frac{\hbar}{i}\nabla e^{i[(k-\frac{\omega}{c}\hat{n})\cdot r]}\right)^{\star} e^{-Zr/a_0}r^2 dr d\Omega$$

$$= \hbar k\cdot\hat{\varepsilon}\int e^{[i(\frac{\omega}{c}\hat{n}-k)\cdot r - Zr/a_0]}r^2 dr d\Omega. \tag{4.164a}$$

If we define the vector $q = k - \frac{\omega}{c}\hat{n}$, the integral over θ and ϕ in the above matrix element M_{mk} becomes

$$\int e^{-iq\cdot r}d\Omega = -2\pi\int_0^n e^{-iqr\cos\theta}d(\cos\theta)$$

$$= -2\pi\left[\frac{e^{iqr} - e^{-iqr}}{-iqr}\right] = 4\pi\frac{\sin qr}{qr}. \tag{4.164b}$$

Equations (4.164a) and (4.164b) imply that

$$M_{mk} = \hbar k\cdot\hat{\varepsilon}4\pi\int_0^\infty \frac{\sin qr}{qr}e^{-Zr/a_0}r^2 dr$$

$$= \hbar k\cdot\hat{\varepsilon}\frac{4\pi}{q}\int_0^\infty r\sin qr\, e^{-Zr/a_0}dr$$

$$= -\hbar k\cdot\hat{\varepsilon}\frac{4\pi}{q}\frac{d}{dq}\int_0^\infty \cos qr\, e^{-Zr/a_0}dr. \tag{4.165}$$

Using the standard integral

$$\int_0^\infty e^{-ax}\cos bx\, dx = \frac{a}{a^2 + b^2}\quad (a > 0), \tag{4.166}$$

the derivative of the integral in Equation (4.165) becomes

$$\frac{d}{dq}\frac{Z/a_0}{\left[\left(\frac{Z}{a_0}\right)^2 + q^2\right]} = \frac{-2qZ/a_0}{\left[\left(\frac{Z}{a_0}\right)^2 + q^2\right]^2}. \tag{4.167}$$

Equation (4.167) implies that the matrix element M_{mk} in Equation (4.165) is

$$M_{mk} = \frac{\hbar k \cdot \hat{\varepsilon} \, 8\pi \dfrac{Z}{a_0}}{\left[\left(\dfrac{Z}{a_0}\right)^2 + q^2\right]^2}.$$

(4.168)

From Equations (4.163) and (4.168) it follows that

$$\frac{d\sigma(\omega)}{d\Omega} = \frac{32\hbar k(\hat{\varepsilon} \cdot k)^2}{m\omega}\left(\frac{Z}{a_0}\right)^5\left(\frac{Z^2}{a_0^2} + q^2\right)^{-4}.$$

(4.169)

Since $q = k - \dfrac{\omega}{c}\hat{n}$, we have

$$\left(\frac{Z}{a_0}\right)^2 + q^2 = \left(\frac{Z}{a_0}\right)^2 + k^2 + \left(\frac{\omega}{c}\right)^2 - 2k\frac{\omega}{c}\cos\theta.$$

(4.170)

The wave vector k of the ionized electron and frequency of the incident photon ω are related by the expression

$$\frac{\hbar^2 k^2}{2m} = \hbar\omega - \chi = \hbar\omega - \frac{Z^2 e^2}{2a_0}.$$

(4.171)

Equations (4.170) and (4.171) imply that

$$\left(\frac{Z}{a_0}\right)^2 + q^2 = \frac{2m\omega}{\hbar}\left[1 - \frac{\hbar k}{mc}\cos\theta + \frac{\hbar\omega}{2mc^2}\right].$$

(4.172)

Since $\hbar\omega/2mc^2$ is small and

$$\frac{\hbar k}{mc} = \frac{v}{c} = \beta,$$

(4.173)

Equation (4.172) reduces to the equation

$$\left(\frac{Z}{a_0}\right)^2 + q^2 = \frac{2m\omega}{\hbar}[1 - \beta\cos\theta].$$

(4.174)

Therefore, the differential cross section $d\sigma/d\Omega$ in Equation (4.169) becomes

$$\frac{d\sigma}{d\Omega} = 2\alpha k(\hat{\varepsilon} \cdot k)^2 \left(\frac{\hbar}{m\omega}\right)^5\left(\frac{Z}{a_0}\right)^5 (1 - \beta\cos\theta)^{-4}$$

$$= 2\alpha k^3 \left(\frac{\hbar}{m\omega}\right)^5\left(\frac{Z}{a_0}\right)^5 \frac{\cos^2\psi}{(1 - \beta\cos\theta)^4},$$

(4.175)

where ψ is the angle between the direction of polarization $\hat{\varepsilon}$ and the wave vector of the photoionized electron whose momentum $p = \hbar k$. Since the average value of $\cos^2\psi$ over all solid angles is

$$\overline{\cos^2\psi} = \frac{1}{4\pi}\int_0^{2\pi}\int_0^{\pi}\cos^2\psi\sin\psi \, d\psi \, d\phi = \frac{1}{3}$$

(4.176)

and in the nonrelativistic limit $\beta = v/c \ll 1$, the cross section $\sigma(\omega)$ for bound–free emission becomes

$$\sigma(\omega) = \sigma(v) = \int \frac{d\sigma}{d\Omega} d\Omega = \frac{8\pi}{3} \alpha k^3 \left(\frac{\hbar}{m\omega}\right)^5 \left(\frac{Z}{a_0}\right)^5, \tag{4.177}$$

with $\hbar\omega$ equal to the energy of the photoionized electron and $\hbar k$ equal to the momentum of the photoionized electron. Equation (4.177) implies that $\sigma(\omega)$ can be written as

$$\sigma(\omega) = \frac{16\sqrt{2}\pi}{3} \frac{\alpha\hbar^7}{m^{3.5}} \left(\frac{Z}{a_0}\right)^5 \frac{(\hbar\omega - \chi)^{3/2}}{(\hbar\omega)^5}, \tag{4.178}$$

with χ equal to the ionization potential and $\alpha = e^2/\hbar c$.

4.7 Free–free absorption

Bremsstrahlung is electromagnetic radiation produced by electrons that are accelerated by the Coulomb fields of protons or other ions. Free–free absorption, which is also known as inverse bremsstrahlung, occurs when a free electron moving in the Coulomb field of a proton or ion absorbs a photon. The Hamiltonian describing free–free absorption is

$$H = H_0 + H_1' + H_2' = \frac{p^2}{2m} - \frac{Ze^2}{r} + \frac{eA \cdot p}{mc}. \tag{4.179}$$

The unperturbed Hamiltonian H_0 in Equation (4.179) is the kinetic energy of the free electron. The unperturbed wave functions of the incident electron and the electron that has absorbed a photon are

$$e^{i(k_a \cdot r - \omega_a t)}$$

and

$$e^{i(k_b \cdot r - \omega_b t)}, \tag{4.180}$$

respectively.

Plane-wave solutions are inadequate to calculate the cross section for free–free absorption. The Coulomb field of a proton (or ion) perturbs incoming and outgoing electrons. The perturbed Hamiltonians H_1' and H_2' in Equation (4.179) represent the Coulomb potential energy of the proton and the interaction with the incident electromagnetic wave, respectively. We will calculate the wave functions of the incoming and outgoing electrons perturbed by the Coulomb potential of the proton (or ion) and then calculate the cross section for the absorption of radiation.

The time-dependent perturbed wave functions and eigenvalues can be expanded as $u_a = u_{a_0} + \lambda u_a'$ and $E_a = E_{a_0} + \lambda E_a'$, where λ is equal to a small parameter and E_{a_0} is equal to the kinetic energy of the incoming electron. Similarly the wave function and eigenvalue of the outgoing electron can be expanded as $u_b = u_{b_0} + \lambda u_b'$ and $E_b = E_{b_0} + \lambda E_b'$. Expressing the Hamiltonian as $H_0 + \lambda H_1'$, with H_0' and H_1' given in Equation (4.179), and using the above expansions for u_a and E_a implies that terms first order in λ are

$$H_0 u_a' + H_1' u_{a_0} = E u_a' + E_1 u_{a_0}. \tag{4.181}$$

Expanding u_a' in plane-wave solutions u_n, we obtain

$$u_a' = \sum_n a_n u_n. \tag{4.182}$$

Substituting u_a' into Equation (4.181), multiplying through by $u_m^\star$ and integrating over all space gives

$$a_m = \frac{\langle m | H_1' | a_0 \rangle}{E_a - E_m}. \tag{4.183}$$

Similarly, for the outgoing electrons we have

$$u_b'^\star = \sum_n b_n^\star u_n^\star \tag{4.184}$$

and

$$b_m^\star = \frac{\langle b_0 | H_1' | m \rangle}{E_b - E_m}. \tag{4.185}$$

The time-independent wave functions for the incoming and outgoing electrons can now be written as

$$u_a = u_{a_0} + u_a'$$

and

$$u_b = u_{b_0} + u_b'. \tag{4.186}$$

Substituting the unperturbed wavefunctions from Equation (4.180) into Equation (4.53) and then making a Fourier transform of the resultant equation gives

$$
\begin{aligned}
a_b(\infty) &= -\frac{i}{\hbar} \int_{-\infty}^{+\infty} \langle b_0 | \frac{e}{mc} \boldsymbol{A} \cdot \boldsymbol{p} | a_0 \rangle e^{i(\omega_b - \omega_a)t} dt \\
&= -\frac{ie}{\hbar mc} \int_{-\infty}^{+\infty} \int_{-\infty}^{+\infty} \langle b_0 | e^{i\boldsymbol{k} \cdot \boldsymbol{r}} \hat{\boldsymbol{\varepsilon}} \cdot \boldsymbol{p} | a_0 \rangle A(\omega) e^{-i(\omega - \omega_b + \omega_a)t} d\omega dt \\
&= -\frac{2\pi ie}{\hbar mc} A(\omega) \langle b_0 | e^{i\boldsymbol{k} \cdot \boldsymbol{r}} \hat{\boldsymbol{\varepsilon}} \cdot \boldsymbol{p} | a_0 \rangle \\
&= -\frac{2\pi ie}{mc} \frac{A(\omega)}{L^3} \hat{\boldsymbol{\varepsilon}} \cdot \boldsymbol{k}_a \int e^{i(\boldsymbol{k} + \boldsymbol{k}_a - \boldsymbol{k}_b) \cdot \boldsymbol{r}} d^3 r \\
&= -\frac{4\pi^2 ie}{mc} \frac{A(\omega)}{L^3} \hat{\boldsymbol{\varepsilon}} \cdot \boldsymbol{k}_a \delta(\boldsymbol{k} + \boldsymbol{k}_a - \boldsymbol{k}_b),
\end{aligned}
\tag{4.187}
$$

with $\omega = \omega_b - \omega_a$ and L^3 equal to the volume of a large cubic region of space. Equation (4.187) implies that the unperturbed plane-wave solutions can make a nonzero contribution to the transition probability only if both energy and momentum are conserved. However, a free particle cannot absorb a photon and conserve both energy and momentum. Therefore, Equation (4.187) must equal zero, and free–free absorption can only occur in the presence of a proton (or ion). Only matrix elements containing at least one perturbed wave function u'_a or u'_b will give a finite contribution to the cross section for free–free absorption.

Retaining only matrix elements with one unperturbed and one perturbed electron wave function (i.e. neglecting terms of order greater than λ) and performing integrations over time as in Equation (4.70), we obtain

$$a_b(\infty) = -\frac{2\pi i e}{\hbar m c} A(\omega)[\langle b_0 | e^{i\mathbf{k}\cdot\mathbf{r}}\hat{\boldsymbol{\varepsilon}}\cdot\mathbf{p} | a'\rangle + \langle b' | e^{i\mathbf{k}\cdot\mathbf{r}}\hat{\boldsymbol{\varepsilon}}\cdot\mathbf{p} | a_0\rangle]. \tag{4.188}$$

Using the plane-wave expansions for $|a'\rangle$ and $\langle b'|$ given in Equations (4.182) and (4.184), the above expression for $a_b(\infty)$ becomes

$$a_b(\infty) = -\frac{2\pi e i}{\hbar m c} A(\omega) \sum_n \left[\frac{\langle b_0 | e^{i\mathbf{k}\cdot\mathbf{r}}\hat{\boldsymbol{\varepsilon}}\cdot\mathbf{p} | n\rangle\langle n | H'_1 | a_0\rangle}{E_a - E_n} \right.$$
$$\left. + \frac{\langle b_0 | H'_1 | n\rangle\langle n | e^{i\mathbf{k}\cdot\mathbf{r}}\hat{\boldsymbol{\varepsilon}}\cdot\mathbf{p} | a_0\rangle}{E_b - E_n} \right], \tag{4.189}$$

where the summation is over intermediate states n. The eigenfunctions in Equation (4.189) have well-defined momentum, and momentum is conserved in matrix elements involving the electromagnetic perturbation. Therefore, only one intermediate state in each of the two terms inside the summation over n contributes.

Labeling these two intermediate states 1 and 2, respectively, the conservation of momentum implies that

$$\mathbf{p}_1 + \hbar\mathbf{k} = \mathbf{p}_b \tag{4.190}$$

and

$$\mathbf{p}_2 = \mathbf{p}_a + \hbar\mathbf{k}. \tag{4.191}$$

The energies E_1 and E_2 of intermediate states 1 and 2 are determined by Equations (4.190) and (4.191). Squaring these latter equations leads to the equations

$$p_1^2 = p_b^2 - 2\hbar\mathbf{p}_b\cdot\mathbf{k} + \hbar^2 k^2 \tag{4.192}$$

and

$$p_2^2 = p_a^2 + 2\hbar\mathbf{p}_a\cdot\mathbf{k} + \hbar^2 k^2. \tag{4.193}$$

For nonrelativistic electrons and energies E_a and E_b that appear in Equation (4.189) are

$$E_a = \frac{p_a^2}{2m} \tag{4.194}$$

and

$$E_b = \frac{p_b^2}{2m} = E_a + \hbar\omega, \tag{4.195}$$

respectively. Therefore, in the nonrelativistic approximation, and assuming that $\hbar k \ll p_a$ (and p_b), we have

$$E_1 = \frac{p_1^2}{2m} = E_a + \hbar\omega \tag{4.196}$$

and

$$E_2 = E_a = E_b - \hbar\omega. \tag{4.197}$$

It follows that the expression for $a_b(\infty)$ in Equation (4.189) becomes

$$a_b(\infty) = -\frac{2\pi ei}{\hbar mc} A(\omega) \left[\frac{\langle b_0 | e^{i\mathbf{k}\cdot\mathbf{r}} \hat{\mathbf{\epsilon}} \cdot \mathbf{p} | 1 \rangle \langle 1 | H_1' | a_0 \rangle}{\hbar\omega} \right.$$
$$\left. - \frac{\langle b_0 | H_1' | 2 \rangle \langle 2 | e^{i\mathbf{k}\cdot\mathbf{r}} \hat{\mathbf{\epsilon}} \cdot \mathbf{p} | a_0 \rangle}{\hbar\omega} \right]. \tag{4.198}$$

Next we will evaluate the matrix elements involving the Coulomb interaction that appear in Equation (4.198). These matrix elements are of the form

$$\langle n | H_1' | m \rangle = - \int \psi_n^\star \frac{Ze^2}{r} \psi_m r^2 dr d\Omega \tag{4.199}$$

with $\psi_m = e^{i\mathbf{k}_m\cdot\mathbf{r}}/L^{3/2}$ and $\psi_n^\star = e^{-i\mathbf{k}_n\cdot\mathbf{r}}/L^{3/2}$. The momentum transferred to the nucleus is

$$\hbar\mathbf{q} = \mathbf{p}_m - \mathbf{p}_n = \hbar(\mathbf{k}_m - \mathbf{k}_n). \tag{4.200}$$

The nucleus is assumed to be sufficiently massive that the recoil energy of the nucleus is negligible. Using Equation (4.200) implies that Equation (4.199) can be expressed as

$$\langle n | H_1' | m \rangle = - 2\pi \int_0^\infty \frac{Ze^2}{r} r^2 dr \int_0^\pi e^{iqr\cos\theta} \sin\theta d\theta. \tag{4.201}$$

Since

$$\int_0^\pi e^{iqr\cos\theta} \sin\theta d\theta = \frac{1}{iqr} \int_{-iqr}^{iqr} e^{ix} dx = \frac{2\sin qr}{qr}, \tag{4.202}$$

Equation (4.201) becomes

$$\langle n | H_1' | m \rangle = - \frac{4\pi}{q} \int_0^\infty \frac{Ze^2}{r} r \sin qr \, dr. \tag{4.203}$$

The integral in Equation (4.203) does not appear to converge. However, if we introduce the screened Coulomb potential

$$V(r) = -\frac{Ze^2}{r}e^{-\kappa r}, \tag{4.204}$$

where $\kappa = 1/\lambda_D$, with λ_D equal to the Debye length, given in Equation (3.289), then Equation (4.203) becomes

$$\langle n|H_1'|m\rangle = -\frac{4\pi Ze^2}{q}\int_0^\infty e^{-\kappa r}\sin qr\,dr$$

$$= -\frac{2\pi}{iq}Ze^2\left[\int_0^\infty e^{-\kappa r+iqr}dr - \int_0^\infty e^{-\kappa r-iqr}dr\right]$$

$$= -\frac{2\pi}{iq}Ze^2\left[\frac{1}{\kappa - iq} - \frac{1}{\kappa + iq}\right] = -\frac{4\pi Ze^2}{q^2 + \kappa^2}. \tag{4.205}$$

Equation (4.205) implies that in the limit $q^2 \gg \kappa^2$, Equation (4.198) becomes

$$a_b(\infty) = \frac{2\pi ei}{\hbar mc}\frac{A(\omega)}{\hbar\omega}\frac{4\pi Ze^2}{q^2}[-\langle b_0|e^{ik\cdot r}\hat{\varepsilon}\cdot p|1\rangle + \langle 2|e^{ik\cdot r}\hat{\varepsilon}\cdot p|a_0\rangle]. \tag{4.206}$$

For electron velocities $v \ll c$, the momentum of the photon satisfies the condition $\hbar k \ll |P_b - P_a|$. Most of the momentum change is caused by scattering with the proton (or ion). Therefore, making the approximation $e^{ik\cdot r} \simeq 1$ implies that the expression for $a_b(\infty)$ in Equation (4.206) becomes

$$a_b(\infty) = \left(\frac{2\pi eiA(\omega)}{\hbar mc\hbar\omega}\right)\left(\frac{4\pi Ze^2}{q^2}\right)(-\hat{\varepsilon}\cdot p_b + \hat{\varepsilon}\cdot p_a)$$

$$= -\frac{2\pi ei}{\hbar mc\omega}A(\omega)\frac{4\pi Ze^2}{q}\cos\theta, \tag{4.207}$$

where we have used the relation $p_b - p_a = \hbar(k_b - k_a) = \hbar q$ and θ is the angle between $\hat{\varepsilon}$ and q. From Equation (4.207) we obtain

$$|a_b(\infty)|^2 = \frac{64\pi^4\alpha^2Z^2e^2\cos^2\theta}{m^2\omega^2q^2}|A(\omega)|^2. \tag{4.208}$$

We assume that the incident radiation is unpolarized. It follows that $\cos^2\theta$ in Equation (4.208) can be chosen equal to

$$\overline{\cos^2\theta} = \frac{\displaystyle\int\cos^2\theta\,d\Omega}{4\pi} = \frac{1}{3}. \tag{4.209}$$

Energy conservation implies that ω, k_a and k_b are related by the equation

$$\hbar\omega = \frac{\hbar^2}{2m}(k_b - k_a)(k_b + k_a). \tag{4.210}$$

Making the approximation $k_b + k_a \simeq 2k_a$ in Equation (4.210), we obtain

$$\omega \simeq \frac{\hbar k_b}{m}(k_b - k_a) = v(k_b - k_a). \tag{4.211}$$

In order to evaluate the cross section for free–free absorption, we multiply Equation (4.208) by $\hbar\omega$, with $\cos^2 \theta$ set equal to $\frac{1}{3}$, divide the resultant equation by the expression for the energy per unit area per unit ω in the pulse implied by Equation (4.87) and then integrate over the density of final states. Therefore, the cross section $\sigma(\omega)$ becomes

$$\sigma(\omega) = \frac{64\pi^4\alpha^2 Z^2 e^2 \hbar c}{3m^2\omega^3\hbar^3(2\pi)^3} \int \frac{\delta(\omega - v(k_b - k_a))\mathrm{d}^3 p_b}{q^2} \tag{4.212}$$

or

$$\sigma(\omega) = \frac{64\pi^4\alpha^2 Z^2 e^2 \hbar c}{3m^2\omega^3(2\pi)^3} \int \frac{k_b^2\delta(\omega - v(k_b - k_a))\mathrm{d}k_b \sin\theta\mathrm{d}\theta}{(k_a^2 + k_b^2 - k_a k_b \cos\theta)}, \tag{4.213}$$

with $p_b = \hbar k_b$. The delta function that appears within the integrand of Equation (4.213) can be omitted from the integration over θ because the scattering of the electron is caused primarily by the Coulomb interaction with the ion, whereas the change in magnitude of k_a is caused by the absorption of the photon. Therefore, the integral over θ in Equation (4.213) is

$$\int_{-1}^{+1} \frac{\mathrm{d}\cos\theta}{k_a^2 + k_b^2 - 2k_a k_b \cos\theta} = \frac{1}{2k_a k_b} \ln\frac{(k_a + k_b)^2}{(k_a - k_b)^2}$$

$$= \frac{1}{k_a k_b} \ln\frac{k_b + k_a}{k_b - k_a}. \tag{4.214}$$

Equation (4.211) implies that

$$\ln\frac{k_b + k_a}{k_b - k_a} \simeq \ln\frac{2k_a v}{\omega} \tag{4.215a}$$

and

$$\frac{\mathrm{d}\omega}{v} = \mathrm{d}k_b. \tag{4.215b}$$

From Equations (4.213–4.215b) we have

$$\sigma(\omega) = \frac{64\pi^4\alpha^2 Z^2 e^2 \hbar c}{3m^2\omega^3 v(2\pi)^3} \int \frac{k_b^2}{k_a k_b} \ln\left(\frac{2k_a v}{\omega}\right) \delta(\omega - k_b v + k_a v)\mathrm{d}\omega. \tag{4.216}$$

The logarithmic term in Equation (4.216) is slowly varying and can therefore be removed from inside the integral. Under conditions of thermodynamic equilibrium the radiation field is that of a blackbody. The Planck function attains its maximum value when $\hbar\omega = 2.8 kT$, and the average energy of a particle is $(3/2) kT$. Since $\hbar k_a = mv$, the logarithmic term can be approximated as

$$\ln\left(\frac{2k_a v}{\omega}\right) \simeq \ln\left(\frac{2mv^2}{\hbar\omega}\right) \simeq \ln\left(\frac{3}{1.4}\right). \tag{4.217}$$

Therefore, since k_a is approximately equal to k_b, Equation (4.216) becomes

$$\sigma(\omega) = \frac{4\alpha^2 Z^2 e^2 hc}{3m^2\omega^3 v} \ln\left(\frac{2mv^2}{\hbar\omega}\right). \tag{4.218}$$

From Equation (4.218) we have

$$\sigma(v) = 2\pi\sigma(\omega) = \frac{4}{3}\frac{Z^2 e^6 \ln\left(\dfrac{2mv^2}{\hbar\omega}\right)}{m^2 chv^3 v}. \tag{4.219}$$

The electron distribution function is Maxwellian. Therefore from Equations (3.13) and (4.219) the free–free absorption cross section caused by electrons in the velocity interval v, $v + dv$ is

$$d\sigma_f(Z, v, v) = \frac{4Z^2 e^6 \ln\left(\dfrac{2mv^2}{\hbar\omega}\right)}{3m^2 chv^3 v} n_e(v) dv, \tag{4.220}$$

with $n_e(v) = 4\pi n_e \left(\dfrac{m}{2\pi kT}\right)^{3/2} \exp\left(-\dfrac{mv^2}{2kT}\right) v^2$. If we let $x = mv^2/2kT$, then the integral of Equation (4.220) over velocity space is

$$\sigma_f = \frac{16\pi Z^2 e^6 n_e}{3m^2 chv^3}\left(\frac{m}{2\pi kT}\right)^{3/2}\left(\frac{kT}{m}\right)\ln\left(\frac{2mv^2}{\hbar\omega}\right)\int_0^\infty \exp(-x)dx$$

$$= \frac{16\pi Z^2 e^6 n_e \ln\left(\dfrac{2mv^2}{\hbar\omega}\right)}{3hc(2\pi m)^{3/2}(kT)^{1/2}v^3}. \tag{4.221}$$

The cross section $\sigma_f(Z, v, T)$ and free–free absorption opacity κ_v^a are related by the equation

$$\rho\kappa_v^a = n_i\sigma_f(Z, v, T), \tag{4.222}$$

with ρ equal to the gas density and n_i equal to the number density of ions. The stellar free–free opacity corrected for stimulated emission becomes

$$\kappa_v^f = \kappa_v^a(1 - e^{-hv/kT}) = \frac{n_i\sigma_f}{\rho}(1 - e^{-hv/kT}). \tag{4.223}$$

Equation (4.223) can be re-expressed as

$$\kappa_v^f = K\frac{n_i n_e(1 - e^{-hv/kT})}{\rho T^{1/2}v^3}, \tag{4.224}$$

with K equal to a constant determined by Equation (4.221).

Stellar opacity is the sum of electron scattering opacity and absorptive opacity caused by absorption lines, bound–free absorption and free–free absorption. The correction for stimulated emission given in Equation (4.224) is applicable for absorption processes but not for electron scattering.

The temperature derivative of U_ν given in Equation (3.87) is

$$\frac{dU_\nu}{dT} = \frac{8\pi h^2 \nu^4}{c^3 kT^2} \frac{\exp(h\nu/kT)}{[\exp(h\nu/kT) - 1]^2}. \tag{4.225}$$

Substituting Equation (4.225) into Equation (2.77) and introducing the variable $x = h\nu/kT$, we obtain

$$\frac{1}{\kappa} = \frac{8\pi h^2 \rho T^{1/2} \left(\dfrac{kT}{h}\right)^8}{c^3 kT^2 K n_i n_e 4aT^3} \int_0^\infty \frac{x^7 \exp 2x \, dx}{[\exp x - 1]^3}$$

$$= C \frac{\rho}{n_i n_e} T^{3.5}, \tag{4.226}$$

with C equal to a constant. From Equation (4.226) we have the Kramers opacity

$$\kappa = C_1 \frac{n_i n_e}{\rho} T^{-3.5} = C_2 \rho T^{-3.5}, \tag{4.227}$$

with $C_1 = 1/C$ and $C_2 = C_1/\mu\mu_e m_p^2$.

4.8   Stellar opacities

The above discussion of atomic absorption processes is limited to hydrogen and hydrogenic ions. In stellar interiors and some stellar atmospheres, absorption by multielectron ions and atoms is the dominant opacity source over certain ranges of temperature and density. The Schrödinger equation must be solved approximately for two or more electron configurations. Variational methods described in standard quantum-mechanics textbooks can be used to obtain accurate energy levels and wave functions for helium and other helium-like ions. We remark that lowest-order variational calculations do not predict a bound state for the H$^-$ negative ion. Higher-order calculations, which were first performed by Chandrasekhar, are required to obtain its bound-state energy of 1.6 eV.

The central-field approximation is the starting point for many-electron atom and ion calculations. In this approximation it is assumed that each electron moves in the effective spherically symmetric potential formed by the nucleus and other electrons. The Pauli exclusion principle asserts that no two electrons can occupy the same quantum state. Atomic subshells are characterized by the quantum numbers n and l with $2(2l + 1)$ equal to the number of states in a particular subshell. Electrons belonging to the same subshell are called equivalent electrons. The total orbital angular momentum L and spin angular momentum S of a closed atomic shell are equal to zero. It is often true that the electrons in an unfilled atomic shell can combine their individual orbital and spin angular momenta consistent with the Pauli exclusion principle so as to obtain more than one value of total L and S. Hund's empirical rules give the ground-state angular-momentum configuration of atoms. These rules state that for a given electron configuration the largest possible S has the lowest energy and for a given value of S the state with the maximum possible L value has the lowest energy.

Accurate stellar-opacity calculations are, in general, quite complex. Tabulated Roseland mean opacities (Cox and Tabor 1976, Iglesias, Rogers and Wilson 1990) are used in performing stellar-interior and hydrodynamic calculations. Stellar opacities depend on statistical physics and assumed chemical compositions as well as atomic physics. Numerical simulations of variable stars, which are discussed further in Chapter 8, are quite sensitive to assumed opacities. β Cepheids (also known as β canis majoris stars) are massive (i.e. $\gtrsim 10 M_\odot$) periodic variable stars that have recently evolved off the main sequence. Unlike Cepheid variables, whose prototype is δ Cepheid, and RR Lyrae variables, they pass through maximum light when their radii are close to minimum because their hydrogen and helium ionization zones are less extensive. Early models of these stars did not provide a physical explanation for their observed variability. Subsequent numerical simulations with higher values for opacities near 10^5 K were able to explain their observed pulsations. Recent improvements in stellar-opacity calculations have given better agreement between numerical simulations and observations of Cepheids. Early numerical hydrodynamic calculations of RR Lyrae variables tended to give somewhat lower mass predictions than obtained by stellar-evolution calculations. Opacity effects are believed to be responsible for these differences. Higher calculated opacities at temperatures of about 10^6 K have increased estimates of the depth of the solar convective envelope.

Electron scattering, bound–free absorption (photoionization), bound–bound absorption and free–free absorption are not the only important opacity sources in stellar interiors. The Auger effect, which is also called autoionization, contributes significantly to the opacity of He and two-electron ions. Previously discussed absorption processes involve singly excited states in which one electron of a two-electron system remains in the ground state, whereas the other electron is excited to a higher state. However, doubly excited states exist in helium and two-electron ions. These discrete states are embedded in the continuum because their energies are higher than the ionization threshold. The $(2s2p)^1P$ state of the helium atom is an example of a state that can be excited by a single ultraviolet photon from the helium ground state (i.e. $(1s)^2\,^1S$) because the selection rules $\Delta L = \pm 1$ and $\Delta S = 0$ are both satisfied. The decay of the $(2s2p)^1P$ state occurs both by radiative and radiationless (autoionizing) transitions. Radiative decay can take place to either a ^1S or ^1D state of the helium atom because the above selection rules apply. In addition, the $(2s2p)^1P$ state of helium can decay by means of a radiationless transition into a He^+ ion in the ground state and a free electron, because its excitation energy is above the helium-atom ionization threshold. The existence of autoionization states of helium can be demonstrated experimentally. They appear as resonances that are superimposed on photoionization absorption which varies more smoothly with frequency.

Problems

1. a. Calculate the frequency differences between the Lyman α and Hα transitions of atomic hydrogen and deuterium.
    b. Positronium consists of an electron and positron. Determine its energy levels, neglecting spin.

2. a. Show that Equation (4.9) can be rewritten as

$$\frac{1}{\rho^2}\frac{d}{d\rho}\left(\rho^2\frac{dR}{d\rho}\right)+\left(\frac{8\pi^2mZe^2}{\alpha\hbar^2\rho}-\frac{1}{4}-\frac{l(l+1)}{\rho^2}\right)R=0$$

with

$$\rho=\alpha r,\alpha^2=\frac{8m|E|}{\hbar^2}\text{ and }\frac{8\pi^2mZe^2}{\alpha\hbar^2}=\frac{Ze^2}{\hbar}\left(\frac{m}{2|E|}\right)^{1/2}.$$

    b. Make the substitution $R(\rho)=F(\rho)e^{-\rho/2}$ into the above equation and find the resultant equation for $F(\rho)$. The solutions $F(\rho)$ are known as the Laguerre polynomials for $l=0$ and the associated Laguerre polynomials L_{n+l}^{2l+1} for $l>0$.

3. Calculate the oscillator strength and line-center cross section of the Hα transition ($n=3$ to $n=2$).

4. Prove that the electron degeneracy of the nth energy level of the hydrogen atom equals $2n^2$.

5. Prove that the expectation value $\langle 1|V(r)|1\rangle$, where $V(r)=-e^2/r$ and $|1\rangle$ denotes the wave function of the lowest energy level of atomic hydrogen, is $-e^2/a_0$. Calculate explicitly the electron kinetic energy and thereby show that the ground-state energy is $-e^2/2a_0$.

6. Infer the rate of radiative recombination to the $n=1$ energy level of hydrogen from the calculated photoionization cross section. *Hint*: Under conditions of thermodynamic equilibrium the number of photoionizations equals the corresponding number of recombinations.

7. Calculate the photoionization cross section from the $n=2$ energy level of atomic hydrogen.

8. Using the principle of detailed balance, infer the bremsstrahlung emissivity from the calculated free–free absorption cross section. Set the logarithmic term equal to unity and compare with the classical formula for bremsstrahlung (free–free) emission.

9. Assume $h\nu\gg\chi$, with χ equal to the ionization potential of hydrogen, and show that under conditions of thermodynamic equilibrium bound–free hydrogenic opacity can be written $\kappa=\kappa_0\rho T^{-7/2}$. Evaluate the constant κ_0.

5 Stellar atmospheres, convective envelopes and stellar winds

5.1 Radiative transfer in stellar atmospheres

The equation of radiative transfer already discussed in Chapter 2 can be expressed as

$$\frac{dI_\nu}{ds} = -\alpha_\nu I_\nu + j_\nu, \tag{5.1}$$

with ds a small element of length. In Equation (5.1) I_ν is the intensity of radiation defined in Equation (2.60), α_ν is in units of cm^{-1} and j_ν has the dimensions of $\mathrm{erg\,cm}^{-3}\,\mathrm{s}^{-1}\,\mathrm{Hz}^{-1}$ steradian^{-1}. The optical depth $\tau_\nu(s)$ is

$$\tau_\nu(s) = \int_0^s \alpha_\nu(s')ds'. \tag{5.2}$$

Equation (5.2) implies that $d\tau_\nu = \alpha_\nu ds$. In describing plane-parallel atmospheres it is convenient to define the vertical distance z so that z decreases inward from its value of $z = 0$ at the outer boundary. It follows that the change in vertical distance z and the vertical optical depth $\tau_\nu(z)$ are related by the equation

$$d\tau_\nu(z) = -\alpha_\nu(z)dz. \tag{5.3}$$

The source function S_ν is defined by the equation

$$S_\nu = \frac{j_\nu}{\alpha_\nu}. \tag{5.4}$$

Under conditions of local thermodynamic equilibrium the source function S_ν is equal to the Planck function B_ν. Equation (5.4) with S_ν equal to B_ν is often referred to as Kirchhoff's law.

If $d\tau_\nu$ instead of ds is chosen as the independent variable, then Equation (5.1) becomes

$$\frac{dI_\nu}{d\tau_\nu} = -I_\nu + S_\nu. \tag{5.5}$$

Equation (5.5) can be written as

$$\frac{d(I_\nu e^{\tau_\nu})}{d\tau_\nu} = e^{\tau_\nu} S_\nu. \tag{5.6}$$

Integrating Equation (5.6), we obtain

$$I_\nu(\tau_\nu) = I_\nu(0)e^{-\tau_\nu} + \int_0^{\tau_\nu} e^{-(\tau_\nu - \tau_\nu')} S_\nu(\tau_\nu') d\tau_\nu'. \tag{5.7}$$

The integral in Equation (5.7) can be readily evaluated if the source function S_ν is independent of optical depth. Solving Equation (5.7) with S_ν equal to a constant gives

$$
\begin{aligned}
I_\nu(\tau_\nu) &= I_\nu(0)e^{-\tau_\nu} + S_\nu(1 - e^{-\tau_\nu}) \\
&= S_\nu + e^{-\tau_\nu}(I_\nu(0) - S_\nu).
\end{aligned} \tag{5.8}
$$

If local thermodynamic equilibrium exists, then $I_\nu(0)$ and S_ν are equal to $B_\nu(T_0)$ and $B_\nu(T)$, respectively, with T_0 the temperature of some background source at optical depth $\tau_\nu = 0$ and T the temperature of a layer of gas between the background source and an observer. Under conditions of local thermodynamic equilibrium $I_\nu = B_\nu(T)$ if τ_ν satisfies the condition $\tau_\nu \gg 1$. Equation (5.8) becomes

$$I_\nu(\tau_\nu) = B_\nu(T_0)(1 - \tau_\nu) + B_\nu(T)\tau_\nu, \tag{5.9}$$

if τ_ν satisfies the condition $\tau_\nu \ll 1$. Equation (5.9) implies that an absorption line is formed if the temperature T is less than T_0 and an emission line is formed if T is greater than T_0.

The flux through the surface of a star integrated over all frequencies is

$$F = 2\pi \int_0^\infty \int_0^{\pi/2} \cos\theta I_\nu \sin\theta d\theta d\nu \equiv \sigma T_{\text{eff}}^4 = \frac{L}{4\pi R^2}. \tag{5.10}$$

Equation (5.10) implies that if $I_\nu = B_\nu(T)$, then

$$F = \pi \int_0^\infty B_\nu(T) d\nu = \frac{ac}{4} T^4. \tag{5.11}$$

5.2   The Eddington approximation

We consider a plane-parallel atmosphere and assume that the absorption coefficient α_ν is independent of frequency. The equation of radiative transfer becomes

$$-\cos\theta \frac{dI_\nu(\theta)}{d\tau} = -I_\nu(\theta) + \frac{j_\nu}{\alpha}, \tag{5.12}$$

where θ is the angle between the normal and the direction considered, and α is the absorption coefficient. The optical depth τ is

$$\tau = \int_z^0 \alpha dz. \tag{5.13}$$

Integrating Equation (5.12) over all frequencies leads to the equation

$$- \cos \theta \frac{\mathrm{d}I(\theta)}{\mathrm{d}\tau} = - I(\theta) + \frac{j}{\alpha}, \tag{5.14}$$

with $I(\theta) = \int I_\nu(\theta)\mathrm{d}\nu$. It is convenient to define the integrals

$$J = \frac{1}{4\pi} \int I(\theta)\mathrm{d}\Omega, \tag{5.15}$$

$$H = \frac{1}{4\pi} \int I(\theta) \cos \theta \, \mathrm{d}\Omega \tag{5.16}$$

and

$$K = \frac{1}{4\pi} \int I(\theta) \cos^2 \theta \, \mathrm{d}\Omega, \tag{5.17}$$

with $\mathrm{d}\Omega = 2\pi \sin \theta \mathrm{d}\theta$. If Equation (5.14) is integrated over all solid angles, we obtain

$$\frac{\mathrm{d}H}{\mathrm{d}\tau} = J - \frac{j}{\alpha}. \tag{5.18}$$

Multiplying Equation (5.14) by $\cos \theta$ and integrating over all solid angle implies that

$$\frac{\mathrm{d}K}{\mathrm{d}\tau} = H. \tag{5.19}$$

Because the flux through a stellar atmosphere is constant, Equation (5.18) reduces to the equation

$$j = \alpha J, \tag{5.20}$$

and the integral of Equation (5.19) becomes

$$K = H\tau + \text{constant}. \tag{5.21}$$

We assume that $I(\theta)$ can be approximated as

$$I = I_1 \quad (0 < \theta < \tfrac{1}{2}\pi), \tag{5.22}$$

$$I = I_2 \quad \left(\tfrac{1}{2}\pi < \theta < \pi\right). \tag{5.23}$$

It can readily be shown that Equations (5.15–5.17), (5.22) and (5.23) imply that

$$J = \frac{1}{2}(I_1 + I_2), \tag{5.24}$$

$$H = \frac{1}{4}(I_1 - I_2) \tag{5.25}$$

and

$$K = \frac{1}{3}J.$$

(5.26)

Because there is no inflow of radiation at the surface, we have $I_2 = 0$ at $\tau = 0$. It follows from Equations (5.24) and (5.25) that $J = 2H$ at $\tau = 0$. Equation (5.26) and the above expression for J at $\tau = 0$ imply that Equation (5.21) can be written as

$$J = H(2 + 3\tau).$$

(5.27)

Under conditions of local thermodynamic equilibrium and α_v independent of frequency we have

$$j_v = \alpha_v B_v(T) = \alpha B_v(T),$$

(5.28)

with B_v equal to the Planck function. Equations (5.20) and (5.28) imply that

$$J = \frac{j}{\alpha} = \int B_v(T)dv = \frac{\sigma T^4}{\pi}.$$

(5.29)

Equations (5.27) and (5.29) imply that

$$\frac{\sigma T^4}{\pi} = H(2 + 3\tau).$$

(5.30)

Since the flux F is

$$F = 4\pi H = \sigma T^4_{\text{eff}},$$

(5.31)

Equation (5.30) becomes

$$T^4 = \frac{T^4_{\text{eff}}}{2}\left(1 + \frac{3}{2}\tau\right).$$

(5.32)

Equation (5.32) implies that the temperature T is equal to the effective temperature T_{eff} when $\tau = 2/3$ and that the temperature T_0 at $\tau = 0$ is $T_0 = T_{\text{eff}}/2^{1/4}$.

Continuum opacity sources such as bound–free absorption from the H^- negative ion are wavelength dependent, and therefore the Eddington approximation does not give an accurate description of a stellar atmosphere even when absorption-line opacities do not contribute appreciably to determining atmospheric structure. The thickness of a stellar atmosphere is very small compared to the radius of a star. It follows that the physical properties of a stellar atmosphere are determined by effective temperature (T_{eff}), photospheric gravity (g) and abundances of elements, which may exist as atoms, ions or molecules. Stellar atmospheres generally depend more sensitively on T_{eff} than gravity because the ionization states of elements depend sensitively on temperature. The electrons that combine with atomic hydrogen to form H^- in the solar atmosphere are primarily from Fe and other metals rather than hydrogen because ionization potentials of metals are lower. Early model calculations assumed that the structure of a stellar atmosphere depends on continuum opacity only. In Section 5.5 below, we show how the equivalent width of a spectral line can be calculated approximately in an atmosphere

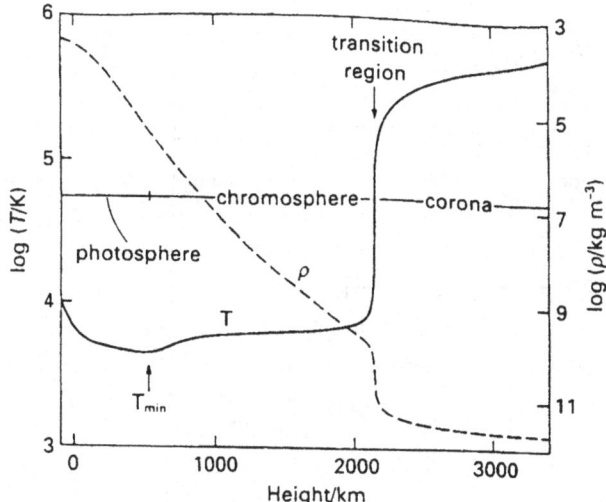

Figure 5.1. Temperature and density are shown as a function of height above the base of the solar photosphere.

whose structure is determined by known continuum opacity. Numerical solutions that include both line and continuum opacities are required to construct accurate models of stellar atmospheres. Such models usually assume local thermodynamic equilibrium (LTE). However, in some hot stars non-LTE effects become important. In Section 5.8 we describe a non-LTE calculation of millimeter-wavelength emission from CO molecules present in strong stellar winds that emanate from luminous red giants.

The structure and elemental abundances of the solar atmosphere are more accurately determined than in any other star. The distribution of temperature and density in the solar atmosphere are shown in Figure 5.1. Figure 5.2 gives the locations of various atoms and ions that cause absorption lines in the photosphere and emission lines in the chromosphere and corona.

Because the temperatures of stellar atmospheres are high, weak absorption lines have Doppler line shapes. However, strong absorption lines are affected by the Lorentzian contribution to the Voigt profile, which implies that near line center the shape is Doppler, whereas in the wings it is Lorentzian. As discussed in Problem 6 at the end of this chapter collisions as well as radiative decay make a Lorentzian contribution to the Voigt profile.

5.3   Line broadening by the Doppler effect

Line radiation from a gas at temperature T is observed along the z-axis. From Equation (3.1) it follows that the fraction of atoms with speeds in the interval $v, v + dv$ that move within the solid angle $d\Omega = 2\pi \sin \theta d\theta$ is

$$2\pi \left(\frac{m}{2\pi kT} \right)^{3/2} e^{-\frac{1}{2}\frac{mv^2}{kT}} v^2 dv \sin \theta d\theta. \tag{5.33}$$

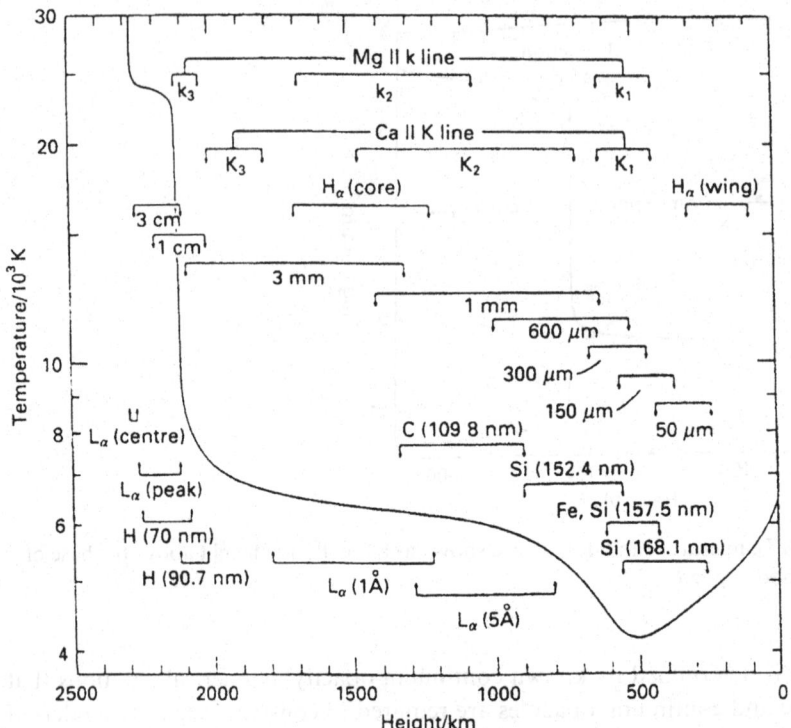

Figure 5.2. The temperature is shown as a function of height in the solar atmosphere. Lines with arrows indicate where various spectral lines come from. The atoms and ions shown are used to determine the temperature and density structure of the chromosphere and lower corona. From Athay (1986).

If v_0 is the frequency of the emitted radiation in the frame of reference of the atom, then the observed Doppler-shifted frequency v follows from the equation

$$v_0 - v = v_0 \frac{v}{c} \mu, \tag{5.34}$$

where $\mu = \cos \theta$. From Equation (5.34) we obtain

$$dv = - v_0 \frac{\mu dv}{c}. \tag{5.35}$$

Substituting Equations (5.34) and (5.35) into Expression (5.33) gives

$$2\pi \left(\frac{m}{2\pi kT} \right)^{3/2} e^{-\frac{a(v-v_0)^2}{\mu^2}} \frac{c^3}{v_0^3} \frac{(v-v_0)^2}{\mu^3} d\mu dv, \tag{5.36}$$

with

$$a = \frac{1}{2} \frac{mc^2}{kTv_0^2}.$$

If we let $x = 1/\mu$, then we have $dx = -d\mu/\mu^2$. Introducing the variable x onto Expression (5.36) gives

$$2\pi \left(\frac{m}{2\pi kT}\right)^{3/2} e^{-a(v-v_0)^2 x^2} \frac{c^3}{v_0^3}(v - v_0)^2 x dx dv. \tag{5.37}$$

The integration over x becomes

$$2\pi \left(\frac{m}{2\pi kT}\right)^{3/2} \frac{c^3}{v_0^3}(v - v_0)^2 dv \int_1^\infty e^{-a(v-v_0)^2 x^2} x dx. \tag{5.38}$$

Making the variable change $y = x^2$ implies that the integral in Equation (5.38) becomes

$$\frac{1}{2}\int_1^\infty e^{-a(v-v_0)^2 y} dy = \frac{e^{-a(v-v_0)^2}}{2a(v - v_0)^2}. \tag{5.39}$$

Expression (5.38), Equation (5.39) and the definition of the constant a given with Expression (5.36) imply that the Doppler line-profile function ϕ_v is

$$\phi_v = \left(\frac{mc^2}{2\pi kT v_0^2}\right)^{1/2} e^{-\frac{mc^2(v-v_0)^2}{2kT v_0^2}}. \tag{5.40}$$

The function ϕ_v in Equation (5.40) is normalized so that

$$\int_0^\infty \phi_v dv = 1. \tag{5.41}$$

5.4   The Voigt line profile

A line profile is caused by the radiative and collisional lifetimes of an energy level as well as the Doppler effect. Consider radiation emitted at frequency v and received by an atom at frequency v'. From Equations (4.131) and (4.150) it follows that the absorption coefficient $\alpha'(v)$ can be expressed as

$$\alpha'(v) = \frac{\pi e^2 f}{mc} \frac{\delta}{\pi} \frac{1}{(v' - v_0)^2 + \delta^2}, \tag{5.42}$$

with $\delta = \gamma/2\pi$ and

$$v' - v_0 = v - v_0\left(1 + \frac{v}{c}\cos\theta\right) - (v - v_0)\frac{v}{c}\cos\theta.$$

$$\simeq v - v_0\left(1 + \frac{v}{c}\cos\theta\right). \tag{5.43}$$

The final approximate relation in Equation (5.43) is a consequence of the condition $v \ll c$. Equations (5.42) and (5.43) imply that $\alpha'(v)$ can be written as

$$\alpha'(v) = \frac{\pi e^2 f}{mc} \frac{\delta}{\pi} \frac{1}{(v - v_0')^2 + \delta^2}, \tag{5.44}$$

with $v_0' = v_0(1 + v\cos\theta/c)$.

Expression (5.33) gives the fraction of atoms with speed in the interval $v, v + dv$ that are moving within the solid angle $d\Omega = 2\pi \sin\theta d\theta$. Equation (5.44) and Expression (5.33) imply that the absorption coefficient caused by a Maxwell–Boltzmann distribution of atoms is

$$\alpha(v) = \frac{\pi e^2 f}{mc}\frac{\delta}{\pi} 2\pi \left(\frac{m}{2\pi kT}\right)^{3/2} \int_0^\infty \int_0^\pi \frac{e^{-\frac{mv^2}{2kT}}\frac{1}{2}v^2 \sin\theta dv d\theta}{\left[v - v_0\left(1 + \frac{v}{c}\cos\theta\right)\right]^2 + \delta^2}. \tag{5.45}$$

Using the standard integral

$$\int \frac{dx}{a^2 + x^2} = \frac{1}{a}\tan^{-1}\frac{x}{a}, \tag{5.46}$$

making a change in variables and integrating over θ, we find that Equation (5.45) is equal to

$$\alpha(v) = \frac{\pi e^2 f}{mc}\frac{2\delta}{\pi^{3/2}b^2}\int_0^\infty e^{-y^2} y\left(\tan^{-1}\frac{w+y}{a} - \tan^{-1}\frac{w-y}{a}\right)dy, \tag{5.47}$$

with

$$a = \delta/b, \, b = \frac{v_0 2kT}{mc}, \, y = \frac{v}{\left(\frac{2kT}{m}\right)^{1/2}} \text{ and } w = \frac{v - v_0}{b}.$$

After an integration by parts, Equation (5.47) becomes

$$\alpha(v) = \frac{\pi e^2 f}{mc}\frac{2\delta}{\pi^{3/2}b^2}\left[\int_0^\infty \frac{e^{-y^2}dy}{a^2 + (w+y)^2} + \int_0^\infty \frac{e^{-y^2}dy}{a^2 + (w-y)^2}\right]$$

$$= \frac{\pi e^2 f}{mc}\frac{2\delta}{\pi^{3/2}b^2}\int_{-\infty}^{+\infty} \frac{e^{-y^2}dy}{a^2 + (w-y)^2} \tag{5.48}$$

If $a = \delta/b$ is small, the integral in Equation (5.48) is large when $y \simeq w$, and therefore setting $x = w - y$, we obtain

$$\int_{-\infty}^\infty \frac{e^{-y^2}dy}{a^2 + (w-y)^2} \simeq e^{-w^2}\int_{-\infty}^\infty \frac{dx}{a^2 + x^2} = \frac{\pi}{a}e^{-w^2}. \tag{5.49}$$

When Equation (5.49) is applicable, Equation (5.48) becomes

$$\alpha(v) \simeq \frac{\pi e^2 f}{mc}\frac{2\delta}{\pi^{3/2}b^2}\frac{\pi}{a}e^{-w^2}. \tag{5.50}$$

If y is assumed to be $\ll w = (v - v_0)/b$, then the integral in Equation (5.48) becomes

$$\int_{-\infty}^{+\infty} \frac{e^{-y^2} dy}{a^2 + (w - y)^2} \simeq \frac{1}{a^2 + w^2} \int_{-\infty}^{+\infty} e^{-y^2} dy = \frac{1}{a^2 + w^2} \sqrt{\pi}, \tag{5.51}$$

and the expression for the absorption coefficient given in Equation (5.48) becomes

$$\alpha(v) = \frac{\pi e^2 f}{mc} \frac{2\delta}{\pi^{3/2} b^2} \frac{\sqrt{\pi}}{a^2 + w^2}. \tag{5.52}$$

5.5   The equation of radiative transfer for spectral-line radiation

The absorption coefficient α_v includes both absorption of radiation and stimulated emission. From Equation (4.91) it follows that the cross section $\sigma(v)$ at frequency v is

$$\sigma(v) = \sigma_t \phi_v = \frac{h v_0}{4\pi} \phi_v B_{12}, \tag{5.53}$$

where ϕ_v is the line-profile function, v_0 is the frequency at the center of the line and B_{12} is the Einstein coefficient for absorption. If stimulated emission is neglected, then the absorption coefficient becomes

$$(\alpha_v)_{abs} = n_1 \sigma(v) = \frac{h v_0}{4\pi} \phi_v n_1 B_{12}. \tag{5.54}$$

If stimulated emission is included, then the absorption coefficient is

$$\alpha_v = \frac{h v_0}{4\pi} \phi_v (n_1 B_{12} - n_2 B_{21}). \tag{5.55}$$

The equation of radiative transfer can now be expressed as

$$\frac{dI_v}{ds} = -\alpha_v I_v + j_v$$

$$= -\frac{h v_0}{4\pi} (n_1 B_{12} - n_2 B_{21}) \phi_v I_v + \frac{h v}{4\pi} n_2 A_{21} \phi_v. \tag{5.56}$$

Using the relation between B_{12} and B_{21} given in Equation (4.92), we find that Equation (5.56) reduces to the equation

$$\frac{dI_v}{ds} = -\frac{h v}{4\pi} n_1 B_{12} (1 - g_1 n_2 / g_2 n_1) \phi_v I_v + \frac{h v}{4\pi} n_2 A_{21} \phi_v. \tag{5.57}$$

The relations between the Einstein coefficients given in Equations (4.92) and (4.93) imply that the source function can be written as

$$S_v = \frac{j_v}{\alpha_v} = \frac{n_2 A_{21}}{n_1 B_{12} - n_2 B_{21}} = \frac{2h v^3}{c^2} \left(\frac{g_2 n_1}{g_1 n_2} - 1 \right)^{-1}. \tag{5.58}$$

The Boltzmann relation given in Equation (3.72) implies that under conditions of local thermodynamic equilibrium we have

$$\alpha_v = \frac{h v_0}{4\pi} n_1 B_{12} (1 - e^{-hv/kT}) \phi_v \tag{5.59}$$

and

$$S_v = B_v(T), \tag{5.60}$$

with $B_v(T)$ equal to the Planck function.

The absorption coefficient given in Equation (5.57) is equal to zero when $g_2 n_1 = g_1 n_2$. Under such physical conditions the source function S_v cannot be defined. Population inversion is said to occur if the condition $g_1 n_2 > g_2 n_1$ is satisfied. When the above inequality holds, the absorption coefficient α_v is negative, and amplification of line radiation (i.e. maser emission) takes place. Maser emission has been observed at microwave frequencies from the expanding circumstellar envelopes of stars and also from dense interstellar clouds.

We assume that scattering is negligible and that the absorption coefficient,

$$\chi_v = \alpha_c + \alpha_v, \tag{5.61}$$

is independent of optical depth. In Equation (5.61) α_c is the continuum absorption coefficient and α_v is the absorption coefficient caused by a spectral line. We assume that the source function $S(\tau_v)$ can be expressed as

$$S(\tau_v) = a + b\tau, \tag{5.62}$$

with a and b constants and τ the vertical optical depth given in Equation (5.3) at frequencies such that line absorption is negligible.

If we solve the radiative-transfer equation given in Equation (5.1) with the optical depth defined by Equation (5.3), the intensity of radiation at $\tau_v = 0$ becomes

$$I_v(\mu) = \int_0^\infty S(\tau_v) e^{-\tau_v/\mu} \frac{d\tau_v}{\mu}. \tag{5.63}$$

The factor μ appears in Equation (5.63) because τ_v is the vertical optical depth.

Equations (5.62) and (5.63) imply that

$$I_v(\mu) = a \int_0^\infty e^{-\tau_v/\mu} \frac{d\tau_v}{\mu} + b \int_0^\infty \tau e^{-\tau_v/\mu} \frac{d\tau_v}{\mu}. \tag{5.64}$$

Since the optical depth τ_v at frequencies that include the spectral line is

$$\tau_v(z) = -\int_z^0 (\alpha_c + \alpha_v) dz' = (\alpha_c + \alpha_v) z, \tag{5.65}$$

and the continuum optical depth is

$$\tau(z) = -\int_z^0 \alpha_c dz = \alpha_c z, \tag{5.66}$$

we have

$$\tau_v = \frac{\alpha_v + \alpha_c}{\alpha_c}\,\tau.$$

(5.67)

Equations (5.61) and (5.67) imply that Equation (5.64) becomes

$$I_v(\mu) = a\int_0^\infty e^{-\tau/\mu}\frac{d\tau_v}{\mu} + \frac{b\alpha_c\mu}{\chi_v}\int_0^\infty \frac{\tau_v}{\mu}e^{-\tau_v/\mu}\frac{d\tau_v}{\mu}$$

$$= a + \frac{\alpha_c b\mu}{\chi_v}.$$

(5.68)

It follows from Equation (5.68) that the flux F_v at the surface (i.e. $\tau = 0$) is

$$F_v = 2\pi\int_0^1 I_v(\mu)\mu d\mu = 2\pi\left(\frac{a\mu^2}{2}\Big|_0^1 + \frac{\alpha_c b\mu^3}{3\chi_v}\Big|_0^1\right) = \pi\left(a + \frac{2}{3}\frac{\alpha_c b}{\chi_v}\right).$$

(5.69)

If we define the dimensionless variable $\eta_v = \alpha_v/\alpha_c = \eta_0\phi_v$, with η_0 a constant and ϕ_v equal to the line-profile function, the expression for the flux given in Equations (5.69) becomes

$$F_v = \pi\left(a + \frac{2}{3}\frac{\alpha_c}{\chi_v}b\right) = \pi\left(a + \frac{2}{3}\frac{b}{1 + \eta_v}\right).$$

(5.70)

At frequencies off the spectral line, $\eta_v = 0$ and therefore from Equation (5.70) we have

$$F_c = \pi\left(a + \frac{2}{3}b\right).$$

(5.71)

Equations (5.70) and (5.71) imply that the absorption profile of the spectral line is

$$A_v = \frac{F_c - F_v}{F_c} = \frac{2}{3}b\frac{\left(1 - \dfrac{1}{1 + \eta_v}\right)}{\left(a + \dfrac{2}{3}b\right)}$$

$$= \frac{\eta_v}{(\eta_v + 1)}\frac{1}{\left(1 + \dfrac{3}{2}\dfrac{a}{b}\right)}.$$

(5.72)

The equivalent width of the spectral line is

$$W_v = \int_0^\infty A_v dv.$$

(5.73)

In agreement with observations, Equation (5.72) shows that A_v is always less than unity for any v because a and b are positive constants.

5.6   Convective envelopes

The presence of convective zones in the outer layers of red giants and main-sequence stars with masses $M \lesssim 1.5\,M_\odot$ is a consequence of the temperature and density dependence of the opacity. By reducing the adiabatic temperature gradient below that of either a fully ionized or atomic gas the occurrence of regions of partially ionized hydrogen (ionization zones) also influences the formation of convective zones. The H^- negative ion, which consists of a proton and two surrounding electrons, is the dominant continuum source of opacity in the outermost layers of low-mass main-sequence stars and some red giants. Since the number of H^- ions depends on the electron number density, which increases very rapidly as a function of increasing optical depth in cool stars, it follows that the opacity caused by H^- also increases rapidly, and, as will be shown below, turbulent convection results.

The Eddington relation between temperature T and optical depth τ given in Equation (5.32) implies that

$$4 \ln T = \ln\left(\frac{1}{2} T_{\text{eff}}^4\right) + \ln\left(1 + \frac{3}{2}\tau\right). \tag{5.74}$$

From Equation (5.74) we obtain

$$\frac{dT}{d\tau} = \frac{3T}{8 + 12\tau}. \tag{5.75}$$

Because $T^{\Gamma_1} P^{1-\Gamma_1}$ is a constant during the adiabatic perturbation of an ideal, non-degenerate gas, we have

$$\left(\frac{dT}{d\tau}\right)_{\text{ad}} = \frac{\Gamma_1 - 1}{\Gamma_1} \frac{T}{P} \frac{dP}{d\tau}. \tag{5.76}$$

Hydrostatic equilibrium implies that

$$\frac{dP}{d\tau} = \frac{g}{\kappa}. \tag{5.77}$$

Integrating Equation (5.77), with κ assumed equal to a constant gives

$$P = \frac{g\tau}{\kappa}. \tag{5.78}$$

Equations (5.77) and (5.78) imply that

$$\frac{1}{P} \frac{dP}{d\tau} = \frac{1}{\tau}. \tag{5.79}$$

Equations (5.76) and (5.79) can be combined to give

$$\left(\frac{dT}{d\tau}\right)_{\text{ad}} = \frac{\Gamma_1 - 1}{\Gamma_1} \frac{T}{\tau}. \tag{5.80}$$

Equations (5.75) and (5.80) imply that the condition for turbulent convection

$$\frac{dT}{d\tau} > \left(\frac{dT}{d\tau}\right)_{ad} \tag{5.81}$$

can be expressed as

$$\frac{3\tau}{8 + 12\tau} > \frac{\Gamma_1 - 1}{\Gamma_1}. \tag{5.82}$$

The maximum value of the left-hand side of Equation (5.82) occurs at $\tau \to \infty$. It follows that when κ is independent of T and ρ (or P), the condition for convection to occur is

$$\Gamma_1 < \frac{4}{3}. \tag{5.83}$$

Since Γ_1 is less than $\frac{4}{3}$, within hydrogen ionization zones turbulent convection can exist even if the opacity is nearly independent of T and ρ.

If the opacity κ can be expressed as $\kappa = a + bP^x$, with a, b and x equal to positive constants, then Equation (5.77) becomes

$$(a + bP^x)\frac{dP}{d\tau} = g. \tag{5.84}$$

Integrating Equation (5.84), we obtain

$$aP + \frac{bP^{x+1}}{x+1} = g\tau. \tag{5.85}$$

At optical depths such that the constant a can be neglected, Equation (5.85) becomes

$$P = \frac{g\tau(x+1)}{bP^x}. \tag{5.86}$$

Equations (5.86) and (5.84), with $a = 0$, imply that

$$\frac{1}{P}\frac{dP}{d\tau} = \frac{1}{(x+1)\tau}. \tag{5.87}$$

From Equation (5.87) it follows that the adiabatic temperature gradient in Equation (5.76), with $\Gamma \equiv \Gamma_1$, becomes

$$\left(\frac{dT}{d\tau}\right)_{ad} = \frac{\Gamma - 1}{\Gamma} \frac{T}{(x+1)\tau}. \tag{5.88}$$

Equations (5.75) and (5.88) imply that the condition for convection is

$$\frac{3\tau}{8 + 12\tau} > \frac{\Gamma - 1}{(x+1)\Gamma}. \tag{5.89}$$

The maximum value of the left-hand side of Equation (5.89) is $\frac{1}{4}$, and therefore convection occurs when

$$\frac{x+1}{4} > \frac{\Gamma-1}{\Gamma}. \tag{5.90}$$

Equation (5.90) implies that if Γ equals $\frac{5}{3}$, then the condition for convection is $x > \frac{3}{5}$.

Convective envelopes contain extensive ionization zones. Thermodynamic variables such as c_p and Γ are modified appreciably as a consequence of partial ionization. We consider a partially ionized hydrogen gas. Let n_H, n_p and n_e equal the number densities of hydrogen atoms, protons and electrons, respectively. Using the ideal gas equation $\beta P = nkT$, with β equal to the gas pressure divided by the total pressure P, and $n = n_H + n_p + n_e$, the Saha equation, given as Equation (3.182) becomes

$$\frac{x_H}{x_p x_e} = \beta P \frac{g_H}{g_p g_e} \left(\frac{m_H}{m_p m_e}\right)^{3/2} (2\pi)^{3/2} \frac{h^2}{(kT)^{5/2}} e^{\chi/kT}, \tag{5.91}$$

with $x_H = n_H/n$, $x_p = n_p/n$, $x_e = n_e/n$ and χ equal to the ionization potential of hydrogen. The gas density ρ can be expressed as

$$\rho = \frac{\beta P \mu m_H}{kT},$$

with the molecular weight $\mu = n_H/n + 0.5\, n_p/n$.

From the first law of thermodynamics the specific heat c_p becomes

$$c_p = T\left(\frac{dS}{dT}\right)_P = \left(\frac{\partial U}{\partial T}\right)_P + P\left(\frac{\partial V}{\partial T}\right)_P$$

$$= \left(\frac{\partial U}{\partial T}\right)_P + \frac{P\delta}{\rho T}, \tag{5.92}$$

with $V = 1/\rho$ and

$$\delta \equiv -\left(\frac{\partial \ln \rho}{\partial \ln T}\right)_P.$$

The internal energy per gram is

$$U = \frac{1}{\mu m_H} \frac{3}{2} kT + \frac{\chi n_H}{\mu m_H n} + \frac{aT^4}{\rho}, \tag{5.93}$$

with $\chi = 13.6\,\text{eV}$, and the adiabatic temperature gradient ∇_{ad} is

$$\nabla_{ad} \equiv \left(\frac{\partial \ln T}{\partial \ln P}\right)_S = \frac{P}{T}\left(\frac{\partial T}{\partial P}\right)_S$$

$$= -\frac{P}{T}\frac{\left(\frac{\partial S}{\partial P}\right)_T}{\left(\frac{\partial S}{\partial T}\right)_P} = -\frac{P}{c_p}\left(\frac{\partial V}{\partial T}\right)_P.$$

$$= -\frac{P}{c_p \rho T}\left(\frac{\partial \ln \rho}{\partial \ln T}\right)_P = \frac{P\delta}{c_p \rho T}. \tag{5.94}$$

The luminosity L_r is a constant throughout a static model for a convective envelope. It follows that the structure of a convective envelope is determined by the equations of hydrostatic equilibrium and mass conservation given by Equations (2.27) and (2.28), respectively, and if superadiabatic gradients are small, by the equation

$$\frac{dT}{dr} = \nabla_{ad}\frac{T}{P}\rho\frac{GM_r}{r^2}, \tag{5.95}$$

with ∇_{ad} defined in Equation (5.94). In the convective envelopes of luminous red giants, temperature gradients become appreciably superadiabatic and the adiabatic mixing-length theory discussed in Section 2.5 must be modified so as to account for radiative losses from rising and falling convective fluid elements.

Nonadiabatic mixing-length theory

Upwards-moving convective elements (blobs of gas) have an excess thermal energy relative to the surrounding gas, and conversely, downward-moving elements have a deficiency of thermal energy. It follows that rising and falling convective elements will radiatively exchange energy with the surrounding gas. In developing a theory of non-adiabatic convection we will need in addition to the adiabatic temperature gradient, ∇_{ad}, given in Equation (5.94), the temperature gradient of the surrounding gas ∇, the temperature gradient of convective elements, ∇', and the radiative temperature gradient that would exist if convection were suppressed, ∇_R. These four temperature gradients satisfy the inequalities

$$\nabla_R \geq \nabla \geq \nabla' \geq \nabla_{ad}. \tag{5.96}$$

From the equation of state of an ideal, nondegenerate gas we have

$$\ln \rho = \ln P - \ln T + \ln \mu - \ln k. \tag{5.97}$$

From Equation (5.97) we obtain

$$d\ln \rho = d\ln P - Q d\ln T, \tag{5.98}$$

with

$$Q = 1 - \left(\frac{\partial \ln \mu}{\partial \ln T}\right)_P.$$

Because there is pressure equilibrium between a convective element and the surrounding gas, the density difference becomes

$$\delta\rho = -Q\rho\frac{\delta T}{T}. \tag{5.99}$$

The buoyancy force per unit volume f_b is $-g\delta\rho$, and therefore Equation (5.99) implies that

$$f_b = gQ\rho\frac{\delta T}{T} = \frac{gQ\rho}{T}\left[-\frac{dT}{dr} - \left(-\frac{dT}{dr}\right)'\right]\Delta r, \tag{5.100}$$

where Δr is the displacement of a convective element, dT/dr is the temperature gradient of the surrounding gas, and $(dT/dr)'$ is the temperature gradient of a convective element. The temperature gradients can be written as

$$\frac{dT}{dr} = \frac{P}{\lambda}\frac{dT}{dP} = \frac{T}{\lambda}\nabla \tag{5.101}$$

and

$$\left(\frac{dT}{dr}\right)' = \frac{T}{\lambda}\nabla' \tag{5.102}$$

with $\lambda = -P(dr/dP)$.

Equations (5.101) and (5.102) imply that the buoyancy force given in Equation (5.100) is

$$f_b = gQ\rho(\nabla - \nabla')\frac{\Delta r}{\lambda}. \tag{5.103}$$

Integrating f_b over a mixing length l, we obtain

$$\int_0^l f_b d(\Delta r) = \frac{gQ\rho}{2}(\nabla - \nabla')\frac{l^2}{\lambda}. \tag{5.104}$$

Equating the work performed on a convective element given in Equation (5.104) to the kinetic energy, as already done in the derivation of convective flux under adiabatic conditions, given in Section 2.5, we find that

$$v_c = \left[\frac{gQ}{\lambda}(\nabla - \nabla')\right]^{1/2}l. \tag{5.105}$$

From Equation (5.105) the convective flux becomes

$$F_c = \rho c_p v_c \delta T = \rho c_p v_c\left[-\frac{dT}{dr} - \left(-\frac{dT}{dr}\right)'\right]\Delta r$$

$$= \rho c_p T v_c(\nabla - \nabla')l/\lambda. \tag{5.106}$$

Substituting the value for the mean convective velocity v_c, given in Equation (5.105), into Equation (5.106) gives

$$F_c = (gQ\lambda)^{1/2}\rho c_p T(\nabla - \nabla')^{3/2}\left(\frac{l}{\lambda}\right)^2. \tag{5.107}$$

The temperature gradient ∇_R, which is the temperature gradient that would exist in the absence of convection, can be obtained from the equation

$$\frac{L}{4\pi r^2} = F_c + F_R = \frac{16\sigma T^4}{3\kappa\rho\lambda}\nabla_R, \tag{5.108}$$

where F_c and F_R are equal to the convective and radiative fluxes, respectively. Using the equation of hydrostatic equilibrium and Equation (5.108) we have

$$\nabla_R = \frac{3\kappa LP}{16\pi acGM_r T^4}. \tag{5.109}$$

Next we will express the temperature gradients ∇ and ∇' in terms of the temperature gradients ∇_R and ∇_{ad}. A rising convective element will lose energy to its surroundings by radiation. We define the parameter γ to be the ratio of the excess energy content of a convective element after moving through a mixing length l divided by the energy lost by radiation during its lifetime. The excess energy content of a convective element is proportional to $\nabla - \nabla'$. If the convective element had moved adiabatically, the excess energy content would be proportional to $\nabla - \nabla_{ad}$. It follows that the energy loss caused by radiation is proportional to

$$(\nabla - \nabla_{ad}) - (\nabla - \nabla') = \nabla' - \nabla_{ad}. \tag{5.110}$$

Because the excess energy content is proportional to $\nabla - \nabla'$ and the radiative energy loss is proportional to $\nabla' - \nabla_{ad}$, we have

$$\gamma = \frac{\nabla - \nabla'}{\nabla' - \nabla_{ad}}. \tag{5.111}$$

For a convective element of volume V and excess temperature δT, the excess energy content is $\rho c_p \delta TV$. The lifetime of a rising convective element is l/v_c. We assume that the rising element is optically thick so that radiative losses can be calculated by the diffusion approximation. If the surface area of a convective element is A and we approximate $-dT/dr \simeq \delta T/l$ with l equal to the mixing length, then we obtain

$$\gamma = \frac{\rho c_p \delta TV}{(16\sigma T^3/3\kappa\rho)(\delta T/l)A(l/v_c)}$$

$$= (\rho c_p v_c/16\sigma T^3)3\kappa\rho\left(\frac{V}{A}\right). \tag{5.112}$$

Equating the expressions for γ, given in Equations (5.111) and (5.112), and setting the ratio $V/A = l/3$ gives

$$\frac{\kappa\rho^2 c_p Tv_c l}{16\sigma T^4} = (\nabla - \nabla')/(\nabla' - \nabla_{ad}). \tag{5.113}$$

The radiative flux F_R can be expressed as

$$F_R = \frac{16}{3} \frac{\sigma T^4}{\kappa \rho \lambda} \nabla. \tag{5.114}$$

From Equations (5.108), (5.114) and (5.107), we have

$$\frac{16\sigma T^4}{3\kappa \rho \lambda}(\nabla_R - \nabla) = (gQ\lambda)^{1/2} \rho c_p T (\nabla - \nabla')^{3/2} \left(\frac{l}{\lambda}\right)^2. \tag{5.115}$$

It is convenient to write Equation (5.115) as

$$U(\nabla_R - \nabla) = (\nabla - \nabla')^{3/2}, \tag{5.116}$$

with

$$U = \frac{16\sigma T^4 \lambda}{3\kappa \rho^2 l^2 (gQ\lambda)^{1/2} c_p T}.$$

Substituting the value for v_c given in Equation (5.105) into Equation (5.113) gives

$$\nabla' - \nabla_{ad} = 3U(\nabla - \nabla')^{1/2}. \tag{5.117}$$

Equation (5.117) can be written as

$$[(\nabla - \nabla')^{1/2}]^2 + 3U(\nabla - \nabla')^{1/2} - (\nabla - \nabla_{ad}) = 0. \tag{5.118}$$

Solving the above quadratic equation for $(\nabla - \nabla')^{1/2}$, we obtain

$$(\nabla - \nabla')^{1/2} = -\frac{3U}{2} + \frac{1}{2}\sqrt{9U^2 + 4(\nabla - \nabla_{ad})}. \tag{5.119}$$

Substituting Equation (5.119) into Equation (5.116) gives

$$\nabla - \nabla_R + \frac{1}{U}\left[\sqrt{\nabla - \nabla_{ad} + 9U^2/4} - \frac{3}{2}U\right]^3 = 0. \tag{5.120}$$

If we let $y = \sqrt{\nabla - \nabla_{ad} + 9U^2/4}$, then Equation (5.120) can be readily expressed as a cubic equation for y and an explicit expression for y obtained from the standard formula for solving a cubic equation. Because ∇_{ad} and U are known, the unknown variable ∇ is determined once the solution for y has been found.

If P_g is the gas pressure, P_R the radiation pressure and L_R the radiative luminosity, then the equations of hydrostatic equilibrium and radiative transfer can be expressed as

$$\frac{dP_g}{dr} + \frac{dP_R}{dr} = -\frac{GM_r \rho}{r^2} \tag{5.121}$$

and

$$\frac{L_R}{4\pi r^2} = -\frac{c}{\kappa \rho} \frac{dP_R}{dr}, \tag{5.122}$$

respectively. Equations (5.121) and (5.122) can be combined to give the equation

$$\frac{P_g}{\rho}\frac{d\rho}{dr} - \frac{P_g}{\mu}\frac{d\mu}{dr} = \frac{1}{r^2}\left[\frac{\kappa\rho L_R}{4\pi c}\left(\frac{P_g}{4P_R}+1\right) - GM_r\rho\right].$$ (5.123)

If molecular-weight gradients ($d\mu/dr$) can be neglected, then Equation (5.123) can be written as

$$\frac{d\rho}{dr} = \frac{1}{r^2}\frac{\mu m_H \rho}{kT}\left[\frac{\kappa L_R}{4\pi c}\left(\frac{P_g}{4P_R}+1\right) - GM_r\right].$$ (5.124)

If P_R is large compared to P_g, then the density gradient $d\rho/dr$ given in Equation (5.124) will satisfy the condition $d\rho/dr \geq 0$, when L_R exceeds the critical luminosity

$$(L_R)_c = \frac{4\pi cGM_r}{\kappa} = \frac{1.32 \times 10^4}{\kappa}\frac{M_r}{M_\odot}L_\odot$$ (5.125)

and hydrostatic equilibrium is not possible. The critical Eddington luminosity $(L_R)_c$ is 0.75 times the Eddington luminosity, given in Equation (2.138). In stellar interiors the total luminosity, which is the sum of the radiative and convective luminosities, can exceed $(L_R)_c$. Thermonuclear runaways in helium-burning shells can cause the critical luminosity L_c to be exceeded in the interior regions of luminous red giants because convection cannot transport most of the energy flux. Convection must become inefficient when convective velocities exceed the sound velocity because shock waves which dissipate energy rapidly are produced by the supersonic motions of convective elements. It follows that the maximum convective flux $(L_c)_{max}$ is

$$(L_c)_{max} \simeq 4\pi r^2 \rho U v_s,$$ (5.126)

where v_s is the local sound velocity and U is the internal energy per gram. In ionization zones and also in the outermost layers of a red giant where molecular hydrogen is dissociated, molecular-weight gradients are present, and therefore from Equation (5.123) it can readily be shown that $d\rho/dr$ can become positive and the star remain in hydrostatic equilibrium.

The equation of state of a mixture of nondegenerate gas and radiation can be expressed as

$$\ln\rho = \ln\beta + \ln P - \ln T + \ln\mu - \ln k,$$ (5.127)

with $\beta = P_g/P$. The local polytropic index n is defined by the equation

$$1 + \frac{1}{n} \equiv \frac{d\ln P}{d\ln\rho}.$$ (5.128)

Equations (5.127) and (5.128) imply that

$$n + 1 = \left(\nabla - \frac{d\ln\beta}{d\ln P} - \frac{d\ln\mu}{d\ln P}\right)^{-1}.$$ (5.129)

From Equation (5.129) it follows that n is small if temperature gradients ∇ become sufficiently superadiabatic. As the polytropic index n becomes small, the density distribution in the convective envelope becomes relatively uniform.

5.7   Stellar winds

Stellar winds have been observed from a number of different types of stars. In a main-sequence star with a convective envelope the motions of convective elements produce waves that dissipate in the corona. Although the wave-energy input is much less than the energy radiated from the photosphere, the resultant heating of the corona is sufficient to produce temperatures of 1–2×10^6 K in a $1M_{\odot}$ star because coronal densities are low, and consequently radiative losses are also low. In the solar corona, energy transport is primarily by means of electron thermal conduction.

Under conditions of spherical symmetry and steady outflow the equations of continuity (mass conservation), momentum and energy are

$$\nabla \cdot (\rho \boldsymbol{v}) = r^{-2} \frac{dr^2 \rho v}{dr} = 0, \tag{5.130}$$

$$\rho v \frac{dv}{dr} = -\frac{dP}{dr} - \rho \frac{GM}{r^2}, \tag{5.131}$$

and

$$\nabla \cdot \left[\rho \boldsymbol{v} \left(\frac{1}{2} v^2 + U + \frac{P}{\rho} \right) \right] = \frac{1}{r^2} \frac{d}{dr} r^2 \rho v \left(\frac{1}{2} v^2 + U + \frac{P}{\rho} \right)$$

$$= -\rho v \frac{GM}{r^2} + \frac{1}{r^2} \frac{d}{dr} r^2 \kappa \frac{dT}{dr}, \tag{5.132}$$

with U equal to the internal energy density, κ equal to the electron thermal conductivity given in Equation (3.364) and $U + P/\rho$ the enthalpy of the gas.

Integrating Equation (5.130), we obtain

$$r^2 \rho v = \text{constant}. \tag{5.133}$$

Because ρv is a function of r, Equation (5.132) can also be integrated, and we obtain

$$4\pi r^2 \rho v \left(\frac{1}{2} v^2 + U + P/\rho - \frac{GM}{r} \right) - 4\pi r^2 \kappa \frac{dT}{dr} = \text{constant}. \tag{5.134}$$

Equations (5.131), (5.133) and (5.134) determine the outflow if boundary conditions are known. Four boundary conditions are required to solve Equations (5.131), (5.133) and (5.134) because they contain two constants and Equations (5.131) and (5.134) are first-order nonlinear differential equations. If the mass-loss rate

$$\dot{M} = 4\pi r^2 \rho v = \text{constant} \tag{5.135}$$

is specified, then the solutions of Equations (5.131) and (5.134) are determined if the temperature T and density ρ (or pressure P) are known at some radial distance r and if the additional boundary condition $dv/dr \to 0$ as $r \to \infty$ is also imposed.

Equation (5.133) is equivalent to the differential equation

$$\frac{2}{r} + \frac{1}{\rho}\frac{d\rho}{dr} + \frac{1}{v}\frac{dv}{dr} = 0. \tag{5.136}$$

If the molecular weight μ is assumed independent of r, then from the equation of state of an ideal, nondegenerate gas, we have

$$\frac{1}{P}\frac{dP}{dr} = \frac{1}{\rho}\frac{d\rho}{dr} + \frac{1}{T}\frac{dT}{dr}. \tag{5.137}$$

Using Equation (5.137) to eliminate the dP/dr term from Equation (5.131) and then using Equation (5.136) to eliminate $d\rho/dr$, we have

$$\left(v^2 - \frac{P}{\rho}\right)\frac{dv}{dr} = \frac{2vP}{r\rho} - \frac{vGM}{r^2} + \frac{vP}{\rho T}\frac{dT}{dr} \tag{5.138}$$

where dT/dr is given by Equation (5.134). The derivative dv/dr in Equation (5.138) is not determined when the condition

$$v^2 = \frac{P}{\rho} \tag{5.139}$$

is satisfied (i.e. when the flow velocity v is equal to the isothermal sound velocity $v = \sqrt{P/\rho}$). The radial distance r_c at which this condition (5.139) is satisfied is known as the critical radius of the stellar-wind solution. When the condition (5.139) is satisfied, then the right-hand side of Equation (5.138) must also equal zero.

Equations (5.138), (5.134), (5.135) and (5.137), the boundary conditions and the two conditions between the dependent variables that apply at the critical point determine a particular stellar-wind solution. Because Equation (5.138) contains a critical point, it cannot simply be integrated from some radial distance r_1 in the corona to $r = \infty$ but must be expanded about the critical point and integrations performed from r_1 to some radial distance r_- slightly interior to r_c and from some radial distance r_+ slightly exterior to r_c to $r = \infty$.

The discussion of stellar-wind solutions can be simplified somewhat if it is assumed that the relationship between pressure and density has the form

$$P = \kappa\rho^\Gamma, \tag{5.140}$$

with Γ equal to a constant. For adiabatic outflow of an ionized gas, $\Gamma = \frac{5}{3}$. Clearly Γ must satisfy the inequality $\frac{5}{3} \geq \Gamma \geq 1$, where $\Gamma = 1$ represents isothermal outflow. Substituting Equation (5.140) into Equation (5.131), we have

$$\rho v\frac{dv}{dr} + \Gamma\kappa\rho^{\Gamma-1}\frac{d\rho}{dr} + \rho\frac{GM}{r^2} = 0. \tag{5.141}$$

Dividing Equation (5.141) by ρ and re-expressing the second term, we obtain

$$v\frac{dv}{dr} + \frac{\Gamma}{\Gamma-1}\kappa\frac{d\rho^{\Gamma-1}}{dr} + \frac{GM}{r^2} = 0 \quad (\Gamma \neq 1). \tag{5.142}$$

Integrating Equation (5.142) gives Bernoulli's equation,

$$\frac{1}{2}v^2 + \frac{\Gamma}{\Gamma-1}\frac{P}{\rho} - \frac{GM}{r} = \frac{1}{2}v_\infty^2 \quad (\Gamma \neq 1), \tag{5.143}$$

where we have assumed that the pressure P at $r = \infty$ is equal to 0.

Using Equation (5.136) to eliminate $d\rho/dr$ from Equation (5.141), we have

$$\left[v^2 - \Gamma\frac{P}{\rho}\right]\frac{dv}{dr} = \frac{2\Gamma Pv}{\rho r} - \frac{GMv}{r^2}. \tag{5.144}$$

Equation (5.144) has a critical point at a radial distance $r = r_c$ when

$$v^2 = \Gamma\frac{P}{\rho}. \tag{5.145}$$

At the critical point r_c the right-hand side of Equation (5.144) is also equal to zero, and therefore we have

$$r_c = \frac{GM\rho}{2\Gamma P}. \tag{5.146}$$

Stellar winds from cool main-sequence stars do not lead to significant amounts of mass loss. The mass-loss rate caused by the solar wind is $\simeq 10^{-13} - 10^{-14}\,M_\odot/\text{yr}$. Because gas densities are low, the collisional mean free paths between electrons, protons and ions become much larger than the dimension of the solar system. A two-fluid model in which the stellar-wind equations are solved separately for electron and proton outflows is required to describe the physical behavior of the solar wind at radial distances such that electron–proton collisional mean free paths are large. Because the solar wind contains magnetic fields, and proton and electron gyroradii are small compared to density scale lengths, the fluid equations remain applicable. Electrons transport heat in the solar wind, and consequently the electron temperature distribution as a function of radial distance from the Sun is more nearly isothermal than the corresponding proton temperature distribution. In the absence of collisionless interactions between electrons and protons the proton temperature would decrease adiabatically (i.e. $T\rho^{-(\Gamma_1-1)} = \text{constant}$) when gas densities become sufficiently low that energy exchange between electrons and protons can be neglected.

Very massive stars on and off the main sequence have sufficiently high luminosities that radiation pressure causes their outermost atmospheric layers to expand supersonically. Such stellar winds are driven primarily by spectral-line absorption at ultraviolet wavelengths.

The momentum flux per unit time of an atmospheric radiation field is

$$\frac{L}{4\pi r^2 c} = \frac{1}{c} \int F_\nu d\nu, \tag{5.147}$$

with F_ν equal to the radiative flux. It follows that the outward radiative force per gram on the gas is

$$g_R = \frac{1}{\rho c} \int \alpha_\nu F_\nu d\nu, \tag{5.148}$$

where α_ν is the absorption coefficient and ρ the gas density. If g_R exceeds the inward gravitational acceleration g, then the outer layers of the stellar atmosphere cannot remain in hydrostatic equilibrium and a stellar wind results. Equation (5.148) is valid at frequencies such that the optical depth τ_ν is $\lesssim 1$.

In stellar winds, most of the acceleration of the gas occurs close to the critical point which separates subsonic from supersonic outflow velocities. Except at large distances from the star, the outflow velocity changes as a function of radial distance r because the gas is accelerated. Close to the critical radius r_c we have, approximately,

$$\frac{dv}{dr} \sim \frac{v_s}{r_c}, \tag{5.149}$$

where v_s is the speed of sound at the critical radius r_c. Consider a spectral line whose line center occurs at a frequency ν_l and for which the optical depth at line center is large. For strong lines, radiation is absorbed within a distance

$$r_a = \Delta v \frac{dr}{dv}, \tag{5.150}$$

where Δv is the line width in units of velocity. The velocity line width Δv and line width $\Delta \nu$ are related by the expression $\Delta v/c = \Delta \nu/\nu$. From Equation (5.150) the radiative force per cm^2 on the gas becomes

$$\rho g_R^l r_a \simeq \rho g_R^l \Delta v \frac{dr}{dv} = \frac{F_\nu \Delta \nu}{c}, \tag{5.151}$$

where g_R^l is the acceleration caused by a spectral line. Equations (5.151) and (5.149) and the relation between velocity width and frequency width imply that

$$g_R^l \simeq \frac{\nu_l F_\nu v_s}{\rho c^2 r_c}. \tag{5.152}$$

5.8   Molecular-line emission from stellar winds

Molecular-line emission has been observed from luminous red giants that are undergoing rapid mass loss. Long-period variables are luminous red giants that pulsate with periods ranging from approximately three months to more than two years. The physical basis for the excitation of pulsations will be discussed in Chapter 8. Long-period variables include stars with normal [C]/[O] abundance ratios and carbon stars, which have [C]/[O] abundance ratios > 1. Stellar winds from long-period variables can be produced in the following manner. Pulsations produce shock waves that eject mass from the photo-spheres of the stars. As the ejected mass cools, dust formation occurs, and radiation pressure acting on these dust particles leads to a steady outflow of gas at radial distances much greater than the photospheric radius. The physical properties of these stellar winds can be inferred from a comparison between theoretical models and infrared continuum radiation from dust and molecular line emission. In the discussion below we will derive the equations that describe the rotational line emission of the CO molecule, which next to the H_2 molecule is the most abundant molecule in stellar winds from red giants. Rotational line emission from the CO molecule has been observed at both millimeter and submillimeter wavelengths. The $J = 1 \rightarrow 0$ rotational transition of CO occurs at a wavelength of 2.6 mm. The wavelengths of rotational transitions between energy levels with higher J values can be readily inferred from the description of rotational energy levels given in Section 3.5.

Calculation of CO emission line strengths requires solution of the equations of radiative transfer for CO line emission, the equations of statistical equilibrium for the CO-level populations and an energy-conservation equation. The radiative-transfer equation for a particular transition of frequency $v = v_{J+1,J}$ is

$$\frac{dI_v}{ds} = -\alpha_v I_v + j_v, \tag{5.153}$$

with α_v and j_v given by Equations (5.55) and (5.56), respectively, and at a small distance along some particular direction.

Equation (5.153) can be solved if the populations of the rotational energy levels of the CO molecule, gas temperature and the ratio n_{CO}/n_{H_2} are given. The populations of the rotational energy levels of CO and gas temperature are determined by the rate equations for the excitation and deexcitation of the rotational energy levels of CO and the energy-conservation equation. Because both the equations of statistical equilibrium and energy conservation depend on the mean intensity of radiation and the intensity of radiation is determined by solving Equation (5.153), an iterative method of solution is required.

Statistical equilibrium exists when the number of radiative and collisional transitions out of each energy level equals the corresponding number of transitions into that energy level. Statistical equilibrium exists in the outflowing molecular gas; however, the more stringent condition of local thermodynamic equilibrium does not exist. The rate equations describing statistical equilibrium between CO rotational energy levels are

$$n_{H_2} \sum_{J=0}^{J_{max}-1} \sum_{J'=J+1}^{J_{max}} [n_{J'} \langle \sigma v \rangle_{J,J'} - n_J \langle \sigma v \rangle_{J',J}]$$

$$+ \sum_{J=0}^{J_{max}} \{ [A_{J+1,J} + \bar{I}_{J+1,J} B_{J+1,J}] n_{J+1} - \bar{I}_{J+1,J} B_{J,J+1} n_J \} = 0 \tag{5.154}$$

where $\langle \sigma v \rangle_{J,J'}$ is the collisional rate coefficient between the J' and J energy levels and

$\bar{I}_{J+1,J}$ is the mean intensity of the radiation field at the frequency $\nu_{J+1,J}$. The highest rotational energy level $J_{\max}$ that must be included in the calculations depends on the temperature of the gas. The rate coefficient $\langle \sigma v \rangle_{J,J'}$ can be expressed as

$$\langle \sigma v \rangle_{J,J'} = 4\pi \int \nu \sigma_{J,J'}(v) f(v) v^2 dv, \tag{5.155}$$

with $\sigma_{J,J'}$ equal to the cross section for a collisional transition between the J' and J levels, and $f(v)$ equal to the Maxwell–Boltzmann distribution function. Collisional excitation and de-excitation of the CO rotational energy levels are caused primarily by collisions between H_2 and CO molecules.

It follows from the principle of detailed balance that under conditions of local thermodynamic equilibrium the collisional rate coefficients for excitation and de-excitation by levels J' and J satisfy the equation

$$n_J \langle \sigma v \rangle_{J',J} = n_{J'} \langle \sigma v \rangle_{J,J'}. \tag{5.156}$$

Under conditions of local thermodynamic equilibrium the ratio $n_{J'}/n_J$ is given by the Boltzmann relation, and therefore Equation (5.156) becomes

$$\langle \sigma v \rangle_{J',J} = \frac{g_{J'}}{g_J} \langle \sigma v \rangle_{J,J'} e^{-\frac{h\nu_{J,J'}}{kT}} \quad (J' > J). \tag{5.157}$$

Equation (5.157) implies that if a particular collisional rate coefficient has been calculated, then the corresponding collisional de-excitation rate coefficient is also known and vice versa.

Most of the outflowing gas from a red giant is molecular hydrogen and consequently the gas density ρ is $\rho \simeq n_{H_2} m_{H_2}$. The heat capacity at constant volume per molecule of a molecular hydrogen gas is

$$c_v(T) = \frac{3}{2}k + c_{vib}(T) + c_{rot}(T), \tag{5.158}$$

where the translational, vibrational and rotational heat capacities given in Equation (5.158) have been discussed in Section 3.5. The energy-conservation equation for steady-state outflow is

$$v\frac{dU}{dr} + Pv\frac{d\left(\frac{1}{\rho}\right)}{dr} = \frac{\Gamma - \Lambda}{\rho}, \tag{5.159}$$

where Γ and Λ are the rates of energy gain and loss, respectively, in units of $\text{erg cm}^{-3}\text{s}^{-1}$. At sufficiently large radial distances from the star the radial velocity of the gas is approximately independent of r, and therefore from the equation of mass conservation given in Equation (5.136) we have

$$\frac{dn_{H_2}}{n_{H_2}} = -2\frac{dr}{r}. \tag{5.160}$$

Equation (5.160) implies that Equation (5.159) can be written as

$$c_v n_{H_2} v \frac{dT}{dr} = -\frac{2n_{H_2} vkT}{r} + \Gamma - \Lambda, \tag{5.161}$$

where the first term on the right-hand side of Equation (5.161) represents expansive cooling, Γ represents heating caused by gas–dust collisions and Λ represents cooling caused by radiative and collisional excitation and de-excitation of the CO molecules. The rate of energy exchange, $-\Lambda = -\Lambda_{CO}$ between the CO molecule, surrounding gas and radiation field is

$$-\Lambda_{CO} = n_{H_2} \sum_{J=0}^{J_{max}-1} \sum_{J'=J+1}^{J_{max}} [n_J \langle \sigma v \rangle_{J',J} - n_{J'} \langle \sigma v \rangle_{J,J'}] h\nu_{J',J}$$

$$- \sum_{J=0}^{J_{max}} [n_{J+1}(A_{J+1,J} + \bar{I}_{J+1} B_{J+1,J}) - n_J \bar{I}_{J+1} B_{J,J+1}] h\nu_{J+1,J}, \tag{5.162}$$

where $A_{J+1,J}$, $B_{J+1,J}$ and $B_{J,J+1}$ are the Einstein coefficients and $\bar{I}_{J+1}$ the mean intensities of the radiation field.

Radiation from the central star and inner region of the expanding circumstellar envelope accelerates dust particles (grains) with respect to the outflowing molecular gas. The grains reach a terminal velocity with respect to the surrounding molecular gas when the radiative force becomes equal in magnitude and opposite in sign to the drag force caused by collisions with the surrounding gas. If m_g, r_g and v_g are the mass, radius and grain drift velocity, respectively, then the force on a grain is

$$m_g \frac{dv_g}{dt} = \frac{Le^{-\tau(r)}}{4\pi r^2 c} \pi r_g^2 Q_p - \frac{v_g}{t_s}, \tag{5.163}$$

where Q_p is the fraction of incident photon momentum that is transferred to a grain, $e^{-\tau(r)}$ is the fraction of radiation from star and inner region of outflowing gas that reaches radius r and t_s, the stopping time, is the timescale for the grain to slow down as a result of collisions with the surrounding gas. Since the optical depth τ and Q_p depend on the frequency ν, we can divide the frequency coordinate ν into intervals centered at frequencies ν_i with $i = 1, 2, \ldots, n$ and let

$$Le^{-\tau(r)} Q_p = \sum_{i=1}^{n} L(\nu_i) e^{-\tau_i(r)} Q_p(\nu_i), \tag{5.164}$$

where $L(\nu_i)$ is the power emitted in some frequency interval $\Delta\nu_i$ by the central star and inner region of outflowing dust.

If we assume that the drift velocity of a grain v_g is 0 at some time $t = 0$ and that the grain attains its limiting drift velocity before traveling a distance comparable to r, then the solution to Equation (5.163) becomes

$$v_g = v_D(1 - e^{-t/t_s}), \tag{5.165}$$

where

$$v_D = \frac{Le^{-\tau(\tau)}r_g^2 Q_p t_s}{4r^2 c}.$$

In stellar winds from red giants the stopping time t_s, which will be derived below, is sufficiently short that grains move at the limiting drift velocity v_D, given in Equation (5.165).

To calculate the force on a grain and determine t_s we must make some assumptions about the geometry of the grain. We assume that the grains are cylinders of circular radius r_g and that the axes of symmetry of these cylinders are aligned along the direction of gas outflow. From considerations of symmetry it follows that net acceleration occurs only as a result of collisions on the front and back side of the cylinder. Therefore, the force on a grain caused by elastic collisions with the surrounding gas is

$$m_g \frac{dv_D}{dt} \equiv m_g \frac{v_D}{t_s}$$

$$= n_{H_2} m_{H_2} \pi r_g^2 \int_{-v_D}^{\infty} \int_{-\infty}^{+\infty} \int_{-\infty}^{+\infty} 2(v_z + v_D)^2 f(v_x, v_y, v_z) dv_x dv_y dv_z$$

$$- n_{H_2} m_{H_2} \pi r_g^2 \int_{v_D}^{\infty} \int_{-\infty}^{+\infty} \int_{-\infty}^{+\infty} 2(v_z - v_D)^2 f(v_x, v_y, v_z) dv_x dv_y dv_z, \qquad (5.166)$$

where z is the direction of grain motion and $f(v_x, v_y, v_z)$ the Maxwell–Boltzmann distribution function, given in Equation (3.13). The first of the two integrals on the right-hand side in Equation (5.166) is the force on the front side of the grain and the second integral is the force on the back side of the grain. In obtaining Equation (5.166) we have assumed that the mean free path of a molecule is appreciably larger than the dimension of a grain.

Equation (5.166) can be re-expressed as

$$m_g \frac{v_D}{t_s} = 2n_{H_2} m_{H_2} \pi r_g^2 \left(\frac{m_{H_2}}{2\pi kT}\right)^{1/2} \left\{ 4 \int_0^\infty v_z v_D e^{-\frac{m_H v_z^2}{2kT}} dv_z \right.$$

$$\left. + \int_{-v_D}^{v_D} (v_z^2 + v_D^2) e^{-\frac{m_H v_z^2}{2kT}} dv_z \right\}. \qquad (5.167)$$

In the limit of large and small drift velocities, Equation (5.167) reduces to the equations

$$m_g \frac{v_D}{t_s} = 2n_{H_2} m_{H_2} \pi r_g^2 v_D^2 \quad (v_D \gg \bar{v}) \qquad (5.168)$$

and

$$m_g \frac{v_D}{t_s} = 8\pi r_g^2 n_{H_2} m_{H_2} v_D \left(\frac{8kT}{\pi m_{H_2}}\right)^{1/2} \quad (v_D \ll \bar{v}), \qquad (5.169)$$

with $\bar{v} = \left(8kT/\pi m_{H_2}\right)^{1/2}$.

If v_z is the z-component of the velocity of a molecule as measured in the reference frame of the gas, then the change in the kinetic energy of a molecule as a result of elastic collisions with the front and back sides of the grain are

$$\Delta E_1 = \frac{1}{2} m_{H_2} (4v_z v_D + 4v_D^2) \tag{5.170}$$

and

$$\Delta E_2 = \frac{1}{2} m_{H_2} (-4v_z v_D + 4v_D^2), \tag{5.171}$$

respectively. The energy gain caused by collisions with cylindrical grains is

$$\Gamma_g = n_{H_2} n_g \pi r_g^2 \left\{ \int_{-v_D}^{\infty} \int_{-\infty}^{\infty} \int_{-\infty}^{\infty} (v_z + v_D) \Delta E_1 f(v_x, v_y, v_z) dv_x dv_y dv_z \right.$$

$$\left. + \int_{v_D}^{\infty} \int_{-\infty}^{\infty} \int_{-\infty}^{\infty} (v_z - v_D) \Delta E_2 f(v_x, v_y, v_z) dv_x dv_y dv_z \right\}, \tag{5.172}$$

where ΔE_1 and ΔE_2 are given by Equations (5.170) and (5.171), respectively, and $f(v_x, v_y, v_z)$ is the Maxwell–Boltzmann distribution function. The expression for Γ_g given in Equation (5.172) can be readily reduced to the equation

$$\Gamma_g = n_{H_2} n_g \pi r_g^2 \left(\frac{m_{H_2}}{2\pi kT} \right)^{1/2} \left\{ 16 v_D^2 \int_0^{\infty} v_z e^{-\frac{m_{H_2} v_z^2}{2kT}} dv_z \right.$$

$$\left. + 4 v_D^3 \int_{-v_D}^{v_D} e^{-\frac{m_{H_2} v_z^2}{2kT}} dv_z + 4 v_D \int_{-v_D}^{v_D} v_z^2 e^{-\frac{m_{H_2} v_z^2}{2kT}} dv_z \right\}. \tag{5.173}$$

In the limits of large and small drift velocities, Equation (5.173) becomes

$$\Gamma_g = n_{H_2} n_g 4\pi r_g^2 v_D^3 \quad (v_D \gg \bar{v}) \tag{5.174}$$

and

$$\Gamma_g = 8\pi r_g^2 n_{H_2} n_g v_D^2 \left(\frac{8kT}{\pi m_{H_2}} \right)^{1/2} \quad (v_D \ll \bar{v}). \tag{5.175}$$

In the limit $v_D \gg \bar{v} = (8kT/\pi m_{H_2})^{1/2}$, the drift velocity is highly supersonic, and from Equations (5.165) and (5.168) we obtain

$$v_D = \frac{L e^{-\tau(r)} r_g^2 Q_p t_s}{4 r^2 c}, \tag{5.176}$$

with

$$t_s = \frac{m_g}{2 n_{H_2} m_{H_2} \pi r_g^2 v_D}.$$

It follows that v_D is

$$v_D = \left(\frac{L e^{-\tau(r)} Q_p m_g}{8\pi r^2 c n_{H_2} m_{H_2}} \right)^{1/2}.$$ (5.177)

Problems

1. The opacity, effective temperature and gravity of a pure hydrogen stellar atmosphere are $\kappa = 0.4$, $T_{eff} = 10^4$ and $g = 2GM_\odot/R_\odot^2$, respectively. Use the Eddington approximation to determine T and ρ at optical depths $\tau = 0, 1/2, 2/3, 1, 2$. The density ρ equals zero at $\tau = 0$ and κ is in units of $cm^2\,g^{-1}$.

2. The Poynting vector (power per unit area) of a monochromatic electromagnetic wave moving in the $\hat{n}$ direction is

$$c \frac{E \times H}{4\pi},$$

with $H = \hat{n} \times E$ and $E = E_0 \sin \omega t$. What is the time-averaged intensity of radiation of the electromagnetic wave?

3. A star has radius R and uniform brightness $B = \int B_\nu d\nu$, with B_ν equal to the Planck function. Show that the flux F at a distance r from the center of the star is

$$F = \pi B \left(\frac{R}{r} \right)^2.$$

4. A spherical, optically thin cloud of radius R, temperature T and distance $D \gg R$ emits thermal radiation with an emissivity of j_ν ($erg\,cm^{-3}\,s^{-1}\,Hz^{-1}\,ster^{-1}$). Calculate the intensity of radiation I_ν at the cloud center, the effective temperature and the flux F_ν from the cloud. Compare I_ν with the Planck function B_ν if the optical thickness is τ_ν.

5. A stellar absorption line of line-center frequency ν_0 is formed in a thin, cool layer of thickness L. The optical depth is $\tau_\nu = nL\alpha_0\phi_\nu$ with n equal to the number density of absorbers and ϕ_ν equal to the line-profile function.
   a. For an optically thin line show that the equivalent width equals $nL\alpha_0$.
   b. Assume that a strong Doppler-broadened line is uniformly absorbing from the line center out to $3\Delta\nu_D$, where the optical depth is unity. Determine how the equivalent width depends on the number density of absorbers.
   c. For a Lorentzian line that is optically thick ($\tau \geq 1$) to three times the half-width find how the equivalent width depends on the number density of absorbers.

6. Spectral lines are broadened by collisions. Assume that an electron in an atom is radiating at a frequency ν_0 until interrupted by a collision. If t_0 is the time interval during which the electron radiates without a collision occurring, then from Fourier analysis the intensity distribution of the pulse emitted between collisions is

$$\left| \int_{-t_0/2}^{t_0/2} \exp[-2\pi i(\nu_0 - \nu)t]dt \right|^2 \propto \left(\frac{\sin \pi(\nu - \nu_0)t_0}{\pi(\nu - \nu_0)} \right)^2.$$

If the mean time interval between collisions is τ_c then the probability that the atom undergoes a collision in the time interval t_0, $t_0 + dt_0$ is $\exp(-t_0/\tau_c)\,dt_0/\tau_c$. Show that the intensity distribution of a collisionally broadened line is Lorentzian with $\gamma_c = 2/\tau_c$. It follows that the damping constant of a line broadened by radiative and collisional broadening is $\gamma = 1/\tau + 2/\tau_c$, with τ equal to the mean radiative lifetime.

7.   A hydrogen gas has a temperature of 10^4 K, and therefore most atoms are ionized. At what density ρ is the mean lifetime of the 2P state equal to the mean time interval for an atom to be collisionally excited by an electron into the 2P state from the 1S state if the collisional cross section is assumed equal to 5×10^{-17} cm^2. Find the corresponding mean lifetime of the 2P state due to collisional de-excitation. What is the combined natural and collisional width γ of the Lyman α transition at density ρ?

8.   A star's luminosity L exceeds the critical luminosity $L_{crit} = 4\pi GcM/\kappa$. An optically thin gas cloud is initially stationary at a radial distance r. Show that the terminal velocity of the cloud is

$$v^2 = \frac{2GM}{r}\left(\frac{\kappa L}{4\pi GMc} - 1\right).$$

9.   The adiabatic exponent Γ_1 is defined by the equation

$$\frac{dP}{P} + \Gamma_1 \frac{dV}{V} = 0.$$

Find Γ_1 for a mixture of gas and radiation pressure. It is convenient to define $\beta = P_{gas}/P$.

10. For a partially ionized hydrogen gas derive expressions for c_v and c_p (neglect radiation pressure). Show that under conditions of partial ionization c_v and c_p can have values much greater than those of a fully ionized gas.

11. The maximum convective flux is determined by the condition that the convective velocity cannot exceed the local sound velocity. Find the maximum convective fluxes of a hydrogen gas at $T = 10^6$ K and $\rho = 10^{-7}$ g cm^{-3} and a helium gas at $T = 10^7$ K and $\rho = 10^2$ g cm^{-3}. What are the corresponding limiting convective luminosities if the hydrogen and helium gases are at radial distances of 10^{11} cm and 10^9 cm, respectively?

12. Write down the stellar-wind equations for a two-fluid (electron and proton) spherical outflow. Assume that the protons expand adiabatically whereas the electrons expand isothermally.

13. Repeat the derivation of Equation (5.177) for spherical grains of radius a.

14. Consider two small, counterstreaming plane-parallel gas layers of thickness L and uniform density ejected from a star. Both gas layers move along the line of sight. A spectral line is redshifted if emitted from the receding layer and blueshifted if emitted from the approaching layer. The number density of

radiating molecules is n. The temperature distributions of molecules satisfy the relation $T = (T_2 - T_1)(r - R_1)/L + T_1$, with $R_2 - R_1 = L$, where R_2 and R_1 are the outer and inner radial distances of the plane-parallel layers. Assume that the molecular line frequency and temperature T satisfy the condition $h\nu/kT \ll 1$, and the line-center optical depth is ~ 1. Show that for the assumed plane-parallel geometry more redshifted line radiation, which is emitted along a positive temperature gradient, is observed than blueshifted line radiation, which is emitted along a negative temperature gradient.

15. a. Derive the well-known barometric equation

$$P(z) = P(z_0)e^{-mg(z-z_0)/kT},$$

which holds in an isothermal atmosphere with g equal to a constant.

b. In Part a, the pressure $P(z)$ obviously goes to zero as $z \to \infty$. Show that if we take into account the fact that g varies inversely as $1/z^2$, the pressure in an isothermal atmosphere approaches a constant value as $z \to \infty$.

6 Thermonuclear reactions and nucleosynthesis

6.1 Nonresonant thermonuclear reactions

The strong interaction is a short-range force, and therefore nuclear reactions can occur only if nuclei come sufficiently close to each other. Thermonuclear reactions take place when relative particle velocities are high enough for Coulomb barrier penetration between positively charged nuclei to occur with significant probability.

The radial Schrödinger equation that describes the interaction between two nuclei with orbital angular momentum quantum number l is

$$-\frac{\hbar^2}{2\mu}\left(\frac{d^2}{dr^2} + \frac{2}{r}\frac{d}{dr}\right)\psi_l(r) + \left[\frac{l(l+1)\hbar^2}{2\mu r^2} + V(r)\right]\psi_l(r) = E\psi_l(r), \tag{6.1}$$

where $\mu = \dfrac{m_1 m_2}{m_1 + m_2}$.

In Equation (6.1) the potential $V(r)$ is assumed to have radial symmetry. If we substitute $\psi_l(r) = \chi_l(r)/r$, then Equation (6.1) becomes

$$-\frac{\hbar^2}{2\mu}\frac{d^2\chi_l}{dr^2} + \left[\frac{l(l+1)\hbar^2}{2\mu r^2} + V(r) - E\right]\chi_l(r) = 0. \tag{6.2}$$

In stellar interiors, interacting nuclei usually have sufficiently low energy that $l = 0$. Therefore, Equation (6.2) reduces to the equation

$$-\frac{\hbar^2}{2\mu}\frac{d^2\chi(r)}{dr^2} + [V(r) - E]\chi(r) = 0. \tag{6.3}$$

If $V(r)$ is taken to be the long-range Coulomb potential, then Equation (6.3) can be solved by means of the WKB approximation and also by transforming the above equation into parabolic coordinates and solving the resultant confluent hypergeometric equation. These methods of solution will be discussed in Section 6.2. In this section we will describe a simpler but more approximate method of solution which is closely related to the WKB approximation. We assume that $\chi(r)$ in Equation (6.3) can be expressed as

$$\chi(r) = Ae^{i\phi(r)/\hbar}. \tag{6.4}$$

Substituting the above expression for $\chi(r)$ into Equation (6.3) leads to the equation

$$i\hbar \frac{d^2\phi(r)}{dr^2} - \left(\frac{d\phi(r)}{dr}\right)^2 + 2\mu[E - V(r)] = 0. \tag{6.5}$$

If we assume that the term containing $\hbar$ in Equation (6.5) is negligible, then the solution to Equation (6.5) becomes

$$\phi(r) = \pm (2\mu)^{1/2} \int_{r_0}^{r} [E - V(r')]^{1/2} dr'. \tag{6.6}$$

The potential $V(r)$ in Equation (6.6) is the Coulomb potential $V(r) = Z_1 Z_2 e^2/r$, with Z_1 and Z_2 equal to the number of protons in each of the two interacting nuclei. The turning point r_1 is defined to be the radial distance between two interacting nuclei at which $E = V(r_1)$. At radial distances r close to r_1 the solution of the Schrödinger equation implied by Equations (6.6) and (6.4) is not valid. The integrand in Equation (6.6) is a real number for $r > r_1$ and imaginary for $r < r_1$.

If the wave function $\chi(r)$ has been normalized throughout all space, then $\chi^*(r)\chi(r)$ gives the probability per unit radial distance that the incoming nucleus is at radial distance r. It is clear that the physically relevant solution of Equation (6.3) must give a decreasing value of $\chi^*(r)\chi(r)$ as r approaches the radius of the nucleus (i.e. range of the nuclear force). The penetration factor $P_l(E)$ between some radial distance $r = r'$ and $r = r_0$ with r_0 equal to the range of the attractive nuclear force is

$$P_l(E) = \frac{\chi^*(r_0)\chi(r_0)}{\chi^*(r')\chi(r')}. \tag{6.7}$$

From Equations (6.5) and (6.6) it follows that the penetration factor varies as

$$\exp\left(-2 \int_{r_0}^{r'} \left[\frac{2\mu}{\hbar^2}(V(r) - E) \right]^{1/2} dr \right). \tag{6.8}$$

In obtaining Expression (6.8) we have used the solution with the positive sign given in Equation (6.6). For the low-energy interacting nuclei that we consider r_0, the range of the nuclear potential, is much less than the radial distance r_1 at which $E = V(r_1)$, and therefore r_0 can be set equal to zero in Expression (6.8). It follows that for $r' < r_1$ the integral in Expression (6.8) becomes

$$\frac{(2\mu E)^{1/2}}{\hbar} \int_0^{r'} (r_1 - r)^{1/2} r^{-1/2} dr, \tag{6.9}$$

with $r_1 = Z_1 Z_2 e^2/E$. If r' is chosen to be approximately equal to the turning point r_1, then integrating Expression (6.9), we obtain

$$\frac{(2\mu E)^{1/2}}{\hbar} \frac{Z_1 Z_2 e^2}{E} \frac{\pi}{2} = \pi \frac{Z_1 Z_2 e^2}{\hbar v}. \tag{6.10}$$

Expression (6.8) and Equation (6.10) imply that the penetration factor is proportional to

$$\exp\left[-\frac{2\pi Z_1 Z_2 e^2}{hv}\right].$$ (6.11)

Expression (6.11) is known as the Gamow factor.

The cross section σ for a nuclear reaction is

$$\sigma = \frac{\text{number of reactions s}^{-1}}{\text{number of incident particles cm}^{-2}\,\text{s}^{-1}}$$ (6.12)

The value of σ given in Equation (6.12) depends on the relative velocity between the two nuclei as measured in the center-of-mass frame of reference. If n_1 and n_2 are the number densities of two different nuclei 1 and 2, which have relative velocity v in the center-of-mass frame of reference, then the number of reactions $\text{s}^{-1}\,\text{cm}^{-3}$ is

$$r = n_1 n_2 v\sigma(v).$$ (6.13)

If the two nuclei are similar particles with zero or integer spin, then the reaction rate becomes

$$r = \frac{n_1^2}{2}v\sigma(v).$$ (6.14)

The factor of $\frac{1}{2}$ appears in Equation (6.14) because n_1^2 counts pairs of nuclei twice. If the interacting nuclei are identical particles with half-integer spin, then the reaction rate r is

$$r = \frac{n_1^2}{8}v\sigma(v).$$ (6.15)

The factor of $\frac{1}{8}$ appears in Equation (6.15) because the total wave function of the identical half-integer spin nuclei must be antisymmetric with respect to changes in position and spin coordinates. Only if the wave function is symmetric with respect to changes in position coordinates and antisymmetric with respect to changes in spin coordinates will there be a significant probability for the two nuclei to approach each other to within the range of the strong interaction. Since the degeneracy of the $S = 0$ antisymmetric spin state is 1, and the degeneracy of the $S = 1$ symmetric spin state is 3 a factor of $\frac{1}{4}$ appears in Equation (6.15) in addition to the factor of $\frac{1}{2}$ that appears in Equation (6.14).

In stellar interiors interacting nuclei have Maxwell–Boltzmann distributions. If n_1 and n_2 are the number densities and v_1 and v_2 the velocities of interacting nuclei, then the product

$$n_1(v_1)\mathrm{d}v_{1x}\mathrm{d}v_{1y}\mathrm{d}v_{1z}n_2(v_2)\mathrm{d}v_{2x}\mathrm{d}v_{2y}\mathrm{d}v_{2z}$$

$$= n_1\left(\frac{m_1}{2\pi kT}\right)^{3/2}\mathrm{e}^{-\frac{m_1 v_1^2}{2kT}}\,\mathrm{d}v_{1x}\mathrm{d}v_{1y}\mathrm{d}v_{1z}n_2\left(\frac{m_2}{2\pi kT}\right)^{3/2}\mathrm{e}^{-\frac{m_2 v_2^2}{2kT}}\,\mathrm{d}v_{2x}\mathrm{d}v_{2y}\mathrm{d}v_{2z}$$

$$= n_1 n_2\frac{(m_1 m_2)^{3/2}}{(2\pi kT)^3}\mathrm{e}^{-\left(\frac{m_1 v_1^2 + m_2 v_2^2}{2kT}\right)}\mathrm{d}^3 v_1\mathrm{d}^3 v_2$$ (6.16)

is the number density of nuclei 1 in the volume of velocity space $\mathrm{d}^3 v_1$ times the number density of nuclei 2 in the volume of velocity space d^3v_2. If V is the velocity of the center of mass and $v = v_2 - v_1$ the relative velocity between two interacting nuclei, then from Equation (4.4) we have

$$v_1 = V - \frac{m_2}{m_1 + m_2}v \tag{6.17}$$

$$v_2 = V + \frac{m_1}{m_1 + m_2}v. \tag{6.18}$$

Equations (6.17) and (6.18) imply that the right-hand side of Equation (6.16) becomes

$$n_1(v_1)d^3v_1 n_2(v_2)d^3v_2 = n_1 n_2 \frac{(m_1 m_2)^{3/2}}{(2\pi kT)^3} e^{-\frac{(m_1+m_2)V^2}{2kT} - \frac{\mu v^2}{2kT}} d^3v_1 d^3v_2, \tag{6.19}$$

where μ is equal to the reduced mass of the two nuclei. The elements of velocity space d^3v_1 and d^3v_2 are related to the corresponding elements d^3v and d^3V by the transformation

$$d^3v_1 d^3v_2 = J_x J_y J_z d^3V d^3v. \tag{6.20}$$

In Equation (6.20) the Jacobian

$$J_x = \begin{vmatrix} \dfrac{\partial v_{1x}}{\partial V_x} & \dfrac{\partial v_{1x}}{\partial v_x} \\[2mm] \dfrac{\partial v_{2x}}{\partial V_x} & \dfrac{\partial v_{2x}}{\partial v_x} \end{vmatrix} = \begin{vmatrix} 1 & \dfrac{-m_2}{m_1 + m_2} \\[2mm] 1 & \dfrac{m_1}{m_1 + m_2} \end{vmatrix} = 1. \tag{6.21}$$

Similarly the Jacobians J_y and J_z are also equal to 1. Equations (6.19–6.21) imply that the right-hand side of Equation (6.16) becomes

$$n_1 n_2 \left[\left(\frac{m_1 + m_2}{2\pi kT} \right)^{3/2} e^{-\frac{(m_1+m_2)V^2}{2kT}} d^3V \right] \left[\left(\frac{\mu}{2\pi kT} \right)^{3/2} e^{-\frac{\mu v^2}{2kT}} d^3v \right]. \tag{6.22}$$

The integral over velocity of the first bracketed term in Expression (6.22) is equal to unity, and therefore if the nuclei are distinguishable, the reaction rate r becomes

$$r = n_1 n_2 \int v\sigma(v) \left(\frac{\mu}{2\pi kT} \right)^{3/2} e^{-\frac{\mu v^2}{2kT}} d^3v. \tag{6.23}$$

It is convenient to express the cross section $\sigma(E)$ as

$$\sigma(E) = \frac{S(E)}{E} e^{-\frac{2\pi Z_1 Z_2 e^2}{\hbar v}}, \tag{6.24}$$

where $S(E)$ is a slowly varying function of energy E and the exponential term is the barrier penetration factor (i.e. Gamow factor). Because $v = \sqrt{2E/\mu}$, Equation (6.24) implies that

$$\sigma(E) = \frac{S(E)}{E} e^{-bE^{-1/2}}, \tag{6.25}$$

with $b = \dfrac{\pi\sqrt{2\mu}Z_1Z_2e^2}{\hbar}$.

Expressing the Maxwell–Boltzmann distribution function, given in Equation (3.13), as a function of energy E, we have

$$f(v)4\pi v^2 dv = \frac{2}{\sqrt{\pi}}\frac{E^{1/2}e^{-E/kT}}{(kT)^{3/2}}dE. \tag{6.26}$$

Equations (6.23), (6.25) and (6.26) imply that the reaction rate per pair of interacting nuclei is

$$\lambda = \langle\sigma v\rangle = \int_0^\infty \sigma(E)vf(v)4\pi v^2 dv$$

$$= \left(\frac{8}{\mu\pi}\right)^{1/2}\frac{1}{(kT)^{3/2}}\int_0^\infty S(E)e^{(-E/kT - bE^{-1/2})}dE. \tag{6.27}$$

Although $S(E)$ is a slowly varying function of energy, the exponential factor within the integral of Equation (6.27) depends strongly on energy. The maximum value of $\exp(-E/kT - bE^{-1/2})$ is obtained by setting

$$\frac{d}{dE}(E/kT + bE^{-1/2}) = \frac{1}{kT} - \frac{b}{2}E_0^{-3/2} = 0. \tag{6.28}$$

If we let $T_6 = T/10^6$, then Equation (6.28) and the definition of b imply that the exponential factor in Equation (6.27) is maximum when

$$E_0 = \left(\frac{bkT}{2}\right)^{2/3} = 1.22(Z_1^2 Z_2^2\mu T_6^2/m_\mu)^{1/3}\text{ keV}. \tag{6.29}$$

Because the function

$$f_1 = e^{-E/kT - bE^{-1/2}} \tag{6.30}$$

is the product of a function that decreases rapidly with energy and one that increases rapidly, it is sharply peaked. Therefore, we can approximate $f_1(E)$ as a Gaussian function, which is equal to f_1 at E_0 and whose first and second derivatives are also equal to those of f_1 at $E = E_0$. We let

$$f_1(E) = e^{-E/kT - bE^{-1/2}} \simeq Ce^{-\left[\frac{E - E_0}{\Delta/2}\right]^2} = f_2(E). \tag{6.31}$$

It follows from Equations (6.29) and (6.31) that the requirement $f_1(E_0) = f_2(E_0)$ implies that

$$C = e^{-E_0/kT - bE_0^{-1/2}} = e^{-3E_0/kT}. \tag{6.32}$$

Evaluating the second derivatives of $f_1(E)$ and $f_2(E)$ at $E = E_0$ gives

$$\frac{d^2 f_1(E_0)}{dE^2} = -(3/4)bE_0^{-5/2}f_1(E_0) \tag{6.33}$$

and

$$\frac{d^2 f_2(E_0)}{dE^2} = -\frac{2 f_2(E_0)}{(\Delta/2)^2}. \tag{6.34}$$

Since $f_1(E_0) = f_2(E_0)$ the requirement that Equations (6.33) and (6.34) are equal at $E = E_0$ implies that

$$\frac{3b}{4} E_0^{-5/2} = \frac{2}{(\Delta/2)^2}. \tag{6.35}$$

Equations (6.29) and (6.35) lead to the equation

$$\Delta = \frac{4}{\sqrt{3}} \sqrt{kTE_0}. \tag{6.36}$$

From Equations (6.29), (6.31) and (6.36) it follows that the expression for the reaction rate per pair of nuclei given in Equation (6.27) can be written

$$\lambda = \left(\frac{8}{\mu\pi}\right)^{1/2} \frac{1}{(kT)^{3/2}} e^{-3E_0/kT} S(E_0) \int e^{-\left[\frac{E-E_0}{(\Delta/2)}\right]^2} dE. \tag{6.37}$$

If we make the variable change $y = (E - E_0)/(\Delta/2)$, then Equation (6.37) becomes

$$\lambda = \left(\frac{2}{\mu\pi}\right)^{1/2} \frac{\Delta}{(kT)^{3/2}} e^{-3E_0/kT} S(E_0) \int_{-2E_0/\Delta}^{\infty} e^{-y^2} dy. \tag{6.38}$$

Since $2E_0/\Delta$ satisfies the condition $2E_0/\Delta \gg 1$ we can replace the lower limit of the integral in Equation (6.38) with $-\infty$. Therefore the integral is equal to $\sqrt{\pi}$ and Equation (6.38) becomes

$$\lambda = \left(\frac{2}{\mu}\right)^{1/2} \frac{\Delta e^{-3E_0/kT}}{(kT)^{3/2}} S(E_0), \tag{6.39}$$

with Δ given by Equation (6.36). Defining the variable $\tau = 3E_0/kT$, we have

$$\Delta = \frac{4}{\sqrt{3}} \sqrt{kTE_0} = \frac{4}{3} \tau^{1/2} kT \tag{6.40}$$

and

$$(kT)^{3/2} = \left(\frac{3E_0}{\tau}\right)^{3/2}. \tag{6.41}$$

From Equations (6.40) and (6.41) we obtain

$$\frac{\Delta}{(kT)^{3/2}} = \frac{4\tau^2}{3^{5/2}} \frac{kT}{E_0^{3/2}} = \frac{\tau^2}{3^{5/2} b}. \tag{6.42}$$

The final equality in Equation (6.42) follows because $E_0^{3/2} = 4bkT$. Equation (6.42) implies that Equation (6.39) can be expressed as

$$\lambda = \frac{1}{3^{5/2}}\left(\frac{2}{\mu}\right)^{1/2}\frac{\tau^2}{b}S(E_0)e^{-\tau},$$

(6.43)

where $\tau = 3E_0/kT = 42.46(Z_1^2 Z_2^2 \mu / m_\mu T_6)^{1/3}$ and b is defined in Equation (6.25).

The nuclear cross section $\sigma(E)$ can be expressed as the product $\sigma(E) = P_l(E)\sigma_n(E)$ where $P_l(E)$ is the penetration factor given in Equation (6.7) and $\sigma_n(E)$ is the nuclear cross section with the Coulomb interaction neglected.

The Coulomb interaction is modified because an electron gas surrounds nuclei in the dense interiors of stars. If kT exceeds the average Coulomb energy, then from Equation (3.295) the Coulomb potential V_c about a nucleus becomes

$$V_c = \frac{Z_1 e}{r}e^{-r/\lambda_D},$$

(6.44)

with λ_D equal to the Debye radius. Expanding the exponential term in Equation (6.44) gives

$$V_c = \frac{Z_1 e}{r} - \frac{Z_1 e}{\lambda_D}.$$

(6.45)

Using Equation (6.45) it can be shown that the thermonuclear reaction rate given by Equation (6.27) is increased by a factor

$$f = \exp\left[\frac{Z_1 Z_2 e^2}{kT\lambda_D}\right].$$

(6.46)

6.2   The penetration factor

The WKB approximation

Expanding $\phi(r)$ in Equation (6.4) in powers of $\hbar$, we obtain

$$\phi(r) = \phi_0(r) + \hbar\phi_1(r).$$

(6.47)

Substituting the expression for $\phi(r)$ given in Equation (6.47) into Equation (6.5) and equating equal powers of $\hbar$ leads to the two equations

$$-\left(\frac{d\phi_0}{dr}\right)^2 + 2\mu(E - V(r)) = 0$$

(6.48)

and

$$i\frac{d^2\phi_0}{dr^2} - 2\frac{d\phi_0}{dr}\frac{d\phi_1}{dr} = 0.$$

(6.49)

Integrating Equations (6.48) and (6.49), we obtain

$$\phi_0(r) = \pm \int_{r_0}^{r} f(r')dr' \tag{6.50}$$

and

$$\phi_1(r) = \frac{i}{2}\ln f(r), \tag{6.51}$$

with

$$f(r) = \left(\frac{2\mu}{\hbar^2}\right)^{1/2} (E - V(r))^{1/2} \tag{6.52}$$

and $V(r)$ equal to the Coulomb potential $Z_1 Z_2 e^2/r$. The solutions ϕ_0 and ϕ_1 given as Equations (6.50–6.52) are not valid in the vicinity of the turning point r_1 (i.e. when $E = V(r_1)$). At radial distances r such that $V(r) < E$ Equations (6.50) and (6.52) imply that the wave function $\chi(r)$ in Equation (6.4) is

$$\chi(r) = A\frac{e^{\pm i \int_{}^{r} f(r')dr'}}{f(r)^{1/2}}. \tag{6.53}$$

Similarly if $V(r) > E$, the wave function χ becomes

$$\chi(r) = B\frac{e^{\pm \int_{}^{r} g(r')dr'}}{g(r)^{1/2}}, \tag{6.54}$$

with $g(r) = \left(\frac{2\mu}{\hbar^2}\right)^{1/2} (V(r) - E)^{1/2}$.

In order to simplify the matching of the solutions given in Equations (6.53) and (6.54) which apply at large and small radial distances, respectively, we assume that the potential $V(r)$ is the Coulomb potential for radial distances $r \leq 0.5\,r_1$, with r_1 given in Equation (6.9), and $V(r) = 0$ for $r > 0.5\,r_1$. At radial distances $r > 0.5\,r_1$ and $l = 0$, the Schrödinger equation given in Equation (6.3) becomes

$$\frac{d^2\chi}{dr^2} + k^2\chi = 0, \tag{6.55}$$

with $k^2 = \dfrac{2\mu E}{\hbar^2}$.

The solution of Equation (6.55) consists of the two independent solutions e^{ikr} and e^{-ikr}, and therefore $\chi(r)$ can be expressed as

$$\chi(r) = A\sin(kr + \delta) = \frac{Ae^{-i\delta}}{2i}(e^{2i\delta}e^{ikr} - e^{-ikr}), \tag{6.56}$$

where δ is the phase shift, e^{-ikr} represents an incoming wave and e^{ikr} represents an outgoing wave.

The plane-wave solution for an incoming particle moving along the z-axis can be represented by the expansion

$$e^{ikz} = \sum_{l=0}^{\infty} (2l + 1)i^l j_l(kr) P_l(\cos \theta), \tag{6.57}$$

where j_l are the spherical Bessel functions and P_l the Legendre polynomials. At low energies only the $l = 0$ term needs to be retained, and consequently the plane wave solution given in Equation (6.57) is

$$j_0(kr) = \frac{\sin kr}{kr} = \frac{1}{2ikr}(e^{ikr} - e^{ikr}). \tag{6.58}$$

If we chose $A = e^{i\delta}/k$, then the wave function $\chi(r)$ in Equation (6.56) becomes

$$\chi = \frac{1}{2ik}(e^{2i\delta} - 1)e^{ikz} + \frac{1}{2ik}(e^{ikr} - e^{ikr}). \tag{6.59}$$

The last two terms in Equation (6.59) divided by r are equal to the spherical Bessel function $j_0(kr)$ given in Equation (6.58). The first two terms in Equation (6.59) can be written

$$\chi_{sc} = \frac{e^{i\delta}}{k} \sin \delta e^{ikr}, \tag{6.60}$$

where χ_{sc} represents the scattered outgoing wave, which is zero in the absence of the potential $V(r)$.

The boundary conditions are that $\chi(r)$ and $\chi'(r)$ are continuous at $r = 0.5r_1$. Applying these boundary conditions with Equations (6.54), (6.59) and (6.60), we obtain

$$\frac{B}{\left(\dfrac{2\mu E}{\hbar^2}\right)^{1/4}} = \frac{e^{i\delta}}{k} \sin \delta e^{i\left(\frac{kr_1}{2}\right)} + \frac{\sin \dfrac{kr_1}{2}}{k} \tag{6.61}$$

and

$$\left(\frac{2\mu E}{\hbar^2}\right)^{1/4} B - \frac{B}{r_1} \frac{1}{\left(\dfrac{2\mu E}{\hbar^2}\right)^{1/4}} = ie^{i\delta} \sin \delta e^{i\left(\frac{kr_1}{2}\right)} + \cos\left(\frac{kr_1}{2}\right). \tag{6.62}$$

Eliminating the phase shift δ from Equations (6.61) and (6.62) leads to the equation

$$(B_r + iB_i)(K_2 - ik) = -iK_1 \sin\left(\frac{kr_1}{2}\right) + K_1 \cos\left(\frac{kr_1}{2}\right), \tag{6.63}$$

with $K_1 = \left(\dfrac{2\mu E}{\hbar^2}\right)^{1/4}$, $K_2 = K_1^2 - \dfrac{1}{r_1}$ and B_r and B_i equal to the real and imaginary components of B. Equation (6.63) reduces to the two equations

$$B_r(K_2^2 + k^2) = K_1 K_2 \cos\left(\frac{kr_1}{2}\right) + kK_1 \sin\left(\frac{kr_1}{2}\right) \tag{6.64}$$

and

$$B_i(K_2^2 + k^2) = kK_1 \cos\left(\frac{kr_1}{2}\right) - K_1 K_2 \sin\left(\frac{kr_1}{2}\right). \tag{6.65}$$

Equations (6.64) and (6.65) determine B_r and B_i as functions of k and r_1. From Equation (6.61) we have

$$\frac{e^{i\delta}}{k}\sin\delta e^{i\left(\frac{kr_1}{2}\right)} = \frac{B_r + iB_i}{K_1} - \frac{\sin\left(\frac{kr_1}{2}\right)}{k}. \tag{6.66}$$

Solving for δ in Equation (6.66) gives

$$\frac{\sin^2\delta}{k^2} = \left| \frac{B}{K_1} - \frac{\sin\left(\frac{kr_1}{2}\right)}{k} \right|^2$$

$$= \left(\frac{B_r}{K_1} - \frac{\sin\left(\frac{kr_1}{2}\right)}{k}\right)^2 + \left(\frac{B_i}{K_1}\right)^2. \tag{6.67}$$

The sign of the phase shift δ in Equation (6.67) is negative because the potential is repulsive.

The above solution is approximate because we have neglected the effect of the Coulomb potential at radial distances $r > r_1/2$ where r_1 is the turning point. To obtain a more accurate solution within the framework of the WKB approximation we must solve the Schrödinger Equation (6.3) in the vicinity of the turning point and then match boundary conditions with solutions (6.53) and (6.54), which are valid for $V(r) < E$ and $V(r) > E$, respectively.

Equation (6.3) can be expressed as

$$\frac{d^2\chi}{dr^2} + \frac{2\mu}{\hbar^2}\left[E - \frac{Z_1 Z_2 e^2}{r}\right]\chi = 0. \tag{6.68}$$

For values of r close to the turning point r_1, we have

$$\frac{1}{r} \simeq \frac{1}{r_1} - \frac{r - r_1}{r_1^2} = \frac{2}{r_1} - \frac{r}{r_1^2}. \tag{6.69}$$

Since by definition

$$E = \frac{Z_1 Z_2 e^2}{r_1}, \tag{6.70}$$

at the turning point r_1 it follows from Equation (6.69) that close to the turning point, Equation (6.68) becomes

$$\frac{d^2\chi}{dr^2} + \frac{2\mu Z_1 Z_2 e^2}{r_1^2 \hbar^2}(r - r_1)\chi = 0. \tag{6.71}$$

If we define $x = \alpha(r - r_1)$, with

$$\alpha = \left(\frac{2\mu Z_1 Z_2 e^2}{r_1^2 \hbar^2}\right)^{1/3}, \tag{6.72}$$

then Equation (6.71) reduces to the equation

$$\frac{d^2\chi}{dx^2} + x\chi = 0. \tag{6.73}$$

It can readily be shown by direct substitution that

$$\chi = \sqrt{x} J_{\pm 1/3}\left(\frac{2}{3}x^{3/2}\right) \tag{6.74}$$

is a solution of Equation (6.73) for $r > r_1$. In Equation (6.74) $J_{\pm 1/3}$ are solutions of Bessel's equation,

$$x^2\frac{d^2y}{dx^2} + x\frac{dy}{dx} + (x^2 - n^2)y = 0, \tag{6.75}$$

with $n = \pm\frac{1}{3}$. Because $J_{-1/3}$ diverges and χ in Equation (6.74) remains finite as $x \to 0$ it follows that the general solution of Equation (6.68) close to $r = r_1$ (i.e. $x = 0$) and satisfying the condition $r > r_1$ is

$$\chi = C\alpha\sqrt{r - r_1} J_{-1/3}\left(\frac{2}{3}\alpha^{3/2}(r - r_1)^{3/2}\right), \tag{6.76}$$

with C a constant.

Since $I_{-1/3}$ diverges and χ remains finite as $x \to 0$, the solution of Equation (6.68) for values of r such that $r = r_1$ and $r < r_1$ is

$$\chi = C'\alpha\sqrt{r - r_1} I_{-1/3}\left[\frac{2}{3}\alpha^{3/2}(r_1 - r)^{3/2}\right], \tag{6.77}$$

where the functions $I_{-1/3}$ are related to the Bessel functions $J_{-1/3}$ by the relation $I_p(x) = i^{-p}J_p(ix)$. It follows that χ and χ' are continuous at $r = r_1$ if $C = i^{1/3}C'$. The constants in Equations (6.76) and (6.77) must satisfy the boundary conditions that χ and $d\chi/dr$ are continuous with the WKB solution given in Equation (6.53) at some value of r such that $r \simeq r_1$ and $r > r_1$ and with the WKB solution given in Equation (6.54) at some value of r such that $r \simeq r_1$ and $r < r_1$.

Barrier penetration in parabolic coordinates

The Schrödinger equation for an assumed Coulomb potential can be expressed as

$$\nabla^2 \psi + \left(k^2 - \frac{2kn}{r} \right) \psi = 0,$$ (6.78)

with

$$k = \frac{mv}{\hbar}, \ n = \frac{Z_1 Z_2 e^2}{\hbar v} \ \text{and} \ 2kn = \frac{2m}{\hbar^2} Z_1 Z_2 e^2.$$

Parabolic coordinates are related to spherical polar coordinates by the equations

$$\xi = r - z = r(1 - \cos \theta) = 2r \sin^2 \frac{\theta}{2},$$ (6.79)

and

$$\eta = r + z = r(1 + \cos \theta) = 2r \cos^2 \frac{\theta}{2}$$ (6.80)

and

$$\phi = \phi.$$ (6.81)

The form of the parabolic coordinates suggests that we introduce the factored wave function

$$\psi = e^{ikz} y(\xi).$$ (6.82)

Substituting the expression for ψ given in Equation (6.82) into Equation (6.78) gives

$$\nabla^2 y + 2ik \frac{\partial y}{\partial z} - \frac{2kn}{r} y = 0.$$ (6.83)

From the definitions of ξ and η given in Equations (6.79) and (6.80) we obtain the relations

$$\nabla^2 y = \frac{4}{\xi + \eta} \frac{d}{d\xi} \left(\xi \frac{dy}{d\xi} \right)$$ (6.84)

and

$$\frac{\partial y}{\partial z} = \frac{-2\xi}{\xi + \eta} \frac{dy}{d\xi},$$ (6.85)

with $r = \frac{1}{2}(\xi + \eta)$.

Substituting Equations (6.84) and (6.85) into Equation (6.83) leads to the equation

$$\xi \frac{d^2 y}{d\xi^2} + (1 - ik\xi) \frac{dy}{d\xi} - kny = 0.$$ (6.86)

Equation (6.86) is a confluent hypergeometric equation and has two singular points. As discussed in Appendix 3 the standard form for the confluent hypergeometric equation is

$$zy'' + [c - z]y' - ay = 0. \tag{6.87}$$

The solution of Equation (6.87) that is regular at $z = 0$ is

$$y = C_1F_1(a, c; z), \tag{6.88}$$

with C equal to the constant and $_1F_1(a, c; z)$ equal to the confluent hypergeometric function. From the theory of confluent hypergeometric functions the asymptotic (i.e. $z \to \infty$) formula for $_1F_1(a, c; z)$ is

$$_1F_1(a, c; z) \simeq \frac{\Gamma(c)}{\Gamma(c - a)}(-z)^{-a} + \frac{\Gamma(c)}{\Gamma(a)}e^z z^{a-c}, \tag{6.89}$$

where Γ, the well-known gamma function, has the integral representation

$$\Gamma(z) = \int_0^\infty e^{-t}t^{z-1}dt \quad (\text{Re } z > 0). \tag{6.90}$$

A comparison between Equations (6.86) and (6.87) shows that $a = -in$, $c = 1$ and $z = ik\xi$. Therefore, for large z Equation (6.89) becomes

$$_1F_1(-in, 1; ik\xi) \simeq \frac{e^{\frac{n\pi}{2}}}{\Gamma(1 + in)}[(k\xi)^{in} - ne^{i(k\xi + 2\eta_0)}(k\xi)^{-(1+in)}] \tag{6.91}$$

with $e^{2i\eta_0} = \dfrac{\Gamma(1 + in)}{\Gamma(1 - in)}$ and $i = e^{\frac{i\pi}{2}}$.

Expressing the solution ψ_c of the Schrödinger equation as

$$\psi_c = Ce^{ikz}{}_1F_1(-in, 1; ik(r - z)), \tag{6.92}$$

and using Equation (6.91), we find that as $|r - z| \to \infty$, ψ_c approaches the function

$$\psi_c \simeq \frac{Ce^{\frac{n\pi}{2}}}{\Gamma(1 + in)}\left[e^{i[kz + n\ln k(r - z)]} - \frac{n}{k(r - z)}e^{i[kr - n\ln k(r - z) + 2\eta_0]} \right]$$

$$= \frac{Ce^{\frac{n\pi}{2}}}{\Gamma(1 + in)}\left[e^{i(kz + n\ln k(r - z))} + \frac{f_c(\theta)}{r}e^{i(kr - n\ln kr)} \right] \tag{6.93}$$

with

$$f_c(\theta) = -\frac{ne^{i[-n\ln(1 - \cos\theta) + 2\eta_0]}}{k(1 - \cos\theta)}.$$

In deriving Equation (6.93) we have used the identity

$$\Gamma(1 - in) = -in\Gamma(-in). \tag{6.94}$$

The differential scattering cross section implied by Equation (6.93) is

$$\sigma_c(\theta) = |f_c(\theta)|^2 = \left(\frac{n}{2k \sin^2 \theta/2}\right)^2$$

$$= \left(\frac{Z_1 Z_2 e^2}{2\mu v^2}\right)^2 \mathrm{cosec}^4\left(\frac{\theta}{2}\right). \tag{6.95}$$

The expression for $\sigma_c(\theta)$ given in Equation (6.95) is the well-known Rutherford scattering cross section already given in Equation (3.332). If the wave function ψ is normalized to unit flux, then the constant C in Equation (6.93) becomes

$$C = v^{-1/2}\Gamma(1 + in)e^{-\frac{1}{2}n\pi}, \tag{6.96}$$

and therefore the Coulomb wave function ψ_c is equal to

$$\psi_c = v^{-1/2}\Gamma(1 + in)e^{-\frac{1}{2}n\pi}e^{ikz}{}_1F_1(- in, 1; ik\xi) \tag{6.97}$$

with $\xi = r - z = r(1 - \cos\theta) = 2r\sin^2\frac{1}{2}\theta$.

For small values of $z = ik\xi$ the confluent hypergeometric function ${}_1F_1(a, c; z)$ can be expanded about $z = 0$ by means of the power series

$${}_1F_1(a, c; z) = \sum_{m=0}^{\infty} \frac{\Gamma(a + m)\Gamma(c)z^m}{\Gamma(a)\Gamma(c + m)}$$

$$= 1 + \frac{az}{c1!} + \frac{a(a + 1)z^2}{c(c + 1)2!} + \cdots \tag{6.98}$$

Equation (6.98) implies that ${}_1F_1(a, c; 0) = 1$. Because the range of the strong interaction is sufficiently small that the position of the nucleus can be taken as $r = 0$, Equations (6.97–6.98) and the equality

$$\Gamma(1 + in)\Gamma(1 - in) = \frac{in\pi}{\sin in\pi} = \frac{2n\pi}{e^{n\pi} - e^{-n\pi}} \tag{6.99}$$

imply that $|\psi_c|^2$ evaluated at $r = 0$ is

$$|\psi_c(0)|^2 = v^{-1}\frac{2n\pi}{e^{2n\pi} - 1}. \tag{6.100}$$

The equality given in Equation (6.99) is derived in Appendix 3.

For small values of v we have $n \gg 1$, since $n = Z_1 Z_2 e^2/\hbar v$, and therefore Equation (6.100) becomes

$$|\psi_c(0)|^2 = \frac{2\pi n}{v}e^{-2n\pi} = \frac{2\pi Z_1 Z_2 e^2}{\hbar v^2}e^{-2\pi Z_1 Z_2 e^2/\hbar v}. \tag{6.101}$$

The exponential penetration factor in Equation (6.101) is the same as the Gamow factor obtained in Section 6.1. For an assumed incident unit flux the probability that the incident nucleus penetrates the Coulomb barrier and reaches the target nucleus, becomes $v|\psi_c(0)|^2$.

6.3   Scattering and resonant reactions

For an assumed spherically symmetric potential $V(r)$ the radial Schrödinger equation describing the interaction between two nuclei is

$$\left[\frac{d^2}{dr^2} + \frac{2}{r}\frac{d}{dr} - \frac{l(l+1)}{r^2} - \frac{2\mu}{\hbar^2}V(r) + k^2\right]u_l(r) = 0 \tag{6.102}$$

with

$$k^2 = 2\mu E/\hbar^2.$$

For $V(r) = 0$ the solutions of Equations (6.102) are the spherical Bessel functions

$$j_l(kr) = \left(\frac{\pi}{2kr}\right)^{1/2} J_{l+1/2}(kr) \tag{6.103}$$

where $J_{l+1/2}$ are the half-integer Bessel functions. The $l = 0$ spherical Bessel function is

$$j_0(kr) = \frac{\sin kr}{kr}. \tag{6.104}$$

As $r \to \infty$ the spherical Bessel functions become

$$j_l(kr) = \frac{\sin(kr - l\pi/2)}{kr}. \tag{6.105}$$

The wave function for an incident flux of nuclei moving along the z-direction is

$$\psi_{in} = Ae^{ikz}. \tag{6.106}$$

In the discussion below the normalization constant A will be chosen equal to unity. The plane-wave solution $\exp(ikz)$ can be expanded as

$$e^{ikz} = \sum_{l=0}^{\infty} A_l j_l(kr) P_l(\cos\theta). \tag{6.107}$$

As shown in Appendix 3 the constants A_l in Equation (6.107) are

$$A_l = (2l + 1)i^l. \tag{6.108}$$

From Equations (6.107), (6.108) and the asymptotic expression (6.105) it follows that

$$e^{ikz} = \sum_{l=0}^{\infty} (2l + 1)i^l P_l(\cos\theta)\frac{\sin(kr - l\pi/2)}{kr}$$

$$= \frac{1}{2kr}\sum_{l=0}^{\infty} (2l + 1)i^{l+1}P_l(e^{-i(kr - l\pi/2)} - e^{i(kr - l\pi/2)}). \tag{6.109}$$

Because there is axial symmetry the most general solution of Equation (6.102) is

$$\psi(r,\theta) = \sum_{l=0}^{\infty} B_l u_l(r) P_l(\cos\theta), \tag{6.110}$$

where B_l are constants. For large values of r the wave function $\psi(r,\theta)$ has the asymptotic form

$$\psi = e^{ikz} + f(\theta)\frac{e^{ikr}}{r}, \tag{6.111}$$

because the scattered wave can contain only outgoing waves, e^{ikr}. If the potential $V(r)$ approaches zero sufficiently rapidly as $r \to \infty$ then the asymptotic behaviour of the spherical Bessel functions j_l, which are given in Equation (6.105) and the eigenfunctions u_l can differ only by a phase shift δ_l. Therefore, it follows from Equations (6.105) that

$$u_l(kr) = \frac{1}{kr}\sin(kr - l\pi/2 + \delta_l), \tag{6.112}$$

where the phase shift δ_l is due to the potential $V(r)$. Equations (6.111) and (6.112) imply that asymptotically

$$\psi = \sum_l P_l(\cos\theta) B_l \frac{1}{kr}\sin(kr - l\pi/2 + \delta_l)$$

$$= (2ikr)^{-1}\sum_l P_l(\cos\theta)B_l[e^{i(kr-l\pi/2)}e^{i\delta_l} - e^{-i(kr-l\pi/2)}e^{-i\delta_l}]. \tag{6.113}$$

The wave function for the scattered wave is

$$\psi_{sc} = \psi - e^{ikz}. \tag{6.114}$$

The asymptotic equation (6.113), Equation (6.114) and the condition that the scattered wave can contain only outgoing waves imply that the constants B_l in Equation (6.113) are

$$B_l = i^l(2l+1)e^{i\delta_l}. \tag{6.115}$$

Equations (6.113) and (6.115) imply that the function $f(\theta)$ given in Equation (6.111) is

$$f(\theta) = \frac{1}{2ik}\sum_l (2l+1)(e^{2i\delta_l} - 1)P_l(\cos\theta) \tag{6.116}$$

and the asymptotic wave functions ψ and ψ_{sc} are

$$\psi(r,\theta) = \frac{1}{2kr}\sum_{l=0}^{\infty} (2l+1)i^{l+1}P_l[e^{-i(kr-l\pi/2)} - \eta_l e^{i(kr-l\pi/2)}] \tag{6.117}$$

and

$$\psi_{sc}(r,\theta) = \frac{f(\theta)}{r}e^{ikr} = \frac{1}{2kr}\sum_{l=0}^{\infty} (2l+1)i^{l+1}P_l(1-\eta_l)e^{i(kr-l\pi/2)} \tag{6.118}$$

with $\eta_l = e^{2i\delta_l}$ and the phase shift δ_l complex if absorption occurs.

The probability current density for scattered particles is

$$J_{sc}(\theta) = \frac{\hbar}{2i\mu}(\psi_{sc}^\star \nabla \psi_{sc} - \psi_{sc} \nabla \psi_{sc}^\star). \tag{6.119}$$

Equation (6.119) implies that the number of particles scattered per second is

$$N_{sc} = \int_0^\pi J_{sc}(\theta) r^2 2\pi \sin\theta d\theta. \tag{6.120}$$

Retaining terms proportional to $1/r$ only we obtain

$$\frac{\partial \psi_{sc}}{\partial r} = ik\psi_{sc} \tag{6.121}$$

$$\frac{\partial \psi_{sc}^\star}{\partial r} = -ik\psi_{sc}^\star. \tag{6.122}$$

Since derivatives with respect to θ give no contribution to N_{sc} Equations (6.118–6.122) imply that

$$N_{sc} = \frac{\hbar k}{\mu} \int_0^\pi \psi_{sc}^\star \psi_{sc} r^2 2\pi \sin\theta d\theta$$

$$= \frac{\hbar\pi}{\mu k} \sum_{l=0}^\infty (2l+1)|1 - \eta_l|^2 \tag{6.123}$$

The incident flux is $\hbar k/\mu$ and therefore from Equation (6.123) we have

$$\sigma_{sc} = \frac{N_{sc}}{(\hbar k/\mu)} = \frac{\pi}{k^2} \sum_{l=0}^\infty (2l+1)|1 - \eta_l|^2$$

$$= \pi\lambda^2 \sum_{l=0}^\infty (2l+1)|1 - \eta_l|^2. \tag{6.124}$$

Equation (6.124) implies that the cross section for each l value is

$$\sigma_{sc,l} = \pi\lambda^2(2l+1)|1 - \eta_l|^2. \tag{6.125}$$

The number of reactions per second is

$$N_R = \frac{\hbar}{2i\mu} \int_0^\pi \left(\psi^\star \frac{\partial\psi}{\partial r} - \psi \frac{\partial\psi^\star}{\partial r}\right) 2\pi r^2 \sin\theta d\theta \tag{6.126}$$

and the absorption cross section is equal to

$$\sigma_{abs} = \frac{N_R}{\left(\frac{\hbar k}{\mu}\right)}. \tag{6.127}$$

From Equations (6.117), (6.126) and (6.127) the absorption cross section becomes

$$\sigma_{abs} = \frac{\pi}{k^2} \sum_{l=0}^{\infty} (2l + 1)(1 - \eta_l^* \eta_l) = \pi \lambda^2 \sum_{l=0}^{\infty} (2l + 1)(1 - |\eta_l|^2).$$ (6.128)

From Equation (6.128) it follows that

$$\sigma_{abs,l} = \pi \lambda^2 (2l + 1)(1 - |\eta_l|^2) < (2l + 1)\pi \lambda^2.$$ (6.129)

Equations (6.124) and (6.128) imply that the total cross section is

$$\sigma_t = \sigma_{sc} + \sigma_{abs} = \frac{2\pi}{k^2} \sum_{l=0}^{\infty} (2l + 1)(1 - \mathrm{Re}\,\eta_l).$$ (6.130)

Equations (6.116) and (6.130) lead directly to the equation

$$\mathrm{Im} f(0) = \frac{k\sigma_t}{4\pi}.$$ (6.131)

The above equation, which relates the imaginary part of the forward scattering amplitude and the total cross section, is known as the optical theorem.

The maximum nuclear cross sections given in Equation (6.129) have the following physical interpretation. The angular momentum of an incoming particle is quantized so that $L = l\hbar$. If p is the momentum of an incoming particle and b the impact parameter then we have

$$L = pb = \frac{\hbar b}{\lambda} = l\hbar$$ (6.132)

with $b = \dfrac{l\lambda}{2\pi} = l\lambda$. The semi-classical interpretation of Equation (6.129) is that

$$\sigma_{r,l}(\max) = \pi[(l + 1)\lambda]^2 - \pi[l\lambda]^2 = (2l + 1)\pi\lambda^2.$$ (6.133)

Measured cross sections for nuclear interactions show resonant behavior. The existence of resonances at particular energies is associated with the formation of a long-lived state (i.e. compound nucleus) consisting of two interacting particles (e.g. proton and ^{12}C nucleus). At a particular energy the compound nuclear state of a proton and ^{12}C nucleus is an excited state of the ^{13}N nucleus. The energy width Γ of the compound nuclear state and the mean lifetime are related by the equation

$$\Gamma = \hbar/\tau.$$ (6.134)

Equation (6.134) implies that for level widths of 1 keV the lifetime of the compound nucleus is

$$\tau = \frac{\hbar}{\Gamma} \simeq 10^{-12}\,\mathrm{s}$$ (6.135)

which is very long as compared to a nuclear crossing time ($\simeq 10^{-21}$ s).

In the theory of the compound nucleus it is assumed that a particular nuclear reaction can be represented as the two-step process

$$A + B \rightarrow C^{\star} \rightarrow D + E, \tag{6.136}$$

where $C^{\star}$ is the compound nuclear state (i.e. excited energy level of nucleus), and D and E represent a particular decay mode. The various decay modes of a compound nucleus are known as channels. The energy width Γ is

$$\Gamma = \Gamma_p + \Gamma_n + \Gamma_\alpha + \Gamma_\gamma + \Gamma_{e^\pm} \tag{6.137}$$

where Γ_p, Γ_n, Γ_α, Γ_γ and $\Gamma_{e^\pm}$ are the energy widths associated with proton, neutron, α-particle, γ-ray, and e^- (or e^+) decay modes respectively. Because it is assumed that the decay mode is independent of the mode of formation of the nucleus the cross section for the formation of a compound nucleus through some entrance channel α and decay through some channel β can be expressed as

$$\sigma_{\beta\alpha} = \sigma_c(\alpha)\frac{\Gamma_\beta}{\Gamma}, \tag{6.138}$$

where Γ_β is the partial width for decay through some channel β and Γ is the total width

$$\Gamma = \sum_\beta \Gamma_\beta. \tag{6.139}$$

The range of the nuclear force is assumed to be such that interacting particles do not interact beyond some channel radius R. We define ρ_l as the logarithmic derivative of the lth particle wave multiplied by R, i.e.

$$\rho_l = \left(\frac{R}{u_l}\frac{du_l}{dr}\right)_{r=R} \tag{6.140}$$

where $u_l = r\psi_l$. Equation (6.117) implies that for s-wave scattering we have

$$u_0(r) = \frac{i}{2k}(e^{-ikr} - \eta_0 e^{ikr}) \qquad (r \geq R). \tag{6.141}$$

Equations (6.140) and (6.141) lead to the equation

$$\eta_0 = \frac{\rho_0 + ikR}{\rho_0 - ikR}e^{-2ikR}. \tag{6.142}$$

Equation (6.142) implies that if ρ_0 is real then $|\eta_0| = 1$, and therefore no reaction has occurred. However, if $\text{Im}\rho_0 < 0$ then $|\eta_0| < 1$ and reactions do occur. From Equations (6.125), (6.128) and (6.142) it follows that the s-wave scattering and absorption cross sections are

$$\sigma_{sc} = \frac{\pi}{k^2}|1 - \eta_0|^2 = \frac{\pi}{k^2}\left|e^{2ikR} - 1 - \frac{2ikR}{\rho_0 - ikR}\right|^2 \tag{6.143}$$

and

$$\sigma_{abs} = \frac{\pi}{k^2}(1 - |\eta_0|^2) = \frac{\pi}{k^2}\left[\frac{-4kR\mathrm{Im}\rho_0}{(\mathrm{Re}\rho_0)^2 + (\mathrm{Im}\rho_0 - kR)^2}\right]. \tag{6.144}$$

Because $\mathrm{Im}\rho_0 < 0$ and σ_{abs} has its maximum value when $\mathrm{Re}\rho_0 = 0$ we can expand ρ_0 about the resonance energy as

$$\rho_0 = -a(E - E_r) - ib, \tag{6.145}$$

where a and b are real constants and E_r is the resonance energy. The expansion given in Equation (6.145) applies only close to $E = E_r$.

Equations (6.143) and (6.145) imply that the scattering cross section close to resonance is

$$\sigma_{sc} = \frac{\pi}{k^2}\left|e^{2ikR} - 1 + \frac{2ikR/a}{(E - E_r) + i(b + kR)/a}\right|^2. \tag{6.146}$$

Equation (6.146) shows that the scattering amplitude consists of a nonresonant component

$$A_{nr} = \frac{1}{2ik}(e^{2ikR} - 1), \tag{6.147}$$

and a resonant component whose amplitude is

$$A_r = \frac{1}{2ik}\left[\frac{2ikR/a}{(E - E_r) + i(b + kR)/a}\right]. \tag{6.148}$$

The resonant amplitude A_r represents the re-emission of the absorbed nucleus by the compound nucleus. The partial width for the compound nucleus to re-emit the incident nucleus through channel α is

$$\Gamma_\alpha = 2kR/a \tag{6.149}$$

and the total width is

$$\Gamma = 2(b + kR)/a. \tag{6.150}$$

From Equations (6.149) and (6.150) the reaction width becomes

$$\Gamma_r = \sum_{i \neq \alpha}\Gamma_i = \Gamma - \Gamma_\alpha = 2b/a. \tag{6.151}$$

Equations (6.146), (6.148), (6.149) and (6.150) imply that the s-wave scattering cross section is

$$\sigma_{sc,\alpha} = \frac{\pi}{k^2}\frac{\Gamma_\alpha^2}{(E - E_r)^2 + \left(\frac{1}{2}\Gamma\right)^2}. \tag{6.152}$$

Similarly from Equations (6.144), (6.145), (6.150) and (6.151) the absorption cross section is

$$\sigma_{abs} = \frac{\pi}{k^2} \frac{\Gamma_r \Gamma_\alpha}{(E - E_r)^2 + \left(\frac{1}{2}\Gamma\right)^2}. \tag{6.153}$$

Equations (6.151), (6.152) and (6.153) imply that the total cross section involving the formation of a compound nucleus through channel α becomes

$$\sigma_t = \sigma_{abs} + \sigma_{sc,\alpha} = \frac{\pi}{k^2} \frac{\Gamma \Gamma_\alpha}{(E - E_r)^2 + \left(\frac{1}{2}\Gamma\right)^2}. \tag{6.154}$$

From Equations (6.151) and (6.154) it follows that the cross section for the compound nucleus to be formed through channel α and then decay through channel β is

$$\sigma_{\alpha\beta} = \frac{\pi}{k^2} \frac{\Gamma_\beta \Gamma_\alpha}{(E - E_r)^2 + \left(\frac{1}{2}\Gamma\right)^2}. \tag{6.155}$$

Because Γ_β and Γ_α are $< \Gamma$ the cross section $\sigma_{\alpha\beta}$ in Equation (6.155) is $< 4\pi/k^2$ as required by Equation (6.128) for s-wave absorption. Equations (6.152–6.155) are referred to as the Breit–Wigner formulae for resonant reactions. Nonresonant nuclear reactions can also be caused by resonances, when reaction energies are not close to a particular resonance.

6.4   Nuclear energy levels

A nucleus like an atom contains interacting particles that occupy discrete energy levels. The deuteron, which consists of one neutron and one proton, is the simplest nucleus besides the proton. We assume that the neutron and proton interact by means of a spherically symmetric potential and that the orbital angular momentum is equal to zero so that there is no spin–orbit interaction. If we let $u = r\psi$ then the Schrödinger equation for a pure s-wave interaction is

$$\frac{d^2u}{dr^2} + k^2(r)u = 0 \tag{6.156}$$

with

$$k(r) = \frac{1}{\hbar}\sqrt{2\mu(E - V(r))}.$$

Because the masses of the neutron and proton are approximately equal we have

$$\mu = \frac{m_p m_n}{m_p + m_n} \simeq \frac{m_n}{2}. \tag{6.157}$$

The binding energy of the deuteron is known to be

$$|E| = 2.23 \text{ MeV}, \tag{6.158}$$

and therefore the energy eigenvalue of Equation (6.156) is also known. The range of the strong interaction between the neutron and proton is $r_0 \simeq 1.3 \times 10^{-13}$ cm. We will assume a particular shape for the nuclear potential and calculate the depth of the potential using the known binding energy of the deuteron given in Equation (6.158).

If it is assumed that the nuclear potential is a spherical potential of range r_0 and depth V_0 with V_0 negative then the solution u of Equation (6.156) is

$$u = C_1 \sin kr, \qquad (r < r_0) \tag{6.159}$$

with

$$k = \frac{1}{\hbar}\sqrt{m_n(E - V_0)} \qquad (|V_0| > |E|)$$

and

$$u = C_2 e^{-r/R}, \qquad (r > r_0), \tag{6.160}$$

where C_1 and C_2 are constants. The continuity of the logarithmic derivative

$$\frac{1}{u}\frac{du}{dr}$$

at $r = r_0$ implies

$$k \cot kr_0 = -\frac{1}{R}. \tag{6.161}$$

If we assume $R \gg r_0$ then u has its maximum value when kr is slightly less than kr_0 and therefore

$$kr_0 \simeq \frac{1}{\hbar}\sqrt{m_n(E - V_0)}r_0 \simeq \frac{\pi}{2}. \tag{6.162}$$

If the nuclear potential $V(r)$ between the neutron and proton is assumed to be a Yukawa potential,

$$V(r) = -|V_0|e^{-r/r_0} = V_0 e^{-r/r_0}, \tag{6.163}$$

then the radial Schrödinger equation becomes

$$\frac{d^2u}{dr^2} + \frac{m_n}{\hbar^2}(E - V_0 e^{-r/r_0})u = 0. \tag{6.164}$$

Introducing the variable

$$x = e^{-r/2r_0}$$

we find that Equation (6.164) reduces to the equation

$$x^2 \frac{d^2u}{dx^2} + x \frac{du}{dx} + (\gamma^2 x^2 - n^2)u = 0 \tag{6.165}$$

with

$$\gamma^2 = -\frac{4m_n r_0^2 V_0}{\hbar^2} \tag{6.166a}$$

and

$$n^2 = -\frac{-4m_n r_0^2 E}{\hbar^2}. \tag{6.166b}$$

Introducing the further transformation $z = \gamma x$ implies that Equation (6.165) reduces to the Bessel equation

$$z^2 \frac{d^2u}{dz^2} + z \frac{du}{dz} + (z^2 - n^2)u = 0. \tag{6.167}$$

The general solution of Equation (6.167) is

$$u = C_1 J_n(z) + C_2 Y_n(z). \tag{6.168}$$

The constant C_2 in Equation (6.168) must be zero because $Y_n(z)$ diverges as $z \to 0$. It follows that the solutions of Equation (6.165) are

$$u = C_1 J_n(z). \tag{6.169}$$

Since $\psi = u/r$ must remain finite u must equal 0 at $r = 0$. Because the variable x defined in Equation (6.164) is equal to unity when $r = 0$, the condition for an eigenvalue becomes

$$J_n(\gamma) = 0$$

where γ and n are given in Equations (6.166a) and (6.166b), respectively.

In a multi-electron atom the electrons move in the self-consistent Coulomb potential produced by the nucleus and all of the other electrons in the atom. Because electrons are spin-$\frac{1}{2}$ particles they obey the Pauli exclusion principle and occupy the available quantum states according to the degeneracy factors of the various energy levels. Lower energy states are filled first. In nuclei protons and neutrons move in the potential caused by other nucleons, and because they are also spin-$\frac{1}{2}$ particles they fill available energy levels according to the degeneracy factors of these energy levels. Although the precise form of the nuclear potential is not known it is reasonable to assume that the shape of the nuclear potential is intermediate between a potential with $V(r)$ constant and negative out to some radial distance R and the potential of a three-dimensional harmonic oscillator. Because they are different particles protons and neutrons separately fill their respective energy levels. The p–p, n–p and n–n strong interactions are assumed to be the same except for considerations of particle symmetry.

A nucleon within an atomic nucleus is assumed to be enclosed in a spherical region where the potential $V(r)$ has some constant and negative value V_0, and it is assumed that at some radius R this spherical region is surrounded by infinitely high potential walls. The solutions of the Schrödinger equation can be separated and the wave function ψ expressed as

$$\psi(r, \theta, \phi) = R_l(r) Y_{l,m}(\theta, \phi) \tag{6.170}$$

with $R_l(r)$ the solution of the equation

$$\frac{d^2 R_l}{dr^2} + \frac{2}{r}\frac{dR_l}{dr} + \left[k^2 - \frac{l(l+1)}{r^2} \right] R_l = 0 \qquad (0 \le r \le R) \tag{6.171}$$

where

$$k^2 = \frac{2m(E - V_0)}{\hbar^2}.$$

Because the potential $V(r)$ is infinitely high at the boundary of the sphere $R_l(r)$ is equal to 0 for values of r such that $r > R$. If we let $z = kr$ and define $y(z)$ by the equation

$$R_l = z^{-1/2} y(z) \tag{6.172}$$

then Equation (6.171) becomes

$$y'' + \frac{1}{z} y' + \left[1 - \frac{\left(l + \frac{1}{2}\right)^2}{z^2} \right] y = 0. \tag{6.173}$$

Equation (6.173) is the Bessel equation whose solutions are the half-integer Bessel functions, $J_{\pm(l+\frac{1}{2})}(kr)$. It follows that the general solution of Equation (6.171) is

$$R_l(r) = \sqrt{\frac{\pi}{2kr}} [C_1 J_{l+\frac{1}{2}}(kr) + C_2 J_{-(l+\frac{1}{2})}(kr)] \tag{6.174}$$

with C_1 and C_2 equal to constants. The constant C_2 in Equation (6.174) must be equal to zero because otherwise the solution $u_l(r)$ would diverge at $r = 0$. The solutions in Equation (6.174) are the spherical Bessel functions

$$j_l(z) = \sqrt{\frac{\pi}{2z}} J_{l+\frac{1}{2}}(z). \tag{6.175}$$

From Equations (6.174) and (6.175) it follows that the solutions of Equation (6.171) are

$$j_l(kR) = \sqrt{\frac{\pi}{2kR}} J_{l+\frac{1}{2}}(kR) = 0. \tag{6.176}$$

For $l = 0$, 1, and 2 the spherical Bessel functions are

$$j_0(kr) = \frac{\sin kr}{kr} \tag{6.177a}$$

$$j_1(kr) = \frac{\sin kr}{(kr)^2} - \frac{\cos kr}{kr} \tag{6.177b}$$

$$j_2(kr) = -3\frac{\cos kr}{(kr)^2} + \left(\frac{3}{(kr)^3} - \frac{1}{kr}\right)\sin kr. \tag{6.177c}$$

Equation (6.177a) implies that the solutions of Equation (6.171) with $l = 0$ occur when the condition $kR = n\pi$ is satisfied. Equations (6.177b) and (6.177c) imply that the $l = 1$ and $l = 2$ solutions of Equation (6.171) are determined by the conditions $\tan kR = kR$ and $\tan kR = 3kR/[3 - (kR)^2]$.

As an additional approximation for a nucleon potential we consider a nucleon moving in the spherical harmonic oscillator potential

$$V(r) = \frac{1}{2}m\omega^2 r^2. \tag{6.178}$$

Substituting the potential $V(r)$ in Equation (6.178) into the Schrödinger equation we obtain

$$\frac{d^2\psi}{dr^2} + \frac{2}{r}\frac{d\psi}{dr} + \left[\frac{2mE}{\hbar^2} - \frac{m^2\omega^2 r^2}{\hbar^2} - \frac{l(l+1)}{r^2}\right]\psi = 0. \tag{6.179}$$

Expressing the wave function $\psi(r)$ as

$$\psi(r) = r^l e^{-\frac{1}{2}\alpha r^2} y(r) \tag{6.180}$$

implies that Equation (6.179) reduces to the equation

$$\frac{d^2y}{dr^2} + 2\left[\frac{l+1}{r} - \alpha r\right]\frac{dy}{dr} - \left[2\alpha\left(l + \frac{3}{2}\right) - \lambda\right]y = 0 \tag{6.181}$$

with α and λ equal to constants.

Making the variable change $t = \alpha r^2$ we find that Equation (6.181) reduces to the equation

$$\frac{td^2y}{dt^2} + \left(l + \frac{3}{2} - t\right)\frac{dy}{dt} - \frac{1}{2}\left(l + \frac{3}{2} - \frac{\lambda}{2\alpha}\right)y = 0. \tag{6.182}$$

Equation (6.182) can be solved by expanding $y(t)$ in a polynomial series. Introducing the polynomial expansion

$$y(t) = \sum_{p=0}^{n_r} a_p t^p \tag{6.183}$$

into Equation (6.182) and arranging terms in order of increasing powers of t gives

$$\sum_{p=0}^{n_r-1}\left[(p+1)\left(p + l + \frac{3}{2}\right)a_{p+1} - \frac{1}{2}\left(2p + l + \frac{3}{2} - \frac{\lambda}{2\alpha}\right)a_p\right]t^p$$

$$-\frac{1}{2}\left(2n_r + l + \frac{3}{2} - \frac{\lambda}{2\alpha}\right)a_{n_r}t^{n_r} = 0. \tag{6.184}$$

Equation (6.184) must hold for all values of t, and consequently the coefficients of different powers of t must be independently equal to zero. For values of p such that $0 \leq p \leq n_r - 1$ the coefficient of t^p is equal to zero if

$$a_{p+1} = \frac{2p + l + 3/2 - \lambda/2\alpha}{2(p + 1)(p + l + 3/2)} a_p. \tag{6.185}$$

The condition that the coefficient of t^{n_r} in Equation (6.184) equal zero implies that

$$\frac{\lambda}{2\alpha} = 2n_r + l + 3/2. \tag{6.186}$$

Substituting the expressions for λ and α implied by Equations (6.179–6.181) into Equations (6.186) gives

$$E(n_r, l) = \left(2n_r + l + \frac{3}{2} \right) h\nu. \tag{6.187}$$

If we define the principal quantum number n as

$$n = 2n_r + l \tag{6.188}$$

Equation (6.187) becomes

$$E(n_r, l) = \left(n + \frac{3}{2} \right) h\nu. \tag{6.189}$$

The quantum numbers n_r and l can be combined in different ways to obtain the same energy eigenvalue, and therefore energy levels with particular values of n are degenerate. If n is an even integer then l can equal any of the $n/2 + 1$ possible even values between 0 and n. For every l value there exist $2l + 1$ states that correspond to eigenfunctions with different values of the azimuthal quantum number m. The degeneracy of energy level $E_n (n = 1, 2, 3, 4)$ is

$$g_n = 2(n + 1)(n/2 + 1). \tag{6.190}$$

The energy levels of nuclei like those of atoms have orbital and spin angular momentum. Nucleons in nuclei occupy certain quantum states and fill certain shells (and subshells) that depend on the nuclear potential and spin–orbit interactions. In nuclei closed shells (and subshells) exist for both neutrons and protons. The total number of protons (or neutrons) required to fill all nuclear energy levels through a particular nuclear shell is known as a magic number. The nuclear magic numbers are 2, 8, 20, 28, 50, 82 and 126 (for neutrons). The nuclei ^{4}He and ^{16}O are called doubly magic nuclei because both protons and neutrons fill closed shells. The binding energies per nucleon of ^{4}He and ^{16}O are high as compared to isotopes having nearly the same number of nucleons. Both square-well and harmonic-oscillator nuclear potentials correctly predict closed shells at magic numbers 2 and 8. Magic numbers greater than 8 are not predicted correctly by either the square-well or harmonic-oscillator potentials because the nuclear potential is not described accurately by either potential and because there is a significant spin–orbit interaction. The spin–orbit interaction in nuclei is caused by nucleons outside closed shells. Because of pairing forces all even–even nuclei such as ^{4}He and ^{16}O have spin equal to zero.

The spin–orbit interaction in nuclei is often caused by a single unpaired nucleon, and therefore the spin–orbit interaction can be expressed as

$$V = - al \cdot s, \tag{6.191}$$

where l and s are the orbital and spin angular momentum respectively of the unpaired nucleons. Since the total angular momentum $j = l + s$ we have

$$j \cdot j = (l + s) \cdot (l + s) = l \cdot l + 2s \cdot l + s \cdot s. \tag{6.192}$$

Equation (6.192) implies that

$$l \cdot s = \frac{1}{2}[j(j + 1) - l(l + 1) - s(s + 1)]. \tag{6.193}$$

If the spin–orbit interaction is caused by a single nucleon then the total angular-momentum quantum number must be either $j = l + 1/2$ or $j = l - 1/2$. From equations (6.191) and (6.193) we find that the change in energy caused by the spin–orbit interaction is

$$\Delta E = \frac{-a}{2} l \qquad (j = l + 1/2) \tag{6.194}$$

and

$$\Delta E = \frac{a}{2}(l + 1) \qquad (j = l - 1/2). \tag{6.195}$$

Equations (6.194) and (6.195) imply that a $P_{3/2}$ subshell will have a lower energy than a $P_{1/2}$ subshell, and therefore nucleons fill the $P_{3/2}$ subshell before the $P_{1/2}$ subshell.

The parity operator P is defined so that if $f(x, y, z)$ is a function of x, y, z then $Pf(x, y, z) = f(- x, - y, - z)$. The parity operator causes a reflection of the spatial coordinates but does not affect the spin coordinates. If the operator is applied to both sides of the Schrödinger equation given in Equation (6.1) it follows immediately that if ψ is an eigenfunction of Equation (6.1) then $P\psi$ is also a solution. The eigenvalues of the parity operator P are $+ 1$ or $- 1$ because the parity operator P satisfies the relation $P^2\psi = \psi$. The above relations show that the parity of a particular quantum state of a nucleus (e.g. the lowest energy level) can be either $+ 1$ or $- 1$ depending on whether $P\psi = \psi$ or $P\psi = - \psi$. Parity is conserved in strong interactions but not in weak interactions.

The wave function ψ of a particular nuclear energy level can be expressed as

$$\psi(r, \theta, \phi) = u_l(r)N_nP_l^m(\cos \theta)e^{im\phi} \tag{6.196}$$

where N_n is the normalization factor and P_l^m an associated Legendre function. In spherical coordinates the parity operator produces the transformation $(r, \theta, \phi) \rightarrow (r, \pi - \theta, \phi + \pi)$ and therefore we have

$$Pu_l(r) = u_l(r) \tag{6.197a}$$

$$PP_l^m(\cos \theta) = (-1)^{l+m} \tag{6.197b}$$

$$Pe^{im\phi} = (-1)^m e^{im\phi}. \tag{6.197c}$$

Equations (6.197a–6.197c) imply that

$$P\psi = (-1)^l \psi \tag{6.198}$$

and therefore the parity of a nuclear energy level depends on whether the orbital angular momentum quantum number l is odd or even. Consider two interacting nuclei that have orbital angular momentum quantum numbers l_1 and l_2 respectively. If the orbital angular momentum quantum number of the interacting particles is l then the parity is

$$(-1)^{l_1 + l_2 + l}. \tag{6.199}$$

Nuclear energy levels are determined by the shape and depth of the nuclear potential, and the spin–orbit interaction. The existence of shells (and subshells) of protons and neutrons inside nuclei explains why certain nuclei are particularly stable. In nuclei as in atoms the $1S_{1/2}$ state with a degeneracy of 2 is the lowest energy level. The next nuclear energy level has a degeneracy of 6 because in order of increasing energy it includes the $2P_{3/2}$ and $2P_{1/2}$ states. It follows that the first two nuclear shells are completed if a nucleus contains 8 or more protons (or neutrons). The third nuclear shell contains in order of increasing energy the $1D_{5/2}$, $2S_{1/2}$ and $1D_{3/2}$ states and therefore has a total of 12 protons (or neutrons). Although the 1D state is of lower energy in both spherical harmonic oscillator and uniform spherical well potentials described previously the spin–orbit interaction is sufficient to give the $1D_{3/2}$ state a higher energy in actual nuclei than the 2S state. In a uniform spherical well potential the energy level above the $1D_{3/2}$ state is a 1F state whereas in a spherical harmonic oscillator potential the 2P and 1F energy levels are degenerate. In actual nuclei the $1F_{7/2}$ state forms the fourth nuclear shell, which therefore contains 8 nucleons and is filled in nuclei containing ≥ 28 protons (or neutrons). The fifth nuclear shell consists of the $2P_{3/2}$, $1F_{5/2}$, $2P_{1/2}$ and $1G_{9/2}$ states. Therefore, it contains 22 protons (or neutrons) and is completed if the number of protons (or neutrons) is ≥ 50. The sixth nuclear shell, which is filled if the proton (or neutron) number is ≥ 82, contains the $1G_{7/2}$, $2D_{5/2}$, $1H_{11/2}$, $3S_{1/2}$ and $2D_{3/2}$ states. The seventh and final nuclear shell is completed with a total of 126 neutrons. Filled nuclear shells with 126 protons are not known to exist.

In the discussion above we have seen that a spin–orbit coupling term (i.e. $\boldsymbol{al \cdot s}$) must be added to the simple harmonic oscillator potential to describe nuclei. A term proportional to the square of the orbital angular momentum (i.e. l^2) is also required. Figure 6.1 shows how these two terms affect nuclear systematics. Figure 6.2 gives in the neutron number (N) versus proton number (Z) plane the locations of nuclear isotopes. Black squares denote isotopes sufficiently stable to be found in nature. Figure 6.3 shows how nucleon

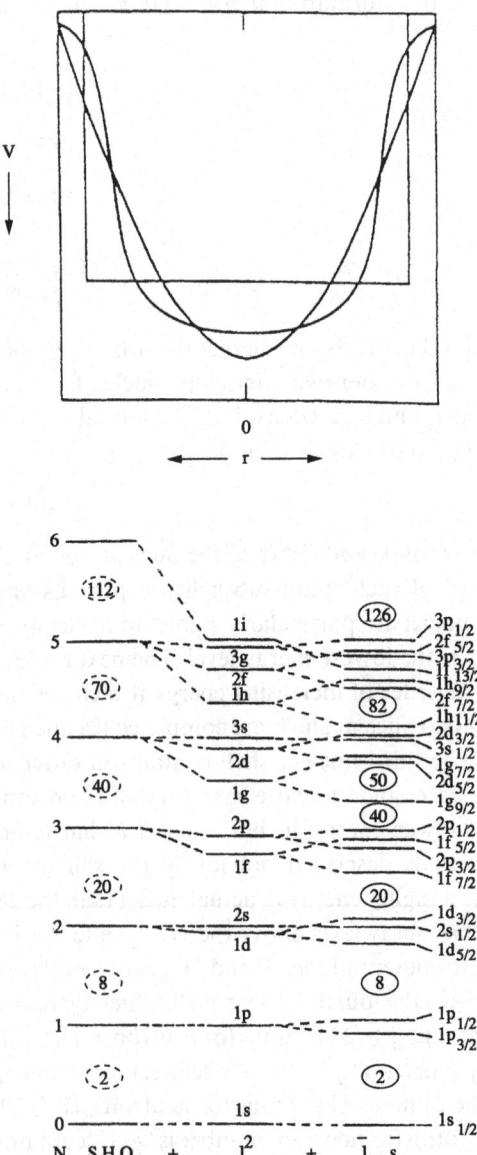

Figure 6.1. The upper figure is a schematic representation of three shell model potentials, a simple harmonic oscillator potential, a square-well potential and an intermediate-shape or modified harmonic oscillator potential. The latter is approximately equivalent to adding to the nuclear potential a term proportional to l^2 (i.e., the square of the orbital angular momentum).

The lower figure gives single-particle energies for a simple harmonic oscillator (SHO), a modified harmonic oscillator with l^2 term, and realistic shell-model potential with l^2 and spin orbit (i.e., l, s terms). From Casten (1990).

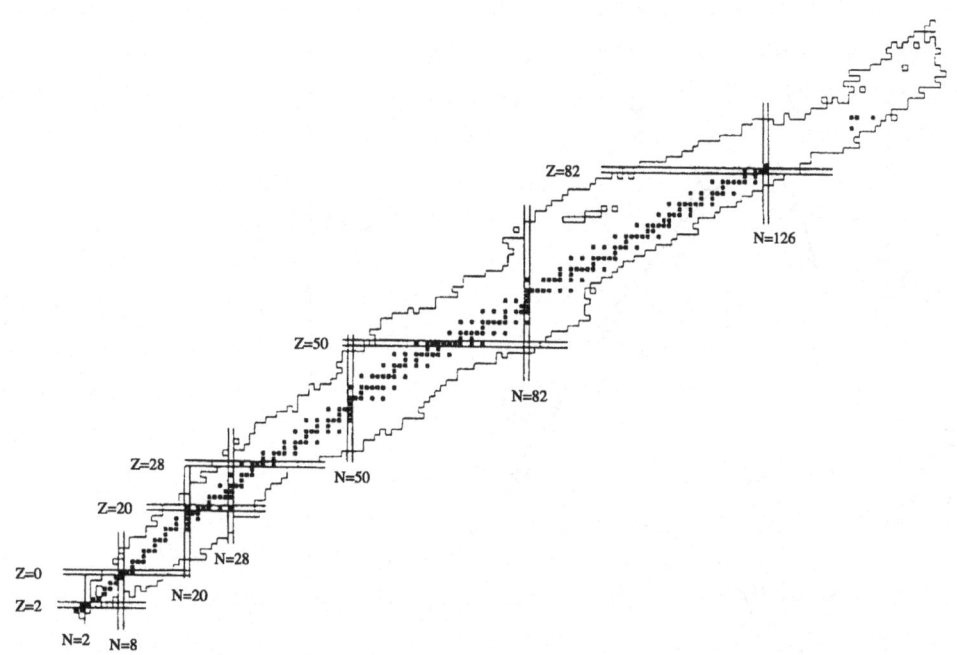

Figure 6.2. Nuclear isotopes shown as a function of neutron number N on the horizontal axis and proton number Z on the vertical axis. Nuclei sufficiently stable to be found in nature are indicated by black squares. Other experimentally observed nuclei are bounded by the thin line. Magic numbers are shown as double lines. From Siemens and Jensen (1987).

(i.e. proton or neutron) magic numbers predicted by the nuclear shell model cause nuclear binding energies to deviate from a smooth curve. Solar and cosmic ray abundances are compared in Figure 6.4.

The principle of charge independence of nuclear interactions states that the strong interaction between nucleons in the same state is the same. Therefore, except for electromagnetic and weak interactions, protons and neutrons are equivalent and we can consider the proton and neutron different charge states of a single particle. In describing the charge independence of nuclear interactions it is convenient to introduce the concept of isospin. Neutrons and protons are each assumed to have an isospin of $\frac{1}{2}$. The third component of isospin is

$$T_3 = Q - \frac{1}{2} \tag{6.200}$$

where Q is the charge of the nucleon in units of electron charge. From Equation (6.200) it follows that protons have $T_3 = +\frac{1}{2}$ and neutrons have $T_3 = -\frac{1}{2}$. Nuclear interactions involving nucleons in the same isospin state are identical.

Although the deuteron is a stable nucleus consisting of a proton and neutron stable nuclei containing two protons (or two neutrons) do not exist. Moreover, the deuteron

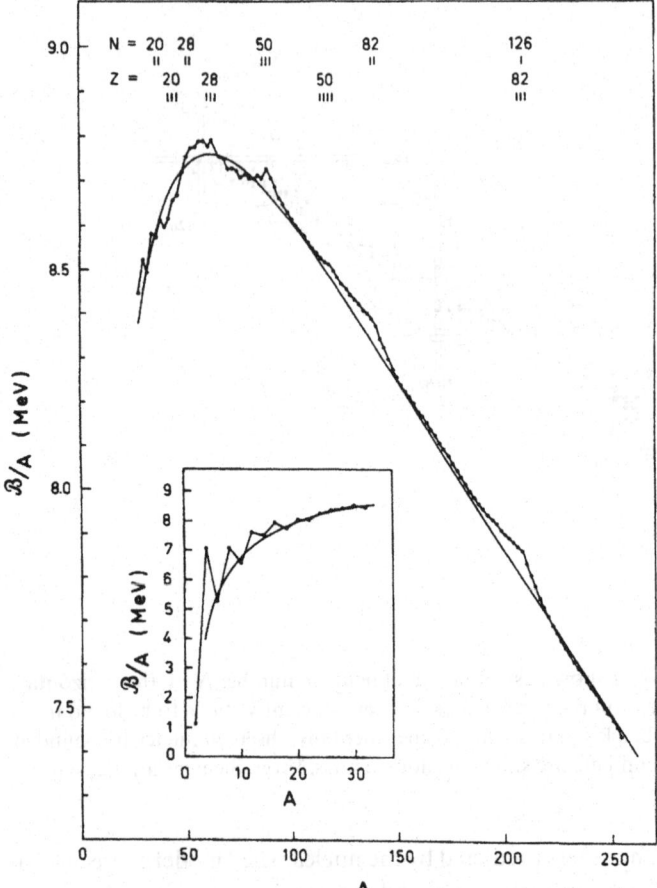

Figure 6.3. Experimental nuclear binding energies. The smooth curve represents the semi-empirical mass formula, which is discussed in standard nuclear physics textbooks. From Siemens and Jensen (1987).

does not have a bound excited state. Two interacting nucleons each with isospin $\frac{1}{2}$ can have either isospin $T = 1$ or 0. An isospin state of $T = 1$ has a degeneracy of $2T + 1 = 3$ whereas an isospin state of $T = 0$ has a degeneracy of $2T + 1 = 1$. The deuteron is a stable nucleus because its isospin state is $T = 0$. If n–n and p–p formed nuclei and a bound first excited state of the deuteron existed then they would have the same energy except for electromagnetic interactions and therefore would constitute a $T = 1$ isobar with p–p, p–n and n–n components designated as $T_3 = +1, 0$ and -1 respectively. Evidence that the deuteron is an isospin singlet is provided by the spin of the deuteron. The measured spin of the deuteron is $S = 1$. Because of the Pauli exclusion principle the deuteron wave function must be antisymmetric with respect to changes in coordinate positions, spin and isospin. It follows that the deuteron wave function must be antisymmetric with respect to changes in isospin coordinates and therefore be an isospin singlet state (i.e. $T = 0$) because the angular momentum is even and spin is equal to 1.

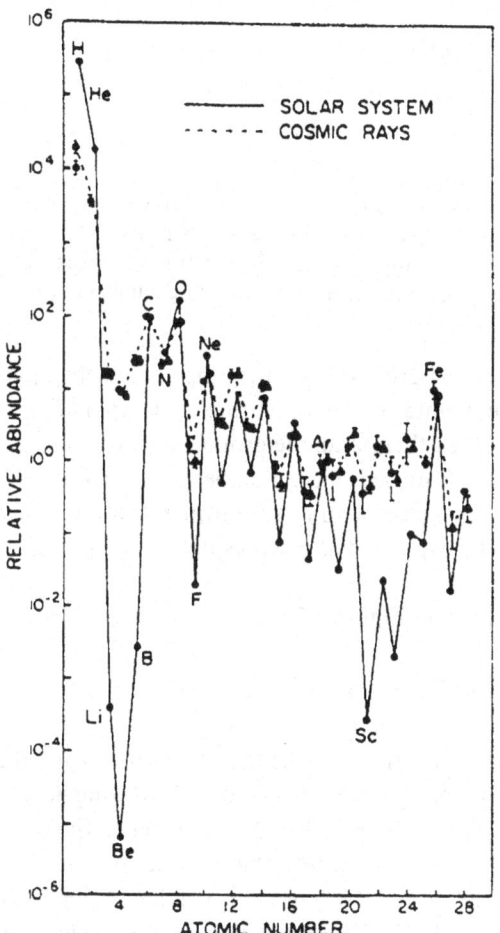

Figure 6.4. Solar abundances of elements compared to cosmic ray abundances as measured above the Earth's atmosphere. The higher abundances of Li, Be and B in cosmic rays are caused by cosmic ray interactions with interstellar gas. Curves are normalized to carbon $= 10^2$. From Rolfs and Rodney (1988).

The existence of nuclear isobars and the utility of the concept of isospin is illustrated by the ^{12}B, ^{12}C and ^{12}N nuclei. The ^{12}C nucleus is stable; however, the ^{12}B nucleus (5 protons and 7 neutrons) and the ^{12}N nucleus (7 protons and 5 neutrons) decay to ^{12}C via the weak interactions

$$^{12}\text{B} \rightarrow {}^{12}\text{C} + \text{e}^- + \bar{\nu}_\text{e}$$
$$^{12}\text{N} \rightarrow {}^{12}\text{C} + \text{e}^+ + \nu_\text{e}. \qquad (6.201)$$

The ground-state energies of the ^{12}B and ^{12}N nuclei are approximately 15 MeV higher than the ground-state energy of the ^{12}C nucleus. Moreover, there exists an excited state of the ^{12}C nucleus with energy approximately equal to the energies of the ground states of

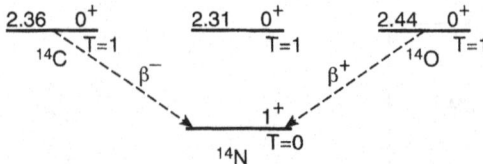

Figure 6.5. Lowest energy levels of ^{14}C, ^{14}N and ^{14}O with excitation energies in MeV shown on the left and spin and parity on the right for each nucleus. ^{14}N has a lower-energy ground state than ^{14}C or ^{14}O because $T = 0$ is allowed. It has more excited states because the number of n–p interactions exceeds the corresponding number of p–p or n–n interactions.

the ^{12}B and ^{12}N nuclei. It follows that the ^{12}N ground state the, approximately equal energy, excited state of ^{12}C and the ground state of ^{12}B constitute an isospin ($T = 1$) triplet with $T_3 = 1, 0, -1$ respectively. On the other hand, the ground state of ^{12}C is an isospin singlet with $T = 0$ and $T_3 = 0$. The ^{12}C nucleus has a number of excited nuclear states besides the particular excited state that is part of an isospin triplet. It follows that most of the excited states of nuclei cannot be explained as purely single-nucleon excitations.

The ^{14}C, ^{14}N, ^{14}O isospin multiplet is shown in Figure 6.5.

6.5   Helium burning

As already discussed in Chapter 2, helium burning involves the triple α reactions and the nuclear reaction ^{12}C(α, γ) ^{16}O. The ^{16}O(α, γ) ^{20}Ne reaction also takes place during helium burning but at much lower rate than the ^{12}C(α, γ) ^{16}O reaction. The first step in the triple α reactions is the formation of ^{8}Be. The ^{8}Be nucleus is unstable and decays into two α-particles with a natural lifetime equal to 2.6×10^{-16} s. The energy difference between two α-particles and the ^{8}Be nucleus is only 92 keV. Helium burning occurs at temperatures $T \gtrsim 10^8$ K and densities sufficiently high that thermodynamic equilibrium exists between ^{4}He and ^{8}Be nuclei. Therefore, from Equation (3.182) the number density of ^{8}Be nuclei is

$$n_{Be} = n_\alpha^2 \frac{h^3}{(2\pi\mu kT)^{3/2}} \exp(-\chi_8/kT) \tag{6.202}$$

with the reduced mass $\mu = 0.5 M_\alpha$ and $\chi_8 = 92$ keV.

The ^{12}C nucleus has a resonance at 7.654 MeV above its ground state. The energy of this excited state of ^{12}C is 278 keV more than the combined rest mass energy of ^{4}He and ^{8}Be. Thermodynamic equilibrium exists between the 7.644 MeV energy level of ^{12}C and ^{4}He (and ^{8}Be) nuclei (i.e. ^{4}He $+$ ^{8}Be $\leftrightarrow$ $^{12}\star$C). Electromagnetic decay to the ground state is forbidden because the 7.644 MeV level and ground state both have zero angular momentum. However, electromagnetic decay to the 4.433 MeV energy level, which has $J = 2$, can occur by the emission of two γ-rays. Decay of the 7.644 MeV excited state of ^{12}C into ^{4}He and ^{8}Be nuclei occurs much more frequently than electromagnetic decay to the 4.433 MeV state. Therefore, from Equation (3.182) the number of ^{12}C nuclei in the 7.644 MeV state becomes

$$n_{12\star C} = n_\alpha n_{Be} \frac{h^3}{(2\pi\mu'kT)^{3/2}} \exp(-\chi_{7.6}/kT) \tag{6.203}$$

with $\chi_{7.6} = 278$ keV and $\mu' = \dfrac{m_\alpha m_{8Be}}{m_{8Be} + m_\alpha}$.

Equations (6.202) and (6.203) imply

$$n_{12\star C} = n_\alpha^3 \frac{h^6 \exp[-(\chi_8 + \chi_{7.6})/kT]}{(\mu'\mu)^{3/2}(2\pi kT)^3}. \tag{6.204}$$

From Equation (6.204) it follows that the 3α reaction rate is

$$r_{3\alpha \to {}^{12}C} = n_{12\star C} \frac{\Gamma_\gamma}{\hbar} \, \text{cm}^{-3} \text{s}^{-1}, \tag{6.205}$$

where $\Gamma_\gamma \simeq 3.6 \times 10^{-3}$ eV and $\Gamma_\gamma/\hbar$ is the decay rate of ^{12*}C.

The ^{12}C(α, γ) ^{16}O reaction occurs simultaneously with the 3α reactions. The α-particle interaction energy E_0 given in Equation (6.29) is approximately 300 keV. The reaction rate of the ^{12}C(α, γ) ^{16}O reaction is primarily the result of two subthreshold resonances, which occur at energies of 7.117 MeV and 6.917 MeV above the ground state of the ^{16}O nucleus. The ^{12}C(α, γ) ^{16}O reaction proceeds primarily through the high-energy tails of these subthreshold resonances located -45 keV and -245 keV below the α-particle threshold. The angular momentum and parity (J^π) of the 7.117 MeV and 6.917 MeV states are 1^- and 2^+ respectively. Therefore, since the ground state is a 0^+ state, electromagnetic decay occurs via single γ-ray (E1) and two γ-ray (E2) decays. The low-energy tail of a $J_\pi = 2^+$ resonance located 2.418 MeV above the α-particle threshold and direct E2 capture to the ground state also contribute to the ^{12}C(α, γ) ^{16}O reaction rate.

Important helium-burning energy levels are shown in Figure 6.6.

The main products of helium burning are ^{12}C and ^{16}O. Because helium burning takes place at lower densities in massive stars than low-mass stars and the 3α reaction rate given in Equation (6.205) varies as the third power of the α-particle number density, more ^{16}O is produced in massive stars than low-mass stars. The 3α reaction rate is more temperature sensitive than the ^{12}C(α, γ) ^{16}O reaction. When thermonuclear runaways occur in helium-burning shells temperatures of $\simeq 3 \times 10^8$ K are attained and approximately 80–90% of the synthesized helium is ^{12}C. Helium-shell flashes occur in low- and intermediate-mass ($M \lesssim 9 \, M_\odot$) red giants. Carbon stars are formed because ^{12}C is mixed into the surfaces of red giants as the result of helium-shell flashes.

6.6   Nucleosynthesis during hydrogen-burning stages

The interstellar and solar system abundance of ^{2}D, which is formed as the result of big-bang nucleosynthesis, is $[^2D]/[H] \simeq 2 \times 10^{-5}$ where $[^2D]$ and $[H]$ are the number

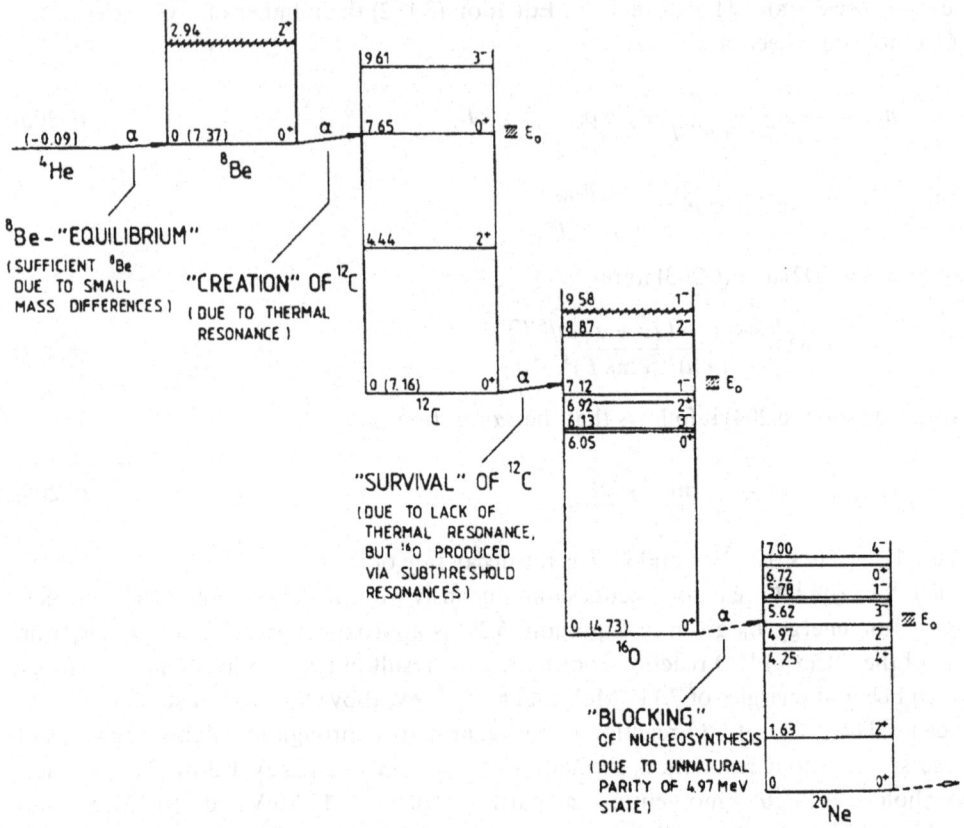

Figure 6.6. Relevant energy levels of nuclei involved in helium-burning reactions. Notice how the parity of the 4.97 MeV energy level affects ^{20}Ne synthesis. From Rolfs and Rodney (1988).

densities of deuterons and protons respectively. Primordial nucleosynthesis also includes the synthesis of large amounts of ^{4}He (i.e. [^{4}He]/[H] $\simeq 0.09$) and significant abundances of ^{3}He and ^{7}Li. Deuterium is depleted during pre-main-sequence evolution because of the ^{2}D(p, γ) ^{3}He reactions in the proton–proton chain. The high burning rate of ^{2}D implies that the abundance of ^{2}D in the hydrogen-burning core (or shell) of a star is always very low. Significant amounts of ^{3}He are present in the hydrogen-burning cores of low-mass stars because the ^{3}He (^{3}He, 2p) ^{4}He and ^{3}He (α, γ) ^{7}Be reactions are slow compared to other nuclear (strong) interactions in the p–p chains. The ^{7}Li (p, α) ^{4}He reaction rate is relatively high and consequently significant abundances of ^{7}Li are not present in hydrogen-burning regions. Although significant abundances of ^{3}He exist within the cores of main-sequence stars ^{3}He is unlikely to reach the stellar surface. Estimated cosmic abundances of both ^{3}He and ^{7}Li are consistent with the predictions of big-bang nucleosynthesis; however, ^{7}Li is synthesized in some giants.

The CNO cycle reactions given by reactions (2.124a–2.124b) are not complete because the ^{17}O(p, γ) ^{18}F reaction competes with the ^{17}O(p, α) ^{14}N reaction. The additional reactions in the CNO cycle are

$$\begin{aligned} ^{17}\text{O} + \text{p} &\rightarrow {}^{18}\text{F} + \gamma, \\ ^{18}\text{F} &\rightarrow {}^{18}\text{O} + \text{e}^+ + \nu, \\ ^{18}\text{O} + \text{p} &\rightarrow {}^{15}\text{N} + {}^4\text{He}. \end{aligned} \qquad (6.206)$$

Reactions (2.124a), (2.124b) and (6.206) include three different ways in which the CNO cycle can be completed. A fourth branch in the CNO cycle is produced by the reactions ^{18}O(p, γ)^{19}F(p, α)^{16}O.

The CN part of the CNO cycle involves the stable isotopes ^{12}C, ^{13}C, ^{14}N and ^{15}N. Because the ^{15}N(p, α) ^{12}C reaction rate is generally higher than the ^{15}N(p, γ) ^{16}O reaction rate the CN part of the CNO cycle goes into equilibrium first. When CN reaction equilibrium is established the relative abundances of the stable isotopes ^{12}C, ^{13}C, ^{14}N and ^{15}N are determined by the thermonuclear reaction rates. Since the ^{15}N(p, γ) ^{15}O reaction rate is high the ^{15}N abundance is depleted and conversely the ^{14}N abundance increased as compared to its solar system value because the ^{14}N(p, γ) ^{15}O reaction is slow. The most important consequence of the CNO cycle from the point of view of nucleosynthesis is the production of ^{14}N. The isotope ^{14}N is therefore synthesized primarily from ^{12}C which has been synthesized during the helium-burning stage of an earlier generation of stars. The isotope ^{12}C is depleted by the CNO cycle and the [^{12}C]/[^{13}C] abundance ratio lowered appreciably as compared to its solar-system value of 90.

Although abundance enhancement of ^{17}O can occur as the result of low temperature ($T < 10^8$ K) hydrogen burning the isotopes ^{15}N and ^{18}O are generally depleted. At high temperature ($T > 10^8$ K) the characteristic reaction time of ^{13}N(p, γ) ^{14}O becomes less than the mean hydrogen-burning timescale and the CNO cycle occurs under nonequilibrium conditions. The β-unstable nucleus ^{14}O which is synthesized at high temperatures decays to ^{14}N. It is likely that ^{15}N is synthesized at high temperatures ($T \simeq 1$–2×10^8 K) in novae or red giants. At such temperatures abundance enhancement of the isotope ^{17}O can occur. The abundance of ^{18}O is limited by the ^{18}O(α, γ) ^{22}Ne reaction. At temperatures $< 10^8$ K the equilibrium value of the abundance ratio [^{13}C]/[^{12}C] is $\simeq 1/4$. Under nonequilibrium conditions the [^{13}C]/[^{12}C] abundance ratio can exceed unity; however at temperatures $> 10^8$ K the ^{13}C isotope is bypassed by the ^{13}N(p, γ)^{14}O reaction.

6.7 Carbon, oxygen and silicon burning

The isotope ^{56}Fe has high binding energy per nucleon. Elements heavier than the iron-group elements Mn, Fe, Co and Ni are synthesized by neutron capture rather than by thermonuclear reactions involving charged nuclei. Carbon, oxygen and silicon burning are the principal thermonuclear burning stages that lead to the formation of elements with mass numbers $20 \leq A \leq 60$.

Carbon burning, which is the next major nuclear burning stage after helium burning, occurs when (and if) the interior temperature of a star exceeds $\simeq 4\text{--}8 \times 10^8$ K. In massive stars ($M \gtrsim 9\,M_\odot$) carbon is ignited at relatively low densities and burning proceeds under conditions of hydrostatic equilibrium. Carbon burning can take place explosively in intermediate-mass stars ($1.4\,M_\odot \lesssim M \lesssim 9\,M_\odot$) because neutrino-emission processes prevent the core from reaching carbon ignition temperatures until after it has become highly degenerate. The interior temperatures of stars less massive than the white dwarf mass limit are kept sufficiently cool by neutrino emission that they evolve into white dwarfs without burning carbon. Moreover, most intermediate-mass stars lose sufficient mass on the asymptotic giant branch that they also become white dwarfs.

The ^{12}C $+$ ^{12}C nuclear reaction proceeds through excited states of the ^{24}Mg nucleus. Its principal reaction channels are ^{12}C(^{12}C, α) ^{20}Ne and ^{12}C(^{12}C, p) ^{23}Na, which occur with approximately equal branching ratios. An appreciable fraction of the ^{23}Na formed by the ^{12}C $+$ ^{12}C reaction is transformed into ^{20}Ne by the reaction ^{23}Na(p, α) ^{20}Ne. The isotope ^{23}Na is also transformed into ^{24}Mg by the ^{23}Na(p, γ) ^{24}Mg reaction. Since ^{16}O is abundant in a stellar core after helium burning the most abundant isotopes upon completion of hydrostatic carbon burning are ^{16}O, ^{20}Ne, ^{23}Na and ^{24}Mg. Important additional nuclear reactions involving the ^{20}Ne nucleus are ^{20}Ne(γ, α) ^{16}O and ^{20}Ne(α, γ) ^{24}Mg(α, γ) ^{28}Si.

In massive stars ^{16}O burning follows carbon burning when interior temperatures exceed $1\text{--}2 \times 10^9$ K. The principal channels of the ^{16}O $+$ ^{16}O reactions are

$$^{16}\text{O} + {}^{16}\text{O} \rightarrow {}^{32\star}\text{S} \rightarrow \alpha + {}^{28}\text{Si}$$
$$\text{p} + {}^{31}\text{P}$$
$$\text{n} + {}^{31}\text{S} \rightarrow {}^{31}\text{P} + \text{e}^+ + \nu_e. \tag{6.207}$$

Nuclear reactions involving the end products of the above reactions also occur and significant abundances of nuclei up to iron-group elements are synthesized.

Silicon burning occurs in massive stars when stellar core temperatures exceed $\simeq 3\text{--}5 \times 10^9$ K. Statistical equilibrium between nuclei with mass numbers $A = 28\text{--}60$ exists as a consequence of α-particle, neutron, proton and photodisintegration reactions at high temperatures. In the cores of massive stars silicon burning takes place under conditions of hydrostatic equilibrium. However, when carbon burning occurs explosively under physical conditions of high-degeneracy silicon burning is also explosive. Because stellar interior densities are not sufficiently high for neutrino trapping to occur statistical equilibrium does not exist between electron capture and β-decay reactions. Therefore, ^{56}Fe is not generally the most abundant iron-peak isotope synthesized by silicon burning. The α-particle reactions and inverse reactions in silicon burning are: ^{28}Si(α, γ)^{32}S(α, γ)^{36}A(α,γ)^{40}Ca(α, γ)^{44}Ti(α, γ)^{48}Cr(α, γ)^{52}Fe(α, γ)^{56}Ni. Alpha-particle reactions from heavier to lighter nuclei can also occur (e.g. ^{28}Si(γ, α)^{24}Mg(γ, α)^{20}Ne(γ, α)^{16}O). In explosive silicon burning the unstable isotope ^{56}Ni is the most abundant end product.

^{56}Ni is subsequently transformed into ^{56}Fe by the electron captures, positron decays and
γ-ray emission:

$$^{56}\text{Ni}(e^-, v)^{56}\star\text{Co and } ^{56}\text{Ni} \rightarrow {}^{56}\star\text{Co} + e^+ + v_e$$
$$^{56}\star\text{Co} \rightarrow {}^{56}\text{Co} + \gamma$$
$$^{56}\text{Co}(e^-, v)^{56}\star\text{Fe and } ^{56}\text{Co} \rightarrow {}^{56}\star\text{Fe} + e^+ + v_e$$
$$^{56}\star\text{Fe} \rightarrow {}^{56}\text{Fe} + \gamma. \tag{6.208}$$

The observed linear magnitude versus time decline of type I supernova light curves is a
consequence of heating caused by absorption of γ-rays emitted by radioactive decays such
as those given by reactions 6.208. Approximately $0.5\,M_\odot$ of ^{56}Ni and other iron-group
elements are formed when a type I supernova undergoes explosive carbon, oxygen and
silicon burning. Observations of γ-rays from the ejecta of Supernova 1987A, which
occurred in the Large Magellanic Cloud, show that ^{56}Co has been synthesized. Therefore,
formation of iron group elements occurs in type II supernovae. However, the amount of
iron-group nucleosynthesis is much less than predicted in a type I supernova outburst.
The estimated amount of ^{56}Ni synthesized by Supernova 1987A is $\simeq 0.07\,M_\odot$.

6.8   Neutron-capture nucleosynthesis

Isotopes with mass number $A \geq 60$ are synthesized by neutron capture on existing iron-
group isotopes. In s-process nucleosynthesis neutron capture occurs in a timescale that is
long as compared to relevant β-decay lifetimes and only isotopes up to ^{209}Bi can be
synthesized. The capture of a neutron by an isotope of charge Z and mass number A is
given by the reaction

$$(Z, A) + n \rightarrow (Z, A + 1) + \gamma. \tag{6.209}$$

If $(Z, A + 1)$ is a stable isotope and exposed to a neutron flux it will eventually capture a
neutron and the isotope $(Z, A + 2)$ will be synthesized. On the other hand, if the isotope
$(Z, A + 1)$ is unstable to β-decay and exposed only to a weak neutron flux it will undergo
β-decay to the isotope $(Z + 1, A + 1)$ before neutron capture can occur. If the positions of
isotopes are plotted in a Z versus $N = A - Z$ plane then the s-process path through this
plane from lighter to heavier isotopes is uniquely determined for most elements. Some
stable isotopes of elements lighter than Pb are bypassed by s-process nucleosynthesis, and
therefore must be synthesized under conditions of rapid neutron capture (r-process).
Elements heavier than Bi can only be synthesized under conditions of very high neutron
flux. Most isotopes along the s-process path(s) in the (Z, N) plane can be synthesized by
r-process nucleosynthesis as well as s-process nucleosynthesis because unstable isotopes
produced by rapid neutron capture undergo β-decays which form isotopes along the
s-process path(s). However, some isotopes along the s-process path(s) in the (Z, N) plane
can only be synthesized under physical conditions of low neutron flux. If N_A is the
number density of a stable s-process isotope then the time rate of change of N_A is
determined by the equation

$$\frac{\mathrm{d}N_A}{\mathrm{d}t} \simeq n(t)\bar{v}[-\sigma_A N_A + \sigma_{A-1} N_{A-1}]. \tag{6.210}$$

In Equation (6.210) $n(t)$ is the neutron number density given as a function of time t, $\bar{v}$ is the mean velocity of a neutron, and σ_A and σ_{A-1} are the neutron capture cross sections of isotopes A and $A - 1$ respectively. For physically relevant gas temperatures the neutron capture cross section of an isotope varies inversely as the energy of the neutron. Under conditions of statistical equilibrium the left-hand side of Equation (6.210) will be small and therefore we have

$$\sigma_A N_A \simeq \sigma_{A-1} N_{A-1}. \tag{6.211}$$

Equation (6.211) implies that σN values of pure s-process isotopes of the same element should be approximately equal. The sets of isotopes (^{122}Te, ^{123}Te, ^{124}Te), (^{134}Ba, ^{136}Ba) and (^{148}Sm, ^{149}Sm, ^{150}Sm) are examples of pure s-process isotopes of the same element. Existing experimental values of neutron capture cross sections and abundances indicate that the $\sigma_A N_A$ values of isotopes within each of the above sets of isotopes are approximately equal. For pure s-process isotopes the product $\sigma_A N_A$ varies slowly as a function of A except close to closed nuclear shells where neutron capture cross sections vary sharply. s-process nucleosynthesis ends at ^{209}Bi which is synthesized after the doubly magic nucleus ^{208}Pb ($Z = 82$, $N = 126$).

s-process nucleosynthesis occurs in red giants as a consequence of helium-shell flashes. During peak burning phases the temperatures within helium-burning shells are $\simeq 3 \times 10^8$ K and neutrons can be formed by the ^{22}Ne(α, n) ^{25}Mg and ^{13}C(α, n)^{16}O reactions. Following a helium-shell flash neutron capture isotopes can be dredged to the surface of a red giant by convection. After a sufficient number of helium-shell flashes a carbon star can be formed if the abundance of ^{12}C in the convective envelope exceeds the abundance of ^{16}O. Examples of s-process elements found in the spectra of some red giants are Zr in the form of ZrO and the unstable isotope ^{99}Tc.

Neutron production via the ^{22}Ne(α, n)^{25}Mg reaction occurs in the following manner. The convective zone that develops as the result of a helium-shell flash extends outwards. The isotope ^{14}N whose abundance has been increased by the CNO cycle is mixed into the helium-burning shell. The isotope ^{22}Ne, which is a source of neutrons, is synthesized by the reactions ^{14}N(α, γ) ^{18}F(e^+, v) ^{18}O(α, γ) ^{22}Ne. Neutron production via the ^{13}C(α, n)^{16}O reaction will occur if protons are mixed into the helium-burning shell during a shell flash. The entropy of the gas within the hydrogen-burning shell is higher than the entropy within the core. The discontinuity in entropy at the boundary layer separating the hydrogen-burning shell and core prevents convection from penetrating into the hydrogen-burning shell. The convective zone produced by a helium-shell flash extends to within less than a pressure scale height of the entropy barrier separating core from hydrogen-burning shell. It is plausible to assume that a small fraction of a pressure scale height of protons are mixed into the helium-burning shell. Protons mixed into the core will rapidly interact with ^{12}C and generate neutrons via the ^{13}C(α, n)^{16}O reaction. Convective overshoot is one possible mechanism for causing protons to be mixed into the core.

Abundances of r-process isotopes that can be formed by both s-process and r-process nucleosynthesis can be estimated by subtracting predicted s-process abundances from measured total abundances. Measured r-process abundance peaks at $A \simeq 80$, $A \simeq 130$ and $A \simeq 195$ are caused by the closure of neutron shells at $N \simeq 50$, 82 and 126. Overabundant r-process isotopes are not the same isotopes synthesized in r-process nucleosynthesis but isotopes to which initially synthesized unstable, neutron-rich isotopes have decayed. s-process abundance peaks are formed near $A = 90$, $A = 140$ and $A = 208$. The abundance peaks of r-process isotopes occur at values of A about 10 atomic mass units less than the corresponding s-process abundance peaks because highly neutron-rich isotopes are initially formed in r-process nucleosynthesis, and consequently the magic neutron numbers $N = 50, 82$ and 126 are reached at lower values of mass number A.

Some r-process nucleosynthesis probably occurs in luminous red giants that are undergoing helium-shell flashes. Supernovae are obvious plausible sites for r-process nucleosynthesis because high temperature and nonequilibrium conditions exist at the time of a supernova outburst. The propagation of a supernova shock wave through the helium zone of a massive star is a plausible mechanism for r-process nucleosynthesis. However, it is not entirely certain that neutron fluxes sufficiently high to synthesize very heavy elements such as Th and U can be caused by such a shock wave. The $^{22}Ne(\alpha, n)^{25}Mg$ and $^{13}C(\alpha, n)^{16}O$ reactions are the principal neutron sources for r-process nucleosynthesis caused by supernova shocks or helium-shell flashes. Intense $^{13}C(\alpha, n)^{16}O$ neutron fluxes can be generated only if the number density of protons is sufficiently low that the $^{13}C(\alpha, n)^{16}O$ reaction is much faster than the $^{13}C(p, \alpha)^{14}N$ reaction because ^{14}N has a large neutron capture cross section.

The reaction $^{22}Ne(\alpha, n)^{25}Mg$ that occurs during helium-shell flashes generates sufficient neutron density (i.e. $\sim 10^{10}\ cm^{-3}$) to explain s-process nucleosynthesis. Although it is possible that some r-process nucleosynthesis in red giants could result from the $^{13}C(\alpha, n)^{16}O$ reaction, synthesis of Th, U and Pu which lie beyond ^{209}Bi require much higher neutron densities ($\gtrsim 10^{18}\ cm^{-3}$) because isotopes with very short half-lives must be synthesized before these relatively stable elements can be formed by neutron capture. The original suggestion (1957) that r-process isotopes are formed in mass just outside the pre-neutron star core and then ejected during a supernova outburst along with other elements synthesized by hydrostatic thermonuclear reactions has not been supported by detailed calculations. The principal reason for obtaining negative results for r-process nucleosynthesis is that previously neglected weak interactions such as coherent neutrino scattering (see Chapter 7) increase the time interval during which neutrinos remain in the layer above the neutron star. Therefore, the number of $v + n \rightarrow e^- + p$ reactions is increased and neutron densities lowered. The physical process (i.e. coherent neutrino scattering) which makes the neutrino transport supernova model more plausible (see Chapter 11) also makes the above r-process scenario less plausible. Nucleosynthesis calculations have indicated that to achieve agreement between solar r-process abundances and those predicted by calculations it is necessary to eject such a large mass of r-process enriched mass that a considerable excess production of r-process elements is also predicted. The possibility of nonspherical mass ejections, however, makes the above criticism not entirely conclusive and therefore the boundary layer of a pre-neutron star is still a possible site for r-process nucleosynthesis.

In Figure 6.7 cosmic abundances are plotted versus atomic weight. The narrow s-process abundance peaks in Figure 6.7 result because measured neutron capture cross sections shown in Figure 6.8 are low at neutron closed shells (i.e. at neutron magic

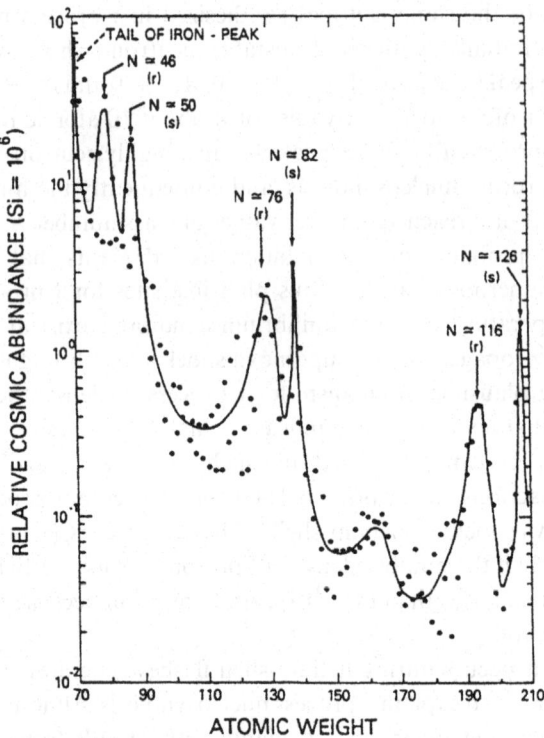

Figure 6.7. Cosmic abundances of heavy elements from Fe to Bi are shown as a function of atomic weight.

numbers). Neutron capture paths for both s- and r-process nucleosynthesis are given in Figure 6.9. The above figures show quite convincingly that elements between iron and bismuth have been synthesized by both s- and r-process nucleosynthesis. As discussed above the most widely accepted interpretation of these results is that s-process nucleosynthesis takes place primarily in asymptotic giant stars whereas r-process nucleosynthesis occurs in supernova explosions. However, this scenario does not explain why s- and r-process abundance peaks are nearly the same. It might seem coincidental that physical processes occurring in very different types of stars should lead to almost the same amount of element production (see Figure 6.7). A possible explanation for comparable synthesis at s- and r-process abundance peaks between iron and bismuth is their production in asymptotic giant-branch (AGB) stars by means of the $^{13}C(\alpha, n)^{16}O$ reaction discussed above. Cowan and Rose (1977) introduced the term I-process (I for intermediate) to describe nucleosynthesis of these elements because neutron fluxes sufficiently high to synthesize them might be generated by mixing protons into the helium-burning region of an AGB star during a shell flash whereas the trans-bismuth actinide rare earths such as Th and U require neutron fluxes much too high to be synthesized in AGB stars. The latter scenario has the advantage that similar abundances might result because both s- and r-process nucleosynthesis occur in similar types of stars but with very different numbers of

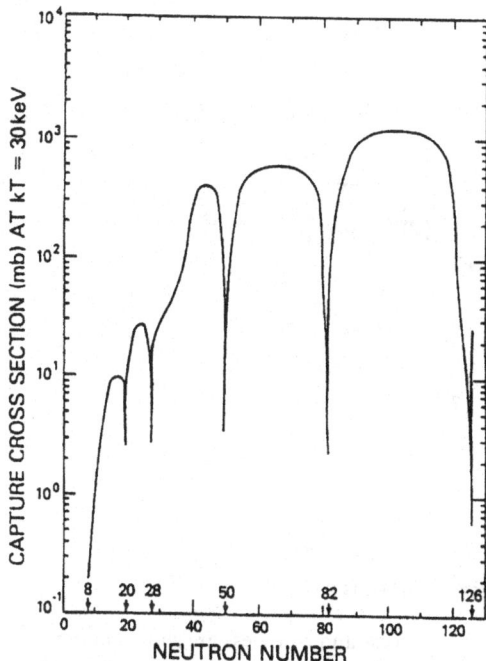

Figure 6.8. Measured average neutron capture cross sections $\langle \sigma \rangle$ are shown as a function of neutron number N. From Rolfs and Rodney (1988).

protons mixed into helium-burning zones during helium-shell flashes. During some flashes proton mixing may not occur at all and therefore only classical s-process can take place. If we accept the conventional point of view that r-process peak elements between iron and bismuth and trans-bismuth elements are synthesized by the classical r-process during supernova explosions then the similar s- and r-process abundance peaks must be considered a coincidence.

Important evidence about nucleosynthesis can be obtained from analysis of meteorites. For example the reaction $^{18}O(\alpha, \gamma)^{22}Ne$ discussed above in this section may explain abundance excesses of ^{22}Ne found in some meteorites. The isotope ^{26}Al, which can be synthesized in supernovae, undergoes the β-decay.

$$^{26}Al \rightarrow {}^{26}Mg + e^+ + \nu_e.$$

with a half-life of 7.1×10^5 years. Excess amounts of ^{26}Mg have been found in some meteorites. One interpretation of this result is that ^{26}Al from a supernova entered the solar system close to the time of its formation (i.e. about 4.55×10^9 years ago). It has been conjectured that a supernova outburst triggered the formation of the solar system.

The excess amounts of Li, Be and B found in cosmic rays as compared to cosmic abundances (see Figure 6.4) are caused by interstellar cosmic-ray spallation reactions between abundant elements such as C, N, O and protons. It has been argued that the entire production of ^6Li, Be and B can be explained by cosmic-ray interactions. As discussed above ^7Li is synthesized during big-bang nucleosynthesis and in some giants as

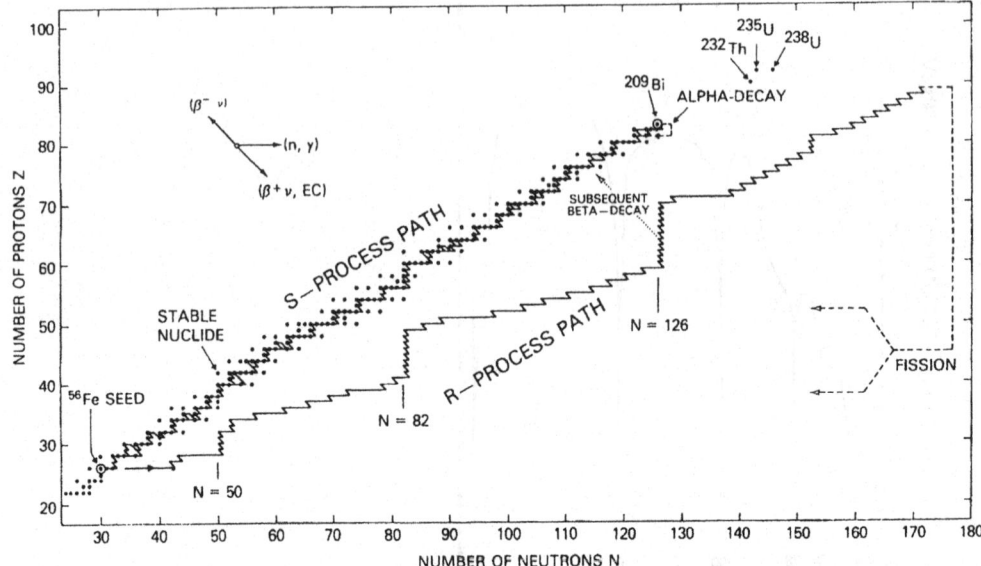

Figure 6.9. Neutron capture paths for the s-process and r-process are shown in the (N,Z)-plane. Both paths start with iron-peak nuclei as seeds (mainly ^{56}Fe). The s- process follows a path along the β-decay stability line and terminates at ^{209}Bi. The r-process produces isotopes far to the neutron-rich side of the stability line and neutron capture continues until β-decayed and neutron-induced fission occur. Although ^{209}Bi undergoes α-particle decay, its half life is $\geq 2 \times 10^{18}$ years. From Rolfs and Rodney (1988).

well as by cosmic-ray interactions. Therefore, it has a higher cosmic abundance. The element Li is known to be depleted in the solar atmosphere.

Problems

1. The p + ^{12}C → ^{13}N + γ reaction occurs in a hydrogen gas of density ρ and temperature T. Evaluate E_0 in Equation (6.29), Δ in Equation (6.36) and the screening factor f in Equation (6.46).

2. Derive Equation (6.46) by repeating the derivation of Equation (6.27) but with the electron Coulomb potential included.

3. Equation (6.46) is referred to as the weak screening limit of the Coulomb correction to the thermonuclear reaction rate. In the strong screening (i.e. high electron density) approximation it is assumed that each interacting nucleus (.e.g ^{12}C) is surrounded by a uniform density electron cloud of total charge equal to the charge of the nucleus (e.g. ^{12}C). The penetration enhancement is calculated by finding the difference in electrostatic potential energies between two isolated ion–electron spheres of charges Z_1e and Z_2e and a single ion–electron sphere of charge $(Z_1 + Z_2)e$. The radii of the electron clouds are equal to the mean distance between nuclei. Find the correction to the thermonuclear reaction rate in the strong screening approximation.

4. Consider a high-density gas of ^{12}C nuclei undergoing nuclear reactions at temperature $T = 5 \times 10^8$ K with E_0 given by Equation (6.29). Find the gas density required for the turning radius r_1 defined by the condition $E_0 = 36\,e^2/r_1$ to equal the mean distance between ^{12}C nuclei.

5. Prove Equations (6.84–6.86).

6. Evaluate the Coulomb penetration probability $v\,|\,\psi_c(0)\,|^2$ with $|\,\psi_c(0)\,|^2$ given by Equation (6.101) for the hydrogen-burning reactions $p + {}^2D \rightarrow {}^3He + \gamma$, ${}^3He + {}^3He \rightarrow {}^4He + 2p$ and $p + {}^{12}C \rightarrow {}^{13}N + \gamma$.

7. The $p + {}^{12}C \rightarrow {}^{13}N^\star \rightarrow {}^{13}N + \gamma$ reaction is a resonant nuclear reaction. Show that $l = 1$ partial waves do not make a significant contribution to the reaction cross section at $T = 30 \times 10^6$ K.

8. a. For the special case of pure elastic resonant scattering the input and output channels are identical. Show that under such physical conditions Equation (6.152) can be derived by assuming that the line shape is Lorentzian about the resonance energy E_r and the scattering cross section equal to the maximum scattering cross section for the $l = 0$ partial wave given by Equation (6.125).

   b. Interacting nuclei have spins of S_1 and S_2, and the angular momentum of the resonant excited state (compound nucleus) is J. Show that for s-wave (i.e. $l = 0$) scattering Equation (6.152) is unaffected by the spin of the incoming nuclei.

   c. Assume that particle energies are sufficiently high for partial waves with $l \neq 0$ to be important. Use arguments similar to those used in part a to show that the predicted resonant scattering cross section is

   $$\frac{2J + 1}{(2S_1 + 1)(2S_2 + 1)} \frac{\pi}{k^2} \frac{\Gamma^2}{(E - E_r)^2 + \left(\dfrac{\Gamma}{2}\right)^2}$$

   with $J = S + l$ and $S = S_1 + S_2$.

7 Weak interactions in stellar interiors

7.1 Introduction

There exist two types of fermions: hadrons and leptons. Hadrons, which include protons, neutrons and mesons, interact with each other via the strong interaction as well as the electromagnetic and weak interactions. Electrons, muons, tau leptons and their respective neutrinos are the six known leptons. Since an antiparticle exists for each lepton the number of leptons and antileptons is 12. Hadrons interact with leptons via the weak interaction. In free space neutrons undergo the weak decay

$$n \to p + e^- + \bar{\nu}_e \tag{7.1}$$

with a half-life of about 10.6 minutes. In a degenerate gas the weak decay of the neutron is inhibited by the Pauli exclusion principle because final states into which the products of the decay must be formed are already occupied with a high degree of probability. Neutrons and protons are also products of the weak interaction and inverse reaction

$$p + e^- \leftrightarrow \nu_e + n. \tag{7.2}$$

Leptons and antileptons undergo weak decay and interaction with each other. Muons (μ^- or μ^+), which are formed during supernova implosions, undergo the weak decays

$$\mu^- \to e^- + \bar{\nu}_e + \nu_\mu \tag{7.3a}$$

$$\mu^+ \to e^+ + \nu_e + \bar{\nu}_\mu. \tag{7.3b}$$

The theory of the weak decay of the neutron and muon are similar except that neutrons interact via the strong interaction and therefore cannot be accurately described as point particles whose wave functions are solutions of the Dirac equation.

Electromagnetic interactions are transmitted by the exchange of photons. For the small time interval Δt during which an interaction between two charged particles occurs the uncertainty principle implies a degree of energy nonconservation ΔE such that

$$\Delta E \Delta t \sim \hbar. \tag{7.4}$$

Because photons travel at the speed of light c the range R implied by expression (7.4) is

$$R \sim c\Delta t. \tag{7.5}$$

Photons have zero mass and therefore can be formed with arbitrarily low energy. Since

ΔE can approach zero in expression (7.4) the range of the electromagnetic interaction given by Equation (7.5) is infinite. Except at high energies the strong interaction is transmitted by the exchange of mesons, and consequently the range r_0 of the strong interaction is approximately equal to the Compton wavelength of a π meson:

$$r_0 = \frac{\hbar}{m_\pi c} \simeq 1.4 \times 10^{-13}\,\text{cm}. \tag{7.6}$$

The range of the weak interaction is much less than that of the strong interaction, and therefore the particles that transmit the weak interaction must be much more massive than π mesons. The weak interaction is the result of the exchange of two massive charged bosons, W^+ and W^-, and the neutral Z boson. The masses of the W and Z bosons have been experimentally determined to be $m_w c^2 \simeq 80\,\text{GeV}$ and $m_z c^2 \simeq 91\,\text{GeV}$ respectively, and consequently the range of the weak interaction is predicted to be $\simeq 2 \times 10^{-16}\,\text{cm}$.

7.2   Solar neutrinos

As previously described in Chapter 2 hydrogen burning can occur either as a consequence of the proton–proton chain or the CNO cycle. At relatively low temperatures ($T \lesssim 18 \times 10^6\,\text{K}$) the proton–proton chain is the dominant energy source whereas at higher temperatures the CNO cycle dominates. Weak decays and interactions occur in both the proton–proton chain and CNO cycle. Because neutrinos have no charge they can interact with matter only by means of the weak interaction. The weak interaction cross section is approximately

$$\sigma = \sigma_0 \left(\frac{E}{mc^2}\right)^2 \tag{7.7}$$

with $\sigma_0 \simeq 1.7 \times 10^{-44}\,\text{cm}^2$, E equal to the energy of the neutrino and m equal to the electron mass. The cross section given by Equation (7.7) is very small, and therefore neutrinos traverse the interiors of most stars with negligible probability of interacting with other particles. Although neutrinos emitted from the hydrogen-burning core of the Sun escape directly from the interior it is obviously difficult to detect solar neutrinos because even a massive detector will capture only a very small fraction of incident neutrinos.

The weak interaction

$$p + p \rightarrow {}^2D + e^+ + \nu_e \tag{7.8}$$

is the initial reaction in the proton–proton chain. The deuteron can also be formed by the weak interaction

$$p + e^- + p \rightarrow {}^2D + \nu_e \tag{7.9}$$

which occurs approximately 400 times less frequently than the p–p reaction. The energy of the neutrino emitted in reaction (7.8) is $\simeq 0.3\,\text{MeV}$, and therefore from Equation (7.7) its interaction cross section is quite low. Moreover, the energy of the neutrino emitted in the p–p reaction is below the threshold energy for the reaction

$$^{37}\text{Cl} + \nu \rightarrow {}^{37}\text{Ar} + e^- \tag{7.10}$$

which is the reaction used in the solar neutrino detector that will be described below. The weak interactions

$$^{7}\text{Be} + e^- \rightarrow {}^{7}\text{Li} + \nu_e \tag{7.11}$$

and

$$^{8}\text{B} \rightarrow {}^{8}\text{Be} + e^+ + \nu_e \tag{7.12}$$

occur in the proton–proton chain that was described in Chapter 2. Neutrinos emitted from the weak decay of ^{8}B are more energetic and therefore more readily detected than those emitted by reactions (7.8), (7.9) or (7.11). If two ^{3}He nuclei collide and produce $^{4}\text{He} + 2p$ then the weak decays (7.11) and (7.12) do not occur. The $^{3}\text{He} + {}^{4}\text{He}$ reaction occurs with higher relative probability than the $^{3}\text{He} + {}^{3}\text{He}$ reaction at high temperatures and high ^{4}He abundances, and consequently more neutrinos are emitted by reactions (7.11) and (7.12) at the center of the Sun than in the outer region of the hydrogen-burning core. Likewise the higher the temperature the higher the relative probability of reaction (7.12) occurring as compared to reaction (7.11). Since neutrinos emitted in reaction (7.12) are more energetic than those emitted in reaction (7.11) it follows that the detectable neutrino flux is sensitive to the central temperature of the Sun.

The CNO cycle described in Chapters 2 and 6 contains the β-decays

$$^{13}\text{N} \rightarrow {}^{13}\text{C} + e^+ + \nu_e, \tag{7.13a}$$

$$^{15}\text{O} \rightarrow {}^{15}\text{N} + e^+ + \nu_e \tag{7.13b}$$

and

$$^{17}\text{F} \rightarrow {}^{17}\text{O} + e^+ + \nu_e. \tag{7.13c}$$

The latter β-decay occurs with relatively low probability because the $^{12}\text{C} + {}^{4}\text{He}$ reaction channel of the $^{15}\text{N} + p$ reaction is produced much more frequently than the $^{16}\text{O} + \gamma$ reaction channel. Most of the neutrinos emitted in reactions (7.13a–17.13c) have energies that are above the threshold energy of 0.81 MeV for the $\nu_e + {}^{37}\text{Cl}$ reaction. Moreover, the neutrino flux from reactions (7.13a) and (7.13b) depends only on the total rate of nuclear energy generation that is a consequence of the CNO cycle.

In the ^{37}Cl solar neutrino detector a tank of $\simeq 100\,000$ gallons of C_2Cl_4 is placed approximately 1 mile underground. The detector must be placed underground to reduce the effect of secondary cosmic-ray protons, which can produce ^{37}Ar by means of the reaction $^{37}\text{Cl}(p, n)\,^{37}\text{Ar}$. The isotope ^{37}Ar which is produced by solar neutrinos interacting with ^{37}Cl has a 35 day radiative half-life. In the solar neutrino experiment the C_2Cl_4 is left undisturbed for about a month. During this time interval some ^{37}Ar is formed by incident solar neutrinos. Argon, which does not react chemically, can be removed from

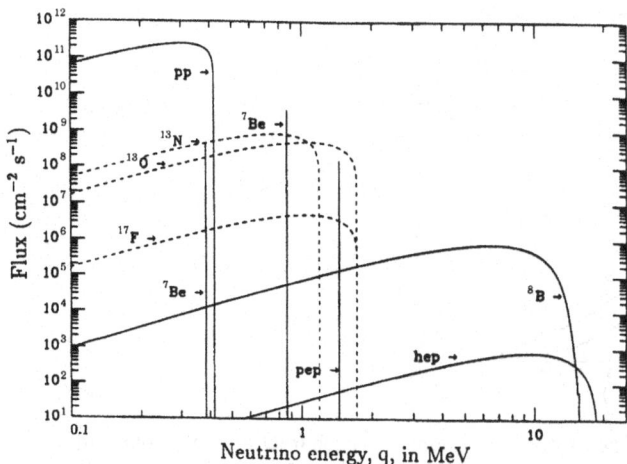

Figure 7.1. The figure shows the neutrino spectrum predicted by a standard solar model. The neutrino fluxes from continuum sources such as p–p and ^{8}B are given in units of number per cm^2 per second per MeV at one astronomical unit. The line fluxes (pep and ^{7}Be) are given in number per cm^2 per second. The neutrino spectra from the pp chain are drawn with solid lines whereas the CNO spectra are drawn with dotted lines. From Bahcall (1989).

the liquid C_2Cl_4 with helium gas and then separated from the helium gas. Electron capture of K shell ^{37}Ar electrons produces 2.8 keV Auger electrons which are formed simultaneously with ^{37}Ar electron capture. The emitted Auger electrons can be detected with a proportional counter. The flux of p–p neutrinos incident on the Earth can be readily estimated. At the top of the Earth's atmosphere the solar constant is $S \simeq 1.37 \times 10^6 \, \mathrm{erg\,cm^{-2}\,s^{-1}}$. If the Sun's luminosity is entirely a consequence of the proton–proton chain then one p–p reaction neutrino is emitted for each 13 MeV of energy in the form of solar radiation. It follows that the neutrino flux from the p–p reaction is

$$\phi_\nu = \frac{S}{13\,\mathrm{MeV}} \simeq 6.7 \times 10^{10} \, \mathrm{neutrino\,cm^{-2}\,s^{-1}}. \tag{7.14}$$

The relative fluxes of other neutrinos emitted in the proton–proton chain and CNO cycle depend on the precise temperature distribution and composition of the solar interior. Neutrinos emitted from the solar interior have been detected. The measured neutrino flux, which is insensitive to p–p reaction neutrinos because they are below threshold, is approximately a factor of three less than predicted by solar interior models.

Figure 7.1 gives the predicted neutrino fluxes based on an interior model of the Sun. Note that fluxes are in units of cm^{-2} s^{-1} for neutrinos emitted at a fixed energy; however, the units of flux are cm^{-2} s^{-1} MeV^{-1} for neutrinos whose spectrum is continuous in energy. Equation (7.7) implies that relatively high-energy neutrinos such as those emitted in the weak decay of ^{8}B have a much higher capture cross section than lower-energy

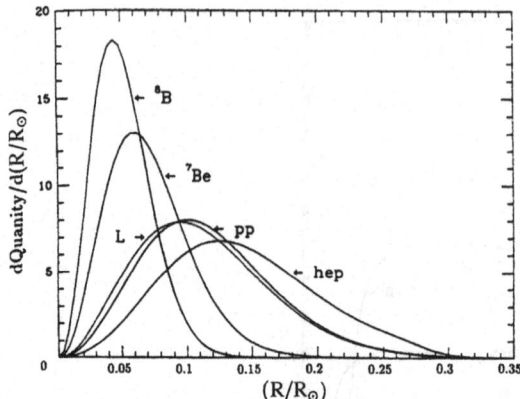

Figure 7.2. The fraction of neutrinos that originate in each fraction of the solar radius is [dFlux/d($R/R_\odot$)]. The figure gives the production fractions for ^{8}B, ^{7}Be, pp, and hep neutrinos for a standard solar model. From Bahcall (1989).

neutrinos which can be below the detection threshold. Because the various nuclear reactions involved in the p–p chain depend sensitively on temperature and also density neutrinos from different sources are preferentially emitted from different regions of the hydrogen-burning core of the Sun. Figure 7.2 shows the predicted distribution with respect to the solar center of different neutrino sources. The ^{8}B neutrino flux is highly dependent on the precise temperature at the center of the Sun. We note that if the CNO cycle were the relevant nuclear energy source then predicted neutrino flux would depend only on the solar luminosity and therefore be independent of calculated solar models. In Figure 7.3 a comparison is made between a computed solar interior model and the neutrino experiment of Davis and collaborators.

7.3   Neutrino emission and stellar evolution

The presence of neutrino emission processes in the proton–proton chain and CNO cycle imply that a small fraction of the total thermonuclear energy release does not contribute to maintaining hydrostatic or thermal equilibrium, and therefore the main-sequence lifetime of a star is slightly less than it would be if all thermonuclear energy release were available as heat input. In the hot cores of evolved stars neutrino-emission processes remove thermal energy from the gas in the absence of thermonuclear reactions. The three most important of these neutrino-emission processes are electron pair annihilation into a neutrino–antineutrino pair, the plasma neutrino process and the photoneutrino process.

At temperatures $\gtrsim 10^9$ K electron–positron pairs are produced in stellar interiors because they are in equilibrium with the radiation field. This equilibrium is described by the reaction and inverse reaction,

$$e^- + e^+ \leftrightarrow \gamma + \gamma. \tag{7.15}$$

Electron–positron pairs can annihilate via the weak interaction

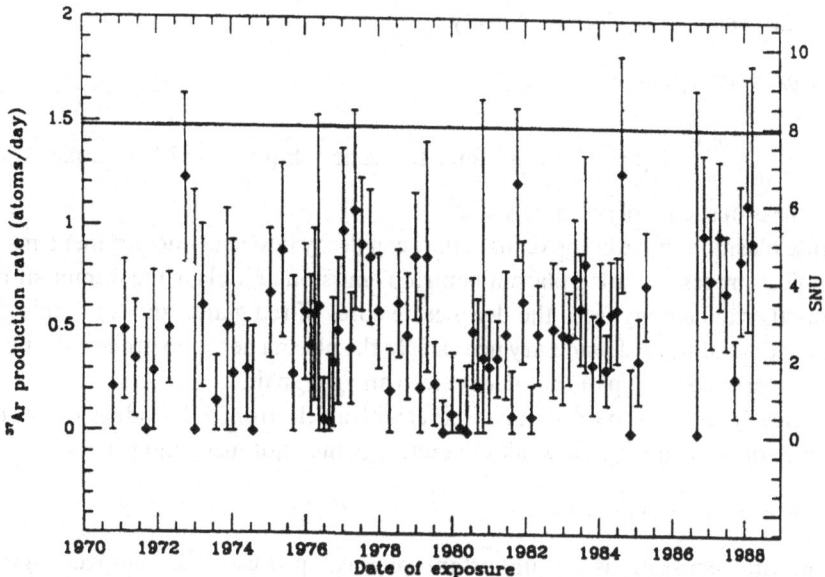

Figure 7.3. The experimental results of the solar-neutrino experiment of R. Davis and collaborators is shown for the period 1971–88. The line at 7.9 SNU is the ^{37}Ar production rate predicted by a standard solar model. From Bahcall (1989).

$$e^- + e^+ \rightarrow v_e + \bar{v}_e. \tag{7.16}$$

The reaction rate of the interaction given in Equation (7.16) can be readily estimated if the temperature and density of the gas is specified. The average rate of energy generation in units of $\mathrm{erg\,cm^{-3}\,s^{-1}}$ becomes

$$Q_{\mathrm{pair}} \simeq n_+ n_- \overline{\sigma v E} \tag{7.17}$$

where n_+ and n_- are the number densities of positrons and electrons respectively, σ is the weak interaction cross section given in Equation (7.7), $\bar{v} \simeq c$ is the relative velocity between positrons and electrons and $\bar{E}$ is the average energy of a neutrino–antineutrino pair. The calculation of n_+ and n_- in Equation (7.17) was described in Chapter 3. The pair annihilation process can become important in the evolved cores of very massive stars; however, the plasma neutrino and photoneutrino processes, which are described below, play the dominant role in affecting the evolution of low- and intermediate-mass stars.

Because it has zero rest mass a photon in free space cannot decay into an electron–positron pair or neutrino–antineutrino pair and conserve both momentum and energy. Consider the possibility of the reverse process occurring (i.e. the annihilation of an e^+–e^- pair or v_e–$\bar{v}_e$ pair into a photon). In the center-of-mass frame of either an e^-–e^+ pair or v_e–$\bar{v}_e$ pair the momentum and energy of an emitted photon would have to be zero. It follows that the inverse processes namely photon decay into e^-–e^+ pairs or v_e–$\bar{v}_e$ pairs cannot take place. However, photons in a dense gas have an effective mass and are referred to as plasmons. The dispersion relation of a transverse plasmon in a nondegenerate, nonrelativistic gas of electron density n_e is

$$\hbar^2\omega^2 = \hbar^2\omega_p^2 + k^2c^2 \tag{7.18}$$

with $\omega_p = \sqrt{\dfrac{4\pi n_e e^2}{m}}$ equal to the plasma frequency. Equation (7.18) implies that the effective mass of such a plasmon is $\hbar\omega_p/c^2$.

Because plasmons have effective mass they can decay into neutrino–antineutrino pairs. The physical process is called plasma neutrino emission and plays the dominant role in removing thermal energy from the degenerate cores of red giants and hot white dwarfs such as the central stars of planetary nebulae. In the plasma neutrino process the plasmon produces a virtual e^--e^+ pair that decays into an $\nu_e-\bar{\nu}_e$ pair.

A sufficiently energetic photon can interact with an electron and produce two electrons and a positron. The analogous weak interaction is the photoneutrino process:

$$\gamma + e^- \rightarrow e^- + \nu_e + \bar{\nu}_e. \tag{7.19}$$

Photoneutrino emission is an important physical process that removes significant amounts of thermal energy from stars with hot, degenerate cores.

In a nondegenerate gas the removal of thermal energy causes the star to contract. The virial theorem implies that as the density increases the temperature of the gas must also increase. However, in a degenerate gas the pressure is nearly independent of temperature and consequently the removal of thermal energy will cause the temperature to decrease. In the degenerate cores of red giants less massive than $\simeq 9\,M_\odot$ the effect of plasma neutrino emission and (to a lesser extent photoneutrino emission) is to prevent the temperature from becoming sufficiently high for carbon burning to occur until after the core mass becomes approximately equal to the white-dwarf mass limit and the central core density $\simeq 1\text{--}2 \times 10^9\,\mathrm{g\,cm^{-3}}$. Under such conditions of very high electron degeneracy carbon burning is explosive and a supernova outburst results. Neutrino emission prevents carbon burning from occurring in stars with white-dwarf end states. Another important effect of neutrino emission is to cause the luminosities of the central stars of planetary nebulae to decrease more rapidly than they would if the plasma and photoneutrino processes did not occur. The observationally inferred lifetimes of the central stars of planetary nebulae are consistent with those obtained with stellar evolutionary models in which the plasma neutrino and photoneutrino processes are included but not with stellar models calculated with these neutrino emission processes neglected. It follows that observations of the central stars of planetary nebulae provide evidence that the plasma neutrino emission process and the physically similar photoneutrino and pair annihilation processes are real physical processes. Figure 7.4 shows in the $\log\rho$ versus $\log T$ diagram where neutrino processes discussed above are dominant.

7.4   Weak interactions in presupernova stars, supernovae and neutron stars

In massive white dwarfs and the cores of red giants with masses close to the white-dwarf

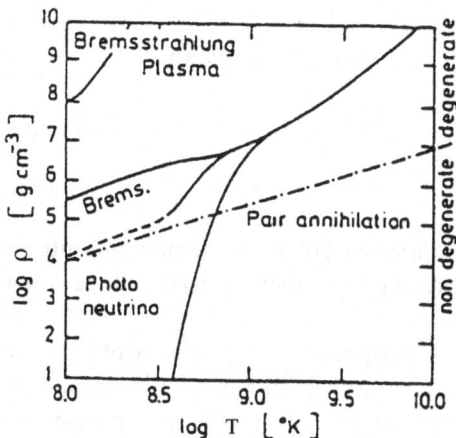

Figure 7.4. The lines in the $\log \rho$ versus $\log T$ diagram separate it into regions where different neutrino processes dominate. From Festa and Ruderman (1969).

mass limit the energies of electrons are sufficiently high that some nuclei can capture free electrons. If the degenerate core of a red giant consists of ^{56}Fe then at a density of $1.14 \times 10^9 \, \mathrm{g \, cm^{-3}}$ the threshold kinetic energies are sufficiently high for the following reactions to occur:

threshold energy

$$^{56}\mathrm{Fe} + \mathrm{e}^- \to {}^{56}\mathrm{Mn} + \nu \quad (3.7 \, \mathrm{MeV}) \tag{7.20a}$$

$$^{56}\mathrm{Mn} + \mathrm{e}^- \to {}^{56}\mathrm{Cr} + \nu \quad (< 3.7 \, \mathrm{MeV}). \tag{7.20b}$$

The second of the above electron captures occurs immediately after the first because the threshold energy is lower. Further electron capture does not occur until the central density equals $1.9 \times 10^{10} \, \mathrm{g \, cm^{-3}}$. Consider a degenerate core that consists primarily of ^{16}O and ^{12}C. At densities $\gtrsim 1.9 \times 10^{10} \, \mathrm{g \, cm^{-3}}$ the electron capture reactions

threshold energy

$$\mathrm{e}^- + {}^{16}\mathrm{O} \to {}^{16}\mathrm{N} + \nu_\mathrm{e} \quad (10.4 \, \mathrm{MeV}) \tag{7.21a}$$

$$\mathrm{e}^- + {}^{16}\mathrm{N} \to {}^{16}\mathrm{C} + \nu_\mathrm{e} \quad (< 10.4 \, \mathrm{MeV}). \tag{7.21b}$$

occur. At the somewhat higher density of $3.9 \times 10^{10} \, \mathrm{g \, cm^{-3}}$ the kinetic energies of electrons become sufficiently high (i.e. 13.37 MeV) to be above threshold for the reactions

$$\mathrm{e}^- + {}^{12}\mathrm{C} \to {}^{12}\mathrm{B} + \nu_\mathrm{e} \tag{7.22a}$$

$$\mathrm{e}^- + {}^{12}\mathrm{B} \to {}^{12}\mathrm{Be} + \nu_\mathrm{e}. \tag{7.22b}$$

When electron kinetic energies exceed about 20 MeV neutron-rich isotopes will tend to emit free neutrons rather than capture electrons. Such emission of neutrons by nuclei is referred to as neutron drip. Electron capture removes degenerate electrons, and therefore the gas pressure, which is caused primarily by the electron degenerate gas, is reduced.

Electron captures that are followed by β-decays and further electron captures are known as the Urca process. If Z is the number of protons in a nucleus and N the number of neutrons then the Urca process becomes

$$e^- + (Z + 1, N - 1) \rightarrow \nu_e + (Z, N) \tag{7.23a}$$

$$(Z, N) \rightarrow (Z + 1, N - 1) + e^- + \bar{\nu}_e. \tag{7.23b}$$

The above reactions are cyclic as far as the nuclei are concerned but the emitted neutrinos and antineutrinos leave the core without interaction with other particles and the core is cooled.

Neutron stars are formed as the result of the implosion of the evolved core of a massive star. Since the gravitational binding energy of a neutron star is $\simeq 10^{53}$ erg and the binding energy of the core of the presupernova star is $\simeq 3 \times 10^{50}$ erg approximately 10^{53} erg must be released during the implosion. Most of the required energy release is in the form of neutrinos. Recently (Feb. 1987) neutrinos have been observed from a supernova outburst in a massive star within the Large Magellanic Cloud. This supernova outburst is the brightest observed supernova since the time of Kepler. The characteristic implosion timescale of a highly degenerate core is $\simeq 2$ seconds. Since the characteristic collapse timescale varies as $1/\sqrt{\rho}$ the final stages of collapse which occur after the core density becomes greater than the nucleon density inside the atomic nucleus (i.e. 3×10^{14} g cm^{-3}) lasts only about 1 ms. If the temperature inside the core exceeds 10^{11} K π^- mesons are produced. At sufficiently low densities the production of π^- mesons is followed by the weak decays

$$\pi^- \rightarrow \mu^- + \bar{\nu}_\mu \tag{7.24a}$$

$$\mu^- \rightarrow e^- + \bar{\nu}_e + \nu_\mu. \tag{7.24b}$$

In the dense interiors of imploding stars π^- mesons produce muons and neutrinos via the reactions

$$\pi^- + n \rightarrow n + e^- + \nu_e \tag{7.25a}$$

$$\pi^- + n \rightarrow n + \mu^- + \nu_\mu \tag{7.25b}$$

and muons interact via the reaction

$$n + p + \mu^- \rightarrow n + n + \nu_\mu. \tag{7.26}$$

The muon neutrinos $(\nu_\mu, \bar{\nu}_\mu)$ would not be strongly coupled to matter during supernova implosions if it were not for the scattering process

$$\nu_\mu + n \rightarrow \nu_\mu + n \tag{7.27a}$$

$$\nu_\mu + p \rightarrow \nu_\mu + p \tag{7.27b}$$

$$\nu_\mu + (Z, A) \rightarrow \nu_\mu + (Z, A). \tag{7.27c}$$

All three of the above scattering processes are mediated by the Z boson and not by the charged W bosons. At the relevant energies the de Broglie wavelength of the neutrino is greater than the dimension of the nuclei and consequently scattering from heavy nuclei is coherent. Because scattering is coherent the cross section for scattering from a heavy nucleus of nucleon number A is proportional to A^2 rather than A.

In addition to the scattering process

$$\nu_e + n \rightarrow \nu_e + n, \tag{7.28a}$$

$$\nu_e + p \rightarrow \nu_e + p, \tag{7.28b}$$

and

$$\nu_e + (Z, A) \rightarrow \nu_e + (Z, A), \tag{7.28c}$$

which are similar to those of muon neutrinos given in Equations (7.27a–7.27c), electron neutrinos and antineutrinos interact with matter via the reactions:

$$\nu_e + n \rightarrow e^- + p \tag{7.29a}$$

$$\nu_e + e^- \rightarrow \nu_e + e^- \tag{7.29b}$$

$$\bar{\nu}_e + p \rightarrow e^+ + n \tag{7.29c}$$

$$\bar{\nu}_e + p + n \rightarrow n + n + e^+. \tag{7.29d}$$

Reactions (7.29a), (7.29c) and (7.29d) are nucleon absorption reactions whereas reaction (7.29b) is electron–neutrino scattering. The observed kinetic energies of supernova ejecta from massive stars result from the transfer of neutrino momenta to mass that is ejected from the star.

The photospheric temperatures of neutron stars decrease as their central temperatures become lower. Although the neutron density is highest the interiors of neutron stars also contain degenerate protons, degenerate electrons and heavy nuclei. Neutron stars are cooled as the result of the following physical process:

$$e^- + (Z, A) \rightarrow e^- + (Z, A) + \nu_e + \bar{\nu}_e \tag{7.30}$$

$$n + n \rightarrow n + p + e^- + \bar{\nu}_e \tag{7.31}$$

$$n + p \rightarrow n + p + \nu_e + \bar{\nu}_e \tag{7.32}$$

$$n + p + e^- \rightarrow n + n + \nu_e. \tag{7.33}$$

Reaction (7.30) is known as neutrino pair bremsstrahlung and is the weak interaction analogue of $e^- - e^+$ pair production that results when energetic electrons interact with the Coulomb field of a nucleus. Reactions (7.31–7.33) are inhibited by the circumstance that neutrons and protons are degenerate in neutron stars, and consequently most reactions that would occur if final states were unoccupied do not occur because of the Pauli exclusion principle.

If the central density in the interior of a neutron star becomes sufficiently high π^- mesons can be produced by the reactions:

$$n + n \rightarrow n + p + \pi^- \tag{7.34}$$

$$n + e^- \rightarrow n + \pi^- + \nu_e \tag{7.35}$$

$$n + \mu^- \rightarrow n + \pi^- + \nu_\mu. \tag{7.36}$$

The reverse reactions

$$\pi^- + n \rightarrow n + e^- + \bar{\nu}_e \tag{7.37}$$

$$\pi^- + n \rightarrow n + \mu^- + \bar{\nu}_\mu \tag{7.38}$$

also produce neutrinos which escape from the neutron star in a timescale much less than the cooling timescale associated with the flow of electromagnetic radiation. If densities and temperatures are sufficiently high for significant numbers of π^- mesons to exist then reactions (7.34–7.38) will cause rapid cooling. The density inside a neutron star may not be high enough for reactions (7.34–7.36) to be above threshold. Under such physical conditions the cooling of a neutron star might result from the much slower reactions (7.30–7.33).

It has been suggested that pion–nucleon condensations (i.e. bound states consisting of a pion and nucleon) can form at neutron densities too low for reactions (7.34–7.36) to become significant. If such pion–nucleon condensations are formed as the result of strong interactions between nucleons then the pion–nucleon condensation N' will undergo the weak decay

$$N' \rightarrow N + e^- + \bar{\nu}_e. \tag{7.39}$$

In reaction (7.39) N is a free nucleon (most likely a neutron). Neutrinos emitted as the result of the formation and β-decay of pion–nucleon condensations could lead to the rapid cooling of the interior of a neutron star. Upper limits on the surface temperatures of a number of neutron stars have been obtained with surface temperatures of two or three neutron stars actually determined. These measurements indicate that decay of π^- mesons or pion–nucleon condensations have caused the interiors of neutron stars to cool rapidly.

7.5   Weak-interaction decay rates and cross sections

The starting point of the theory of weak interactions is the assumption that a β-decay such as the decay of the neutron can be calculated from time-dependent perturbation theory. From Equation (4.52) the expansion coefficient $a_m(t)$ for an assumed time-dependent perturbed Hamiltonian H'_{mk} becomes

$$a_m = \frac{-iH'_{mk}}{\hbar} \int_0^t e^{i\omega_{mk}t} dt = \frac{H'_{mk}}{\hbar \omega_{mk}} (1 - e^{i\omega_{mk}t})$$

$$= \frac{-2iH'_{mk}}{\hbar \omega_{mk}} e^{i\omega_{mk}t/2} \sin\left(\frac{\omega_{mk}t}{2}\right). \tag{7.40}$$

Equation (7.40) implies that the transition probability from the initial state to some final state m is

$$|a_m|^2 = \frac{4|H'_{mk}|^2}{\hbar^2 \omega_{mk}^2} \sin^2\left(\frac{\omega_{mk}t}{2}\right). \tag{7.41}$$

If the final states can be regarded as continuous then from Equation (7.41) we have

$$\sum_m |a_m|^2 = \int_{-\infty}^{\infty} |a_m(t)|^2 \rho\, dE = \int_{-\infty}^{+\infty} |a_m(t)|^2 \rho \frac{dE}{d\omega}\, d\omega$$

$$= \frac{|H'_{mi}|^2}{\hbar} \int_{-\infty}^{\infty} \rho \frac{\sin^2 \omega t/2}{(\omega/2)^2}\, d\omega = \frac{2\pi}{\hbar} |H'_{mi}|^2 \rho t \tag{7.42}$$

where ρ is the density of final states. Equation (7.42) implies that the transition probability per unit time out of initial state i is

$$\Gamma = \frac{2\pi}{\hbar} |H'_{mi}|^2 \rho. \tag{7.43}$$

From Equation (7.43) it follows that the decay rate of the neutron is

$$d\Gamma = \frac{2\pi}{\hbar} \frac{1}{2} \sum_{\text{spin}} |H'_{mi}|^2 \rho_e \rho_{\bar{\nu}} \delta(E_e + E_{\bar{\nu}} - E_0) dE_e dE_{\bar{\nu}} \tag{7.44}$$

where E_0 is the energy difference including rest mass energy between the initial neutron and final proton, ρ_e and $\rho_{\bar{\nu}}$ are the densities of electron and antineutrino states in the energy intervals E_e, $E_e + dE_e$ and $E_{\bar{\nu}}, E_{\bar{\nu}} + dE_{\bar{\nu}}$ respectively, H'_{mi} is the interaction Hamiltonian matrix element connecting initial and final states. Equation (7.44) includes an average over initial spin states and a summation over final spin states.

In the original Fermi theory of weak interactions H'_{mi} is assumed to have the form

$$H'_{mi} = G_F \int \psi_p^\star \psi_e^\star \psi_{\bar{\nu}} \psi_n d^3 r \tag{7.45}$$

where G_F is a coupling constant, ψ_n is the wave function of the initial neutron and ψ_p, ψ_e and $\psi_{\bar{\nu}}$ are the wave functions of the proton, electron and antineutrino. The integral in the matrix element H'_{mi} given in Equation (7.45) is the probability amplitude that all four particles exist at the same point in space. We can assume that the weak interaction is a point interaction and therefore we have

$$H'_{mi} = \int \Psi_m^\star(r) V_w(r,r') \Psi_i(r') d^3 r d^3 r' \tag{7.46}$$

with $\Psi_m = \psi_p \psi_e \psi_{\bar{\nu}}$, $\Psi_i = \psi_n$ and $V_w(r,r') = G_F \delta(r - r')$. The wave functions of the electron and antineutrino are plane-wave solutions with de Broglie wavelengths much greater than the dimensions of the nucleons. It follows that

$$\psi_e \psi_{\bar{\nu}} = e^{i k_e \cdot r} e^{i k_{\bar{\nu}} \cdot r} \simeq 1 + i(k_e + k_{\bar{\nu}}) \cdot r \simeq 1. \tag{7.47}$$

Equation (7.47) implies that Equation (7.46) becomes

$$H'_{mi} = \int\int \psi_p^\star(r) G_F \delta(r - r') \psi_n(r') \mathrm{d}^3 r \mathrm{d}^3 r'$$

$$= G_F \int \psi_p^\star(r) \psi_n(r) \mathrm{d}^3 r. \qquad (7.48)$$

The overlap integral in Equation (7.48) is approximately equal to unity for neutron decay because the wave function of the neutron and proton are nearly the same. In nuclei overlap integrals similar to those of Equation (7.48) can be quite different than unity. In particular they can be equal to zero if a β-decay is not allowed.

The form of the interaction Hamiltonian given in Equation (7.45) is not unique because there are other Lorentz-invariant ways to formulate the theory of weak interactions. If ψ_1 and ψ_2 are two solutions of the Dirac equation then it can be shown (see Appendix 2) that under a Lorentz transformation with reflections $\bar{\psi}_1\psi_2$ transforms as a scalar, $\bar{\psi}_1\gamma^\lambda\psi_2$ transforms as a vector and $\bar{\psi}_1\gamma^\lambda\gamma^5\psi_2$ transforms as an axial vector. Corresponding products involving $\bar{\psi}_1$ and ψ_2 that transform as tensors and pseudoscalars can also be constructed. Leptons can be regarded as point particles and therefore since they are fermions their wave functions satisfy the Dirac equation.

In the standard theory of weak interactions the weak interaction matrix element for muon decay is

$$M_\mu = \frac{G_F}{\sqrt{2}} (\bar{\psi}_e(1 - \gamma_5)\gamma^\lambda\psi_{\nu_e})(\bar{\psi}_{\nu_\mu}(1 - \gamma_5)\gamma_\lambda\psi_\mu). \qquad (7.49)$$

The matrix element given in Equation (7.49) includes vector and axial vector terms. The theory of weak interactions based on the above matrix element is known as the V–A theory of weak interactions. If neutrons and protons could be regarded as Dirac particles (i.e. particles whose wave functions satisfy the Dirac equation) then in the standard theory of weak interactions the matrix element for neutron decay would be

$$M_n = \frac{G_F}{\sqrt{2}} (\bar{\psi}_p(1 - \gamma_5)\gamma^\lambda\psi_n)(\bar{\psi}_e(1 - \gamma_5)\gamma_\lambda\psi_\nu). \qquad (7.50)$$

Because both the neutron and proton interact via the strong interaction the neutron decay matrix element is expressed as

$$M_n = \frac{G_F}{\sqrt{2}} (\bar{\psi}_p(C_V - C_A\gamma_5)\gamma^\lambda\psi_n)(\bar{\psi}_e(1 - \gamma_5)\gamma_\lambda\psi_\nu) \qquad (7.51)$$

where C_V and C_A are constants whose values must be found from experiment. The values of C_V and C_A have been determined to be $C_V \simeq 1$ and $C_A \simeq 1.25$. The absolute values of weak interaction matrix elements can be inferred from weak interaction decay lifetimes.

In our discussion of weak interactions we will write the neutron decay rate given in Equation (7.44) as

$$\mathrm{d}\Gamma = \frac{2\pi}{\hbar} G_F^2 M^2 \rho_e \rho_{\bar{\nu}} \delta(E_e + E_{\bar{\nu}} - E_0) \mathrm{d}E_e \mathrm{d}E_{\bar{\nu}} \qquad (7.52)$$

with $G_F^2 M^2 = \dfrac{1}{2} \sum_{\text{spin}} |H_{mi}|^2$ and $G_F \simeq 1.4 \times 10^{-49}\,\text{erg cm}^3$.

The term M^2 that appears in Equation (7.52) is determined by the 925 second mean lifetime of the neutron. The corresponding M^2 term in muon decay has a value that is comparable to but not equal to the M^2 term in neutron decay because the neutron is a strongly interacting particle. In nucleon beta decays the M^2 terms are determined by measured mean lifetimes of beta-unstable nuclei.

If neutrons exist in an electron-degenerate gas then the transition rate for neutron decay given by Equation (7.52) becomes

$$d\Gamma = \frac{2\pi}{\hbar} G_F^2 M^2 \rho_e \rho_{\bar{\nu}} \delta(E_e + E_{\bar{\nu}} - E_0)(1 - f_e) dE_e dE_{\bar{\nu}} \tag{7.53}$$

with $f_e = \dfrac{1}{1 + \exp[(E_e - \mu_e)/kT]}$.

The factor $1 - f_e$ in Equation (7.53) is the probability that an electron state is unoccupied. Integrating Equation (7.53) over $dE_{\bar{\nu}}$ we obtain

$$d\Gamma = \frac{2\pi}{\hbar} G_F^2 M^2 \rho_e \rho_{\bar{\nu}} (1 - f_e) dE_e. \tag{7.54}$$

Neglecting spin the density of electron states is

$$\rho_e = \frac{4\pi p^2}{h^3} \frac{dp}{dE_e} = \frac{4\pi p E_e}{c^2 h^3}. \tag{7.55}$$

The neutrino has zero (or very small) mass, and therefore

$$p_{\bar{\nu}} c = E_{\bar{\nu}} = E_0 - E_e. \tag{7.56}$$

It follows that the density of antineutrino states is

$$\rho_{\bar{\nu}} = \frac{4\pi(E_0 - E_e)^2}{h^3 c^3}. \tag{7.57}$$

Substituting the above expressions for ρ_e and $\rho_{\bar{\nu}}$ into Equation (7.54) we obtain

$$d\Gamma = \frac{64\pi^4 G_F^2 M^2}{h^7 c^6} (E_e^2 - m^2 c^4)^{1/2} E_e (E_0 - E_e)^2 (1 - f_e) dE_e. \tag{7.58}$$

Equation (7.58) can be re-expressed as

$$d\Gamma = \frac{64\pi^4 G_F^2 M^2}{h^7} m^5 c^4 (\varepsilon^2 - 1)^{1/2} \varepsilon (\varepsilon_0 - \varepsilon)^2 (1 - f_e) d\varepsilon \tag{7.59}$$

where $\varepsilon = \dfrac{E_e}{mc^2}$, $\varepsilon_0 = \dfrac{E_0}{mc^2} \simeq \dfrac{m_n c^2 - m_p c^2}{mc^2} \simeq 2.53$ and $f_e = \dfrac{1}{e^{y(\varepsilon - x)}}$, $y = \dfrac{mc^2}{kT}$, $x = \dfrac{\mu_e}{mc^2}$.

If the gas is nondegenerate so that $f_e \simeq 0$ then the integral of Equation (7.59) becomes

$$\Gamma = \int d\Gamma = \frac{64\pi^4 G_F^2 M^2 m^5 c^4}{h^7} \int_1^{\varepsilon_0} (\varepsilon^2 - 1)^{1/2}\varepsilon(\varepsilon_0 - \varepsilon)^2 d\varepsilon$$

$$= 1.64 \frac{64\pi^4 G_F^2 M^2 m^5 c^4}{h^7}. \tag{7.60}$$

If the gas is highly electron degenerate then

$$1 - f_e = 0 \left(\varepsilon < \frac{\mu}{mc^2} \right) \tag{7.61a}$$

$$1 - f_e = 1 \left(\varepsilon > \frac{\mu}{mc^2} \right). \tag{7.61b}$$

In the above expressions we have assumed that both E_e and μ include the electron rest mass energy mc^2. If $\mu/mc^2 > \varepsilon_0$ then from Equation (7.59) $\Gamma = 0$. On the other hand, if $\mu/mc^2 < \varepsilon_0$ then Γ becomes

$$\Gamma = \int d\Gamma = \frac{64\pi^4 G_F^2 M^2 m^5 c^4}{h^7} \int_{\mu/mc^2}^{\varepsilon_0} (\varepsilon^2 - 1)^{1/2}\varepsilon(\varepsilon_0 - \varepsilon)^2 d\varepsilon. \tag{7.62}$$

The reaction and inverse reaction

$$e^- + p \leftrightarrow n + \nu_e \tag{7.63}$$

are physically similar to the decay of the neutron. Consider electron capture onto free protons under physical conditions such that initial electrons are degenerate but the other particles are nondegenerate. The electron capture rate per proton is

$$\frac{4\pi}{h^3} \int \sigma_c v f_e p_e^2 dp_e = \frac{4\pi}{h^3 c^2} \int \sigma_c v f_e p_e E_e dE_e$$

$$= \frac{2\pi}{\hbar h^3} \int 4\pi p_e^2 dp_e f_e \delta(E_\nu + E_0 - E_e) G_F^2 M^2 \rho_\nu dE_\nu. \tag{7.64}$$

The density of neutrino final states is

$$\rho_\nu = \frac{4\pi p_\nu^2}{h^3 c} = \frac{4\pi E_\nu^2}{h^3 c^3} \tag{7.65}$$

since $p_\nu c = E_\nu$. Using Equation (7.56) and the relation $p_e dp_e = E_e dE_e/c^2$ Equation (7.64) becomes

$$\frac{4\pi}{h^3 c^2} \int_{E_0}^{\infty} \sigma_c v f_e p_e E_e dE_e$$

$$= \frac{64\pi^4 G_F^2 M^2}{h^7 c^5} \int p_e E_e dE_e f_e E_\nu^2 \delta(E_\nu + E_0 - E_e) dE_\nu. \tag{7.66}$$

Integrating the right-hand side of Equation (7.66) over dE_ν we obtain

$$\frac{4\pi}{h^3c^2}\int_{E_0}^{\infty}\sigma_c vf_ep_eE_edE_e$$

$$=\frac{64\pi^4G_F^2M^2}{h^7c^5}\int p_eE_edE_ef_e(E_e-E_0)^2. \tag{7.67}$$

Equation (7.67) implies that the electron capture cross section of the proton is

$$\sigma_c=\frac{16\pi^3G_F^2M^2(E_e-E_0)^2}{vh^4c^3} \tag{7.68}$$

where v is the velocity of the electron.

The process that is inverse to electron capture by a free proton is absorption of a neutrino by a free neutron. Neutrinos travel at the speed of light c and therefore we have

$$\sigma_a c\rho_v dE_v=\int d\Gamma$$

$$=\frac{2\pi}{\hbar}G_F^2M^2\rho_v dE_v\int \rho_e dE_e(1-f_e)\delta(E_v+E_0-E_e). \tag{7.69}$$

Using Equation (7.55) and integrating over dE_e Equation (7.69) becomes

$$\sigma_a c\rho_v dE_v=\frac{16\pi^3G_F^2M^2}{h^4c^3}\rho_v dE_v\int (E_e^2-m^2c^4)^{1/2}E_e(1-f_e)\delta(E_v+E_0-E_e)dE_e$$

$$=\frac{16\pi^3G_F^2M^2}{h^4c^3}\rho_v dE_v[(E_v+E_0)^2-m^2c^4]^{1/2}$$

$$\times [E_v+E_0]\left[1-\frac{1}{e^{(E_v+E_0-\mu_e)/kT}+1}\right]. \tag{7.70}$$

Equation (7.70) implies that the absorption cross section σ_a of a free neutron follows from the equation

$$\sigma_a c=\frac{16\pi^3G_F^2M^2}{h^4c^3}[(E_v+E_0)^2-m^2c^4]^{1/2}[E_v+E_0]\left[1-\frac{1}{e^{(E_v+E_0-\mu_e)/kT}+1}\right]. \tag{7.71}$$

Neutrinos transfer thermal energy from the interiors of neutron stars by means of the Urca reactions given as Equations (7.31–7.33). Because protons produced by neutrons in the first of the Urca reactions are transformed into neutrons by the remaining two reactions the Urca reactions can occur many times as a neutron star cools. In a highly degenerate gas most of the low-energy states are occupied except those of neutrinos, which we assume to be nondegenerate. It follows that each density of initial states factor must be multiplied by f_j and each density of final states factor except the density of neutrino final states must be multiplied by $1-f_j$ where

$$f_j=\frac{1}{e^{(E_j-\mu_j)/kT}+1}. \tag{7.72}$$

Since f_j in Equation (7.72) is the probability that a particular energy state is occupied the terms $1 - f_j$ represent the probability that a particular energy state is unoccupied.

Labeling the two initial neutrons in Equation (7.31) as 1 and 2, and the final state neutron as 3 it follows that the neutrino luminosity per cm^3 is

$$L_{\bar{\nu}} = \frac{2\pi}{\hbar h^{15}} G_F^2 M'^2 \int d^3p_1 d^3p_2 d^3p_3 d^3p_p d^3p_e d^3p_{\bar{\nu}} \delta(E_f - E_i)$$

$$\times \delta^3(\boldsymbol{p}_f - \boldsymbol{p}_i) f_1 f_2 (1 - f_3)(1 - f_p)(1 - f_e) E_{\bar{\nu}}$$

$$= \frac{2\pi}{\hbar h^{15}} G_F^2 M'^2 \int \prod_{j=1}^{6} p_j^2 dp_j d\Omega_j f_1 f_2 (1 - f_3)(1 - f_4)(1 - f_5)$$

$$\times \delta(E_f - E_i) \delta^3(\boldsymbol{p}_f - \boldsymbol{p}_i) E_{\bar{\nu}} \tag{7.73}$$

where f and i denote final and initial states respectively. In the final expression in Equation (7.73) the labels 4, 5 and 6 stand for proton, electron and antineutrino states respectively. Because of momentum conservation Equation (7.73) contains one less phase space factor h^{-3} than final state particles.

Equation (7.73) can be written as

$$L_{\bar{\nu}} = \frac{2\pi G_F^2 M'^2}{\hbar h^{15}} \int \prod_{j=1}^{6} p_j^2 dp_j d\Omega_j f_1 f_2 (1 - f_3)(1 - f_4)(1 - f_5) E_{\bar{\nu}} \delta(E_f - E_i)$$

$$\times \int \prod_{j=1}^{6} d\Omega_j \delta^3(\boldsymbol{p}_f - \boldsymbol{p}_i). \tag{7.74}$$

The initial and final momenta are

$$\boldsymbol{p}_i = \boldsymbol{p}_1 + \boldsymbol{p}_2 \tag{7.75}$$

and

$$\boldsymbol{p}_f = \boldsymbol{p}_3 + \boldsymbol{p}_4 + \boldsymbol{p}_5 + \boldsymbol{p}_6 \simeq \boldsymbol{p}_3 + \boldsymbol{p}_4 + \boldsymbol{p}_5. \tag{7.76}$$

The last relation in Equation (7.76) is valid because $\boldsymbol{p}_6$ is small. We are considering a highly degenerate gas in which kT is much less than the Fermi energies of neutrons, protons and electrons. In a neutron star the Fermi energies of neutrons are appreciably higher than those of protons or electrons, and therefore $|\boldsymbol{p}_3|$ is much greater than the momenta of final state protons, electrons or neutrinos. Using Equations (7.75) and (7.76) the δ function that expresses the conservation of momentum can be written as

$$\delta^3(\boldsymbol{p}_f - \boldsymbol{p}_i) = \frac{\delta(p_3 - |\boldsymbol{p}_1 + \boldsymbol{p}_2 - \boldsymbol{p}_4 - \boldsymbol{p}_5 - \boldsymbol{p}_6|)}{p_3^2} \delta(\phi_3 - \phi_3')\delta(\cos\theta_3 - \cos\theta_3'). \tag{7.77}$$

Because the integral

$$\int \delta(\phi_3 - \phi_3')\delta(\cos\theta_3 - \cos\theta_3') d\Omega_3 = 1 \tag{7.78}$$

it follows that the integration over solid angles contained in Equation (7.74) becomes

$$\int \prod_{j=1}^{6} d\Omega_j \delta^3(\boldsymbol{p}_{\mathrm{f}} - \boldsymbol{p}_{\mathrm{i}}) = \int \prod_{\substack{j=1 \\ j \neq 3}}^{6} d\Omega_j \frac{\delta(p_3 - |\boldsymbol{p}_2 + \boldsymbol{p}_1 - \boldsymbol{p}_4 - \boldsymbol{p}_5|)}{p_3^2}. \tag{7.79}$$

The next step in the derivation is to integrate over $d\Omega_1$. We choose the z-axis for $\boldsymbol{p}_1$ along $\boldsymbol{p}_2 - \boldsymbol{p}_4 - \boldsymbol{p}_5$ and regard $p_3 = |\boldsymbol{p}_2 + \boldsymbol{p}_1 - \boldsymbol{p}_4 - \boldsymbol{p}_5|$ as a function of $\cos \theta_1$ with θ_1 the angle between $\boldsymbol{p}_1$ and the z-axis. Making the definitions $a = \cos \theta_1$ and $\boldsymbol{p}_s = \boldsymbol{p}_4 + \boldsymbol{p}_5$ the δ function in Equation (7.79) expresses the condition

$$f(a) = p_3 - (p_1^2 + |\boldsymbol{p}_2 - \boldsymbol{p}_s|^2 - 2p_1 |\boldsymbol{p}_2 - \boldsymbol{p}_s| \cos \theta_1)^{1/2} = 0. \tag{7.80}$$

Equation (7.80) implies

$$p_3^2 = p_1^2 + |\boldsymbol{p}_2 - \boldsymbol{p}_s|^2 - 2p_1 |\boldsymbol{p}_2 - \boldsymbol{p}_s| \cos \theta_1. \tag{7.81}$$

Solving for $\cos \theta_1$ in Equation (7.81) we have

$$a = \cos \theta_1 = \frac{p_3^2 - p_1^2 - |\boldsymbol{p}_2 - \boldsymbol{p}_s|^2}{2p_1 |\boldsymbol{p}_2 - \boldsymbol{p}_s|}. \tag{7.82}$$

Regarding f in Equation (7.80) a function of $x = \cos \theta$ with $a = \cos \theta_1$, we obtain

$$\frac{df}{dx}(x)|_{x \to a} = \frac{p_1 |\boldsymbol{p}_2 - \boldsymbol{p}_s|}{(p_1^2 + |\boldsymbol{p}_2 - \boldsymbol{p}_s|^2 - 2p_1 |\boldsymbol{p}_2 - \boldsymbol{p}_s| \cos \theta_1)^{1/2}}. \tag{7.83}$$

The delta function satisfies the equation

$$\delta\{f(x)\} = \sum_{i=1}^{N} \frac{\delta(x - x_i)}{|df/dx|_{x = x_i}}, \quad f(x_i) = 0, i = 1, 2, \ldots, N. \tag{7.84}$$

Equations (7.83) and (7.84) imply that after integrating over $d\Omega_1$ the integral over solid angles in Equation (7.79) becomes

$$\frac{2\pi}{p_3 p_1} \int \frac{d\Omega_2 d\Omega_4 d\Omega_5 d\Omega_6}{|\boldsymbol{p}_2 - \boldsymbol{p}_4 - \boldsymbol{p}_5 - \boldsymbol{p}_6|}. \tag{7.85}$$

Since the momentum $\boldsymbol{p}_2$ satisfies the inequality

$$p_2 \gg p_4 + p_5 + p_6 \tag{7.86}$$

it follows that the integral given as expression (7.85) is approximately equal to

$$\frac{2\pi(4\pi)^4}{p_3 p_1 p_2}. \tag{7.87}$$

For neutrons and protons momentum and energy satisfy the relationship

$$p \, dp = \frac{E \, dE}{c^2} \simeq m \, dE \tag{7.88}$$

whereas the momenta and energies of neutrinos and electrons satisfy the relationship

$$dp = \frac{dE}{c}. \tag{7.89}$$

All p_j except $p_{\bar{\nu}}$ can be set equal to $p_F(j)$ in the integral given in Equation (7.74). Equations (7.88) and (7.89) imply that $L_{\bar{\nu}}$ in Equation (7.74) is proportional to

$$m_n^3 m_p p_F(p) p_F^2(e) \int \prod_{j=1}^{6} dE_j E_{\bar{\nu}}^3 S \delta(E_f - E_i) \tag{7.90}$$

where

$$S = \prod_{j=1}^{5} (1 + e^{x_j})^{-1} \tag{7.91}$$

with

$$\begin{aligned} x_j &= (E_j - \mu_j)/kT \qquad (j = 1, 2) \\ x_j &= -(E_j - \mu_j)/kT \qquad (j = 3, 4, 5). \end{aligned} \tag{7.92}$$

If we let $y = E_\nu/kT$ then expression (7.90) becomes

$$m_n^3 m_p p_F(p) p_F^2(e)(kT)^8 I \tag{7.93}$$

with $I = \int dy\, y^3 J$ and

$$J = \int \prod_{j=1}^{5} dx_j (1 + e^{x_j})^{-1} \delta\left[\sum_{j=1}^{5} x_j - y\right]. \tag{7.94}$$

It can be shown by contour integration that the integral I in Equation (7.93) is

$$I = \frac{11\,513\pi^8}{120\,960}. \tag{7.95}$$

From expression (7.93), Equation (7.95), and expression (7.87) it follows that $L_{\bar{\nu}}$ given by Equation (7.74) is proportional to

$$G_F^2 M'^2 m_n^3 m_p p_F(p) p_F^2(e)(kT)^8 I. \tag{7.96}$$

The p–p reaction

Following our discussion of the neutron decay rate and neglecting Coulomb effects the p–p reaction cross section $d\sigma$ becomes

$$d\sigma = \frac{2\pi}{v\hbar} G_F^2 M_{pp}^2 \rho_{e^+} \rho_\nu \delta(E_e + E_\nu - E_0) dE_{e^+} dE_\nu \tag{7.97}$$

where v is proton velocity in center of mass coordinates and E_0 is the energy difference between the two initial protons and final state deuteron. The proton kinetic energy is small as compared to the difference in rest mass energy between two protons and a deuteron. It is convenient to let

$$M_{pp}^2 = M_{spin}^2 M_{space}^2 \tag{7.98}$$

with $M_{spin}^2 = \frac{3}{2}$ and $M_{space} = \int \psi_p(r)\psi_D(r)d^3r$.

The wave functions $\psi_p(r)$ and $\psi_D(r)$ represent the initial protons and final state deuteron respectively. A factor of $\frac{1}{2}$ arises in M_{spin}^2 because the protons are identical particles. An addition factor of $\frac{1}{4}$ is due to the average over initial spin states of the spin-$\frac{1}{2}$ protons. Since the spin of the deuteron is $S = 1$ the number of final states is $2S + 1 = 3$. The weak interaction matrix element is increased by a factor of 2 and the square of the matrix element by a factor of 4 because each of the two initial state protons can be converted into a neutron in forming a deuteron. Multiplying these above factors it follows that M_{spin}^2 is equal to $\frac{3}{2}$ as given in Equation (7.98).

Integrating Equation (7.97) over e^+ and ν final states as in deriving Equation (7.60) we obtain

$$\sigma = \frac{64\pi^4}{v}\frac{m^5 c^4}{h^7}f(\varepsilon_0)G_F^2 M_{spin}^2 M_{space}^2 \tag{7.99}$$

where $\varepsilon_0 = \dfrac{Q}{mc^2} = \dfrac{2m_p - m_D}{m}$

and $f(\varepsilon_0) = (\varepsilon_0^2 - 1)^{1/2}\left(\dfrac{\varepsilon_0^4}{30} - \dfrac{3\varepsilon_0^2}{20} - \dfrac{2}{15}\right) + \dfrac{\varepsilon_0}{4}\ln[\varepsilon_0 + (\varepsilon_0^2 - 1)^{1/2}]$.

If $P(E)$ is the Coulomb barrier penetration probability implied by Equation (6.101) then the p–p reaction cross section becomes

$$\sigma_{pp} = P(E)\sigma = \frac{2\pi e^2 \sigma}{\hbar v}e^{-2\pi e^2/\hbar v} \tag{7.100}$$

with σ given by Equation (7.99). Because the identical particle factor of $\frac{1}{2}$ and average over initial spin factor of $\frac{1}{4}$ have been included in σ and σ_{pp} the p–p reaction rate does not include the factor of $\frac{1}{8}$ that appears in Equation (6.15).

An approximate cross section for the proton–proton weak interaction can be obtained by the following physical considerations. At energies $\gtrsim 1\,\text{MeV}$ the Coulomb barrier between the interacting protons can be neglected. From low-energy ($\gtrsim 1\,\text{MeV}$) scattering experiments the proton–proton nuclear potential can be inferred and the resultant nuclear scattering cross section readily calculated. This cross section turns out to be about 36 barns (1 barn = $10^{-24}\,\text{cm}^2$). The weak interaction is a short-range force and therefore can occur only when two proton wave functions overlap. The probability that a weak interaction occurs during interaction is equal to a nucleon crossing timescale ($\gtrsim 10^{-23}\,\text{s}$) divided by a beta-decay timescale (925 seconds for neutrons). It is appropriate to use the free neutron β-decay mean lifetime because protons and neutrons are similar particles except for charge which is irrelevant during a β-decay. It follows that at 1 MeV the predicted p–p weak interaction cross section becomes

$$\sigma \simeq 36 \times 10^{-24}\left(\frac{10^{-23}}{925}\right) \simeq 4 \times 10^{-49}\,\text{cm}^2. \tag{7.101}$$

From the above cross section at $1\,\mathrm{MeV}$ we can infer from Equation (6.24) that $S(E) \sim 4 \times 10^{-49}\,\mathrm{MeV\,cm^2}$ at relevant stellar proton energies of $\simeq 1 - 10\,\mathrm{keV}$ where the Gamow factor is important. The p–p weak interaction cross section is too low to be measured directly.

7.6   Neutrino mass and solar neutrinos

Neutrinos (v_e, v_μ, v_τ) and the corresponding antineutrinos are usually assumed to be particles of zero mass and definite helicity with neutrinos left-handed (i.e. spin direction opposite to direction of motion) and antineutrinos right-handed (i.e. spin direction along direction of motion). However, it is possible that neutrinos have small but finite mass. Since electron neutrinos, muon neutrinos and tauon neutrinos are different particles their masses could also be different. Lepton number is known to be conserved in weak interactions, however, lepton-number eigenstates need not be the same as eigenstates associated with neutrino mass. It has been suggested that neutrinos formed with definite lepton number can mix with other eigenstates of definite lepton number. If such neutrino oscillations occur in nature then an electron neutrino emitted from the solar core has a certain probability of being transformed into v_μ or v_τ upon arrival at Earth. A solar electron neutrino transformed into either v_μ or v_τ would not be detected. If maximal mixing of the three types of neutrinos occurred then the factor of 3 disagreement between predicted and measured solar neutrino flux would be explained. However, the very good agreement between predicted and observed neutrino fluxes from Supernova 1987A would be worsened.

For simplicity only oscillations between electron and muon neutrinos will be consider-ed. If mass eigenstates (v_1, v_2) are not the same as lepton number eigenstates (v_e, v_μ) neutrino oscillations can occur. Mass lepton number eigenstates are related by a unitary transformation:

$$\begin{pmatrix} v_e \\ v_\mu \end{pmatrix} = \begin{pmatrix} \cos\theta & \sin\theta \\ -\sin\theta & \cos\theta \end{pmatrix} \begin{pmatrix} v_1 \\ v_2 \end{pmatrix} \text{ or}$$

$$\begin{aligned} v_e &= \cos\theta\, v_1 + \sin\theta\, v_2 \\ v_\mu &= -\sin\theta\, v_1 + \cos\theta\, v_2. \end{aligned} \tag{7.102}$$

If an electron neutrino is formed at $t = 0$ the eigenstate at time t becomes

$$\begin{aligned} \psi(t) &= \exp\left(-i\frac{E_1 t}{\hbar}\right)\cos\theta\,|v_1\rangle + \exp\left(-i\frac{E_2 t}{\hbar}\right)\sin\theta\,|v_2\rangle \\ &= \left[\exp\left(-i\frac{E_1 t}{\hbar}\right)\cos^2\theta + \exp\left(-i\frac{E_2 t}{\hbar}\right)\sin^2\theta\right]|v_e\rangle \\ &\quad + \left[\exp\left(-i\frac{E_2 t}{\hbar}\right) - \exp\left(-i\frac{E_1 t}{\hbar}\right)\right]\sin\theta\cos\theta\,|v_\mu\rangle. \end{aligned} \tag{7.103}$$

Equation (7.103) implies that the probability of a neutrino v_e emitted at time $t = 0$ being transformed into a v_μ at time t is

$$P(v_e \rightarrow v_\mu) = |\langle v_\mu | \psi(t) \rangle|^2 = \frac{\sin^2 2\theta}{2}\left[1 - \cos\left(\frac{E_2 - E_1}{\hbar}\right)t\right]. \tag{7.104}$$

Since neutrino rest mass energies $m_1 c^2$ and $m_2 c^2$ are much less than E_1 and E_2 respectively we have

$$E_1 \simeq pc + \frac{m_1^2 c^3}{2p} \text{ and } E_2 \simeq pc + \frac{m_2^2 c^3}{2p}. \tag{7.105}$$

Inserting Equations (7.105) into Equation (7.104) we obtain

$$P(v_e \rightarrow v_\mu) = \frac{\sin^2 2\theta}{2}\left[1 - \cos\left[\left(\frac{m_2^2 - m_1^2}{2p\hbar}\right)c^3 t\right]\right]. \tag{7.106}$$

Equation (7.106) implies that the neutrino oscillation length is

$$\Lambda = \frac{4\pi p\hbar}{c^2 |m_2^2 - m_1^2|}. \tag{7.107}$$

Equation (7.107) shows that for solar neutrinos the oscillation length is very small unless neutrino masses (or mass differences) are also small. From experiment it is known that $m_{v_e} c^2$ is $< 20 \text{ eV}$. Moreover, experiments with muon neutrinos rule out large mixing angles (i.e. θ in Equation (7.106)) unless neutrino masses are much less than 1 eV.

Neutrino mass is also important from the point of view of cosmology since big-bang cosmology indicates that large number densities of relic electron, muon and tauon neutrinos are present in the expanding universe but cannot be detected because of their low energies. A high number density of relic massive neutrinos would imply that they have the dominant effect on expansion of the universe. The present upper mass limit on the electron neutrino ($\sim 20 \text{ eV}$) is nearly sufficient to exclude them as dominating cosmological expansion. However, if nonzero, the masses of muon neutrinos and tauon neutrinos would presumably be much greater than electron neutrino mass because muons and tauons are much more massive than electrons. If the mixing angle in Equation (7.106) is large enough to explain the difference between predicted and observed solar neutrino fluxes then neutrino masses are probably too small to control cosmological expansion.

In the above discussion the effects of dense matter on neutrino propagation have been neglected. It has been suggested that the effective neutrino mass might depend on the density of surrounding solar mass. Since electrons interact with solar matter via charged current (W-boson mediated) weak interactions as well as neutral current (Z-boson mediated) weak interactions whereas muon neutrinos interact with solar matter only via neutral currents it is reasonable to assume that electron neutrino mass depends more sensitively on solar interior density than muon neutrino effective mass. If electron and muon effective masses were to become equal at some solar density less than the solar central density a resonance condition might become satisfied and solar electron neutrinos transformed into muon neutrinos even if the neutrino mixing angle θ is too small for neutrino vacuum oscillations to appreciably reduce electron neutrino fluxes. This theory for lower than predicted solar neutrino fluxes does not, however, explain why they are a factor of 3 lower than stellar model predictions.

In addition to the previously discussed Homestake solar neutrino experiment of Davis and collaborators the more recent Kamiokande and Sage/Gallex neutrino detectors have measured solar neutrino fluxes. The pattern of experimental results is not easily reconciled with standard solar models. If future solar neutrino experiments show that matter-enhanced neutrino oscillations briefly discussed above are the solution to the solar neutrino problem then the standard model of electroweak interactions will require substantial modification. Mikheyev, Smirnov and Wolfenstein first proposed that the presence of matter could enhance neutrino oscillations and therefore this theory is known as the MSW mechanism.

An upper limit to the mass of the electron neutrino can be obtained from the detection of neutrinos from Supernova 1987A because neutrinos with energies between 7.5 to 35 MeV were detected in a time interval of 12.4 seconds (see Chapter 11 for experimental results) and in addition the distance to the supernova is known (distance $\simeq 50\,\text{kpc} = 1.57 \times 10^{23}\,\text{cm}$). The velocity of a relativistic particle with rest mass m and energy E is

$$v = \frac{pc^2}{E} = \sqrt{E^2 - m^2 c^4}\,\frac{c}{E} = c\sqrt{1 - \frac{m^2 c^4}{E^2}} \simeq c\left(1 - \frac{m^2 c^4}{2E^2}\right) \tag{7.108}$$

where the final approximate equality follows from the binomial theorem and the fact that neutrino energy E is known to be much greater than electron neutrino mass. Equation (7.108) implies that the neutrino time of flight is

$$T(E) \simeq \frac{L}{c}\left(1 + \frac{m^2 c^4}{2E^2}\right). \tag{7.109}$$

From Equation (7.109) it follows that the difference in arrival times of two simultaneously emitted neutrinos with energies E_1 and $E_2 (E_1 > E_2)$ is

$$c\Delta T = c[T(E_2) - T(E_1)] = \frac{m^2 c^4}{2}\left(\frac{1}{E_2^2} - \frac{1}{E_1^2}\right)L. \tag{7.110}$$

Solving Equation (7.110) for mc^2 we obtain

$$mc^2 = \left(\frac{2c\Delta T}{L}\,\frac{E_1^2 E_2^2}{(E_1^2 - E_2^2)}\right)^{1/2}. \tag{7.111}$$

If we substitute $E_1 = 35\,\text{MeV}$, $E_2 = 7.5\,\text{MeV}$ and $\Delta T = 12.4$ seconds into Equation (7.111) then the electron neutrino mass limit becomes

$$m_v c^2 < 17\,\text{eV}. \tag{7.112}$$

The above expression is an inequality because neutrinos of different energies may have been emitted at different times.

The Crab pulsar is known to emit radio, optical, x-ray and gamma ray pulses that are observed in phase (see Figure 10.5). This circumstance allows us to use the same physical reasoning discussed above for electron neutrinos to place a limit on photon mass. The distance to the Crab pulsar is 2 kpc and its periodicity is 33 milliseconds. If we consider radio frequency photons of frequency $v_2 = 2 \times 10^8$ Hz and optical photons of frequency $v_1 = 3 \times 10^{14}$ Hz then since $E_1 = hv_1 \gg E_2 = hv_2$ expression (7.111) becomes

$$m_\gamma c^2 < \left(\frac{2c\Delta T}{L}\right)^{1/2} hv_2 = 2 \times 10^{-13} \, \text{eV} \tag{7.113}$$

where we have assumed $\Delta T = 3$ ms. The above limit on photon mass implies that the range of the Coulomb force satisfies the inequality

$$\frac{\hbar}{m_\gamma c^2} > 10^3 \, \text{km}. \tag{7.114}$$

Problems

1. Consider an election-degenerate gas consisting of ^{12}C. What are the effective masses of plasmons at densities of 10^5, 10^7 and 10^9 g cm^{-3}?

2. The density at the center of a star is 10^6 g cm^{-3}. Estimate the rate of neutrino energy losses caused by electron–positron annihilation in units of erg g^{-1} s^{-1}:
    a. for a pure ^{56}Fe gas at a temperature of $T = 10^9$ K,
    b. for an α-particle gas at $T = 10^{10}$ K.
    c. Compare the thermal-cooling timescales implied by your answers to part a and part b with the gravitational collapse timescale $1/\sqrt{G\rho}$.

3. a. From the known values of G_F and Γ evaluate M^2 in Equation (7.60) for neutron decay.
    b. Using the value of M^2 obtained from Equation (7.60) evaluate the electron capture cross section given in Equation (7.68), assuming an electron velocity $v = 10^9$ cm s^{-1}.
    c. Using the value of M^2 obtained from Equation (7.60) and assuming that the gas is nondegenerate, evaluate the neutron–neutrino absorption cross section σ_a implied by Equation (7.71) for neutrino energies of 1 MeV and 10 MeV.

8 Stellar stability and hydrodynamics

8.1 Pulsational instability

In Chapter 1 we discussed some of the observational properties of periodic variable stars. The instability that drives pulsations in RR Lyrae variables, Cepheids and long-period variables is associated with hydrogen and helium ionization zones. The large heat capacity of these ionization zones causes the phase of maximum luminosity to be delayed by approximately 90° as compared to the phase of minimum radius. Thermonuclear reactions can also cause stars to become pulsationally unstable. Very massive stars and white dwarfs in which thermonuclear runaways are caused by mass accretion from a binary companion become pulsationally unstable as the result of their hydrogen-burning sources. To determine whether a particular star is pulsationally unstable one first determines the structure of the star (i.e. $r = r(M_r), P = P(M_r), \rho = \rho(M_r), L_r = L_r(M_r)$) and then solves the linearized equation of motion for the oscillatory modes. It is usually adequate to assume that stellar oscillations are adiabatic. If the oscillatory modes of a star have been determined we can evaluate a stability integral which will be derived below. The sign of this stability integral determines whether a particular stellar model is unstable to self-excited oscillations at a particular frequency (eigenmode). We are usually interested only in radial modes of oscillation and in most circumstances only the longest period mode is pulsationally unstable. In β Canis Majoris stars (also known as β Cepheid variables) nonradial oscillatory modes can also become excited.

Figure 8.1 shows the locations in the H–R diagram of the most important types of periodic variable stars. Long-period variables include both stars with normal oxygen abundances (i.e. O > C) and the less common late-type carbon stars. Extensive hydrogen ionization zones cause them to become unstable to radial pulsations. Zones of partially ionized He and He^+ play essentially no role in causing instability. The amplitudes of their pulsations, which increase with luminosity and also depend on other physical parameters, become sufficiently large that shock waves are generated in their atmospheres. The standard scenario for producing mass loss from these stars is that shock waves eject mass. As ejected mass cools dust is formed. Radiation pressure on dust particles then generates a stellar wind. As will be discussed below the formation of a planetary nebula is believed to represent the final evolution of most long period variables.

The pulsations of classical Cepheids, which are usually referred to as simply Cepheids, result from both hydrogen and helium ionization zones. The so called second helium

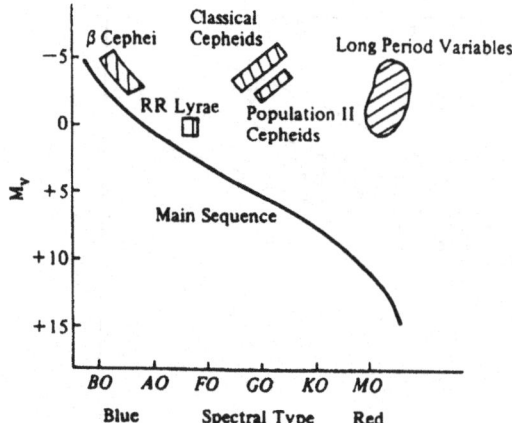

Figure 8.1. The regions of the H–R diagram that are occupied by the most important classes of periodic variable stars are shown schematically.

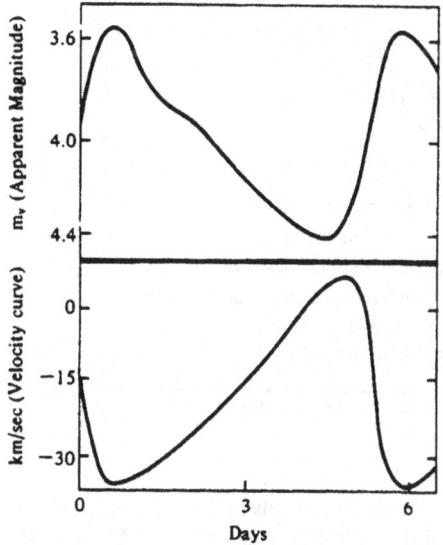

Figure 8.2. A typical Cepheid light curve and velocity curve are shown schematically. Note that maximum luminosity occurs after minimum radius.

ionization zone containing a mixture of He^+ and He^{++} is particularly important. Because of the high luminosities and stable oscillation periods Cepheids play a vital role in estimating distances to galaxies. Figure 8.2 shows the light and velocity curves of a typical Cepheid variable. Pulsation amplitudes are large and therefore the variability shown in Figure 8.2 is not sinusoidal. The fundamental oscillatory period of a variable star such as a Cepheid is approximately equal to the time it takes a sound wave to travel from the center to atmosphere. The empirical period–luminosity relation shown for several galaxies in Figure 8.3 is used to measure distances. Population II Cepheids, which

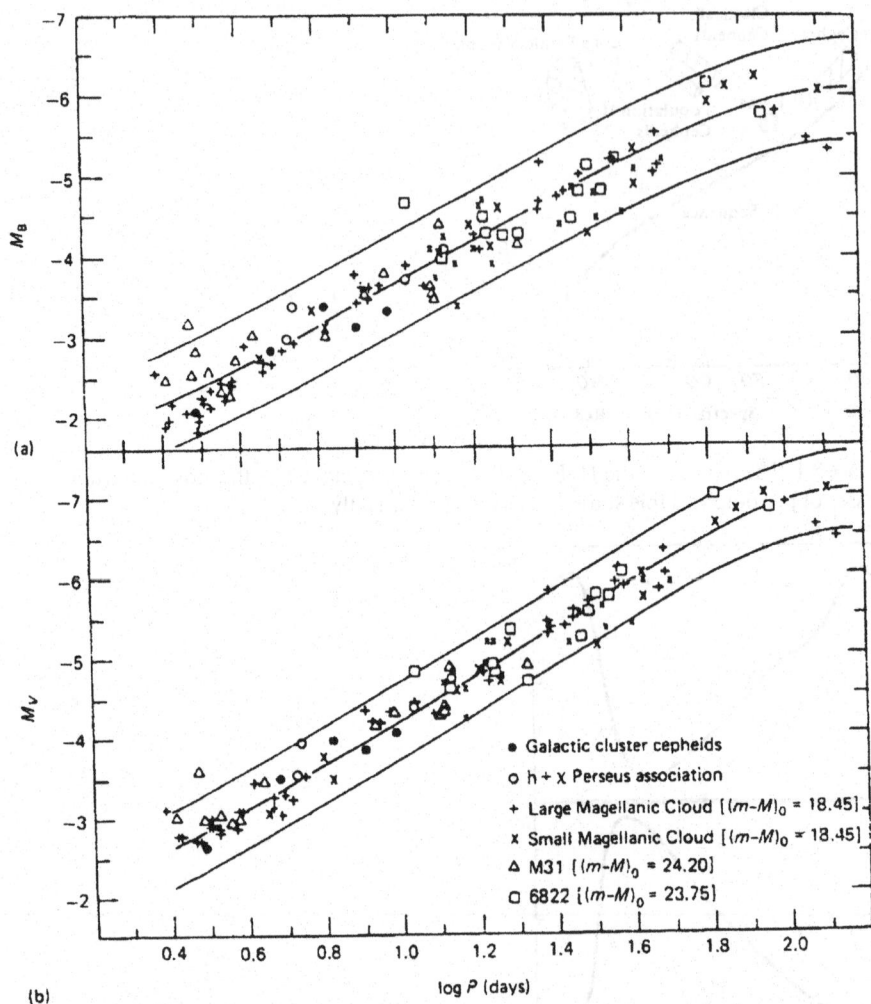

Figure 8.3. The period–luminosity relation for classical Cepheids in the Galaxy, Large Magellanic Cloud, Small Magellanic Cloud, M31 (Andromeda) and NGC 6882. In (a) M_B is the absolute blue magnitude at mean light corrected for interstellar extinction. In (b) M_v is the absolute visual magnitude at mean light. From Sandage (1972).

are also known as W Virginis stars, are much rarer than Cepheids. Their pulsation amplitudes are sufficiently large to generate shock waves. Cepheid and RR Lyrae pulsations are probably not sufficiently strong to cause mass loss.

RR Lyrae variables are located on the horizontal branch and therefore their luminosities are approximately the same. They occur in globular clusters whose nearly spherical distribution about the Galactic center makes it possible to estimate the distance to the Galactic center. This distance is believed equal to 8.5 ± 1 kpc. The helium ionization zone containing both the He^+ and He^{++} is essential in producing RR Lyrae pulsations and

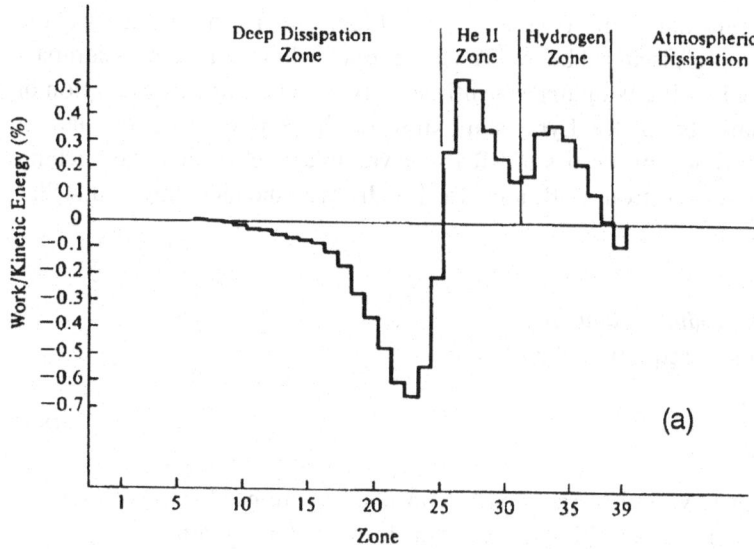

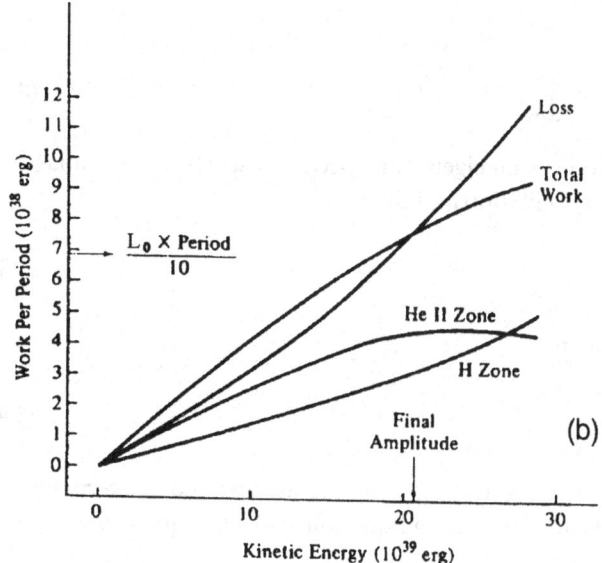

Figure 8.4. (a) The relative amounts of energy production and dissipation are shown for a number of mass shells of a model RR Lyrae variable. (b) The energy production and loss per period are shown as a function of the kinetic energy of pulsation. From Christy (1966).

also in limiting their pulsation amplitudes as shown in Figure 8.4. The linear theory of the excitation of oscillations is discussed below. The presence of helium abundances comparable to those predicted by big-bang nucleosynthesis is required to explain excitation of pulsations. Some relatively hot RR Lyrae stars known as Type C variables oscillate in their first overtone. Physical properties of RR Lyrae variables such as mass and helium abundance can be estimated from both numerical hydrodynamic modeling and stellar evolution calculations.

The equation of radial oscillation

The spherically symmetric equation of motion is

$$\frac{d^2r}{dt^2} = -\frac{1}{\rho}\frac{dP}{dr} - \frac{GM_r}{r^2}. \tag{8.1}$$

Applying the Lagrangian variation operator Δ defined in Equation (2.54) to both sides of Equation (8.1) and assuming periodic changes of the form $\xi e^{i\omega t}$ we obtain

$$\Delta\frac{d^2r}{dt^2} = -\omega^2\xi = -\Delta\left(4\pi r^2\frac{dP}{dM_r} + \frac{GM_r}{r^2}\right)$$

$$= -8\pi r\xi\frac{dP}{dM_r} - 4\pi r^2\frac{d\Delta P}{dM_r} + \frac{2GM_r}{r^2}\frac{\xi}{r} \tag{8.2}$$

where $\xi = \Delta r$ and $\omega^2 = (2\pi/\text{period})^2$ is an eigenvalue. Because oscillations are adiabatic $P\rho^{-\Gamma_1}$ does not change during an oscillation and therefore

$$\frac{\Delta P}{P} = \Gamma_1\frac{\Delta\rho}{\rho}. \tag{8.3}$$

Equations (2.43) and (8.3) imply that

$$\Delta P = -\Gamma_1 P\frac{1}{r^2}\frac{dr^2\xi}{dr}. \tag{8.4}$$

Using the equation of hydrostatic equilibrium to combine the first and last term on the right-hand side of Equation (8.2) and substituting Equation (8.4) into Equation (8.2) gives

$$-\omega^2\rho\xi = -4\frac{\xi}{r}\frac{dP}{dr} - \frac{d\Delta P}{dr}$$

$$= -4\frac{\xi}{r}\frac{dP}{dr} + \frac{d}{dr}\left(\Gamma_1 P\frac{1}{r^2}\frac{dr^2\xi}{dr}\right). \tag{8.5}$$

Equation (8.5) is a second-order linear differential equation for ξ, because the dependent variables P, ρ and r are known as a function of M_r. Equation (8.4) and the first of the two equalities in Equation (8.5) are equivalent to the second-order equation for ξ given by Equation (8.5). It is convenient to introduce the dimensionless variables ξ/r and $\Delta P/P$ and solve the resultant coupled linear first-order differential equations. These coupled equations have the boundary conditions $\xi/r = 0$ at $r = 0$ and $\Delta P/P = 1$ at the surface of the star. Because the equation(s) of radial oscillation are linear they do not determine the limiting pulsational amplitudes. The stellar hydrodynamic equations described in Section 8.4 must be solved to obtain pulsational amplitudes.

A necessary condition that a function f be homogeneous of degree k is

$$f(\lambda x) = \lambda^k f(x). \tag{8.6}$$

Differentiating Equation (8.6), we obtain

$$\frac{df(\lambda x)}{d(\lambda x)}\frac{d(\lambda x)}{d\lambda} = k\lambda^{k-1}f(x). \tag{8.7}$$

If we let $\lambda = 1$ then Equation (8.7) becomes

$$x\frac{df}{dx} = kf(x) \tag{8.8a}$$

and in particular if $k = 1$ Equation (8.8a) is

$$x\frac{df}{dx} = f(x). \tag{8.8b}$$

Assume that $P(r)$, $\rho(r)$ and M_r are homogeneous functions of degree one for a stellar model. From Equation (8.8b) the equations of hydrostatic equilibrium and mass conservation reduce to

$$r\frac{dP}{dr} = P = -\rho\frac{GM_r}{r} \tag{8.9}$$

and

$$r\frac{dM_r}{dr} = M_r = 4\pi r^3\rho. \tag{8.10}$$

Taking the Lagrangian variations of Equations (8.9) and (8.10) gives

$$\frac{\Delta\rho}{\rho} = -3\frac{\Delta r}{r}, \tag{8.11}$$

and

$$\frac{\Delta P}{P} = -4\frac{\Delta r}{r}. \tag{8.12}$$

From the equation of state of an ideal gas we have

$$\frac{\Delta P}{P} = \frac{\Delta\rho}{\rho} + \frac{\Delta T}{T}, \tag{8.13}$$

where the molecular weight has been assumed not to vary. Equations (8.11–8.13) imply that

$$\frac{\Delta T}{T} = -\frac{\Delta r}{r}. \tag{8.14}$$

The condition for pulsational instability

The energy equation is

$$\frac{dQ}{dt} = \frac{dU}{dt} + \frac{dW}{dt} = \varepsilon - \frac{dL_r}{dM_r}, \tag{8.15}$$

with $\dfrac{dW}{dt} = P\dfrac{dV}{dt}$.

For adiabatic oscillations we have

$$T\frac{dS}{dt} = \frac{dQ}{dt} = \Delta\varepsilon - \frac{d\Delta L_r}{dM_r}. \tag{8.16}$$

The constancy of the internal energy density U from cycle to cycle implies that

$$W = \int\int \frac{dQ}{dt} \, dM_r dt, \tag{8.17}$$

where W in Equation (8.17) is the total work done during one cycle (i.e. during the time interval equal to the pulsation period of the particular mode considered).

The entropy density S is a perfect differential and from Equation (8.16) we have

$$dS = \frac{1}{T}\frac{dQ}{dt} \, dt. \tag{8.18}$$

If pulsations occur under adiabatic conditions then Equation (8.18) implies that

$$0 = \int dS = \int \frac{1}{T}\frac{dQ}{dt} \, dt. \tag{8.19}$$

For small variations in T the reciprocal of T can be approximated as

$$\frac{1}{T} \simeq \frac{1}{T_0} - \frac{\Delta T}{T_0^2}. \tag{8.20}$$

Substituting Equation (8.20) into Equation (8.19) and integrating over the various mass shells gives

$$\int\int \frac{dQ}{dt} \, dM_r dt = \int \frac{\Delta T}{T_0}\frac{dQ}{dt} \, dM_r dt. \tag{8.21}$$

Equations (8.16), (8.17) and (8.21) imply that

$$\left\langle \frac{dW}{dt} \right\rangle = \left\langle \int \frac{\Delta T}{T_0} \left(\Delta \varepsilon - \frac{d\Delta L_r}{dM_r} \right) dM_r \right\rangle,$$ (8.22)

where the brackets in Equation (8.22) indicate that an integration has been performed over an entire cycle, and we have made use of the physical condition that U is a constant from cycle to cycle.

The kinetic energy of an oscillatory mode averaged over a cycle is

$$\langle KE \rangle = \frac{\omega^2}{2} \int \left(\frac{\Delta r}{r} \right)^2 r^2 dM_r$$ (8.23)

where

$$\omega^2 = \left(\frac{2\pi}{\text{period}} \right)^2.$$

Because the average gravitational potential of the oscillations equals the average oscillatory kinetic energy the e-folding time for the growth of oscillatory energy in a single mode follows from the equation

$$\frac{1}{\tau} = \frac{\left\langle \dfrac{dW}{dt} \right\rangle}{2\langle KE \rangle},$$ (8.24)

where the numerator and denominator of Equation (8.24) are given by Equations (8.22) and (8.23) respectively.

8.2   Evolution to the red-giant branch and thermonuclear runaways

Numerical solutions of the equations of stellar interiors show that after hydrogen depletion has occurred in their cores main-sequence stars evolve onto the red-giant branch. Low-mass stars ($M \lesssim 1.2\,M_\odot$), which burn hydrogen by means of the proton–proton chain on the main sequence evolve gradually from main sequence to red-giant evolutionary stages. During their rather long lifetimes on the subgiant branch ($\simeq 0.5 \times 10^9$ years for $1\,M_\odot$ stars), hydrogen is transformed into helium by the proton–proton chain in a thick shell surrounding an inert helium core. As evolution continues the helium core mass increases and eventually the temperature in the hydrogen-burning region becomes sufficiently high that the CNO cycle becomes the dominant thermonuclear energy source. The CNO cycle is highly temperature sensitive, and therefore hydrogen burning via the CNO cycle occurs in a thin shell surrounding the helium core. As the mass and density of the inert helium core increase the luminosity and radius of the red giant also increase, and the star evolves to higher luminosity and lower surface temperature as observed in the Hertzsprung–Russell diagram. The cores of stars that are sufficiently massive ($M \geq 1.2\,M_\odot$) to burn hydrogen by means of the CNO cycle on the main sequence contract rapidly (i.e. in a Kelvin–Helmholtz timescale) after hydrogen core exhaustion, and then evolve more rapidly onto the red-giant branch than stars of lower mass. In some stellar clusters a paucity of post-main-sequence stars known as the Hertzsprung gap is observed. These observations provide evidence for rapid core contraction, and therefore evidence that the CNO cycle is the dominant thermonuclear energy source in those cluster stars that are presently evolving off the main sequence.

Because the thermonuclear energy release that results from hydrogen burning is very large ($\simeq 6 \times 10^{18}$ erg g^{-1}) the time-derivative term in Equation (2.131) can be neglected in calculating main-sequence stellar models. If the time-derivative term in Equation (2.131) is neglected in calculating post-main-sequence evolution then the calculated stellar models have isothermal cores that are surrounded by hydrogen-burning shells. Numerical calculations show that isothermal cores consisting of a nondegenerate gas surrounded by a hydrogen-burning shell source do not exist if the core mass exceeds $\simeq 0.1$–0.15 times the mass of the star. These limiting isothermal core masses are referred to collectively as the Schönberg–Chandrasekhar limit. The existence of a limiting isothermal core for a particular initial mass main-sequence star shows that core contraction must occur in post-main-sequence evolution.

The physical basis for a limiting isothermal core mass can be inferred from the virial theorem. In deriving the virial theorem given in Equation (2.35) the surface term that results from a partial integration is zero because the pressure is zero at the stellar surface. If the surface term is evaluated at the boundary radius r_1 of an isothermal core then the virial theorem becomes

$$4\pi P_1 r_1^3 \simeq 3(\Gamma_1 - 1)U_1 + \Omega_1 \tag{8.25}$$

where $U_1 \simeq \dfrac{3}{2}\dfrac{kT_c}{m_p\mu}M_{core}$, $\Omega_1 \simeq \dfrac{-GM_{core}^2}{r_1}$ and $\Gamma_1 = \frac{5}{3}$ for an ionized nondegenerate gas.

We wish to obtain the condition for the pressure at the core boundary P_1 to have a maximum value with T_c and M_{core} given. Because the CNO cycle is highly temperature sensitive, the isothermal core temperature cannot vary appreciably. The maximum surface temperature is obtained by setting the derivative

$$\left(\frac{dP}{dr}\right)_{r=r_1} = 0. \tag{8.26}$$

Multiplying Equation (8.25) by r, differentiating the resultant equation with respect to r and using the condition given in Equation (8.26) gives

$$16\pi P_{Max} r_1^3 \simeq \frac{3kT_c M_{core}}{m_p\mu}. \tag{8.27}$$

Eliminating P_{Max} from (8.26) and (8.27) with $\Gamma_1 = \frac{5}{3}$ we obtain

$$r_1 = \frac{4}{9}\frac{GM_{core}m_p\mu}{kT_c}. \tag{8.28}$$

Using the expression for r_1 given in Equation (8.28) to eliminate r_1 from Equation (8.27) we have

$$P_{\text{Max}} \simeq \frac{3^7}{4^5\pi}\left(\frac{kT_c}{m_p\mu}\right)^4 \frac{1}{G^3 M_{\text{core}}^2}. \tag{8.29}$$

Since T_c does not vary appreciably Equation (8.29) implies that as M_{core} increases P_{Max} decreases. The core must contract if the boundary pressure exceeds P_{Max}. As a consequence of contraction the central temperature becomes appreciably greater than the temperature of the hydrogen-burning shell. Although the cores of red giants become sufficiently dense for electrons to be degenerate, hydrogen- (and helium-) burning shell sources remain at nondegenerate gas densities.

For a nondegenerate ideal gas the local polytropic index n can be found from the equation

$$n + 1 = \frac{d\ln P}{d\ln T}. \tag{8.30}$$

From Equation (8.30) and the equation of hydrostatic equilibrium we obtain

$$\frac{dT}{dr} = \frac{dT}{dP}\frac{dP}{dr} = \frac{-T\rho GM_r}{P(n+1)r^2}. \tag{8.31}$$

In computed red-giant models, n is approximately 3 close to hydrogen-burning shell sources, and therefore the gas density and temperature decrease rapidly for radial distances r greater than the hydrogen-burning shell source. Integrating Equation (8.31) with M_r and n assumed constant gives

$$T - T_1 = \frac{GM_r m_p\mu}{(n+1)k}\left(\frac{1}{r} - \frac{1}{r_1}\right). \tag{8.32}$$

If the conditions $r \gg r_1$ and $T \ll T_1$ are satisfied, then Equation (8.32) becomes

$$T_1 \simeq \frac{GM_r\mu m_p}{(n+1)kr_1}. \tag{8.33}$$

Because the core mass M_r and temperature of the hydrogen-burning shell T_1 do not increase rapidly as the star evolves, Equation (8.33) implies that the radius of the hydrogen-burning shell does not change appreciably either.

Numerical solutions of the equations of stellar interiors show that as the core mass of a red giant increases the luminosity and radius increase by a large factor but the core radius changes by only a small amount. A qualitative explanation for the large radii of red giants is given by the following arguments. From Equation (2.28) we have

$$\frac{d\ln r}{d\ln M_r} = \frac{M_r}{4\pi r^3\rho}. \tag{8.34}$$

The logarithm of the radius of a red giant $\ln R$ can be obtained from the integral

$$\ln R = \int_{M_1}^{M}\frac{d\ln r}{d\ln M_r}d\ln M_r + \ln r_1 \tag{8.35}$$

where M, M_1 and r_1 are the mass of star, mass interior to hydrogen-burning shell and radius of hydrogen-burning shell respectively. Since $\ln r_1$ remains approximately the same as the red giant evolves it follows that small changes in the integrand of Equation (8.35) can lead to large changes in the radius R.

Thermonuclear runaways

If the radial modes of oscillation are obtained from the equation of adiabatic pulsations given in Equation (8.5) then Equation (8.24) can be used to calculate the (linear) timescale for the growth of pulsational energy. If we had taken the energy conservation and energy transport equations into account in calculating the relative amplitudes and eigenvalues of pulsational modes we would have found that the eigenvalues are complex and proportional to $\exp(i\omega t + \delta t/2)$ with δ real and positive for unstable modes. For unstable modes $1/\delta$ gives the characteristic timescale for the growth of pulsational energy. When a star is dynamically unstable $i\omega$ is real and positive and the instability develops in a free-fall timescale. In addition to pulsational and dynamical instabilities a star may also become thermally unstable.

Suppose that $P = P(M_r)$, $r = r(M_r)$, $T = T(M_r)$ and $L_r = L_r(M_r)$ are calculated solutions of the equations of stellar interiors with the time derivative term in the energy equation included. It follows that we can linearize the equations of stellar interiors and solve for solutions of the type

$$P + \Delta P e^{t/\tau_k},$$
$$r + \Delta r e^{t/\tau_k},$$
$$T + \Delta T e^{t/\tau_k},$$
$$L_r + \Delta L_r e^{t/\tau_k}, \tag{8.36}$$

where the eigenvalues τ_k are real. In any stellar model modes with τ_k negative always exist. Such modes describe mathematically how departures from thermal equilibrium are damped. The star is thermally unstable if there exists a thermal mode for which τ_k is positive. The perturbations ΔP, ΔL_r, ΔT and Δr given in Equation (8.36) preserve hydrostatic equilibrium. A thermally unstable star remains in hydrostatic equilibrium, at least, during the early stages of the development of the instability.

Main-sequence stars have thermonuclear sources at their centers. It is of interest to determine whether main-sequence stars can be thermally unstable. The relative amplitudes of the thermal modes of main-sequence stars are given approximately by Equations (8.11), (8.12) and (8.14). The latter equations give

$$\frac{\delta P}{P} = \frac{4}{3}\frac{\delta\rho}{\rho} \tag{8.37}$$

$$\frac{\delta T}{T} = \frac{1}{3}\frac{\delta\rho}{\rho}. \tag{8.38}$$

Equations (8.37) and (8.38) imply that if the rate of thermonuclear energy generation in the core is perturbed above its quiescent value then the gas will expand and the density and temperature decrease. Since the rates of thermonuclear energy generation are highly temperature sensitive the decrease in temperature causes a relatively large reduction in thermonuclear energy generation, and therefore the main-sequence star is thermally stable.

Hydrogen-burning shells are generally thermally stable in red giants. However, if mass is accreted onto a white dwarf thermal instability results. The resultant thermonuclear runaways are responsible for the nova outburst. The 3α process is even more temperature sensitive than the CNO cycle, and red giants that have evolved to the stage in which hydrogen and helium burning occurs in thin shells about an inert carbon and oxygen core have thermally unstable helium-burning shells.

The gravity g at the center of a star is equal to zero; however, the gravity in the vicinity of a helium-burning shell is comparable to that on the surface of a white dwarf and perturbations preserving hydrostatic equilibrium are such that

$$\frac{\delta P}{P} \simeq 0. \tag{8.39}$$

The perturbations δP, $\delta \rho$ and δT are related by Equation (8.13) because the helium-burning shell is nondegenerate. From Equation (8.13) we have

$$\frac{\delta T}{T} \simeq -\frac{\delta \rho}{\rho}. \tag{8.40}$$

Equation (8.40) implies that expansion resulting from an excess amount of thermonuclear energy generation can be accompanied by an increase in the temperature of the gas. Therefore, since the rate of thermonuclear energy generation is highly temperature sensitive a thermonuclear runaway can develop. The e-folding time τ_k associated with the increase in thermonuclear energy generation during a thermonuclear runaway is given by the equation

$$\frac{1}{\tau_k} \simeq \frac{\varepsilon_{He}}{3P/2\rho}, \tag{8.41}$$

where ε_{He} is the rate of energy generation from the 3α process and $\frac{3}{2} P/\rho$ is the thermal energy density of the gas. The limiting amplitude (i.e. highest rate of thermonuclear energy release) of a thermal instability occurs when the pressure in the region of the helium-burning shell begins to decrease rapidly as the result of expansion.

The radiative transfer equation given as Equation (2.80) can be expressed as

$$\frac{L_r}{4\pi r^2} = -\sigma \frac{dT}{dr} \tag{8.42}$$

with $\sigma = \dfrac{4ac}{3\kappa\rho} T^3$.

If the time derivative term is neglected then the energy equation becomes

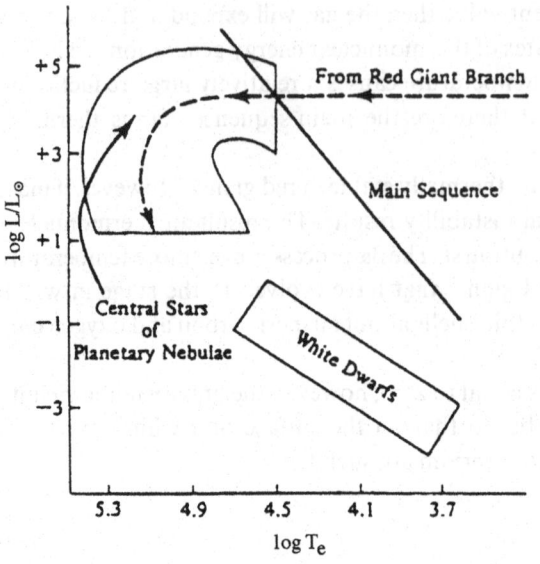

Figure 8.5. The evolution of a star from the red-giant branch into the white-dwarf state is shown in the H–R diagram.

$$0 = -\nabla \cdot \boldsymbol{F} + \rho\varepsilon_{3\alpha}. \tag{8.43}$$

From Equations (8.42) and (8.43) we obtain

$$\frac{1}{r^2}\frac{\partial}{\partial r}\left(r^2\sigma\frac{\mathrm{d}T}{\mathrm{d}r}\right) + \rho\varepsilon_{3\alpha} = 0. \tag{8.44}$$

Suppose that the density and temperature at two boundary radii are given. If a temperature distribution satisfying Equation (8.44) exists then helium burning is thermally stable. Otherwise, the time derivative term must be included and helium burning is unstable. The onset of a thermonuclear runaway in stars is physically similar to reaching the critical condition such that there exists no steady state temperature distribution satisfying Equation (8.44).

Figure 8.5 shows the final evolution of an asymptotic giant branch star.

8.3   Nonradial oscillations

Radial oscillations are described by either a second order linear differential equation or two coupled first order linear differential equations. Nonradial oscillations have an angular dependence proportional to a spherical harmonic Y_{lm} and can have periodicities close to the corresponding periodicities of radial modes of oscillation. The periodic variable β Canis Majoris and some other β Cepheid variables have periodicities that indicate both the longest period radial mode and a $l = 2$ nonradial oscillation with nearly the same periodicity have been excited. Nonradial pulsations divide into two branches: p modes and g modes. The p (pressure) modes have a period distribution such that periodicities vary from a longest period mode, whose period is generally within a factor of

2 of the corresponding fundamental radial mode to arbitrarily shorter periodicities. On the other hand, the g modes where g stands for gravity are characterized by a shortest period mode and a spectrum of modes with longer periodicity. Some white dwarfs are known to have periodicities that appear to be caused by gravity modes whose periods are much longer than the corresponding longest period radial mode.

In deriving the equations of nonradial oscillation we will assume that oscillatory perturbations occur about a spherically symmetric stellar model whose structure is determined from the equations

$$\frac{1}{\rho}\frac{dP}{dr} = \frac{-d\Phi}{dr}, \tag{8.45}$$

$$\nabla^2\Phi = \frac{d^2\Phi}{dr^2} + \frac{2}{r}\frac{d\Phi}{dr} = 4\pi G\rho, \tag{8.46}$$

and the remaining equations of stellar interiors described in Chapter 2. Equations (8.45) and (8.46) lead directly to the identity

$$\frac{1}{\rho}\frac{d^2P}{dr^2} = \frac{1}{\rho^2}\frac{d\rho}{dr}\frac{dP}{dr} - \frac{2}{r\rho}\frac{dP}{dr} - 4\pi G\rho. \tag{8.47}$$

If the Eulerian perturbations δP, $\delta\rho$ and $\delta\Phi$ are added to the unperturbed values of P, ρ and Φ the equation of motion becomes

$$\frac{\partial \boldsymbol{v}}{\partial t} = -\nabla\Phi - \nabla\delta\Phi - \frac{1}{\rho}\nabla(\delta P + P) + \frac{\delta\rho}{\rho^2}\nabla P. \tag{8.48}$$

Because Φ satisfies Equation (8.45) it follows that Equation (8.48) reduces to

$$\frac{\partial \boldsymbol{v}}{\partial t} = -\nabla\delta\Phi - \frac{1}{\rho}\nabla\delta P + \frac{\delta\rho}{\rho^2}\nabla P. \tag{8.49}$$

The equation of continuity can be expressed as

$$\frac{\partial \rho'}{\partial t} + \rho'\nabla\cdot\boldsymbol{v} + \boldsymbol{v}\cdot\nabla\rho' = 0, \tag{8.50}$$

with $\rho' = \rho + \delta\rho$. Since

$$\frac{\partial \rho'}{\partial t} = \frac{\partial \delta\rho}{\partial t}, \tag{8.51}$$

Equation (8.50) becomes

$$\frac{\partial \delta\rho}{\partial t} + \boldsymbol{v}\cdot\nabla\rho = -\rho\nabla\cdot\boldsymbol{v}. \tag{8.52}$$

Because the timescale for radiative diffusion is much longer than the corresponding dynamical timescale the adiabatic condition, $P\rho^{-\Gamma} = $ constant, is satisfied during a Lagrangian perturbation. Therefore, we have

$$\frac{\Delta P}{P} = \Gamma \frac{\Delta \rho}{\rho}. \tag{8.53}$$

If the vector function $a(t)$ with t equal to the time describes the motion of a fixed position r with respect to some mass element then

$$P(r, t) = P(a(t)) + \Delta P(a(t), t). \tag{8.54}$$

Differentiating Equation (8.54) and neglecting second-order terms gives

$$\frac{dP}{dt} = \frac{d\delta P}{dt} = \frac{da(t)}{dt} \cdot \nabla P + \frac{\partial}{\partial t} \Delta P. \tag{8.55}$$

Since v is the velocity of a fluid element with respect to some position we have

$$\frac{da}{dt} = -v. \tag{8.56}$$

Equations (8.55) and (8.56) imply that

$$\frac{\partial \delta P}{\partial t} + v \cdot \nabla P = \frac{\partial}{\partial t} \Delta P. \tag{8.57}$$

Variations in ρ satisfy an equation similar to Equation (8.57), and therefore from Equations (8.53) and (8.57), we obtain

$$\frac{1}{P}\left(\frac{\partial \delta P}{\partial t} + v \cdot \nabla P\right) = \frac{\Gamma}{\rho}\left(\frac{\partial \delta \rho}{\partial t} + v \cdot \nabla \rho\right). \tag{8.58}$$

Equations (8.52) and (8.58) lead directly to the equation

$$\frac{\partial \delta P}{\partial t} + v \cdot \nabla P = -\Gamma P \nabla \cdot v. \tag{8.59}$$

Linearizing Equation (8.46) gives

$$\nabla^2 \delta \Phi = 4\pi G \delta \rho. \tag{8.60}$$

Equations (8.49), (8.52), (8.59), (8.60), two boundary conditions at the center of the star and two at the surface determine the amplitudes and eigenvalues of the oscillatory modes. The above equations can be combined into two coupled linear second-order differential equations.

8.4   Stellar hydrodynamics

In order to simulate the nonlinear oscillations of periodic variables such as Cepheids, RR Lyrae variables and long-period variables, and the hydrodynamics of nova and supernova outbursts it is necessary to solve the gas dynamic equations numerically. Shock-wave formation occurs in stars when departures from hydrostatic equilibrium become large. Therefore, numerical solutions must be able to simulate the formation and propagation of shock waves that result when there are supersonic motions between mass elements.

For an assumed spherically symmetric geometry the stellar hydrodynamic equations, not including an equation of energy transport, are

$$\frac{dr}{dt} = v, \tag{8.61}$$

$$\frac{dv}{dt} = -4\pi r^2 \frac{\partial(P+Q)}{\partial M_r} - \frac{GM_r}{r^2} \tag{8.62}$$

$$\rho = \frac{1}{4\pi r^2}\frac{\partial M_r}{\partial r}, \tag{8.63}$$

$$\frac{dU}{dt} + (P+Q)\frac{d(1/\rho)}{dt} = \varepsilon - \frac{\partial L_r}{\partial M_r}, \tag{8.64}$$

the physical variables in Equations (8.61–8.64) are the same as those that appear in the equations of stellar interiors described in Chapter 2 except for the physical variable Q, which is required for shock-wave simulation. Equations (8.61–8.64) are not complete because they do not include the equation(s) of energy transport. The luminosity $L_r = L_R(M_r) + L_c(M_r)$ with $L_R(M_r)$ equal to the radiative luminosity given by Equation (2.80) and L_c equal to the convective luminosity. The convective luminosity and convective velocity $u_c = u_c(M_r)$ are determined from the equations

$$\frac{\partial L_c}{\partial t} = -(L_c - L_c^0)\frac{u_c}{l} \tag{8.65}$$

and

$$\frac{\partial u_c}{\partial t} = -\frac{u_c^2}{l}, \tag{8.66}$$

where l is the convective mixing length and L_c^0 is the steady-state convective luminosity predicted by the theory of nonadiabatic convection described in Chapter 5. The mixing length l is usually chosen equal to $\simeq 0.5$–1 pressure scale heights.

In stars the thickness of a shock wave is approximately equal to the collisional mean free path between particles. This distance is very small for the physical conditions we consider, and therefore physical variables such as P, T and ρ undergo discontinuous changes across a shock front. In order to solve the hydrodynamic Equations (8.61–8.66) we must replace these differential equations with the corresponding set of difference equations. It follows that the thickness of a shock wave will be approximately equal to the distance between mass points (i.e. $M_r(i + 1)$ and $M_r(i)$ with $i = 1, \ldots, n - 1$ and n equal to the total number of mass points).

The term Q in Equation (8.64) is known as the artificial viscous pressure because it is introduced for purely numerical reasons and can be chosen as any convenient function of P, v and their derivatives provided the thickness of a shock is approximately equal to the distance between neighboring mass points. The effect of terms containing Q must be small outside the shock, and Q must be chosen so that discontinuities are avoided. For the spherical geometry that we consider Q is given by the equations

$$Q = c_0 \rho l^2 \left[\frac{1}{r^2} \frac{\partial}{\partial r} (r^2 v) \right]^2, \qquad \frac{\partial v}{\partial r} < 0 \tag{8.67}$$

(compression)

$$Q = 0, \qquad \frac{\partial v}{\partial r} > 0, \tag{8.68}$$

(expansion)

where c_0 is a dimensionless parameter of order unity, l is the (variable) spacing of mass points and v is the gas velocity. The viscous pressure variable Q defined by Equations (8.67) and (8.68) is a nonzero addition to the gas pressure P only when a region is undergoing compression. The presence of a significant Q in the absence of a shock front would cause unphysical damping of gas motions.

In obtaining solutions of the finite-difference equations a grid of mass points $M_r(1)$, $M_r(2), \ldots, M_r(N) = M$ is chosen. Stellar models are calculated for each time $^1t, {}^2t, {}^3t, \ldots, {}^nt, {}^{n+1}t$ but the mass points are held fixed for an entire calculation. The subscript j on a particular physical variable denotes the particular mass point. In finite-difference form Equation (8.63) can be expressed as

$$\Delta m_{j+\frac{1}{2}} = \frac{4\pi}{3} {}^n\rho_{j+\frac{1}{2}} [({}^n r_{j+1})^3 - ({}^n r_j)^3]. \tag{8.69}$$

If the time intervals $^{n+\frac{1}{2}}\Delta t = {}^{n+1}t - {}^n t$ are chosen the same for all values of n, then in finite-difference form the equation of motion, given as Equation (8.62), becomes

$$\frac{{}^{n+\frac{1}{2}}v_j - {}^{n-\frac{1}{2}}v_j}{\Delta t} = - 4\pi({}^n r_j)^2 \frac{{}^n P_{j+\frac{1}{2}} + {}^n Q_{j+\frac{1}{2}} - {}^n P_{j-\frac{1}{2}} - {}^n Q_{j-\frac{1}{2}}}{\Delta m_j}. \tag{8.70}$$

The finite-difference equation corresponding to the energy-conservation equation, given as Equation (8.64), is

$$\frac{{}^{n+1}U_{j+\frac{1}{2}} - {}^n U_{j+\frac{1}{2}}}{\Delta t} = [{}^{n+\frac{1}{2}}P_{j+\frac{1}{2}} + {}^{n+\frac{1}{2}}Q_{j+\frac{1}{2}}] \frac{\dfrac{1}{{}^{n+1}\rho_{j+\frac{1}{2}}} - \dfrac{1}{{}^n\rho_{j+\frac{1}{2}}}}{\Delta t}$$

$$= {}^{n+\frac{1}{2}}\varepsilon_{j+\frac{1}{2}} - \frac{[{}^{n+\frac{1}{2}}L_j - {}^{n+\frac{1}{2}}L_{j-1}]}{\Delta m_j}. \tag{8.71}$$

The velocity v_j is obtained from the equation

$$^{n+\frac{1}{2}}v_j = \frac{^{n+1}r_j - {}^n r_j}{\Delta t}. \tag{8.72}$$

The method of solution used to solve the hydrodynamic equations involves the following steps. The values of the physical variables must be given at some initial time. The variables $^{n+\frac{1}{2}}v_j$ and $^{n+1}r_j$ can be found from Equations (8.70) and (8.72), respectively. The density $^{n+1}\rho_{j+\frac{1}{2}}$ can then be obtained from Equation (8.69) and the artificial viscous pressure $^{n+\frac{1}{2}}Q_{j+\frac{1}{2}}$ can be calculated from the equation

$$^{n+\frac{1}{2}}Q_{j+\frac{1}{2}} = c_0^{n+\frac{1}{2}}\rho_{j+\frac{1}{2}} \left[\frac{(^{n+\frac{1}{2}}r_{j+1})^2\,{}^{n+\frac{1}{2}}v_{j+1} - (^{n+\frac{1}{2}}r_j)^2\,{}^{n+\frac{1}{2}}v_j}{(^{n+\frac{1}{2}}r_{j+\frac{1}{2}})^2} \right]^2. \tag{8.73}$$

Next we must solve Equation (8.64), which in finite-difference form is given by Equation (8.71). Equations (2.80), (8.65) and (8.66) written in finite-difference form can be used to eliminate the luminosity from the energy Equation (8.71). Since all other physical variables are known at t^{n+1} except the temperature, the energy equation determines the temperature. The resultant energy equation must be solved implicitly (i.e. iteractively) for the temperature values at the various mass points because the difference equations are nonlinear equations. Moreover, the difference equations contain values of the temperature at three different mass points, and consequently a tridiagonal matrix must be inverted in solving for the temperature.

8.5 Helioseismology and stellar seismology

Observations of the solar atmosphere have shown that patterned motions result from large numbers of simultaneously excited oscillatory, interior modes whose periods of oscillation vary from several minutes to several hours. Horizontal wavelengths of modes range from global to as small as several thousand kilometers. Measured oscillations have been classified as p modes (mostly nonradial, $l > 0$) with upper boundaries just below the photosphere. Since the frequencies of excited modes depend on the density and temperature structure of the Sun, accurate frequency determinations can provide detailed information about solar interior layers where amplitudes of particular modes are significant. Identified modes have appreciable amplitudes only in the solar convective zone ($r \gtrsim 0.7\,R_\odot$) and direct inferences about the solar interior have been limited to $r > 0.5\,R_\odot$. Therefore, little can be deduced about the hydrogen-burning core. To determine properties of the solar convective zone it is necessary to measure modes with high l values and to supplement ground based observations with measurements from spacecraft. Most reliable information about solar stratification has been obtained from observations of the five minute p modes. Helioseismic measurements have yielded a helium mass fraction of 0.25 and a solar convective zone depth of $0.287 \pm 0.003\,R_\odot$. Therefore, they provide important confirmation of other independent estimates of these parameters. Excitation of

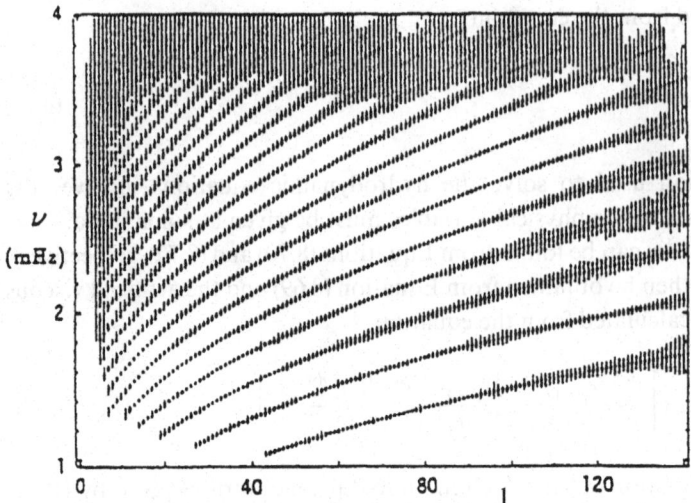

Figure 8.6. Some measured solar p-mode frequencies. Vertical lines represent $\pm 1000\,\sigma$ errors, where σ is the standard error. Sequences of frequencies of modes of fixed order n are approximately parabolic at high l. The lowest-frequency sequence has $n = 1$. From Libbrecht and Woodward (1990). See also Gough and Toomre (1991).

high-frequency p modes is believed to occur in the upper solar convective boundary layer.

A number of white dwarfs, hot pre-white-dwarf stars and central stars of planetary nebulae have been observed to undergo multiperiodic pulsations. These periodicities in light appear to result from the excitation of gravity (g) modes. The interpretation of photometric observations of such stars is referred to as stellar seismology because many nonradial oscillatory modes may be simultaneously excited. From the point of view of stellar evolution, pulsating compact objects are related because the central stars of planetary nebulae cool and become normal white dwarfs in about 10^6 years.

Hot pre-white-dwarf stars are referred to as DO variables (or DOV stars) and their ages are typically 10^5 years after planetary-nebula formation, whereas pulsating helium white dwarfs (DBV stars) have ages of about 10^7 years, are cooler than DOV stars and their cooling is dominated by normal radiative loss instead of plasma neutrino emission. The oldest and coolest (i.e., $T \simeq 11\,000\,\mathrm{K}$) class of compact pulsators are the pulsating hydrogen white dwarfs, which are called either DAV stars or ZZ Ceti stars. They are approximately 10^9 years old and all have about the same photospheric temperatures. Most white dwarfs can be classified as having either nearly pure helium atmospheres or pure hydrogen atmospheres. The latter class of white dwarfs is referred to as spectro-scopic type DA whereas the former is called spectroscopic type DB. DA white dwarfs are approximately four times more abundant than DB white dwarfs. All four classes of pulsators have at least several oscillatory modes simultaneously excited. Because DAV stars show only hydrogen in their spectra and are found only near the well-defined hydrogen opacity maximum it is clear that excitation of pulsations is connected with the development of the partial ionization zone of hydrogen close to the surface layer. Likewise, DBV stars show only helium lines in their spectra and their location in the H–R diagram indicates that unlike nonvariable DB stars they exist only near the HeI opacity maximum. Partial ionization zones of carbon and oxygen are believed responsible for driving pulsations in variable DO stars and the still hotter central stars of planetary nebulae.

Observations of the pulsating DB white dwarf GD 358 have shown that the power spectrum of its light curve contains more than 100 significant peaks. The interpretation of these results leads to a fairly accurate determination of white dwarf mass, luminosity and distance (i.e., $M = 0.6 M_\odot$, $L = 0.05 L_\odot$ and distance = 42 pc). The rate of change of period with time of DAV (i.e. ZZ Ceti) stars depends on the cooling timescale of a relative cool DA white dwarf. Since such stars cool very slowly, changes in their periodicities are also very gradual. ZZ Ceti stars are perhaps the most stable optical clocks since changes in pulsation period can be less than a part in 10^{14} during a time interval of one period.

Measured solar p mode frequencies are shown in Figure 8.6.

9 Binary stars, mass accretion, stellar rotation and meridional circulation

9.1 Binary stars

We consider binary stars with masses M_1 and M_2 that have circular orbits and are separated by a distance a. In the rotating frame of reference the force on each star is zero. The condition that the force on the star with mass M_1 is zero in the rotating frame leads to the equation

$$0 = -\frac{GM_1M_2}{a^2}\hat{r} - M_1\omega \times (\omega \times r_1), \tag{9.1}$$

where $\hat{r}$ is a unit vector along the direction defined by the centers of the two stars and ω is the orbital angular velocity. The position vector r_1 in Equation (9.1) is the position vector of the center of the star with mass M_1 as measured with respect to the center of mass of the binary system. We choose the x-axis of the rotating coordinate system to lie along the direction of r_1. From the definition of the center of mass we have

$$M_1x_1 = -M_2x_2, \tag{9.2}$$

with $x_1 - x_2 = a$. Solving for x_1 and x_2 gives

$$x_1 = \frac{M_2a}{M_1 + M_2}, \tag{9.3a}$$

$$x_2 = \frac{-M_1a}{M_1 + M_2}. \tag{9.3b}$$

From the definition of the rotating coordinate system the second term on the right-hand side of Equation (9.1) becomes

$$-M_1\omega \times (\omega \times r_1) = M_1\omega^2x_1. \tag{9.4}$$

Equations (9.1), (9.3) and (9.4) imply that

$$\omega^2 = \frac{G(M_1 + M_2)}{a^3}. \tag{9.5}$$

Most binary stars are too distant for their angular separation to be measured directly on a photographic plate. However, if spectral lines are observable it is possible to measure spectral line Doppler shifts and thereby infer the projection of the orbital velocity along the line of sight. The projected orbital velocity of a star is

$$v_1 = \omega x_1 \sin i = \frac{2\pi}{P} x_1 \sin i, \tag{9.6}$$

where P is the orbital period of the binary system and the angle of inclination i is the angle between the line of sight and the line perpendicular to the orbital plane that passes through the center of the star. Equations (9.3), (9.5) and (9.6) imply that

$$f_1(M_1, M_2, i) = \frac{(M_2 \sin i)^3}{(M_1 + M_2)^2} = \frac{P v_1^3}{2\pi G}. \tag{9.7}$$

The function f_1, which is known as the mass function, depends on the observable P and v_1. The mass function given in Equation (9.7) can be determined for nearby stars even if orbital velocities are too low for accurate measurement of Doppler shifts because $a_1 \sin i$ and the orbital period P can be observed.

If spectral Doppler shifts can be measured from both stars then the mass function

$$f_2(M_1, M_2, i) = \frac{(M_1 \sin i)^3}{(M_1 + M_2)^2} = \frac{P v_2^3}{2\pi G} \tag{9.8}$$

can also be determined. Equations (9.7) and (9.8) imply that

$$\frac{M_2}{M_1} = \frac{v_1}{v_2}, \tag{9.9}$$

independent of the inclination angle i. In eclipsing binary systems the angle i can be estimated and therefore the masses M_1 and M_2 are determined by Equations (9.9) and (9.8) (or (9.7)).

In the rotating coordinate system of binary stars, the motion of a point mass m is

$$m\frac{d^2 r}{dt^2} = F_1 + F_2 - m\omega \times (\omega \times r) - 2m\left(\omega \times \frac{dr}{dt}\right), \tag{9.10}$$

where F_1 and F_2 are the gravitational forces on m caused by stars of mass M_1 and M_2 respectively. The last two terms in Equation (9.10) represents the centrifugal and coriolis forces. The origin of the rotating coordinate system is chosen at the center of mass of the binary system, the z-axis is taken along the orbital axis ω and the x-axis lies along the line that separates the centers of the two stars. The x and y components of Equation (9.10) in the rotating coordinate system are

$$\ddot{x} = \frac{-GM_1(x - x_1)}{[(x - x_1)^2 + y^2]^{3/2}} - \frac{GM_2(x - x_2)}{[(x - x_2)^2 + y^2]^{3/2}} + \frac{G(M_1 + M_2)x}{a^3} + 2\omega\dot{y} \tag{9.11}$$

$$\ddot{y} = \frac{GM_1 y}{[(x - x_1)^2 + y^2]^{3/2}} - \frac{GM_2 y}{[(x - x_2)^2 + y^2]^{3/2}} + \frac{G(M_1 + M_2)y}{a^3} - 2\omega\dot{x}, \tag{9.12}$$

where the positions of M_1 and M_2 on the x-axis are given by Equations (9.3) and (9.4) and $x_1 - x_2 = a$.

The centrifugal force given in Equation (9.10) can be derived from the centrifugal potential

$$V_c = -\frac{1}{2}m\omega^2(x^2 + y^2) = -\frac{Gm(M_1 + M_2)(x^2 + y^2)}{2a^3}. \tag{9.13}$$

Because it is perpendicular to the direction of motion the coriolis force can do no work on the point mass m.

If m lies in the x-y plane then its energy is

$$E = \frac{1}{2}m(\dot{x}^2 + \dot{y}^2) + V(x, y, 0), \tag{9.14}$$

with

$$V(x, y, 0) = -\frac{GmM_1}{[(x - x_1)^2 + y^2]^{1/2}} - \frac{GmM_2}{[(x - x_2)^2 + y^2]^{1/2}} - \frac{Gm(M_1 + M_2)(x^2 + y^2)}{2a^3}. \tag{9.15}$$

In terms of the dimensionless variables

$$\xi = \frac{x}{a}, \eta = \frac{y}{a}, \xi_1 = \frac{M_2}{M_1 + M_2} \text{ and } \xi_2 = \frac{M_1}{M_1 + M_2}$$

the potential V in Equation (9.15) becomes

$$V(\xi, \eta, 0) = -\frac{Gm(M_1 + M_2)}{a}$$

$$\times \left[\frac{\xi_2}{[(\xi - \xi_1)^2 + \eta^2]^{1/2}} + \frac{\xi_1}{[(\xi - \xi_2)^2 + \eta^2]^{1/2}} + \frac{1}{2}(\xi^2 + \eta^2) \right]. \tag{9.16}$$

The potential $V(\xi, \eta, 0)$ in Equation (9.16) has maxima at three critical points which are known as Lagrangian points along the ξ-axis. Two additional Lagrangian points are located off the ξ-axis. The Lagrangian point L_1, which is located on the ξ-axis between the binary stars, is of particular interest because if one binary star expands sufficiently that part of its surface reaches the Lagrangian point L_1 mass transfer between the two stars will occur. It can be shown by expanding the potential $V(\xi, \eta, 0)$ about the Lagrangian points that motions of particles located close to Lagrangian points are unstable. The equipotential surface about a star that includes the Lagrangian point L_1 is called the critical Roche potential (or Roche lobe). Equipotential surfaces close to either M_1 or M_2 are nearly spherical surfaces about individual stars. The equipotential surfaces surround both stars at distances outside the Roche lobes (see Figures 9.1 and 9.2).

Equations (9.3), (9.4) and (9.6) imply that the angular momentum J of a binary system as measured in the center of mass frame is

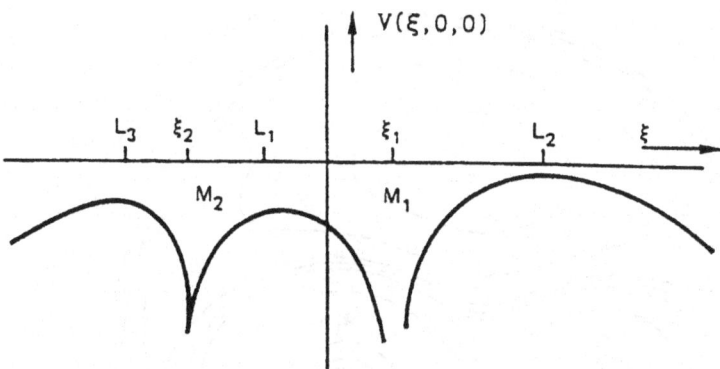

Figure 9.1. $V(\xi, 0, 0)$ is shown as a function of ξ.

$$J = M_1 \omega x_1^2 + M_2 \omega x_2^2 = \frac{M_2 M_2 (Ga)^{1/2}}{(M_1 + M_2)^{1/2}}. \tag{9.17}$$

Solving for a in Equation (9.17) gives

$$a = \frac{MJ^2}{G(M_1 M_2)^2} = \frac{MJ^2}{GM_1^2(M - M_1)^2}, \tag{9.18}$$

with $M = M_1 + M_2$. If the total mass M and angular momentum J are conserved during mass transfer then Equation (9.18) implies that

$$\delta a = \frac{2MJ^2}{GM_1^3(M - M_1)^2} \left(\frac{-M + 2M_1}{M - M_1} \right) \delta M_1, \tag{9.19}$$

with $\delta M_1 + \delta M_2 = 0$. Equation (9.19) gives the relation between the amount of mass transferred and the change in distance between the two stars.

Most supernova explosions occur in massive stars. About one-half of such massive stars have binary companions. As the result of a supernova explosion the binary system may become unbound. For assumed circular orbits the speeds of M_1 and M_2 as measured in the center of mass frame are

$$v_1 = M_2 \left[\frac{G}{a(M_1 + M_2)} \right]^{1/2} \tag{9.20}$$

and

$$v_2 = M_1 \left[\frac{G}{a(M_1 + M_2)} \right]^{1/2}, \tag{9.21}$$

respectively. We assume that a supernova occurs in M_1. The timescale for the supernova outburst to eject mass from the binary system is short as compared to the binary orbital period. Therefore, we can neglect orbital motion of M_1 and M_2 during the supernova outburst. The velocity changes of M_1 and M_2 can also be considered small as measured in

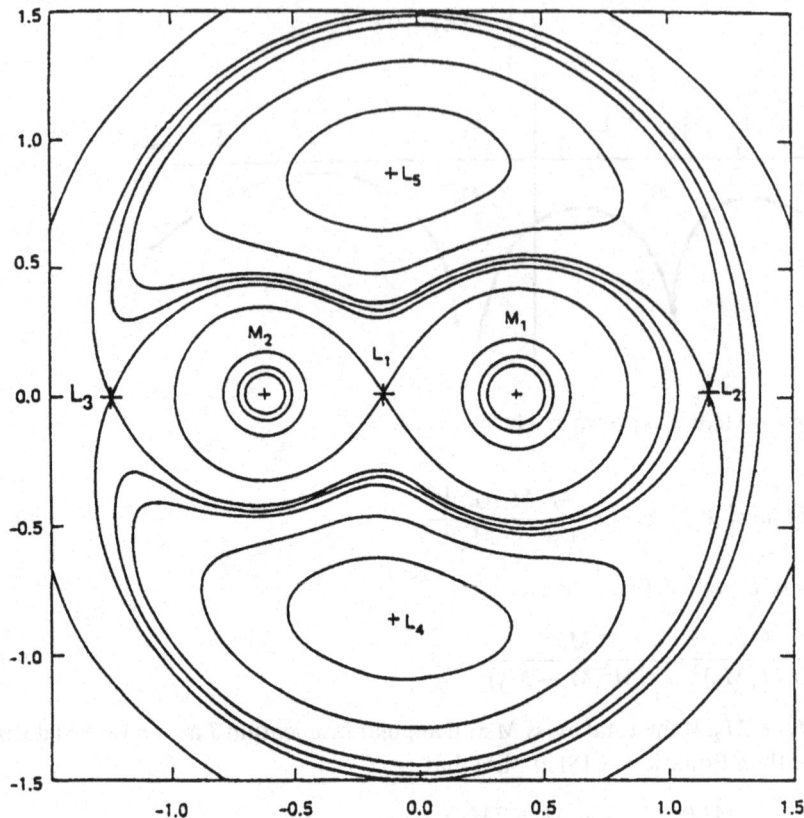

Figure 9.2. A plot of equipotential surfaces. The critical Roche equipotential surface intersects the inner Lagrangian point L_1.

the initial center-of-mass frame. The mass of the stellar remnant (most likely a neutron star) is qM_1 and therefore the ejected mass is $(1 - q)M_1$.

From Equations (9.3) and (9.4) the orbital velocities of M_1 and M_2 in the center-of-mass frame immediately after the supernova outburst are

$$v_1 = \omega x_1 = \frac{\omega M_2 a}{qM_1 + M_2} = \frac{M_2}{(qM_1 + M_2)}\left(\frac{G(M_1 + M_2)}{a}\right)^{1/2} \tag{9.22}$$

and

$$v_2 = \omega x_2 = \frac{-\omega q M_1 a}{qM_1 + M_2} = \frac{-qM_1}{(qM_1 + M_2)}\left(\frac{G(M_1 + M_2)}{a}\right)^{1/2}, \tag{9.23}$$

respectively. The energy of the binary stars in the post-supernova center-of-mass frame is

$$E = \frac{1}{2}qM_1 v_1^2 + \frac{1}{2}M_2 v_2^2 - \frac{GqM_1 M_2}{a}. \tag{9.24}$$

Substituting the expressions for v_1 and v_2 given in Equations (9.22) and (9.23) into Equation (9.24) we obtain

$$E = \frac{GqM_1M_2}{2a}\left(\frac{(M_1 + M_2)M_2}{(qM_1 + M_2)^2} + \frac{(M_1 + M_2)qM_1}{(qM_1 + M_2)^2} - 2\right)$$

$$= \frac{GqM_1M_2}{2a(qM_1 + M_2)}[M_1 + M_2 - 2(qM_1 + M_2)]. \tag{9.25}$$

The condition for the binary stars to be gravitationally unbound after the supernova outburst is $E > 0$. From Equation (9.25) this condition becomes

$$M_1 - M_2 > 2qM_1. \tag{9.26}$$

9.2   Mass accretion

Spherical mass accretion

Spherical mass accretion can occur if a star accretes mass from an interstellar cloud or from a stellar wind emanating from a binary companion. If a stellar wind is the cause of the mass accretion then the accretion will be spherical if the stellar-wind velocity is appreciably less than the free-fall velocity. Spherical accretion is described by the stellar-wind equations discussed in Chapter 5 except that the boundary conditions are different. We assume that the relation between pressure and density is given by Equation (5.140). Integrating Equation (5.142) we obtain the Bernoulli equation

$$\frac{1}{2}v^2 + \frac{\Gamma}{\Gamma - 1}\frac{P}{\rho} - \frac{GM}{r} = \frac{\Gamma}{\Gamma - 1}\frac{P_\infty}{\rho_\infty} = \frac{v_s^2(\infty)}{\Gamma - 1} \qquad (\Gamma \neq 1), \tag{9.27}$$

where $v_s(\infty)$ is the speed of sound at large distances from the accreting star. Equation (9.27) is similar to Equation (5.143) except for the boundary condition. Eliminating $d\rho/dr$ from Equations (5.136) and (5.141) and then eliminating dv/dr from the same equations gives

$$\left[v^2 - \Gamma\frac{P}{\rho}\right]\frac{dv}{dr} = \frac{2\Gamma Pv}{\rho r} - \frac{GMv}{r^2} \tag{9.28}$$

and

$$\left[v^2 - \Gamma\frac{P}{\rho}\right]\frac{d\rho}{dr} = \frac{2v^2\rho}{r} - \rho\frac{GM}{r^2}. \tag{9.29}$$

Equation (9.28) is the same as Equation (5.144). Equations (9.28) and (9.29) imply that as in spherical wind solutions the critical radius occurs when there is a transition from sonic to supersonic flow.

Since the right-hand sides of Equations (9.28) and (9.29) must be equal to zero at the critical radius we obtain the conditions

$$v_c^2 = \Gamma \frac{P_c}{\rho_c} \equiv v_s^2(r_c),$$

(9.30)

$$2\Gamma \frac{P_c}{\rho_c} = \frac{GM}{r_c},$$

(9.31)

$$v_c^2 = \frac{GM}{2r_c}.$$

(9.32)

From Equations (9.27), (9.30), (9.31) and (9.32) we obtain

$$v_c^2 = \frac{2(\Gamma - 1)\Gamma}{(5 - 3\Gamma)} \frac{P_\infty}{\rho_\infty} = \frac{2(\Gamma - 1)}{(5 - 3\Gamma)} v_s(\infty)^2,$$

(9.33)

with $\Gamma < 5/3$ but approximately equal to 5/3 and

$$r_c = \frac{GM(5 - 3\Gamma)\rho_\infty}{4(\Gamma - 1)P_\infty} = \frac{GM(5 - 3\Gamma)\Gamma}{4(\Gamma - 1)v_s(\infty)^2}.$$

(9.34)

Equation (5.140) can be re-expressed as

$$\rho = \rho_\infty \left(\frac{v_s(r)}{v_s(\infty)} \right)^{2/(\Gamma - 1)},$$

(9.35)

where the speed of sound $v_s = \sqrt{\Gamma P / \rho}$.

At the critical radius the infall velocity v_c is equal to the speed of sound. It follows that the mass accretion rate $\dot{M}$ is

$$\dot{M} = 4\pi r^2 \rho v = 4\pi r_c^2 \rho_c v_c$$

$$= 4\pi r_c^2 v_c \rho_\infty \left(\frac{v_c}{v_s(\infty)} \right)^{2/(\Gamma - 1)},$$

(9.36)

where r_c and v_c are given in terms of M, Γ and the boundary condition $v_s(\infty)$ in Equations (9.33) and (9.34). We note that the speed of sound depends only on the temperature, molecular weight and Γ. The value of Γ that most accurately describes a particular set of physical conditions depends on radiative losses, which in turn depend on physical parameters such as gas density and chemical composition.

Mass accretion from a stellar wind

Suppose that a stellar wind from a star in a binary system streams toward its binary companion with a wind velocity much larger than the sound velocity of the gas. We neglect the pressure forces in the stellar wind and consider uniform streaming of the cold gas onto a star of mass M and radius R. Let v_∞ and ρ_∞ equal the gas velocity and density respectively at a large distance from the star of mass M. If p_c is the critical impact parameter such that inflowing gas hits the star then conservation of angular momentum implies

$$p_c v_\infty = R v_\theta(R).$$ (9.37)

From the conservation of energy we have

$$\frac{1}{2} v_\infty^2 = \frac{1}{2} v_\theta(R)^2 - \frac{GM}{R}.$$ (9.38)

Using Equation (9.37) the mass accretion rate becomes

$$\frac{dM}{dt} = \pi p_c^2 \rho_\infty v_\infty = \frac{\pi R^2 v_\theta^2}{v_\infty^2} \rho_\infty v_\infty.$$ (9.39)

Eliminating v_θ^2 from Equations (9.38) and (9.39), we obtain

$$\frac{dM}{dt} = \frac{\pi R^2}{v_\infty^2} \rho_\infty v_\infty \left(v_\infty^2 + \frac{2GM}{R} \right) = \pi R^2 \rho_\infty v_\infty \left(1 + \frac{2GM}{v_\infty^2 R} \right).$$ (9.40)

As inflowing gas flows around the star from both directions shock fronts are produced as the result of counterstreaming gas. As a consequence of such colliding gas streams the incoming gas loses must of its transverse velocity. It follows that gas whose radial velocity after passing through a shock front is less than the escape velocity will free-fall onto the star whereas gas whose radial velocity is greater than the escape velocity will not be accreted onto the star. As a consequence of the formation of shock fronts by counter-streaming gas the capture radius (critical impact parameter) can be much greater than p_c given in Equation (9.37). The new critical impact parameter is $p_c = r_a$ with r_a determined by the equation

$$\frac{1}{2} v_r^2(r_a) = \frac{GM}{r_a}.$$ (9.41)

The radial velocity v_r of a particle of mass m in an inverse square law gravitational field can be found from the energy equation

$$v_r^2 = \frac{2}{m} \left(E - \frac{GmM}{r} \right) - \frac{L^2}{m^2 r^2},$$ (9.42)

with $E = \frac{1}{2} m v_\infty^2$ and the angular momentum L. Equations (9.41) and Equation (9.42) relate the critical impact parameter r_a to the initial particle energy and angular momentum. The mass accretion rate becomes

$$\frac{dM}{dt} = \pi r_a^2 \rho_\infty v_\infty.$$ (9.43)

Before shock-wave formation the equation describing a particle orbit in the gravitational field of the star is

$$r = \frac{p_c^2 v_\infty^2}{GM} \frac{1}{1 + \varepsilon \cos \theta}.$$ (9.44)

Equation (9.44) implies that as $r \to \infty$ $\varepsilon = -1/\cos \theta$ and therefore on the accretion axis we have $\cos\theta = 1/\varepsilon$. Evaluating (9.44) on the accretion axis and using the relation $p_c = r_a$ we obtain

$$r_a = \frac{2GM}{v_\infty^2}.$$ (9.45)

From Equations (9.43) and (9.45) we find that the mass accretion rate is

$$\frac{dM}{dt} = 4\pi\frac{(GM)^2\rho_\infty}{v_\infty^3}.$$ (9.46)

The mass accretion rate given in Equation (9.46) greatly exceeds that given in Equation (9.40) when the condition $GM/v_\infty^2 R \gg 1$ is satisfied.

Accretion disks

Under physical conditions such as Roche lobe overflow mass inflowing toward the surface of a star will have sufficient angular momentum to form an accretion disk. We assume that the amount of mass within the accretion disk is much less than that of the central star, and therefore gravitational forces are caused entirely by the star. The orbits of mass elements within the accretion disk are nearly circular. Equating the radial gravitational force and centrifugal force on a small mass element within the accretion disk gives

$$\omega = \sqrt{\frac{GM}{r^3}},$$ (9.47)

where ω is the angular velocity at radial distance r and M is the mass of a central star.

Equation (9.47) implies that the angular momentum per unit mass within the accretion disk is

$$(GMr)^{1/2}.$$ (9.48)

The gravitational force per unit mass perpendicular to the symmetry plane of the accretion disk is

$$\frac{GMz}{r^3},$$ (9.49)

with z equal to the distance from the symmetry plane. Equation (9.49) implies that in the z-direction the equation of hydrostatic equilibrium can be approximated as

$$\frac{dP}{dz} \sim \frac{P}{h} \sim \rho\frac{GMh}{r^3},$$ (9.50)

with h equal to the half-thickness of the accretion disk.

A mass element must lose most of its angular momentum before being accreted onto the surface of a star or into a black hole. It follows that under steady-state conditions angular momentum must be continuously transported away from the outer boundary of the accretion disk. Moreover, viscous stresses must also act within the accretion disk because the angular momentum per unit mass given by expression (9.48) decreases as r decreases.

The outward rate of angular momentum transport across some radial distance r is

$$\dot{J} = \dot{M}(GMr)^{1/2}. \tag{9.51}$$

The net outflow rate of angular momentum from the inner edge of the accretion disk is

$$\dot{J}_I = \beta\dot{M}(GMr_I)^{1/2}, \tag{9.52}$$

with $\beta \le 1$. The constant β is unity if the central star does not exert stress on the inner edge of the accretion disk. Conservation of angular momentum implies that the torque caused by stress at radial distance r is

$$\text{Torque} = \dot{J} - \dot{J}_I, \tag{9.53}$$

with $\dot{J}$ and $\dot{J}_I$ given by Equations (9.51) and (9.52) respectively. It follows that the stress $t_{r\theta}$ required to conserve angular momentum under conditions of steady-state mass accretion is

$$t_{r\theta} = \frac{-\dot{M}}{4\pi r^2 h}[(GMr)^{1/2} - \beta(GMr_I)^{1/2}]. \tag{9.54}$$

The equation of motion of a viscous gas is

$$\rho\frac{dv_j}{dt} = -\frac{\partial P}{\partial x_j} + \frac{\partial t_{ij}}{\partial x_i} \tag{9.55}$$

where the stress tensor is

$$t_{ij} = t_{ji} = \eta\left(v_{i,j} + v_{j,i} - \frac{2}{3}\delta_{ij}v_{k,k}\right). \tag{9.56}$$

In Equation (9.56) η is the dynamic viscosity and $v_{i,j}$ is the derivative of the i component of velocity with respect to coordinate j. The bulk viscosity can be neglected. The trace of t_{ij} (i.e. t_{ii} with summation over all i understood) is equal to zero.

Next we will calculate the rate at which the gas in the accretion disk is heated as the result of viscous dissipation. The rate of change of energy density is

$$\frac{d\left(\frac{1}{2}v^2 + U\right)}{dt} = \boldsymbol{v}\cdot\frac{d\boldsymbol{v}}{dt} + \frac{dU}{dt}, \tag{9.57}$$

where U is the internal energy per unit mass. Equation (9.55), the equation of continuity given as Equation (2.41) and the first law of thermodynamics imply that Equation (9.57) can be re-expressed as

$$\frac{d}{dt}\left(\frac{1}{2}v^2 + U\right) = -\frac{v_i}{\rho}\frac{\partial P}{\partial x_i} + \frac{v_i}{\rho}\frac{\partial t_{ji}}{\partial x_i} + T\frac{ds}{dt}$$

$$= -\frac{1}{\rho}\frac{\partial}{\partial x_i}(Pv_i - t_{ij}v_j) + T\frac{ds}{dt} - \frac{1}{\rho}t_{ij}v_{i,j}$$

The first term on the right-hand side of Equation (9.58) can be integrated over some finite volume and therefore represents work done by external forces. The sum of the last two terms in Equation (9.58), representing internal heating and entropy change, must be equal to zero. Therefore, we have

$$\rho T\frac{ds}{dt} = v_{i,j}t_{ij}. \tag{9.59}$$

Because the tensor $v_{i,j}$ is symmetric Equation (9.59) can be rewritten as

$$\rho T\frac{ds}{dt} = \frac{1}{2}(v_{i,j} + v_{j,i})t_{ij}. \tag{9.60}$$

Equations (9.56) and (9.60) lead to the equation

$$\rho T\frac{ds}{dt} = \frac{1}{2}\left(\frac{1}{\eta}t_{ij} + \frac{2}{3}\delta_{ij}v_{k,k}\right)t_{ij}$$

$$= \frac{1}{2\eta}t_{ij}t_{ij}. \tag{9.61}$$

The only nonzero components of t_{ij} in cylindrical coordinates are

$$t_{r\theta} = t_{\theta r} = \eta\left(\frac{v_{r,\theta}}{r} + v_{\theta,r} - \frac{v_\theta}{r}\right). \tag{9.62}$$

For circular orbits the only nonzero velocity is

$$v_\theta = r\omega = \left(\frac{GM}{r}\right)^{1/2}. \tag{9.63}$$

Equations (9.62) and (9.63) imply that

$$t_{\theta r} = t_{r\theta} = -\frac{3}{2}\eta\omega = -\frac{3}{2}\eta\left(\frac{GM}{r^3}\right)^{1/2}. \tag{9.64}$$

Using Equations (9.61) and (9.64) the heat generated by viscosity becomes

$$\rho T\frac{ds}{dt} = \frac{(t_{\theta r})^2}{\eta}. \tag{9.65}$$

From Equations (9.54), (9.64) and (9.65) we have

$$\rho T\frac{ds}{dt} = \frac{3\dot{M}}{8\pi r^2 h}\frac{GM}{r}\left(1 - \beta\left(\frac{r_1}{r}\right)^{1/2}\right). \tag{9.66}$$

We assume that heat generated within the accretion disk is radiated away from the surfaces of the accretion disk. Therefore, from Equation (9.66) the radiative flux emitted from the top (or bottom) surface of the accretion disk is

$$F = \frac{3\dot{M}}{8\pi r^2}\frac{GM}{r}\left[1 - \beta\left(\frac{r_1}{r}\right)^{1/2}\right]. \tag{9.67}$$

We note that Equations (9.66) and (9.67) do not contain the dynamic viscosity η.

If the accretion disk is optically thick in the vertical direction and in radiative equilibrium then we have

$$F = \frac{ac}{3\kappa\rho}\frac{\mathrm{d}T^4}{\mathrm{d}z} \sim \frac{ac}{3\kappa\rho}\frac{T^4}{h}, \tag{9.68}$$

where κ is the opacity. For an accretion disk about a neutron star or black hole the opacity is primarily electron scattering and consequently $\kappa = 0.2(1 + X)$ with X equal to the mass fraction in the form of hydrogen. If the mass accretion rate $\dot{M}$ and η are given and the accretion disk is optically thick then the approximate structure of the accretion disk can be determined from the equations

$$P = \rho\frac{GM}{r^3}h^2 \tag{9.69}$$

$$\frac{ac}{3\kappa\rho}\frac{T^4}{h} = \frac{3}{8\pi r^2}\dot{M}\frac{GM}{r}\left(1 - \beta\left(\frac{r_1}{r}\right)^{1/2}\right) \tag{9.70}$$

$$\frac{ac}{3\kappa\rho}\frac{T^4}{h} = \frac{9}{4}\eta h\left(\frac{GM}{r^3}\right) \tag{9.71}$$

$$P = \frac{\rho kT}{\mu m_\mathrm{p}} + \frac{1}{3}aT^4, \tag{9.72}$$

where Equations (9.69) and (9.70) follow from Equations (9.50) and (9.67) and (9.68). Equation (9.71) follows from Equations (9.64) and (9.65). If $\dot{M}$ and η are assumed known Equations (9.69–9.72) determine P, ρ, T and h as a function of r.

It is convenient to express the viscous stress tensor $t_{r\theta}$ as $-\alpha P$ with P equal to the pressure and $\alpha = \alpha(r) \le 1$. Substituting $-\alpha P = t_{r\theta}$ into Equations (9.54) and (9.64) gives

$$\alpha P = \frac{\dot{M}}{4\pi r^2 h}[(GMr)^{1/2} - \beta(GMr_1)^{1/2}] \tag{9.73}$$

and

$$\eta = \frac{2}{3}\alpha P\left(\frac{r^3}{GM}\right)^{1/2}. \tag{9.74}$$

Equations (9.71) and (9.74) lead to the equation

$$\frac{ac}{\kappa\rho}\frac{T^4}{h} = \frac{9}{2}\alpha Ph\left(\frac{GM}{r^3}\right)^{1/2}. \tag{9.75}$$

We assume that radiation pressure is dominant as compared to gas pressure. Substituting the relation $P = \frac{1}{3}aT^4$ into Equation (9.75), we obtain

$$\frac{cP}{\kappa\rho h} = \frac{3}{2}\alpha Ph\left(\frac{GM}{r^3}\right)^{1/2}.$$

(9.76)

Equations (9.69) and (9.76) imply that

$$P = \frac{2}{3}\frac{c}{\alpha\kappa}\left(\frac{GM}{r^3}\right)^{1/2}.$$

(9.77)

Substituting the expression for P given in Equation (9.77) into Equation (9.73) we have

$$h = \frac{3\kappa}{8\pi c}\dot{M}\left[1 - \beta\left(\frac{r_1}{r}\right)^{1/2}\right].$$

(9.78)

Equation (9.78) implies that h is independent of r except close to the inner edge of the accretion disk. Solving Equation (9.76) for ρ we obtain

$$\rho = \frac{2c}{3\alpha\kappa h^2}\left(\frac{r^3}{GM}\right)^{1/2}.$$

(9.79)

If the gas pressure in an accretion disk is much greater than the radiation pressure then $P = \rho kT/\mu m_p$. Substituting the latter expression for the gas pressure into Equation (9.69) gives

$$h = \left(\frac{kT}{\mu m_p}\right)^{1/2}\left(\frac{r^3}{GM}\right)^{1/2} = v_s\left(\frac{r^3}{GM}\right)^{1/2},$$

(9.80)

with v_s equal to the isothermal sound velocity. From Equations (9.73), (9.80) and the above expression for the gas pressure we obtain

$$\rho v_s^3 = \frac{\dot{M}}{4\pi r^2\alpha}\left(\frac{GM}{r^3}\right)^{1/2}[(GMr)^{1/2} - \beta(GMr_1)^{1/2}] = f_1(r).$$

(9.81)

Using Equation (9.80), Equation (9.75) can be rewritten as

$$\frac{ac}{\kappa\rho h}\left(\frac{\mu m_p}{k}\right)^4 v_s^8 = \frac{9h}{2}\alpha P\left(\frac{GM}{r^3}\right)^{1/2}.$$

(9.82)

Equations (9.80), (9.82) and the relation $P = \rho v_s^2$ imply that

$$\frac{v_s^4}{\rho^2} = \frac{9\kappa\alpha}{2ac}\left(\frac{k}{\mu m_p}\right)^4\left(\frac{r^3}{GM}\right)^{1/2} = f_2(r).$$

(9.83)

From Equations (9.81) and (9.83) we obtain

$$v_s^{10} = f_1^2(r)f_2(r)$$

(9.84)

and

$$\rho = \frac{f_1^{2/5}(r)}{f_2^{3/10}(r)}, \tag{9.85}$$

where $f_1(r)$ and $f_2(r)$ are defined in Equations (9.81) and (9.83), respectively.

From Equation (9.67) we have

$$F = \sigma T_{\text{eff}}^4 \simeq \frac{3\dot{M}}{16\pi r^2} \frac{GM}{r} \qquad (r \gg r_1). \tag{9.86}$$

Equation (9.86) implies that

$$T_{\text{eff}} \simeq \left(\frac{3\dot{M}GM}{16\pi\sigma}\right)^{1/4} \frac{1}{r^{3/4}} \qquad (r \gg r_1). \tag{9.87}$$

Further discussion of accretion disks is given in Chapter 13.

9.3   Stellar rotation

The structures of stars are changed by rotation. Because massive main sequence stars rotate rapidly their structures are appreciably affected by rotation. Another consequence of rotation is meridional circulation, which is also known as rotational mixing.

The equation of hydrostatic equilibrium of a rotating star is

$$\frac{1}{\rho}\nabla P = -\nabla\Phi + \omega^2 R, \tag{9.88}$$

where ω is the orbital angular velocity and $R = r\sin\theta$ is the distance to the rotation axis. The gravitational potential Φ is determined by Poisson's equation

$$\nabla^2\Phi = 4\pi G\rho. \tag{9.89}$$

If rotation is uniform on cylinders (i.e. $\omega = \omega(R)$) then the centrifugal force term in Equation (9.88) is derivable from a centrifugal potential Φ_c. The total potential V is

$$V = \Phi + \Phi_c = \Phi - \int_0^R R\omega^2(R)\mathrm{d}R. \tag{9.90}$$

For uniform rotation V becomes

$$V = \Phi - \frac{1}{2}r^2\omega^2\sin^2\theta, \tag{9.91}$$

with θ equal to the angle with respect to the rotation axis.

Equations (9.88) and (9.90) imply that if rotation is uniform on cylinders the equation of hydrostatic equilibrium is

$$\nabla P = -\rho\nabla V. \tag{9.92}$$

Equation (9.92) implies that ∇P and ∇V are parallel. Taking the curl of Equation (9.92) gives

$$\nabla\rho \times \nabla V = 0. \tag{9.93}$$

Equation (9.93) implies that $\nabla\rho$ is parallel to ∇V. Because ∇P is also parallel to ∇V it follows that surfaces of constant ρ, P and V are the same. Therefore, ρ, P and T are constant on equipotential surfaces (i.e. $\rho = \rho(V)$, $P = P(V)$, $T = T(V)$).

In stars for which $P = P(\rho)$ we will show below that the requirement of hydrostatic equilibrium implies that rotation must be uniform on cylinders. Examples of such stars are white dwarfs and stars whose structures can be represented by a single polytropic index n. We assume that the relation between pressure and density has the form $P(\rho)$. We wish to find a function $f(\rho)$ such that

$$\frac{1}{\rho}\nabla P(\rho) = \nabla f(\rho). \tag{9.94}$$

Equation (9.94) can be rewritten as

$$\frac{1}{\rho}\frac{dP}{d\rho}\nabla\rho = \frac{df(\rho)}{d\rho}\nabla\rho. \tag{9.95}$$

A function $f(\rho)$ that satisfies Equation (9.95) is

$$f(\rho) = \int_0^\rho \frac{1}{\rho'}\frac{dP(\rho')}{d\rho'}d\rho'. \tag{9.96}$$

From Equations (9.94) and (9.96) it follows that the equation of hydrostatic equilibrium given in Equation (9.88) becomes

$$\nabla(f(\rho) + \Phi) = \omega^2 R. \tag{9.97}$$

The curl of the left-hand side of Equation (9.97) is equal to zero. Therefore, the right-hand side of Equation (9.97) must be derivable from the centrifugal potential given in Equation (9.90). It follows that rotation is uniform on cylinders.

Equation (9.89) reduces to Laplace's equation outside a star. In spherical polar coordinates Laplace's equation is

$$\frac{1}{r^2}\frac{\partial r^2}{\partial r}\frac{\partial\Phi}{\partial r} + \frac{1}{r^2\sin\theta}\frac{\partial}{\partial\theta}\left(\sin\theta\frac{\partial\Phi}{\partial\theta}\right) + \frac{1}{r^2\sin^2\theta}\frac{\partial^2\Phi}{\partial\phi^2} = 0. \tag{9.98}$$

The solution Φ of Equation (9.98) can be expressed as

$$\Phi(r, \theta, \phi) = R(r)\Theta(\theta)e^{\pm im\phi}. \tag{9.99}$$

Substituting the function Φ given in Equation (9.99) into Equation (9.98) we obtain the two equations

$$\frac{1}{\Theta}\frac{1}{\sin\theta}\frac{d}{d\theta}\left(\sin\theta\frac{d\Theta}{d\theta}\right) - \frac{m^2}{\sin^2\theta} + l(l+1) = 0 \tag{9.100}$$

$$\frac{1}{R}\frac{d}{dr}\left(r^2\frac{dR}{dr}\right) - l(l+1) = 0, \tag{9.101}$$

with $l = 0, 1, 2, \ldots$ Equation (9.100) is the same equation as Equation (4.13) and its solutions are the associated Legendre functions $P_l^m(\cos\theta)$.

Equation (9.101) can be rewritten as

$$r^2\frac{d^2R}{dr^2} + 2r\frac{dR}{dr} - l(l+1)R = 0. \tag{9.102}$$

If we substitute $R = r^\alpha$ into Equation (9.102) we obtain

$$\alpha(\alpha - 1) + 2\alpha - l(l+1) = 0. \tag{9.103}$$

From Equation (9.103) we have $\alpha = l$ or $\alpha = -(l+1)$, and therefore the function $R(r)$ becomes

$$R(r) = C_{1,l}r^l + C_{2,l}r^{-(l+1)}, \tag{9.104}$$

where $C_{1,l}$ and $C_{2,l}$ are arbitrary constants. From Equations (9.99), (9.100) and (9.104) it follows that a solution of Laplace's equation in spherical polar coordinates is

$$\Phi_{lm} = (C_{1,l}r^l + C_{2,l}r^{-(l+1)})P_l^m(\cos\theta)e^{\pm im\phi}, \tag{9.105}$$

with $l = 0, 1, 2, \ldots$ and $m = 0, \pm 1, \pm 2, \ldots$ The general solution of Laplace's equation is the superposition of solutions given in Equation (9.105). The solutions given in Equation (9.105) diverge as $r \to \infty$ and therefore in determining the gravitational potentials outside stars the constants $C_{1,l}$ must be equal to zero. Rotating stars are axially symmetric and therefore only solutions with $m = 0$ are relevant. Because rotating stars are symmetric with respect to the equatorial plane only solutions given in Equation (9.105) with $l = 0, 2, 4, \ldots$ are nonzero, and therefore the general solution to Laplace's equation exterior to a rotating star is the superposition of solutions of the form

$$\Phi_{l,0} = \Phi_l = C_{2,l}r^{-(l+1)}P_l(\cos\theta), \tag{9.106}$$

with $l = 0, 2, 4, \ldots$

If the relation $P = P(\rho)$ is satisfied in a particular rotating star the solutions to Equations (9.88) and (9.89) can be found by means of the following iterative method of solution. The density distribution, which we assume to be known, can be expanded as

$$\rho = \left(\sum_{n=0,1}^{N} c_n r^n\right)\left(\sum_{l=0,2}^{2M} d_l P_l(\cos\theta)\right), \tag{9.107}$$

with c_n and d_l equal to constants. We expand the gravitational potential Φ as

$$\Phi = \left(\sum_{n=0}^{N} a_n r^n\right)\left(\sum_{l=0,2}^{2M} b_l P_l(\cos\theta)\right) \tag{9.108}$$

where a_n and b_l are constants whose values are determined from Poisson's equation and the boundary conditions. At the center of the star the boundary condition is $d\Phi/dr = 0$. In addition, the solutions of Poisson's equation must be such that Φ and $\nabla\Phi$ are continuous with solutions of Laplace's equation on a sphere whose radius is $\gtrsim$ the equatorial radius of the star.

Having solved Poisson's equation for Φ we substitute Φ into the equation of hydrostatic equilibrium, which contains the given axially symmetric distribution of angular velocity $\omega = \omega(r\sin\theta)$, and determine a corrected density distribution. An improved estimate for the gravitational potential is then found by substituting the corrected density distribution into Poisson's equation. Iteration is continued until convergence is achieved.

9.4   Meridional circulation

If the pressure, density and temperature are constant on equipotential surfaces then the radiative flux given in Equation (2.80) can be expressed as

$$F = \left(-\frac{4ac}{3\kappa\rho} T^3 \frac{dT}{dV} \right) \nabla V. \tag{9.109}$$

Equation (9.109) implies that the radiative flux is proportional to the potential gradient ∇V.

Taking the divergence of the radiative flux given in Equation (9.109) we obtain

$$\nabla \cdot F = \frac{d}{dV} \left(-\frac{4ac}{3} \frac{T^3}{\kappa\rho} \frac{dT}{dV} \right) |\nabla V|^2 - \frac{4ac}{3} \frac{T^3}{\kappa\rho} \frac{dT}{dV} \nabla^2 V. \tag{9.110}$$

If the star is assumed to rotate uniformly then Equations (9.89) and (9.91) imply that

$$\nabla^2 V = 4\pi G\rho - 2\omega^2. \tag{9.111}$$

In radiative zones outside the central hydrogen-burning region the condition for thermal equilibrium in main-sequence stars is

$$\nabla \cdot F = 0. \tag{9.112}$$

All terms with the exception of the $|\nabla V|^2$ term are constant on equipotential surfaces in Equation (9.110). In a rotating star the potential gradient cannot be constant on equipotential surfaces because these surfaces are more widely separated in equatorial regions than in polar regions, and consequently the radiative flux is not constant on equipotential surfaces. A rotating star adjusts itself so that Equation (9.112) is satisfied as averaged over an entire equipotential surface. The polar regions of the star receive an excess amount of radiative heating whereas the equatorial regions receive a deficiency of radiation, and therefore are cooled by rotation. It follows that gas in the equatorial regions will sink and be heated as it circulates downwards. On the other hand, the gas in the polar regions that receives an excess radiative heat will tend to cool as it rises. Our next task is to calculate $\nabla \cdot F$ given in Equation (9.110) and then calculate the circulation velocity and corresponding rotational mixing time.

We assume that equipotential surfaces can be represented by the equation

$$r = \bar{r}(1 + c_2(\bar{r})P_2(\cos\theta)), \tag{9.113}$$

where $\bar{r}$ is the mean distance of an equipotential surface from the center of the star and departures from spherical symmetry have been assumed to be small. Using Equation (9.113) we find that the potential gradient $|\nabla V|$ appearing in Equation (9.110) is

$$|\nabla V| = \frac{dV}{d\bar{r}}\frac{d\bar{r}}{dr} = \frac{dV}{d\bar{r}}\left[1 - \frac{d\bar{r}c_2}{d\bar{r}}P_2(\cos\theta)\right], \tag{9.114}$$

where we have assumed that the term containing c_2 is small as compared to unity.

Substituting the expression for $|\nabla V|$ given by Equation (9.114) into Equation (9.110) and retaining only terms that are first order in $c_2(r)$ and dc_2/dr we find that the resultant equation contains terms that depend only on $\bar{r}$ as well as one term that depends on $P_2(\cos\theta)$. Terms that are constant on equipotential surfaces and terms depending only on $\bar{r}$ must add up to zero in the expression for $\mathbf{V}\cdot\mathbf{F}$ given in Equation (9.110) because $\mathbf{V}\cdot\mathbf{F}$ must equal zero as averaged over an equipotential surface. Only the term proportional to $P_2(\cos\theta)$ is responsible for making $\mathbf{V}\cdot\mathbf{F} \neq 0$.

The term proportional to $P_2(\cos\theta)$ contains the factor

$$\frac{d}{dV}\left(-\frac{4ac}{3}\frac{T^3}{\kappa\rho}\frac{dt}{dV}\right)\left(\frac{dV}{d\bar{r}}\right)^2$$

$$= \frac{d}{dr}\left(-\frac{4ac}{3}\frac{T^3}{\kappa\rho}\frac{dT}{dr}\frac{dr}{dV}\right)\left(\frac{dV}{d\bar{r}}\right)$$

$$\simeq \frac{d}{dr}\left(\frac{L}{4\pi r^2}\frac{r^2}{GM_r}\right)\frac{GM_r}{r^2}. \tag{9.115}$$

Because L is independent of r in main-sequence stars except in the hydrogen-burning core the expression in Equation (9.115) becomes

$$\frac{LM_r}{4\pi r^2}\frac{d\left(\dfrac{1}{M_r}\right)}{dr} = -\frac{L\rho}{M_r}, \tag{9.116}$$

where we have used Equation (2.28).

Equations (9.110) and (9.114–9.116) and the condition that we retain only first order terms in $[d(\bar{r}c_2)/d\bar{r}]P_2(\cos\theta)$ imply

$$\mathbf{V}\cdot\mathbf{F} = \frac{2L\rho}{M_r}\frac{d\bar{r}c_2(r)}{d\bar{r}}P_2(\cos\theta). \tag{9.117}$$

The rate of energy loss per cm^3 caused by meridional circulation in a nondegenerate, ionized gas is

$$-\frac{dE}{dt} = P\nabla \cdot \boldsymbol{v} + \nabla \cdot \left(\frac{3}{2}\frac{P}{\rho}\rho\boldsymbol{v}\right), \tag{9.118}$$

where $\boldsymbol{v}$ is the velocity of the gas. Under steady-state conditions the equation of continuity is

$$0 = \nabla \cdot (\rho\boldsymbol{v}) = \rho\nabla \cdot \boldsymbol{v} + \boldsymbol{v} \cdot \nabla\rho. \tag{9.119}$$

Equations (9.118) and (9.119) lead to the equation

$$-\frac{dE}{dt} = P\nabla \cdot \boldsymbol{v} + \frac{3}{2}P\nabla \cdot \boldsymbol{v} + \frac{3}{2}\boldsymbol{v} \cdot \nabla P$$

$$= \frac{5}{2}\frac{P}{\rho}\boldsymbol{v} \cdot \nabla\rho + \frac{3}{2}\boldsymbol{v} \cdot \nabla P. \tag{9.120}$$

If we assume a polytropic relation between P and ρ then we obtain

$$\nabla P = \nabla(K\rho^{1+\frac{1}{n}}) = \frac{P}{\rho}\left(1 + \frac{1}{n}\right)\nabla\rho. \tag{9.121}$$

Equation (9.121) implies that

$$\nabla\rho = \frac{\rho\nabla P}{P\left(1 + \frac{1}{n}\right)}. \tag{9.122}$$

Since rotation is assumed to have only a first-order perturbation on the structure of the star we have

$$\nabla P = -\rho\frac{GM_{\bar{r}}}{\bar{r}^2}. \tag{9.123}$$

Substituting Equations (9.122) and (9.123) into Equation (9.120) and making the approximation

$$\boldsymbol{v} \cdot \nabla P \simeq v\rho\frac{GM_{\bar{r}}}{\bar{r}^2}, \tag{9.124}$$

we obtain

$$-\frac{dE}{dt} = \left(-\frac{5}{2}\frac{n}{n+1} + \frac{3}{2}\right)\boldsymbol{v} \cdot \nabla P \simeq v\rho\frac{GM_r}{\bar{r}^2}\left(-\frac{5}{2}\frac{n}{n+1} + \frac{3}{2}\right). \tag{9.125}$$

Using the relation

$$\nabla \cdot \boldsymbol{F} = \frac{dE}{dt} \tag{9.126}$$

and Equations (9.117), (9.125) and (9.126), we obtain

$$\nabla \cdot F = \frac{2L\rho}{M_r} \frac{d\bar{r}c_2}{d\bar{r}} P_2(\theta)$$

$$\simeq \left(\frac{5}{2}\frac{n}{n+1} - \frac{3}{2}\right) v\rho \frac{GM_r}{\bar{r}^2}, \tag{9.127}$$

where $P_2(\cos\theta) = \frac{1}{2}(3\cos^2\theta - 1)$. Equation (9.127) implies that the circulation velocity is

$$v \simeq \frac{4}{(2n-3)} \frac{L\bar{r}^2}{GM_{\bar{r}}^2} \frac{d\bar{r}c_2(\bar{r})}{d\bar{r}} P_2(\cos\theta). \tag{9.128}$$

We can approximate v in Equation (9.128) as

$$v \simeq \frac{L\bar{r}^2}{GM_{\bar{r}}^2} \lambda P_2(\cos\theta), \tag{9.129}$$

with $\lambda \sim E_{\text{rot}}/E_{\text{B.E.}}$, $E_{\text{rot}} \simeq \frac{1}{2}Mv_\theta^2$ and

$$E_{\text{B.E.}} \sim \frac{1}{2}\frac{3}{5-n}\frac{GM^2}{R}.$$

Substituting the above expressions for E_{rot} and $E_{\text{B.E.}}$ into Equation (9.129) gives

$$\bar{v} \sim \frac{LR^2}{GM^2} \frac{\frac{1}{2}Mv_\theta^2}{\dfrac{GM^2}{R}} = \frac{v_\theta^2 LR^3}{2G^2M^3}. \tag{9.130}$$

The overall rotational mixing timescale is

$$t_{\text{mixing}} \sim R/\bar{v}. \tag{9.131}$$

Low-mass main-sequence stars rotate sufficiently slowly ($v_\theta \simeq 2\,\text{km s}^{-1}$ for the Sun) that rotational mixing timescales given by Equation (9.131) are appreciably greater than main-sequence lifetimes. On the other hand, massive main-sequence stars often have surface rotational velocities $\gtrsim 100\,\text{km s}^{-1}$ and rotational mixing timescales less than their main-sequence lifetimes. Mixing of ^{4}He and ^{14}N enriched gas from the hydrogen-burning core of a main-sequence star to the surface would lengthen the main-sequence lifetime. However, molecular weight gradients formed by hydrogen burning can cause gas circulation currents in the core to be confined to the core and circulation currents in the outer region of the star to be excluded from the hydrogen-burning core.

Circulation currents that are unaffected by molecular weight gradients are shown in Figure 9.3.

The abundance of Li in the solar atmosphere is known to be much less than its cosmic abundance. As mentioned in Chapter 1 Li burning occurs at significantly lower temperatures than the proton–proton chain. However, the temperature at the base of the solar convective envelope (i.e. $\simeq 2 \times 10^6\,\text{K}$) is too low for Li burning to occur. Apparently, depletion of solar Li was caused by meridional circulation currents extending from the

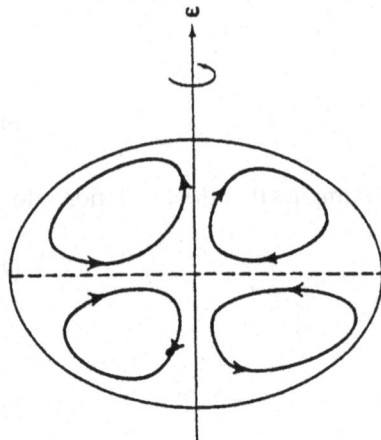

Figure 9.3. Stellar rotation causes circulation currents to develop because of the need to maintain thermal balance.

outer radiative core to the base of the convective zone. An alternate explanation for Li depletion namely that convection extended inwards to sufficiently high temperature ($\simeq 7 \times 10^6$ K) for Li burning to take place during pre-main-sequence evolution is not supported by stellar interior model calculations. However, deuterium depletion is predicted to occur by these same calculations. Although rapid rotation of the solar radiative core has been ruled out by measurements of sphericity, differential rotation sufficient to explain Li depletion is possible. It is known that mixing between the hydrogen burning cores and atmospheres of main-sequence stars does not take place because they evolve onto the red-giant branch. It has been argued that the meridional circulation (i.e. rotational mixing) explanation for Li depletion requires fine tuning of solar physical parameters. We note that molecular weight gradients in a standard solar model do not become appreciable until temperatures exceed 7×10^6 K. It seems plausible, therefore, that circulation currents are inhibited by molecular weight gradients in the inner hydrogen-burning core but not from the outer radiative core where gradients are small.

Problems

1. A black hole has a mass of $10^6\,M_\odot$. Consider the black hole a point Newtonian mass and estimate how close a $1\,M_\odot$ main-sequence star can orbit about it without being disrupted by tidal forces.

2. A sphere is deformed so that

$$r = r_0(1 + \alpha_2 P_2(\cos\theta))$$

with α_2 small. Find the area of the deformed sphere to order α_2^2. *Hint*: An element of surface area is

$$dA = \left[r^2 + \left(\frac{dr}{d\theta} \right)^2 \right]^{1/2} r \sin\theta d\theta d\phi.$$

3. A star is rotating with uniform angular velocity ω. Calculate $\boldsymbol{v} = \omega \times \boldsymbol{r}$ and prove that $\nabla \times \boldsymbol{v} = 2\omega$.

4. A $20\,M_\odot$ star produces a supernova explosion. The remnant star is a $1.4\,M_\odot$ neutron star. How massive must the binary companion be in order for the binary system not to be disrupted? Assume that the binary companion is $2\,M_\odot$, and calculate the velocities of the neutron star and $2\,M_\odot$ companion after the supernova outburst has disrupted the binary system. Assume that the initial binary orbit is circular. If the supernova occurs in the galactic plane estimate how long it takes the neutron star to reach a distance of 10^3 parsecs above the plane.

5. A $1.5\,M_\odot$ main-sequence star is inside an atomic-hydrogen cloud whose density and temperature are $10^2\,\mathrm{cm}^{-3}$ and $50\,\mathrm{K}$, respectively. Assume adiabatic spherical infall and calculate the rate of mass accretion from the cloud.

6. A binary star loses mass at a rate of $10^{-6}\,M_\odot/\mathrm{yr}$ via a stellar wind whose outflow velocity is $10^3\,\mathrm{km\,s}^{-1}$. The binary companion is a $2\,M_\odot$ star and the separation between binary stars is $10\,R_\odot$. Estimate how much of the $10^{-6}\,M_\odot/\mathrm{yr}$ stellar wind is accreted onto the $2\,M_\odot$ companion.

7. A binary system consists of a $1\,M_\odot$ white dwarf and a $0.5\,M_\odot$ subgiant that fills its Roche lobe. The orbit is circular and the binary star separation is $10^{11}\,\mathrm{cm}$. Find the orbital period, position of the first Lagrangian point and orbital velocities of the binary stars. Estimate changes in orbital parameters after $10^{-5}\,M_\odot$ has been transferred from the subgiant to the white dwarf. What is the outer radius of the accretion disk that forms around the white dwarf?

8. Find the equations that replace Equations (9.84) and (9.85) if radiation pressure as well as gas pressure are included.

9. For axial symmetry Laplace's equation in cylindrical coordinates is

$$\frac{1}{R} \frac{\partial}{\partial R} \left(R \frac{\partial \Phi}{\partial R} \right) + \frac{\partial^2 \Phi}{\partial z^2} = 0.$$

Show that $\Phi_\pm(R, Z) = \exp(\pm kz)\, J_0(kR)$ with $J_0(kR)$ equal to the zeroth-order Bessel function are solutions of the above equation.

10 Stellar magnetic fields

10.1 Magnetic fields in stellar winds

If the electric conductivity of an ionized gas is σ and the velocity of the gas is v with respect to some stationary frame of reference then Ohm's law and the equation for the transformation of an electric field imply

$$J = \sigma E' = \sigma\left(E + \frac{v}{c} \times B\right),\tag{10.1}$$

with E' equal to the electric field in the reference frame of the gas, and E (and B) equal to the corresponding electric (and magnetic) field in the stationary frame of reference. From the Maxwell Equation (3.345) and Equation (10.1) we obtain

$$\nabla \times \left(\frac{v}{c} \times B - \frac{J}{\sigma}\right) = \frac{1}{c}\frac{\partial B}{\partial t}.\tag{10.2}$$

The dimensions of the regions we consider are large and therefore the J/σ term in Equation (10.2) can be neglected. It follows that Equation (10.2) becomes

$$\nabla \times \left(\frac{v}{c} \times B\right) = \frac{1}{c}\frac{\partial B}{\partial t}.\tag{10.3}$$

Equation (10.3) remains valid so long as the magnetic diffusion timescale given in Equation (3.350) can be regarded as very long.

Let C be a closed contour of a surface A that is frozen into a gas moving with velocity v through a magnetic field. The total rate of change of the magnetic flux Φ is caused by the explicit time-dependence of the magnetic field and also by the motion through the magnetic field of the surface bounded by the contour C. The change in magnetic flux due to the explicit time dependence of the magnetic field is

$$\frac{\partial \Phi}{\partial t} = \int \frac{\partial B}{\partial t} \cdot dA = \int \nabla \times (v \times B) \cdot dA,\tag{10.4}$$

where the second relation in Equation (10.4) follows from Equation (10.3). The change in the magnetic flux resulting from motion of the surface A through the magnetic field is

$$\int B \cdot (v \times ds) = -\int (v \times B) \cdot ds = -\int \nabla \times (v \times B) \cdot dA.\tag{10.5}$$

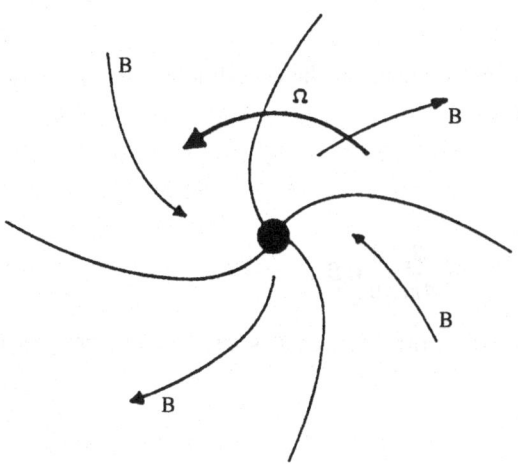

Figure 10.1. Unlike the theoretical model described in the text the actual magnetic field distribution in the solar wind has sector structure (i.e. reversals in the sign of B).

Equations (10.4) and (10.5) imply that the total change in the magnetic flux through the surface A is zero. Magnetic flux constancy is an important physical property of the magnetic fields in ionized gases such as stellar winds from low-mass main-sequence stars.

In the equatorial plane of a low-mass star such as the Sun the velocity $\boldsymbol{v}$ and magnetic field $\boldsymbol{B}$ in a stellar wind can be expressed as

$$\boldsymbol{v} = v_r\hat{\boldsymbol{r}} + v_\phi\hat{\boldsymbol{\phi}} \tag{10.6}$$

and

$$\boldsymbol{B} = B_r\hat{\boldsymbol{r}} + B_\phi\hat{\boldsymbol{\phi}}, \tag{10.7}$$

where r and ϕ are cylindrical coordinates. From considerations of axial symmetry it follows that $\boldsymbol{v}$ and $\boldsymbol{B}$ do not depend on the coordinate ϕ (see Figure 10.1).

Because the outflowing gas can be considered a perfect conductor Equation (10.1) implies $\boldsymbol{E} = -\boldsymbol{v} \times \boldsymbol{B}/c$. Moreover, under steady-state conditions we have $\partial B/\partial t = 0$. The above two relations and the Maxwell Equation (3.345) imply

$$(\boldsymbol{\nabla} \times \boldsymbol{E})_\phi = \frac{1}{r}\frac{d}{dr}[r(v_rB_\phi - v_\phi B_r)] = 0. \tag{10.8}$$

Integrating Equation (10.8) gives

$$r(v_rB_\phi - v_\phi B_r) = \text{constant} = -\omega r_0^2 B_{r_0}, \tag{10.9}$$

with $v_r = 0$ at $r = r_0$. The Maxwell equation $\boldsymbol{\nabla} \cdot \boldsymbol{B} = 0$ becomes

$$\frac{1}{r^2}\frac{d}{dr}r^2B_r = 0, \tag{10.10}$$

and therefore we have

$$r^2 B_r = r_0^2 B_{r_0}. \tag{10.11}$$

Because the physical variables do not depend on the coordinate ϕ the $\hat{\phi}$ equation of motion for steady-state outflow becomes

$$\rho \frac{v_r}{r} \frac{d}{dr}(rv_\phi) = \left(\frac{J}{c} \times B\right)_\phi$$

$$= \frac{1}{4\pi}[(\nabla \times B) \times B]_\phi = \frac{B_r}{4\pi r} \frac{d}{dr}(rB_\phi). \tag{10.12}$$

Equation (10.11) and an assumed constant rate of mass loss ($\rho v_r r^2 = $ constant) imply

$$\frac{B_r}{4\pi \rho v_r} = \frac{B_r r^2}{4\pi \rho v_r r^2} = \text{constant}. \tag{10.13}$$

Using Equation (10.13) the integral of Equation (10.12) becomes

$$rv_\phi - \frac{B_r r B_\phi}{4\pi \rho v_r} = \text{constant} \equiv L. \tag{10.14}$$

Introducing the variable M_A defined by the equation

$$M_A^2 = \frac{4\pi \rho v_r^2}{B_r^2}. \tag{10.15}$$

Equations (10.9), (10.11) and (10.14) lead to the equation

$$v_\phi = \omega r \frac{(M_A^2 L r^{-2} \omega^{-1} - 1)}{(1 - M_A^2)}. \tag{10.16}$$

Equation (10.16) implies that a critical point r_a occurs when the Alfvén velocity,

$$v_A = \left(\frac{B_r^2}{4\pi \rho}\right)^{1/2}, \tag{10.17}$$

equals the flow velocity v_r (i.e. $M_A = 1$). At the critical point r_a the numerator of Equation (10.16) must also equal zero. It follows that the constant L in Equation (10.16) becomes

$$L = \omega r_a^2. \tag{10.18}$$

Because of the conditions $\rho v_r r^2 = $ constant and $B_r r^2 = $ constant the ratio $M_A^2/v_r r^2$ is equal to a constant. Evaluating this constant at the critical point we obtain

$$M_A^2 = \frac{v_r r^2}{v_{r_a} r_a^2} = \frac{\rho_a}{\rho}. \tag{10.19}$$

Equations (10.18) and (10.19) imply that Equation (10.16) can be rewritten as

$$v_\phi = \frac{\omega r}{v_{r_a}} \frac{v_r - v_{r_a}}{1 - M_A^2}. \tag{10.20}$$

Equations (10.9), (10.15), (10.19) and (10.20) lead to the equation

$$B_\phi = B_r \frac{\omega r}{v_{r_a}} \frac{r_a^2 - r^2}{r_a^2(M_A^2 - 1)}. \tag{10.21}$$

The radial equation of motion is

$$\rho v_r \frac{dv_r}{dr} = -\frac{dP}{dr} - \rho \frac{GM}{r^2} + \left(\frac{J}{c} \times B\right)_r + \rho \frac{v_\phi^2}{r}. \tag{10.22}$$

If the gas is assumed to be ionized hydrogen then the pressure P becomes

$$P = 2\frac{\rho kT}{m_p}. \tag{10.23}$$

We assume that pressure P and density ρ are related by the equation

$$P = P_0 \left(\frac{\rho}{\rho_0}\right)^\gamma = P_a \left(\frac{\rho}{\rho_a}\right)^\gamma. \tag{10.24}$$

The radial component of the $\dfrac{J}{c} \times B$ term in Equation (10.22) is

$$\left(\frac{J}{c} \times B\right)_r = -\frac{1}{4\pi r} B_\phi \frac{d}{dr}(rB_\phi). \tag{10.25}$$

Equations (10.23), (10.24) and (10.25) imply that Equation (10.22) becomes

$$\frac{d}{dr}\left[\frac{1}{2}v_r^2 + \frac{\gamma}{\gamma-1}\frac{P_a}{\rho_a}\left(\frac{\rho}{\rho_a}\right)^{\gamma-1} - \frac{GM}{r}\right] = \frac{v_\phi^2}{r} - \frac{rB_\phi}{4\pi\rho r^2}\frac{d}{dr}(rB_\phi), \tag{10.26}$$

where the variables ρ, v_ϕ and B_ϕ can be expressed as functions of v_r and r.

The first term on the right-hand side of Equation (10.26) can be expressed as

$$\frac{v_\phi^2}{r} = -\frac{1}{2}\frac{dv_\phi^2}{dr} + \frac{v_\phi}{r}\frac{d(rv_\phi)}{dr}. \tag{10.27}$$

Equation (10.12) implies that the second term on the right-hand side of Equation (10.27) becomes

$$\frac{v_\phi}{r}\frac{d}{dr}rv_\phi = \frac{v_\phi}{r}\frac{B_r r^2}{4\pi\rho v_r r^2}\frac{drB_\phi}{dr}. \tag{10.28}$$

From Equations (10.27) and (10.28) it follows that the right-hand side of Equation (10.26) is equal to

$$-\frac{1}{2}\frac{dv_\phi^2}{dr} + \frac{1}{4\pi\rho v_r r^2}[rv_\phi B_r - rB_\phi v_r]\frac{drB_\phi}{dr}. \tag{10.29}$$

Equations (10.9), (10.28) and (10.29) imply that the right-hand side of Equation (10.26) can be rewritten as

$$-\frac{1}{2}\frac{dv_\phi^2}{dr} + \frac{\omega r^2 B_r}{4\pi\rho v_r r^2}\frac{drB_\phi}{dr}.$$ (10.30)

The coefficient of $d(rB_\phi)/dr$ in Equation (10.30) is a constant and therefore Equation (10.30) can be integrated. Using Equation (10.30) the integral of Equation (10.26) becomes

$$\frac{1}{2}v_r^2 + \frac{\gamma}{\gamma-1}\frac{P_a}{\rho_a}\left(\frac{\rho}{\rho_a}\right)^{\gamma-1} - \frac{GM}{r} + \frac{1}{2}v_\phi^2 - \frac{B_r B_\phi}{4\pi\rho}\frac{r\omega}{v_r} = \text{constant}.$$ (10.31)

An important consequence of the presence of magnetic fields in stellar winds such as the solar wind is that they enhance the rate of loss of angular momentum. The photosphere of the star corotates with the outflowing gas out to radial distances $r \simeq r_a$ with r_a equal to the critical radius defined in Equation (10.16). It follows that the rate of angular momentum loss per unit mass is

$$\frac{dJ}{dt} \simeq \omega r_a^2,$$ (10.32)

with r_a greater than the photospheric radius. Although the mass-loss rates of the Sun and presumably other solar-mass main-sequence stars are too low (i.e. $\simeq 10^{-13} M_\odot$ yr $- 10^{-14} M_\odot$/yr) to lead to significant amounts of mass loss during their main-sequence lifetimes angular momentum losses are likely to be appreciable. Main-sequence stars with photospheric temperatures too high to produce outer convective regions rotate much more rapidly than lower-mass main-sequence stars. Mechanical motions of convective blobs of gas are the energy sources for stellar winds such as the solar wind.

10.2   Hydrostatic equilibrium in the presence of strong magnetic fields

Large scale magnetic fields in stars are usually too weak to significantly affect the condition for hydrostatic equilibrium. However, magnetic forces are comparable to forces caused by gas pressure in localized regions such as within sunspots. The equation for hydrostatic equilibrium in the presence of a magnetic field is

$$0 = -\nabla P - \rho g + \frac{J}{c} \times B.$$ (10.33)

Equation (10.33) implies that if $J \times B = 0$ then the condition for hydrostatic equilibrium is unaffected by the presence of the magnetic field. Such magnetic field configurations are known as force-free magnetic fields. Using the Maxwell equation (3.342) with the time derivative neglected the condition for a force-free magnetic field becomes

$$J \times B = \frac{1}{4\pi}(\nabla \times B) \times B$$

$$= \frac{1}{4\pi}\left[(B \cdot \nabla)B - \frac{1}{2}\nabla B^2\right] = 0.$$ (10.34)

We assume that a magnetic field is axially symmetric and that in cylindrical coordinates (r, θ, z) the magnetic field $\boldsymbol{B}$ can be expressed as

$$\boldsymbol{B} = B_\theta(r)\hat{\boldsymbol{\theta}} + B_z(r)\hat{\boldsymbol{z}}$$ (10.35)

and

$$\boldsymbol{J} = J_\theta(r)\hat{\boldsymbol{\theta}} + J_z(r)\hat{\boldsymbol{z}},$$ (10.36)

respectively. Because the nonzero components of the curl of $\boldsymbol{B}$ are

$$(\nabla \times \boldsymbol{B})_\theta = \frac{\partial B_r}{\partial z} - \frac{\partial B_z}{\partial r} = -\frac{\partial B_z}{dr}$$ (10.37)

and

$$(\nabla \times \boldsymbol{B})_z = \frac{1}{r}\left[\frac{\partial}{\partial r}(rB_\theta) - \frac{\partial B_r}{\partial \theta}\right] = \frac{1}{r}\frac{drB_\theta}{dr},$$ (10.38)

it follows that the equation of hydrostatic equilibrium given as Equation (10.34) becomes

$$-\frac{dB_z}{dr}B_z - \frac{1}{r}\frac{d}{dr}(rB_\theta)B_\theta = 0.$$ (10.39)

Equation (10.39) can be rewritten as

$$-\frac{1}{2}\frac{d}{dr}[B_z^2(r) + B_\theta^2(r)] - \frac{B_\theta^2(r)}{r} = 0.$$ (10.40)

The Bessel functions $J_n(x)$ satisfy the recursion relation

$$\frac{d}{dx}(x^n J_n(x)) = x^n J_{n-1}(x).$$ (10.41)

In particular Equation (10.41) implies that

$$\frac{dxJ_1(x)}{dx} = xJ_0.$$ (10.42)

Since the derivative $J_0'(x)$ is related to $J_1(x)$ by the equation

$$J_0'(x) = -J_1(x),$$ (10.43)

it follows that $J_0(x)$ and $J_1(x)$ are solutions of the equation

$$-\frac{dJ_0(x)}{dx}J_0(x) - \frac{1}{r}\frac{d(rJ_1(x))}{dx}J_1(x) = 0.$$ (10.44)

Equation (10.44) is identical to Equation (10.39) and therefore we conclude that one set of solutions of Equation (10.39) is

$$B_z(r) = J_0(\alpha r),$$

$$B_\theta(r) = J_1(\alpha r), \tag{10.45}$$

with α equal to a constant.

Another solution of Equation (10.39) (or 10.40) is

$$B_z = \frac{1}{1 + \mu^2 r^2}, \tag{10.46a}$$

$$B_\theta = \frac{\mu r}{1 + \mu^2 r^2}, \tag{10.46b}$$

with μ equal to a constant. One can readily verify that the expressions for B_z and B_θ are solutions of Equation (10.39) by direct substitution. From Equations (10.46a) and (10.46b) we have

$$\frac{B_\theta}{r B_z} = \mu = \text{constant}. \tag{10.47}$$

Equation (10.47) implies that the magnetic field B is twisted with μ equal to a constant. A magnetic field configuration similar to that given in Equation (10.47) might be produced by rotation. In stars force-free magnetic fields are likely to represent only a first approximation to the actual magnetic field configuration.

We assume that a magnetic field is axisymmetric and poloidal (i.e. $B_\theta = 0$). A dipole magnetic field is an example of such a magnetic field. In cylindrical coordinates (r, θ, z) the magnetic field is

$$B = B_r(r, z)\hat{r} + B_z(r, z)\hat{z}. \tag{10.48}$$

A poloidal magnetic field can be expressed as

$$B = \nabla \times A, \tag{10.49}$$

with the magnetic vector potential $A = A_\theta(r, z)\hat{\theta}$. In terms of A_θ the magnetic field components B_r and B_z become

$$B_r = -\frac{\partial A_\theta}{\partial z} \tag{10.50a}$$

and

$$B_z = \frac{1}{r} \frac{\partial r A_\theta}{\partial r}. \tag{10.50b}$$

Equations (10.49) and (3.347) with the time derivative neglected imply

$$\nabla \times (\nabla \times A) = -\nabla^2 A + \nabla(\nabla \cdot A) = \frac{4\pi}{c} J. \tag{10.51}$$

Since $A_\theta = A_\theta(r, z)$ and only the θ component of A is nonzero we have

$$\nabla \cdot \mathbf{A} = \frac{1}{r}\frac{\partial r A_r}{\partial r} + \frac{1}{r}\frac{\partial A_\theta}{\partial \theta} + \frac{\partial A_z}{\partial z} = 0. \tag{10.52}$$

From Equations (10.51) and (10.52) we obtain

$$\nabla^2(A_\theta \hat{\theta}) = \left(\nabla^2 A_\theta - \frac{1}{r^2}A_\theta\right)\hat{\theta} = -\frac{4\pi}{c}J_\theta(r, z)\hat{\theta}. \tag{10.53}$$

Equation (10.53) implies that the only nonzero component of $\mathbf{J}$ is J_θ. Expressing the Laplacian in Equation (10.53) in cylindrical coordinates gives

$$\frac{1}{r}\frac{\partial}{\partial r}\left(r\frac{\partial A_\theta}{\partial r}\right) + \frac{\partial^2 A_\theta}{\partial z^2} - \frac{1}{r^2}A_\theta = -\frac{4\pi}{c}J_\theta. \tag{10.54}$$

Multiplying Equation (10.33) by the vector product $\mathbf{B} \times$ we find that

$$\mathbf{J} = J_\theta \hat{\theta} = \frac{1}{B^2}\mathbf{B} \times (-\nabla P - \rho \mathbf{g}), \tag{10.55}$$

where the direction of $\mathbf{g}$ is along the $\hat{z}$-axis. The above equations can be solved for A_θ, J_θ and $P = P(r, z)$ if the pressure is given in a plane $z = 0$ and a temperature distribution is assumed. An iterative method of solution is required. If the current density J_θ is assumed known then Equation (10.54) can be solved for A_θ. The components of the magnetic field B_r and B_θ follow from Equations (10.50a) and (10.50b). Equation (10.33) can now be used to determine the pressure distribution and then Equation (10.55) used to determine J_θ. If J_θ differs significantly from the initial estimate then Equation (10.54) can be solved for A_θ using the new values of J_θ.

Surface boundary conditions must also be given if a solution is to be found. If P_{ext} and B_{ext} are the external pressure and magnetic field respectively then we have

$$P_{\text{ext}} - P = \frac{1}{8\pi}(B^2 - B_{\text{ext}}^2) = \frac{1}{8\pi}(B - B_{\text{ext}})(B + B_{\text{ext}}). \tag{10.56}$$

From Ampère's law it follows that the current per unit length J_L at the boundary is

$$J_L = \frac{c}{4\pi}(B - B_{\text{ext}}). \tag{10.57}$$

Equations (10.56) and (10.57) imply that

$$J_L = \frac{2}{c}\frac{P_{\text{ext}} - P}{B_{\text{ext}} + B}. \tag{10.58}$$

10.3 Magnetohydrodynamic waves

A static magnetic field $\mathbf{B}_0$ in the y-z plane is given by the equation

$$\mathbf{B}_0 = B_0 \sin\theta \hat{y} + B_0 \cos\theta \hat{z}. \tag{10.59}$$

We assume that waves propagate along the z-direction and that all variables depend on z and the time t only. Let $\delta\rho$, δP and B' be variations in density, pressure and magnetic field, respectively.

The linearized equation of motion is

$$\rho\frac{d\boldsymbol{v}}{dt} = -\nabla\delta P + \frac{1}{4\pi}(\nabla\times\boldsymbol{B}')\times\boldsymbol{B}_0. \tag{10.60}$$

For adiabatic variations in density and pressure we have

$$\delta P = \Gamma_1 P\frac{\delta\rho}{\rho} = v_s^2\delta\rho, \tag{10.61}$$

with v_s equal to the (adiabatic) speed of sound. From Equations (10.60) and (10.61) we have

$$\rho\frac{\partial\boldsymbol{v}}{\partial t} + v_s^2\nabla\delta\rho - \frac{1}{4\pi}(\nabla\times\boldsymbol{B}')\times\boldsymbol{B}_0 = 0. \tag{10.62}$$

The linearized equation of continuity is

$$\frac{\partial\delta\rho}{\partial t} + \rho\nabla\cdot\boldsymbol{v} = 0. \tag{10.63}$$

We assume that magnetic flux is conserved. From Equation (10.3) it follows that the linearized equation expressing the conservation of magnetic flux is

$$\frac{\partial\boldsymbol{B}'}{\partial t} - \nabla\times(\boldsymbol{v}\times\boldsymbol{B}_0) = 0. \tag{10.64}$$

Since all physical variables depend only on z and t we have

$$0 = \nabla\cdot\boldsymbol{B}' = \frac{\partial B_z'}{\partial z}. \tag{10.65}$$

Equation (10.65) implies that $B_z' = 0$. The x- and y-components of Equation (10.64) are

$$\frac{\partial B_x'}{\partial t} = B_0\cos\theta\frac{\partial v_x}{\partial z} \tag{10.66}$$

and

$$\frac{\partial B_y'}{\partial t} = B_0\cos\theta\frac{\partial v_y}{\partial z} - B_0\sin\theta\frac{\partial v_z}{\partial z}. \tag{10.67}$$

The three-component equations of the vector equation (10.62) and Equation (10.63) are

$$\rho\frac{\partial v_x}{\partial t} = \frac{B_0\cos\theta}{4\pi}\frac{\partial B_x'}{\partial z}, \tag{10.68}$$

$$\rho\frac{\partial v_y}{\partial t} = \frac{B_0\cos\theta}{4\pi}\frac{\partial B_y'}{\partial z}, \tag{10.69}$$

$$\rho \frac{\partial v_z}{\partial t} + v_s^2 \frac{\partial \delta\rho}{\partial z} + \frac{B_0 \sin\theta}{4\pi} \frac{\partial B'_y}{\partial z} = 0,$$ (10.70)

$$\frac{\partial \delta\rho}{\partial t} + \rho \frac{\partial v_z}{\partial z} = 0.$$ (10.71)

Equations (10.66) and (10.68) imply that B'_x satisfies the wave equation

$$\frac{\partial^2 B'_x}{\partial t^2} - v_z^2 \frac{\partial^2 B'_x}{\partial z^2} = 0,$$ (10.72)

with $v_z^2 = \dfrac{B_0^2 \cos^2\theta}{4\pi\rho} \equiv v_A^2 \cos^2\theta.$

It follows immediately from Equations (10.66) and (10.68) that v_x also satisfies Equation (10.72). The wave whose propagation is governed by Equation (10.72) is known as the Alfvén wave. If the direction of propagation of the wave is along the magnetic field then the velocity of the wave is the Alfvén velocity v_A.

Equation (10.67) and Equations (10.69–10.71) determine the wave motion of the dependent variables v_y, v_z, B'_y and $\delta\rho$. If we assume that the dependent variables vary as $\exp[i(\omega t - kz)]$ then Equation (10.67) and Equations (10.69–10.71) reduce to the equations

$$kB_0 \cos\theta v_y - kB_0 \sin\theta v_z + \omega B'_y = 0,$$ (10.73)

$$\rho\omega v_y + \frac{kB_0 \cos\theta}{4\pi} B'_y = 0,$$ (10.74)

$$\rho\omega v_z - \frac{kB_0 \sin\theta}{4\pi} B'_y - kv_s^2 \delta\rho = 0,$$ (10.75)

$$- \rho k v_z + \omega\delta\rho = 0.$$ (10.76)

Eliminating $\delta\rho$ from Equations (10.75) and (10.76) we obtain

$$\rho\left(\frac{\omega^2}{k^2} - v_s^2\right)v_z - B'_y \frac{\omega B_0 \sin\theta}{k4\pi} = 0.$$ (10.77)

Similarly eliminating v_y from Equations (10.73) and (10.74) gives

$$\left(\frac{\omega^2}{k^2} - \frac{(B_0 \cos\theta)^2}{4\pi\rho}\right)B'_y - \frac{\omega}{k} B_0 \sin\theta v_z = 0.$$ (10.78)

In Equations (10.77) and (10.78) ω/k is the phase velocity of the wave. Eliminating v_z from Equations (10.77) and (10.78) leads to the equation

$$v^4 - (v_s^2 + v_A^2)v^2 + v_s^2 v_A^2 \cos^2\theta = 0.$$ (10.79)

In the direction along the magnetic field (i.e. $\theta = 0$) Equation (10.79) becomes

$$v^4 - (v_s^2 + v_A^2)v^2 + v_s^2 v_A^2 = 0.$$ (10.80)

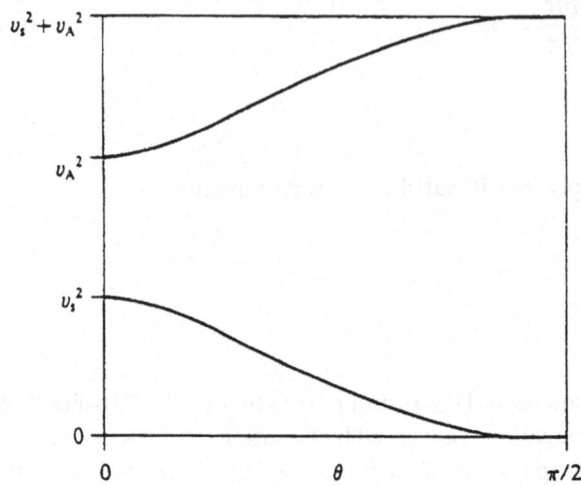

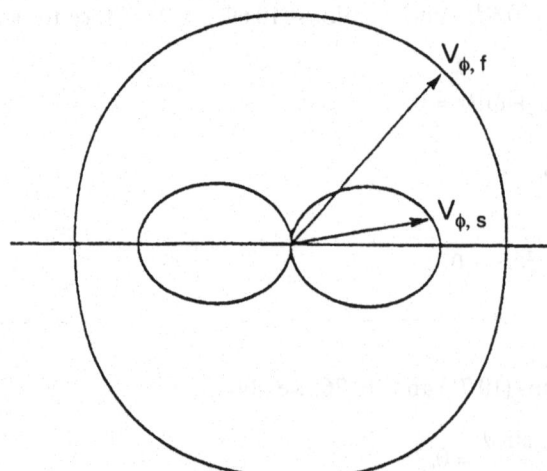

Figure 10.2. The upper figure gives the square of the phase velocity of fast and slow magnetosonic modes as a function of angle between magnetic field vector and direction of propagation (for $v_A > v_s$). The lower figure is a polar diagram showing the phase velocity vectors of fast and slow modes. From Sturrock (1994).

Equation (10.80) can be rewritten as

$$(v^2 - v_s^2)(v^2 - v_A^2) = 0. \tag{10.81}$$

The solutions of Equation (10.81) are $v = \pm v_s$ and $v = \pm v_A$. The first of these solutions is a sound wave and the second solution is a wave that propagates at the Alfvén velocity v_A. In the direction perpendicular to the magnetic field (i.e. $\theta = \pi/2$) only one type of wave can propagate and the velocity of this magnetosonic wave can be either $v = \pm\sqrt{v_A^2 + v_s^2}$. In directions that are not perpendicular to the magnetic field two types of magnetic waves known as the fast and slow waves can propagate in addition to the Alfvén wave described by Equation (10.72). The phase velocities of fast and slow magnetosonic waves are described in Figure 10.2.

10.4 Dynamo magnetic fields

Stars can retain magnetic fields much longer than characteristic magnetic field decay timescales because magnetic fields are generated in stars and planets by turbulent motions and rotation. We consider an axisymmetric magnetic field B consisting of a poloidal component B_p that has field lines in meridional planes with respect to the symmetry axis of the magnetic field and a toroidal component B_t that has field lines encircling the magnetic symmetry axis. Turbulent motions in convective zones can cause the toroidal magnetic field B_t to have a curl that generates a current along the unperturbed direction of the toroidal magnetic field B_t. The resultant toroidal current produces a poloidal magnetic field. The twisting of poloidal magnetic fields generates a current along the initial untwisted poloidal magnetic field. The resultant poloidal current generates a toroidal magnetic field. It follows that turbulence and rotation can amplify an infinitesimal initial magnetic field until magnetic forces impede turbulent motions.

Consider a turbulent gas in a magnetic field B. If v' is the fluctuating component of the gas velocity, v equal to the gas velocity and $\bar{v}$ equal to the mean velocity then $v' = v - \bar{v}$. Because the displacement current can be neglected Maxwell's equations for the mean fields become

$$\nabla \times \bar{E} = -\frac{1}{c}\dot{\bar{B}}, \tag{10.82}$$

$$\nabla \times \bar{B} = \frac{4\pi}{c}\bar{J} \quad (\bar{B} = \bar{H}), \tag{10.83}$$

$$\nabla \cdot \bar{B} = 0. \tag{10.84}$$

Ohm's law for the mean fields is

$$\bar{J} = \eta_t^{-1}(\bar{E} + \bar{v} \times \bar{B} + \overline{v' \times B}), \tag{10.85}$$

where η_t is the turbulent resistivity, which is much higher than the electric resistivity. The mean force $\overline{v' \times B}$ can have a nonzero component along the direction of the mean magnetic field $\bar{B}$. We assume that fluctuating velocities v' lead to the relation

$$\overline{v' \times B} = \alpha\bar{B}, \tag{10.86}$$

where α can be a function of position. Equation (10.85) becomes

$$\bar{J} = \eta_t^{-1}(\bar{E} + \bar{v} \times \bar{B} + \alpha\bar{B}). \tag{10.87}$$

Equations (10.82), (10.83), (10.84) and (10.87) lead to the equation

$$\frac{\partial \boldsymbol{B}}{\partial t} = \bar{\nabla} \times (\bar{\boldsymbol{v}} \times \bar{\boldsymbol{B}}) + \nabla \times (\alpha \bar{\boldsymbol{B}}) + \frac{c^2}{4\pi} \eta_t \nabla^2 \bar{\boldsymbol{B}}. \tag{10.88}$$

We consider a rotating, turbulent region of a star in which the mean magnetic field is axisymmetric. The bar notation used above will be omitted below because only mean fields appear in the equations. The turbulent magnetic diffusity $\eta_D = c^2 \eta_t / 4\pi$ that appears in Equation (10.88) is approximately equal to $\alpha_t l v_c$ where α_t is a constant ≤ 1, l is the convective mixing length and v_c is the convective velocity. The electric conductivities of ionized gas in stars are sufficiently high that radiative regions can be regarded as perfect conductors.

The mean magnetic field consists of a poloidal magnetic field $\boldsymbol{B}_p$ and a toroidal magnetic field $\boldsymbol{B}_t$. In spherical polar coordinates (r, θ, ϕ) the poloidal and toroidal magnetic fields $\boldsymbol{B}_t$ and $\boldsymbol{B}_p$ respectively becomes

$$\boldsymbol{B}_p = \nabla \times \boldsymbol{A}, \tag{10.89}$$

with $\boldsymbol{A} = A_\phi(r, \theta, t)\hat{\phi}$ and

$$\boldsymbol{B}_t = B_\phi(r, \theta, t)\hat{\phi}, \tag{10.90}$$

with t equal to the time. Substituting $\boldsymbol{B} = \boldsymbol{B}_p + \boldsymbol{B}_t$ into Equation (10.88) we obtain the two separate equations

$$\frac{\partial \boldsymbol{B}_p}{\partial t} = \nabla \times (\alpha \boldsymbol{B}_t) + \eta_D \nabla^2 \boldsymbol{B}_p \tag{10.91}$$

and

$$\frac{\partial \boldsymbol{B}_t}{\partial t} = \nabla \times (\boldsymbol{v} \times \boldsymbol{B}_p) + \nabla \times (\alpha \boldsymbol{B}_p) + \eta_D \nabla^2 \boldsymbol{B}_t. \tag{10.92}$$

From Equation (10.89) it follows that Equation (10.91) can be replaced by the equation

$$\frac{\partial \boldsymbol{A}}{\partial t} = \alpha \boldsymbol{B}_t + \eta_D \nabla^2 \boldsymbol{A} + \nabla \psi, \tag{10.93}$$

with ψ equal to some scalar function. We can choose $\nabla \psi$ equal to zero in Equation (10.93) because $\nabla \psi$ can only have a component along the ϕ-direction, and therefore if $\nabla \psi$ were nonzero it would describe a current following along the axis of symmetry.

In Cartesian coordinates the unit vector $\hat{\phi}$ is

$$\hat{\phi} = -\sin \phi \hat{x} + \cos \phi \hat{y}. \tag{10.94}$$

It follows that the Laplacians $\nabla^2 \boldsymbol{A}$ and $\nabla^2 \boldsymbol{B}_t$ that appear in Equations (10.93) and (10.92) respectively become

$$\nabla^2 (A\hat{\phi}) = \left(\nabla^2 A_\phi - \frac{A_\phi}{r^2 \sin^2 \theta} \right) \hat{\phi} \tag{10.95}$$

and

$$\nabla^2(B_\phi \hat{\phi}) = \left(\nabla^2 B_\phi - \frac{B_\phi}{r^2 \sin^2 \theta}\right)\hat{\phi}. \tag{10.96}$$

Using Equations (10.95) and (10.96) and setting $\nabla\psi$ equal to zero implies that the dynamo equations (10.92) and (10.93) can be written in the form

$$\frac{\partial B_\phi}{\partial t}\hat{\phi} - \eta_D\left(\nabla^2 B_\phi - \frac{B_\phi}{r^2 \sin^2 \theta}\right)\hat{\phi} = \nabla \times (\boldsymbol{v} \times \boldsymbol{B}_p) + \nabla \times (\alpha \boldsymbol{B}_p) \tag{10.97}$$

and

$$\frac{\partial A_\phi}{\partial t} - \eta_D\left(\nabla^2 - \frac{1}{r^2 \sin^2 \theta}\right)A_\phi = \alpha B_\phi, \tag{10.98}$$

respectively, where the components of $\boldsymbol{B}_p$ are

$$B_r = \frac{1}{r \sin \theta}\frac{\partial}{\partial \theta}(A_\phi \sin \theta), \tag{10.99}$$

$$B_\theta = -\frac{1}{r}\frac{\partial(rA_\phi)}{\partial r} \tag{10.100}$$

and the velocity $\boldsymbol{v}$ that appears in Equation (10.97) is

$$\boldsymbol{v} = v_\phi(r, \theta)\hat{\phi} = r \sin \theta \omega(r, \theta)\hat{\phi}. \tag{10.101}$$

In stars amplification of B_ϕ is caused primarily by differential rotation and therefore we can neglect the last term in Equation (10.97). Using Equation (10.101) we find that Equation (10.97) reduces to the equation

$$\frac{\partial B_\phi}{\partial t} - \eta_D\left(\nabla^2 - \frac{1}{r^2 \sin^2 \theta}\right)B_\phi = B_r\left(\frac{\partial v_\phi}{\partial r} - \frac{v_\phi}{r}\right) + r \sin \theta \frac{B_\theta}{r}\frac{\partial \omega}{\partial \theta}. \tag{10.102}$$

Equations (10.98) and (10.102) are the dynamo equations. If we choose a particular functional form for $\alpha(r, \theta)$ and v_ϕ then we can reduce Equations (10.98) and (10.102) to an eigenvalue problem by substituting $A_\phi(r, \theta, t) = A_\phi(r, \theta)e^{t/\tau}$ and $B_p(r, \theta, t) = B_p(r, \theta)e^{t/\tau}$ into Equations (10.98) and (10.102) with τ equal to the eigenvalue. If such solutions to Equations (10.98) and (10.102) exist then they describe a magnetic field distribution that grows exponentially with time until magnetic forces affect turbulent motions. The intensity of the magnetic field increases until magnetic stresses become comparable to the coriolis force. An approximate estimate for the balance between these two forces is

$$\langle B_p B_\phi \rangle 4\pi R^{-1} \sim \langle \rho v_c \omega \rangle, \tag{10.103}$$

where R is the dimension of the convective region, v_c is the convective velocity and ω is equal to the angular velocity of rotation. It can also be argued that dynamo magnetic fields grow until the magnetic energy density is approximately equal to the turbulent

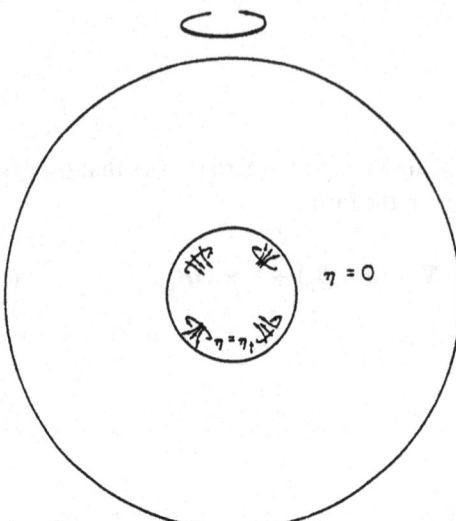

Figure 10.3. An idealized model consisting of a differentially rotating, convective core surrounding by a medium with infinite electrical conductivity (i.e. $\eta = 0$). The resistivity of the core is $\eta_t \sim v_c l$ where v_c is the convective velocity and l is the pressure scale height. Right-handed and left-handed helical motions do not occur with equal probability (the α effect). From Levy and Rose (1974).

kinetic energy density. The latter condition for the growth of the dynamo magnetic field implies

$$\left\langle \frac{1}{8\pi}(B_p^2 + B_\phi^2) \right\rangle \sim \frac{1}{2}\langle \rho v_c^2 \rangle. \tag{10.104}$$

Massive stars develop extensive convective cores during quasi-static carbon, oxygen and silicon burning that occur prior to core collapse. Figure 10.3 shows a schematic model of a rotating star with a turbulent convective core surrounded by a radiative zone. Generation of dynamo magnetic fields in the cores of massive stars provides a plausible explanation for the production of neutron star magnetic fields. Figure 10.4 gives a simplified model for a self-maintaining dynamo where toroidal currents (j_t) generate poloidal magnetic fields (B_p) and vice versa.

It is interesting to note that recent supercomputer simulations of the geodynamo by Glatzmaier and Roberts (*Physics Today*, January 1996) require some differential rotation between the outer liquid iron–nickel core where magnetic fields are produced, and inner solid core and silicate mantle to model the Earth's nearly dipolar magnetic field. These calculations simulate in an approximate manner magnetic field reversals which are known to occur about every 10^6 years. Viscous and magnetic forces are believed responsible for differential rotation. The dynamo equations have for different physical input

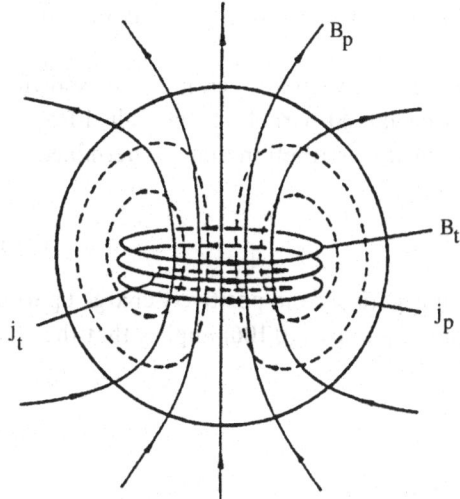

Figure 10.4. A possible dynamo self-maintaining magnetic field configuration with α effect. Ohm's law is $j = \sigma(E + \alpha B)$ with $\sigma = 1/\eta$ finite. The magnetic field B is axisymmetric and consists of toroidal, B_t, and poloidal, B_p, components. B_t field lines encircle the axis of symmetry and as a result of the α effect generate a toroidal current j_t which produces the poloidal magnetic field B_p. The presence of differential rotation can cause a toroidal magnetic field B_t to be generated from B_p and therefore the magnetic field configuration is self-maintaining. From Krause and Rädler (1980).

parameters solutions that predict both periodic and quasi-periodic magnetic field reversals. This is important physically because the solar magnetic field is known to undergo magnetic field reversals with a period of 22 years, which is twice the Sunspot cycle. The Sun's magnetic field has only a weak dipolar component. This is also compatible with the dynamo explanation for the origin of the solar magnetic field.

   The giant planets Jupiter, Saturn, Uranus and Neptune are also known to have large scale magnetic fields. The magnetosphere of Jupiter is enormous and sometimes extends to the orbit of Saturn. The magnetic fields of Jupiter and Saturn are generated in metallic hydrogen cores whereas those of Uranus and Neptune are believed to be produced in cores that are mostly compressed H_2O. Some white dwarfs are known to have strong magnetic fields that are probably dynamo generated.

10.5 Pulsar magnetic fields

Pulsars are rapidly rotating neutron stars with characteristic magnetic fields of 10^{12} gauss. Their interiors have high electrical conductivities and therefore in the corotating reference frame the electric field must equal zero. It follows that in a nonrotating frame we have

$$E + \frac{(\omega \times r) \times B}{c} = 0 \tag{10.105}$$

with ω equal to the rotational angular velocity. For simplicity we assume that the magnetic and rotation axes are aligned.

If it is assumed that a pulsar is surrounded by a vacuum then it can be shown that electric forces are sufficiently strong to cause charged particles to be pulled from their surfaces. The resultant plasma has a charge density distribution that is determined by the Poisson equation

$$\mathbf{V} \cdot [(\omega \times \mathbf{r}) \times \mathbf{B}] = 4\pi ec(n_- - n_+), \qquad (10.106)$$

where n_- and n_+ are the number densities of negatively and positively charged particles, which we have assumed to have unit charge. Equation (10.106) implies that the charge density in the polar regions is

$$e(n_- - n_+) \simeq \frac{\omega \cdot \mathbf{B}}{2\pi c}. \qquad (10.107)$$

A dipolar magnetic field is described by the equation

$$\mathbf{B} = \frac{B_p R^3}{2} \left(\frac{2\cos\theta}{r^3} \hat{r} + \frac{\sin\theta}{r^3} \hat{\theta} \right). \qquad (10.108)$$

If we differentiate

$$r = K \sin^2 \theta \qquad (10.109)$$

with K equal to a constant we obtain

$$\frac{B_r}{B_\theta} = \frac{dr}{rd\theta} = \frac{2\cos\theta}{\sin\theta}. \qquad (10.110)$$

Equation (10.110) shows that Equation (10.109) describes the lines of force of a dipole magnetic field. The light cylinder of a rotating neutron star is defined to be the cylinder whose distance from the rotation axis is

$$r \sin \theta = \frac{c}{\omega}. \qquad (10.111)$$

Magnetic lines of force include open field lines that extend beyond the light cylinder and also field lines that form a corotating magnetosphere because they close within the light cylinder. If the magnetic field is dipolar then open field lines satisfy the inequality

$$\theta < \theta_{max} = \sin^{-1} \left(\frac{\omega R}{c} \right)^{1/2}, \qquad (10.112)$$

where R is the neutron star radius and θ is the angle between the rotation axis and magnetic line of force.

It can be shown that the potential difference between two points on the surface of a pulsar is

$$V_B - V_A = \frac{B_p R^2 \omega}{2c}(\sin^2 \theta_B - \sin^2 \theta_A).$$ (10.113)

Equations (10.112) and (10.113) imply that the potential difference between a magnetic pole and a point with magnetic field direction θ_{max} is

$$\Delta V \simeq \frac{1}{2}\left(\frac{\omega R}{c}\right)^2 RB_p \qquad (\theta_{max} \ll 1).$$ (10.114)

The above equation shows that charged particles may attain energies as high as

$$E_{max} \simeq 3 \times 10^{12}\frac{ZR_6^3 B_{12}}{P^2}\text{ eV},$$ (10.115)

where R_6 is the neutron star radius in units of 10^6 cm, B_{12} is the surface magnetic field in units of 10^{12} gauss and P is the pulsar period in seconds.

The potential of a point within the polar regions of a pulsar cannot in general be the same as the potential at large distance because it varies across the surface. In order to maintain charge neutrality there should exist a critical angle between the magnetic pole, and angle θ_{max} where the surface potential is the same as at large distances. At angles greater than this critical angle the sign of $\omega \cdot B$ in Equation (10.107) changes and therefore the sign of the charge density also changes.

The total magnetic flux that passes beyond the light cylinder is

$$\Phi \simeq \pi r_p^2 B_p,$$ (10.116)

with $r_p = R \sin \theta_{max} \simeq R\left(\frac{\omega R}{c}\right)^{1/2}$. Equations (10.107) and (10.116) imply that the maximum current from a pulsar polar cap is

$$I_{max} \simeq c\pi r_p^2\frac{\omega B_p}{2\pi c} \simeq \frac{\Phi\omega}{2\pi}.$$ (10.117)

Although the current will tend to move along magnetic lines of force voltage differences create a torque. Equations (10.114) and (10.117) indicate that the rate of rotational energy loss satisfies the inequality.

$$\frac{d\left(\frac{1}{2}I\omega^2\right)}{dt} \leq I_{max}\Delta V \simeq \frac{\Phi\omega}{2\pi}\frac{\omega^2 R^3 B_p}{2c^2} \simeq \frac{\omega^4 R^6 B_p^2}{4c^3}.$$ (10.118)

Pulsars slow down at a rate that is often consistent with the above relation. The rate of rotational energy loss predicted by magnetic dipole radiation has a similar dependence on ω, R and B_p.

If a gap in charge density develops close to the surface then an appreciable fraction of the total potential drop within the polar regions of a pulsar may be created across it. A potential drop that is $\gtrsim 10^{12}$eV generates sufficiently energetic electrons (or positrons)

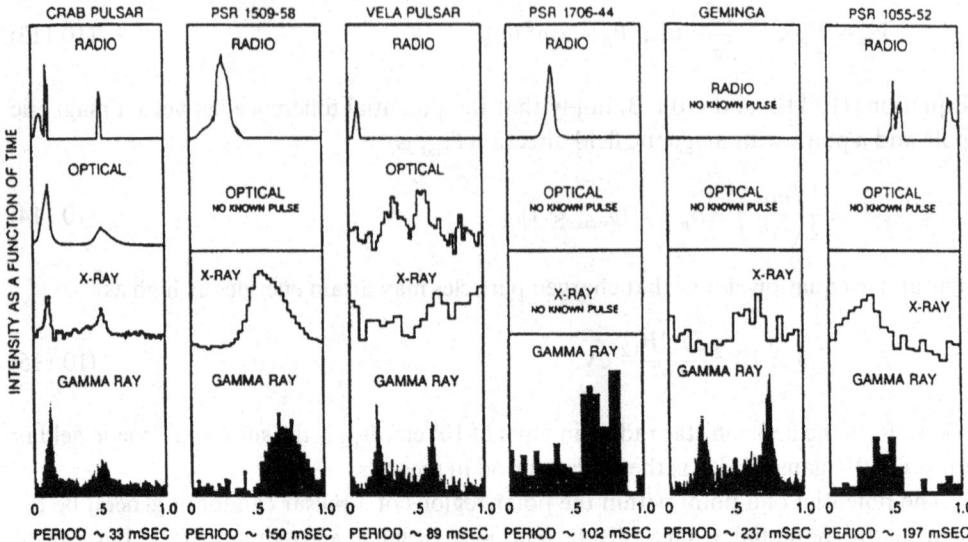

Figure 10.5. Light curves of the six known (1994) gamma-ray pulsars. Note that multiwavelength pulses from the Crab pulsar are in phase and that radio pulses have not been observed from Geminga. From Thompson (1994).

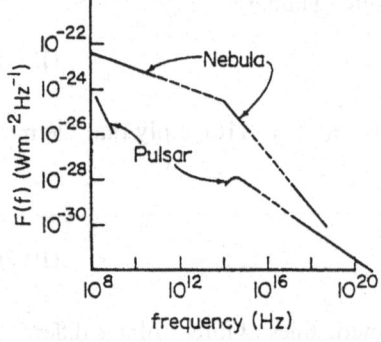

Figure 10.6. Spectra of the Crab pulsar and nebula.

that curvature radiation photons can form electron–positron pairs as they traverse magnetic lines of force. In this manner an electron–positron avalanche can be created and most pulsar radiation is caused by secondary e^-–e^+ pairs, which have much lower energies than the very high-energy primary electrons (or positrons) A coherent radiative process is required to explain the high radio brightness temperatures of pulsars.

Figure 10.5 shows light curves of six pulsars that have been detected at γ-ray wavelengths. Only in the case of the Crab pulsar are radio–γ-ray pulses nearly synchronous. The optical–γ-ray pulses are believed to result from incoherent synchrotron emission. Most of the radiation from the Crab nebular and pulsar (see Figure 10.6) is emitted at x-ray–γ-ray wavelengths. Magnetic fields in the Crab nebula are estimated to

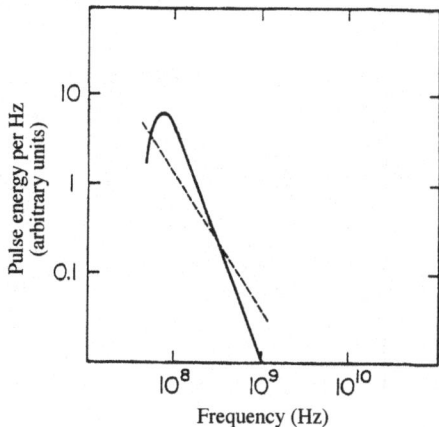

Figure 10.7. Radio spectra of two pulsars. One spectrum shows a low frequency cutoff.

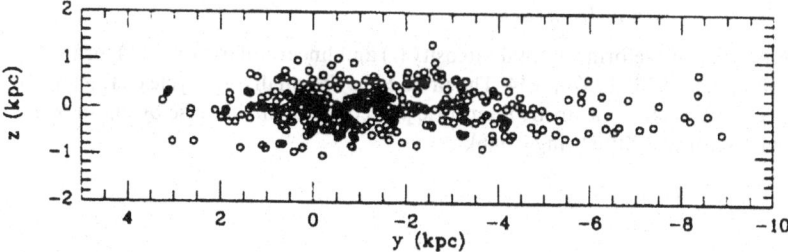

Figure 10.8. The approximate locations of 391 pulsars, projected onto the *y-z* plane of the Galaxy. The Sun lies at (0, 0), and the Galactic center is assumed to be at (−10, 0). From Taylor and Stinebring (1986).

be ~ 10^{-3}–10^{-4} gauss and the high-energy electrons (and positrons) responsible for observed synchrotron radiation have radiative lifetimes shorter than the age of the supernova. The inferred rate of loss of rotational energy is to within an order of magnitude equal to the total radiative output. The Crab and Vela pulsars are the only pulsars with observed optical pulses. A possible explanation for the absence of radio pulses from the nearby pulsar Geminga is that they are not emitted in our direction.

As shown in Figure 10.7 some pulsars have low radio frequency cutoffs whereas others do not. The amount of radio emission is generally much less than inferred rates of rotational energy loss. The positions of 391 pulsars with respect to galactic coordinates are shown in Figure 10.8. A number of pulsars with periods nearly as short as one millisecond have been discovered. Average pulse waveforms from one of these millisecond pulsars are shown in Figure 10.9. The top of Figure 10.10 gives the period distribution of 398 pulsars. Measured period derivatives are shown as a function of period at the bottom of Figure 10.10.

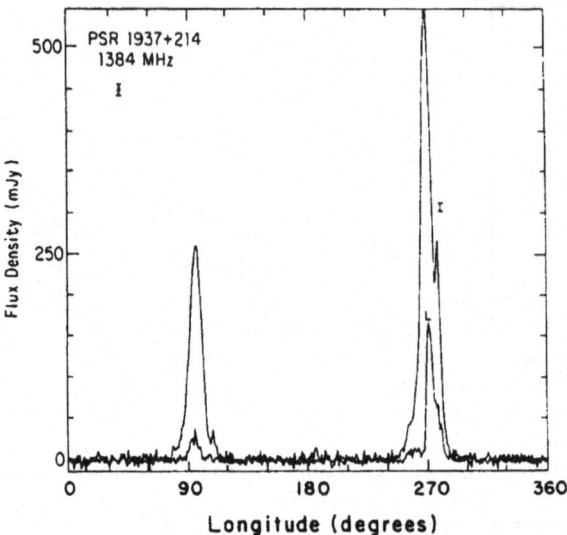

Figure 10.9. Average waveforms in total intensity (I) and linear polarization (L) for the 1.5-millisecond pulsar PSR 1937 + 214. The profile has the small duty cycle characteristic of most pulsars, and it shows a strong interpulse preceding the main pulse by about half a period. From Taylor and Stinebring (1986).

Problems

1.   The mass-loss rate of the solar wind is 10^{-13}–$10^{14}\,M_\odot$/yr, the outflow velocity about $400\,\mathrm{km\,s^{-1}}$ and the coronal temperature $T = 2 \times 10^6\,\mathrm{K}$ at $R = 2\,R_\odot$.

   a. Estimate the radial distance r at which $e^- $–$e^-$, e^-–p and p–p collisional mean free paths become greater than r.

   b. What would the temperature of solar-wind protons be at the orbit of the Earth if they undergo adiabatic expansion? What is the gyroradius of a proton of energy $E = \frac{3}{2}kT$ with $T = 2 \times 10^6\,\mathrm{K}$ in a 1 gauss magnetic field?

   c. Estimate electron thermal energy fluxes from the corona toward the chromosphere. Is the outward electron thermal energy flow sufficient to prevent electron temperatures from declining nearly adiabatically in the solar wind?

2.   Give complete derivations of Equations (10.113–10.115).

3.   Estimate the rate of angular-momentum loss from the Sun if the critical Alfvén radius r_a in Equation (10.32) is $10\,R_\odot$. Assume uniform rotation and estimate how long it will take the Sun to lose half its angular momentum. The rotation velocity of the solar photosphere is $2\,\mathrm{km\,s^{-1}}$.

4.   Prove that the magnetic vector potential A of a uniform magnetic field B equals

$$A = \frac{1}{2}(B \times r).$$

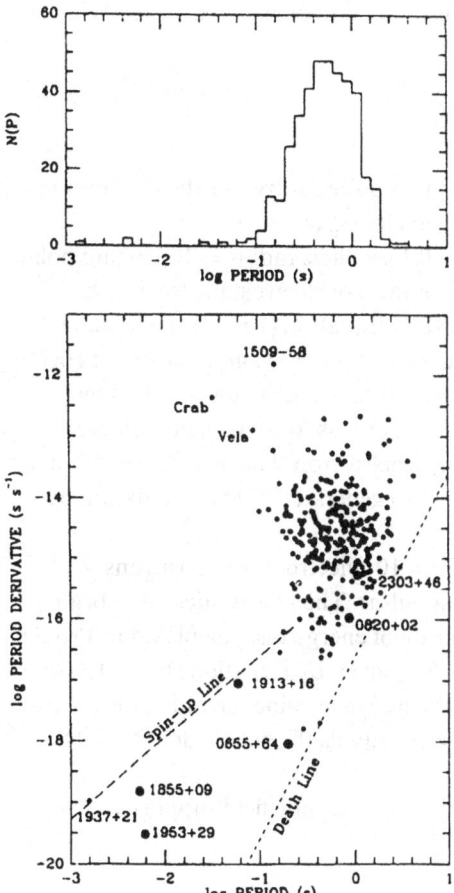

Figure 10.10. (Top) Observed period distribution of 398 pulsars. (Bottom) Distribution of periods and period derivatives of 308 pulsars. Slowdown in rotation and/or decay of magnetic fields cause radio pulses to cease and therefore pulsars do not exist to the right of the 'death line'. Pulsars in the lower left-hand portion of the figure have been spun-up by mass accretion from a binary companion. From Taylor and Stinebring (1986).

5. Prove that the magnetic vector potential

$$A = \frac{m \times r}{r^3}, \qquad m = \frac{BR^3}{2},$$

represents a dipole magnetic field.

6. A pulsar with a magnetic moment $m = BR^3/2$ is surrounded by a vacuum, and the direction of the pulsar magnetic axis is inclined by an angle θ with respect to the rotation axis. The pulsar emits magnetic dipole radiation because rotation

causes its magnetic moment to change as a function of time. The rate of energy loss from magnetic dipole radiation is

$$\frac{dE}{dt} = -\frac{2}{3c^3} \frac{d^2 m_\perp}{dt^2},$$

with $E = \frac{1}{2}I\omega^2$. Show how m depends on t, and then solve the above equation assuming some initial angular rotation velocity ω_0 at $t = 0$.

7.   Consider a pulsar with a rotation period 0.1 seconds, radius $= 10$ km and polar magnetic field $B = 10^{12}$ gauss. What is the ratio of electrostatic force and gravitational force on an electron implied by Equation (10.114)? Determine the maximum energy to which an electron can be accelerated and maximum current from the polar cap. Calculate the maximum rate of energy loss implied by Equation (10.118) and compare with the power loss from magnetic-dipole radiation calculated in Problem 6. Using synchrotron theory with $r_c = 10^7$ cm, find the electron Lorentz factor γ required to generate 10^3 MHz radiation by means of curvature radiation.

8.   A beam consisting of equal numbers of $\gamma = 10^2$ electrons and positrons is emitted from the magnetic polar cap of a pulsar. The relativistic $e^- - e^+$ beam whose energy-loss rate is the maximum rate of energy loss calculated in Problem 7 is confined to a cone angle $\frac{1}{2}\theta_{max}$, with θ_{max} given by Equation (10.112). Using pulsar physical parameters given in Problem 7 determine the radial distance at which the energy density of the $e^- - e^+$ beam equals the energy density of the dipole magnetic field.

9.   The equation of hydrostatic equilibrium for a magnetic field supported by pressure gradients is

$$\frac{1}{4\pi}(\nabla \times B) \times B = \nabla P.$$

In cylindrical coordinates (r, θ, z), a magnetic field has the components $B = (B_r, B_\theta, B_z)$. We assume axial symmetry, so that $\frac{\partial}{\partial \theta} = 0$, and in addition $\frac{\partial}{\partial z} = 0$. Prove that the Maxwell equation $\nabla \cdot B = 0$ implies that $B_r = 0$. Integrate the equation of hydrostatic equilibrium and show that the pressure P satisfies the equation

$$P = \text{constant} - \frac{B_z^2}{8\pi} - \frac{1}{4\pi} \int_0^R \frac{B_\theta}{r'} \frac{d(r' B_\theta)}{dr'} dr'.$$

10.   Rederive the dispersion relationship for the magnetohydrodynamic waves given in Section 10.3, allowing for the effect of finite but small electric conductivity.

11 White dwarfs, novae and supernovae

11.1 White dwarfs

White dwarfs have low luminosities ($\sim 10^{-4} - 1 L_\odot$) but their photospheric temperatures ($T \simeq 6000 - 2 \times 10^4$ K) are comparable to those of main-sequence stars. It follows that the radii of white dwarfs are small ($\simeq 10^9$ cm). The masses and radii of several white dwarfs in binary systems have been determined. The white dwarfs Sirius B and 40 Eri B are known to have masses and radii equal to ($M = 1.05 M_\odot$, $R = 0.0074 R_\odot$) and ($M = 0.48 M_\odot$, $R = 0.0124 R_\odot$) respectively. Because the central densities of white dwarfs are high ($\simeq 10^6$–10^9 g cm^{-3}) and the temperatures of their isothermal cores relatively low ($\sim 10^7$ K), electrons are completely degenerate except in thin surface layers.

The structures of white dwarfs are determined by the equations of hydrostatic equilibrium (Equation (2.27)), mass conservation (Equation (2.28)) and the equation of state of an electron-degenerate gas described in Section 3.2. Unlike main-sequence stars the radii of white dwarfs decrease as a function of increasing mass and in addition there exists an upper limit to the mass of a white dwarf. This mass limit, which is known as the Chandrasekhar mass limit, depends on the electron molecular weight because the electron pressure increases as the number density of electrons increases. The calculated white dwarf mass limit M_c for uniform electron molecular weight μ_e is

$$M_c = \frac{1.44 M_\odot}{(\mu_e/2)^2}. \tag{11.1}$$

Since most white dwarfs consist primarily of fully ionized ^{4}He, ^{12}C and ^{16}O their electron molecular weight is $\mu_e = 2$. The limiting mass of a white dwarf is reduced to $\simeq 1.4 M_\odot$ because inverse beta-decays already discussed in Chapter 7 remove electrons from the gas when densities exceed $\simeq 10^{10}$ g cm^{-3}. General relativistic effects, which will be described in Chapter 12, become important at central densities $\gtrsim 3 \times 10^{10}$ g cm^{-3}.

The existence of a white dwarf mass limit can be readily understood by means of dimensional analysis. Using the dimensional relations (2.132) and (2.133) and assuming that the gas pressure is caused by a relativistic electron gas we obtain

$$\rho \frac{GM}{R^2} \sim \frac{GM^2}{R^5} \sim \frac{P}{R} \sim K_1 \frac{M^{4/3}}{R^5} \tag{11.2}$$

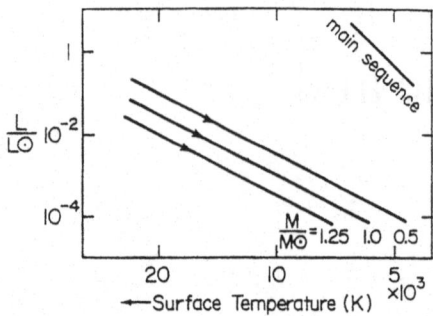

Figure 11.1. The cooling tracks of $0.5M_\odot$, $1.0M_\odot$ and $1.25M_\odot$ white dwarfs. As will be discussed below, crystallization occurs as white dwarfs cool. It takes about 10^{10} years for a white dwarf to reach a luminosity of $\simeq 10^{-4}L_\odot$. At this luminosity, diffusion between different elements such as ^{12}C and ^{16}O may introduce a gravitational energy source that postpones nearly indefinitely further reduction in luminosity.

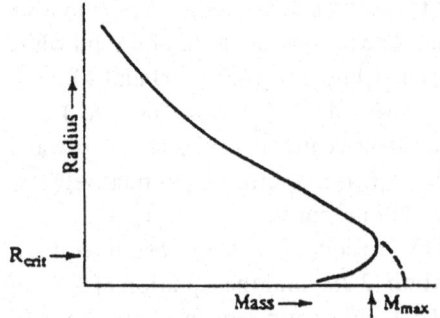

Figure 11.2. The white-dwarf mass–radius relation is shown. The dashed line would apply if inverse beta decay could be neglected. White dwarfs with radii less than R_{crit} do not exist because solutions are unstable.

with K_1 equal to a constant. Because the dependence of the radius R is the same on both sides of relation (11.2) the gravitational term must exceed the pressure gradient term for sufficiently large values of M.

Low mass ($M \lesssim 0.6\,M_\odot$) white dwarfs are polytropes of index $n = 3/2$ because the electron-degenerate gas in their interiors is entirely nonrelativistic, which implies $P = K\rho^{5/3} = K\rho^{1+1/n}$. White dwarfs with mass close to the white dwarf mass limit are polytropes of index $n = 3$ because their electron-degenerate gas is mostly relativistic and therefore $P = K_1\rho^{4/3}$. The white dwarf mass limit can be determined from Equation (2.150) because a massive white dwarf is a polytrope of index $n = 3$ with the equation of state of an ultrarelativistic degenerate electron gas. The cooling tracks of white dwarfs and the mass–radius relation are shown schematically in Figures 11.1 and 11.2 respectively.

Under normal physical conditions an ionized gas will recombine as it cools. The density in the interior of a white dwarf is sufficiently high that the gas remains ionized independent of temperature. The persistence of ionization in a dense, cooling gas is known as pressure ionization. Consider a gas containing nuclei of charge Z. Ions of such nuclei with one bound electron have hydrogenic wave functions, and consequently the most probable distance of an electron from the nucleus is

$$R = Z^{-1/3} a_0 \tag{11.3}$$

with a_0 equal to the Bohr radius. The mean distance between ions is

$$R_i = Z^{1/3} \left(\frac{4\pi n_e}{3}\right)^{-1/3} \tag{11.4}$$

with n_e equal to the number density of electrons. Equations (11.3) and (11.4) imply that

$$\frac{R_i}{R} = Z^{2/3} \left(\frac{4\pi n_e a_0^3}{3}\right)^{-1/3} \lesssim 1 \tag{11.5}$$

is the condition for pressure ionization.

Cooling of white dwarfs

In the interiors of white dwarfs the electrons are strongly degenerate but the ions are nondegenerate. The heat capacity per unit volume c_v of a gas consisting of nondegenerate ions and nonrelativisitic, degenerate electrons is

$$c_v = \frac{3}{2} n_i k + \frac{\pi^2}{2} n_e k \left(\frac{kT}{E_F}\right) \tag{11.6}$$

with E_F equal to the Fermi energy. Equation (11.6), which follows from Equations (2.15) and (3.173), implies that the electrons do not contribute appreciably to the heat capacity in the interior of a white dwarf because the condition for strong degeneracy (i.e. $E_F \gg kT$) is satisfied.

The radii of white dwarfs do not change significantly as they cool, and consequently gravitational contraction does not supply the energy required to maintain their observed luminosities. Thermonuclear energy sources have been exhausted prior to evolution into the white dwarf end state. It follows that the luminosities of white dwarfs are maintained by the thermal energy of ions, and consequently L is given by the expression

$$L = -\frac{d}{dt} \int \int c_v dT dV. \tag{11.7}$$

As the interior of a white dwarf cools the temperature eventually becomes sufficiently low for the ions to crystallize. At relatively high temperatures quantum effects can be neglected and the heat capacity of ions in the crystal is $c_v = 3 n_i k$ with n_i equal to the number density of ions. As will be described below the ionic heat capacity decreases as T^3 in the low-temperature limit. The ionic heat capacity c_v varies from $\frac{3}{2} n_i k$ to $3 n_i k$ as ions pass through a liquid phase prior to crystallization.

Electron conduction is the principal mechanism of energy transport in white dwarfs.

The interiors of white dwarfs are isothermal because the electron thermal conductivity is very high. The temperature change from interior to surface occurs within the thin layer of weakly degenerate and nondegenerate gas that separates the isothermal interior from the photosphere. The mass of this outer layer is much less than the mass of the white dwarf and energy transport is by means of radiation.

The opacity κ of the gas in the outer layer is caused by free–free and bound–free absorption and can be expressed as

$$\kappa = \kappa_0 \rho T^{-3.5} \tag{11.8}$$

with κ_0 equal to the constant that depends on composition.

Equations (2.27), (2.80) and (11.8) can be combined to give the equation

$$\frac{dP}{dT} = \frac{4ac}{3} \frac{4\pi G M T^{6.5}}{\kappa_0 L \rho}. \tag{11.9}$$

Close to the surface the gas is nondegenerate and the pressure is given by Equation (2.6). Eliminating ρ from Equations (2.6) and (11.9) gives

$$PdP = \frac{4ac}{3} \frac{4\pi G M}{\kappa_0 L} \frac{k}{\mu m_p} T^{7.5} dT. \tag{11.10}$$

Since the photospheric temperature of the white dwarf is small as compared to the interior temperature Equation (11.10) can be integrated using the boundary condition $T = 0$ at $P = 0$. Integrating Equation (11.10) and using Equation (2.6) we obtain

$$\rho = \left(\frac{2}{8.5} \frac{4ac}{3} \frac{4\pi G M}{\kappa_0 L} \frac{\mu m_p}{k} \right)^{1/2} T^{3.25}$$

$$= K_1 \left(\frac{M}{L} \right)^{1/2} T^{3.25}. \tag{11.11}$$

Equation (11.11) clearly becomes invalid at densities such that electron degeneracy is important because the equation of state of a nondegenerate gas was used in the derivation. We assume that the boundary density ρ_c of the isothermal core is determined by the condition that the pressure of a nondegenerate electron gas is equal to the pressure of a completely degenerate electron gas with the temperature of the isothermal core. From Equations (2.6), (3.81) and (3.82) this condition becomes

$$\frac{\rho_c k T_c}{\mu m_p} \simeq K \left(\frac{\rho_c}{\mu_e} \right)^{5/3} \tag{11.12}$$

with K equal to a constant, T_c equal to the temperature of the isothermal core and ρ_c equal to the gas density at the core boundary. If we assume that Equation (11.11) is valid when $\rho = \rho_c$ and $T = T_c$ then after eliminating ρ_c from Equations (11.11) and (11.12) we obtain

$$L = K_2 M T_c^{3.5} \tag{11.13}$$

with K_2 equal to a constant determined by Equations (3.81), (3.82), (11.11) and (11.12).

If c_v is equal to the heat capacity per cm^3 in the isothermal core then the total thermal energy of the white dwarf becomes

$$U = \int \frac{c_v T}{\rho} \, dM_r \simeq \frac{\bar{c}_v}{\bar{\rho}} T_c M \tag{11.14}$$

with $\bar{c}_v$ and $\bar{\rho}$ equal to mean values of c_v and ρ. Equations (11.7), (11.13) and (11.14) imply that

$$\frac{-d\left(\frac{\bar{c}_v}{\bar{\rho}} T_c\right)}{dt} = K_2 T_c^{3.5}. \tag{11.15}$$

If $\bar{c}_v$ is assumed to be time independent then the integral of Equation (11.15) from some initial time t_0 is

$$\frac{2}{5} \frac{\bar{c}_v}{\bar{\rho}} (T^{-2.5} - T_0^{-2.5}) = K_2(t - t_0) = K_2 \tau. \tag{11.16}$$

If T satisfies the condition $T \ll T_0$ then Equations (11.13) and (11.16) give

$$\tau = \frac{2}{5} \frac{\bar{c}_v}{\bar{\rho}} \frac{T_c M}{L}. \tag{11.17}$$

Eliminating T_c from Equations (11.13) and (11.17) we find

$$\tau \simeq \frac{2}{5} \frac{\bar{c}_v}{\bar{\rho} K_2^{2/7}} \left(\frac{M}{L}\right)^{5/7}. \tag{11.18}$$

Crystallization and heat capacity of the ionic lattice

In ordinary matter the interaction between electron clouds surrounding nuclei gives rise to crystalline structure. Electrons are not effective in screening nuclei in the very dense gases that exist in the interiors of white dwarfs. As long as the zero point energy of the ions is sufficiently small it is energetically favorable for nuclei to arrange themselves in a regular lattice. In white dwarfs ions are likely to form a body-centered cubic lattice.

We define the ratio

$$\Gamma = \frac{(Ze)^2}{r_i kT} \tag{11.19}$$

with r_i determined by the equation

$$n_i \frac{4\pi r_i^3}{3} = 1. \tag{11.20}$$

The ratio Γ given in Equation (11.19) is the approximate ratio of the Coulomb energy per ion divided by the thermal energy per ion. If Γ given by Equation (11.19) satisfies the inequality $\Gamma \ll 1$ then the Coulomb interaction will not produce a significant correlation between the positions of ions. On the other hand, if Γ satisfies the inequality $\Gamma \gg 1$ the Coulomb interaction will be sufficiently strong for the ions to form a regular lattice.

When ions form a lattice the Hamiltonian describing the vibrations of ions about their equilibrium positions is

$$H = \sum_{i=1}^{3N} \frac{p_i^2}{2M} + V(\xi_1, \xi_2, \ldots, \xi_{3N}) \tag{11.21}$$

where ξ_i $(i = 1, \ldots, 3N)$ are the displacements of ions from their equilibrium positions, $p_i = M\dot{\xi}_i$ are the momenta of the displaced ions and $V(\xi_1, \ldots, \xi_{3N})$ is the potential energy caused by forces between ions. Since the potential V is a minimum when the ions are at their equilibrium positions we have

$$\left(\frac{\partial V}{\partial \xi_i}\right)_0 = 0, \qquad i = 1, \ldots, 3N. \tag{11.22}$$

For small displacements Equations (11.21) and (11.22) imply

$$H = \sum_i \frac{p_i^2}{2M} + \frac{1}{2} \sum_{i,k} V_{ik} \xi_i \xi_k. \tag{11.23}$$

Using Equation (11.23) we find that the equations of motion of the ions are

$$M\ddot{\xi}_i = -\frac{\partial V}{\partial \xi_i} = -\sum_k V_{ik} \xi_k. \tag{11.24}$$

If $V_{ik} = 0$ when $i \neq k$ then Equation (11.24) becomes

$$M\ddot{\xi}_i = -V_{ii}\xi_i \equiv -K\xi_i. \tag{11.25}$$

If solutions to Equation (11.25) have the form $\xi = \xi_0 \cos \omega_0 t$ then

$$\left(\frac{K}{M}\right)^{1/2} = \omega_0 = \left(\frac{4\pi n_i Z^2 e^2}{3M}\right)^{1/2} = \frac{\Omega_p}{3^{1/2}} \tag{11.26}$$

where Ω_p is known as the ion plasma frequency. Each vibrational mode has kT of energy unless the temperature is too low for the mode to be excited. The average kinetic energy of the mode is $\frac{1}{2}kT$ and the average potential energy $\frac{1}{2}kT$. Empirically crystallization occurs when

$$\frac{\xi_i^2}{r_i^2} \lesssim \frac{1}{16}. \tag{11.27}$$

Since $\frac{1}{2}K\langle \xi_i^2 \rangle = \frac{1}{2}kT$ Equation (11.26) gives

$$\langle \xi_i^2 \rangle = \frac{3kT}{M\Omega_p^2}. \tag{11.28}$$

Equation (11.20), inequality (11.27) and Equation (11.28) imply that the condition for crystallization is

$$\frac{\langle \xi_i^2 \rangle}{r_i^2} = \frac{3^{1/3}kT}{M\Omega_p^2}(4\pi n_i)^{2/3}$$

$$= \frac{kTr_i}{Z^2e^2} \lesssim \frac{1}{16}. \tag{11.29}$$

The heat of crystallization is approximately kT per ion and therefore it adds significantly to the cooling time of a white dwarf.

We consider a cubic region of the crystal that contains N ions and has dimension L. We assume plane-wave solutions of the form $\exp(i\mathbf{k} \cdot \mathbf{r})$ and impose periodic boundary conditions. As in our discussion of solutions of the Dirac equation in free space given in Section 3.2 plane-wave solutions $\exp(i\mathbf{k} \cdot \mathbf{r})$ must satisfy the conditions

$$k_x L = 2\pi n_x$$
$$k_y L = 2\pi n_y$$
$$k_z L = 2\pi n_z \tag{11.30}$$

with n_x, n_y and n_z equal to positive integers. Since the dimension L is chosen large as compared to the spacing between ions the number of modes with wave vector less than k is

$$\frac{\frac{4\pi}{3}k^3}{\left(\frac{2\pi}{L}\right)^3}. \tag{11.31}$$

In a solid there are three polarizations (two transverse and one longitudinal). Expression (11.31) implies that the number of modes per unit volume is

$$\frac{k^3}{2\pi^2}. \tag{11.32}$$

Therefore the number of modes of a particular polarization with wave vector between k and $k + dk$ becomes

$$n_1(k)dk = \frac{1}{3}d\left(\frac{k^3}{2\pi^2}\right) = \frac{1}{2\pi^2}k^2 dk. \tag{11.33}$$

The number of such modes with angular frequency in the interval ω, $\omega + d\omega$ can be expressed as

$$g_1(\omega)d\omega = n_1(k)\left(\frac{dk}{d\omega}\right)_1 d\omega \tag{11.34}$$

with $(d\omega/dk)_1$ equal to the sound velocity of a particular mode. The sound velocities of the two transverse polarizations are equal and therefore from Equations (11.33) and (11.34) the total number of modes with ω in the interval ω, $\omega + d\omega$ becomes

$$g(\omega) = \frac{\omega^2}{2\pi^2}\left(\frac{1}{v_l^3} + \frac{2}{v_t^3}\right) \tag{11.35}$$

with v_l and v_t equal to the longitudinal and transverse sound velocities respectively. If for simplicity we assume that v_l and v_t are independent of frequency and $v_l = v_t$ then Equation (11.35) reduces to the equation

$$g(\omega) = \frac{3\omega^2}{2\pi^2 v_s^3} \tag{11.36}$$

with v_s equal to the sound velocity.

The total number of modes per unit volume is $3n$ with n equal to the number of ions per unit volume. Therefore, using Equation (11.32) and recalling that there are three different polarizations it follows that the maximum wave vector k_{max} is

$$k_{max} = (6\pi^2 n)^{1/3} \tag{11.37}$$

and the high frequency cutoff occurs at

$$\omega_{max} = v_s(6\pi^2 n)^{1/3}. \tag{11.38}$$

The energy of a particular mode (phonon) is $\hbar\omega$ and the relative probability that a mode of energy $\hbar\omega$ is present follows from the same equation used for electromagnetic waves under equilibrium conditions (i.e. Equation (3.68)). Equations (3.68) and (11.36) imply that the wave energy per unit volume is

$$U = \frac{3\hbar}{2\pi^2 v_s^3}\int_0^{\omega_{max}} \frac{\omega^3}{e^{\hbar\omega/kT} - 1}\,d\omega. \tag{11.39}$$

Defining the dimensionless parameter $x = \hbar\omega/kT$ Equation (11.39) becomes

$$U = \frac{3k^4 T^4}{2\pi^2 \hbar^3 v_s^3}\int_0^{x_{max}} \frac{x^3 dx}{e^x - 1} \tag{11.40}$$

where $x_{max} = \hbar\omega_{max}/kT \equiv \theta/T$. The Debye temperature θ can be expressed as

$$\theta = (\hbar v_s/k)(6\pi^2 n)^{1/3}. \tag{11.41}$$

Differentiating U given in Equation (11.39) with respect to T we obtain

$$c_v = 9nk(T/\theta)^3 \int_0^{x_{max}} \frac{e^x x^4 dx}{(e^x - 1)^2}. \tag{11.42}$$

At high temperatures the variable x in Equation (11.42) is small within the range of integration and therefore approximating e^x as $e^x \simeq 1 + x$ the heat capacity c_v becomes

$$c_v \simeq 9nk(T/\theta)^3 \int_0^{x_{max}} x^2 dx = 3nk. \tag{11.43}$$

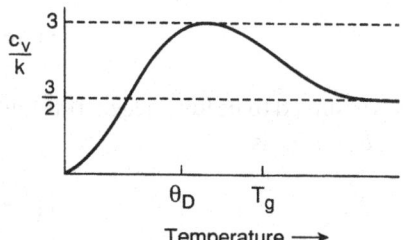

Figure 11.3. The specific heat capacity, c_v, of ions divided by the Boltzmann constant, k. T_g is the temperature above which vibrational energy becomes sufficiently high to dissolve the crystal. At temperatures $\ll \theta_D$ heat capacity c_v varies as T^3.

Equation (11.43) shows that at temperatures $T \gg \theta$, c_v approaches its classical value. At low temperatures (i.e. $T \ll \theta$) we can approximate Equation (11.43) by assuming $x_{max} = \infty$. Performing a partial integration on the integral given in Equation (11.42) we have

$$\int_0^\infty \frac{e^x x^4 dx}{(e^x - 1)^2} = 4 \int_0^\infty \frac{x^3 dx}{e^x - 1} = \frac{4\pi^4}{15}. \tag{11.44}$$

Therefore, in the low-temperature limit Equation (11.42) is

$$c_v = \frac{12\pi^4}{5} nk \left(\frac{T}{\theta}\right)^3 \qquad (T \ll \theta). \tag{11.45}$$

Equation (11.45) implies that the heat capacity of the ion lattice approaches zero at low temperature.

In our calculation of heat capacity we have not distinguished between longitudinal and transverse modes. At low values of ω the dispersion relation of longitudinal modes is such that $\omega = v_l k$ with v_l very large. Since low-frequency modes are dominant at low temperatures it follows from Equation (11.35) that longitudinal modes do not contribute to the heat capacity at low temperatures. In Figure 11.3 ionic heat capacity is shown schematically as a function of temperature.

Coulomb corrections to the equation of state

In Section 3.6 we derived an expression for the Coulomb correction to a degenerate electron gas assuming that the electrons are uniformly distributed about ions. The Thomas–Fermi method is a more accurate method of calculating Coulomb corrections in an electron-degenerate gas consisting of free electrons and ions because it takes into account the nonuniformity of electron charge distributions about ions. A Wigner–Seitz cell is a spherical region about an ion of volume $1/n_i$ with n_i equal to the number density of ions. Let $V(r)$ be the potential in each Wigner–Seitz cell. In the Thomas–Fermi method it is assumed that the Fermi energy E_F of an electron satisfies the condition

$$E_F = -eV(r) + \frac{p_F^2}{2m} = \text{constant} \tag{11.46}$$

throughout each Wigner–Seitz cell. The electron gas is assumed to be fully degenerate and therefore from Equation (3.81) the number density of electrons is

$$n_e = \frac{8\pi}{3h^3} p_F^3 = \frac{8\pi}{3h^3} [2m(E_F + eV(r))]^{3/2}. \tag{11.47}$$

The potential $V(r)$ surrounding an ion is determined by the equation

$$\nabla^2 V = \frac{1}{r^2} \frac{d}{dr} r^2 \frac{d}{dr} V(r) = 4\pi n_e e, \qquad r \neq 0. \tag{11.48}$$

Equation (11.48) has the boundary conditions

$$\lim_{r \to 0} r V(r) = Ze \tag{11.49}$$

and

$$\left. \frac{dV}{dr} \right|_{r_0} = 0 \tag{11.50}$$

with r_0 equal to the boundary radius of the Wigner–Seitz cell.

Equations (11.47) and (11.48) imply that

$$\frac{1}{r^2} \frac{d}{dr} r^2 \frac{d}{dr} V(r) = \frac{32\pi^2 e}{3h^3} [2m(E_F + eV(r))]^{3/2}. \tag{11.51}$$

Making the variable change $r = \mu x$ with

$$\mu = \left(\frac{9\pi^2}{128Z} \right)^{1/3} \frac{\hbar^2}{me^2} = \left(\frac{9\pi^2}{128Z} \right)^{1/3} a_0 \tag{11.52}$$

gives

$$E_F + eV(r) = \frac{Ze^2}{r} \phi(x). \tag{11.53}$$

Equations (11.51–11.53) lead to the equation

$$\frac{d^2\phi}{dx^2} = \frac{\phi^{3/2}}{x^{1/2}}. \tag{11.54}$$

Equations (11.49) and (11.53) imply that

$$\phi(0) = 1. \tag{11.55}$$

At the boundary of the Wigner–Seitz cell $x = x_0 = r_0/\mu$. Taking the derivative of Equation (11.53) and using Equation (11.50) we obtain

$$\phi'(x_0) = \frac{\phi(x_0)}{x_0}. \tag{11.56}$$

Equations (11.55) and (11.56) are the two boundary conditions required to solve Equation (11.54).

From Equation (3.82) the pressure at $r = r_0$ becomes

$$P = \frac{8\pi}{15h^3 m} p_F^5(r_0). \tag{11.57}$$

Equations (11.53) and (11.56) imply that the boundary pressure is

$$P = \frac{1}{10\pi} \frac{Z^2 e^2}{\mu^4} \left(\frac{\phi(x_0)}{x_0}\right)^{5/2}. \tag{11.58}$$

From Equation (11.58) it follows that the pressure P is determined by solving Equation (11.54) numerically for $\phi(x_0)$. The average mass density within a Wigner–Seitz cell is

$$\rho_0 = \frac{A m_u}{4\pi \mu^3 x_0^3 / 3} \tag{11.59}$$

where A is the number of protons and neutrons within a nucleus.

It is of interest to show how the pressure at the Wigner–Seitz boundary varies as a function of density at relatively low densities. At low densities we can assume $x \to \infty$ except in the vicinity of $x = 0$. We assume that the low density asymptotic solution of Equation (11.54) is

$$\phi(x) \sim \frac{C}{x^3}, \tag{11.60}$$

with C equal to a constant. Substituting into Equation (11.54) we find that $\phi(x)$ given in Equation (11.60) is an asymptotic solution of Equation (11.54) if C is set equal to 144. This asymptotic solution does not satisfy the inner boundary condition of Equation (11.54). Equations (11.58–11.60) imply that the outer boundary (i.e. $x = x_0$) pressure satisfies the proportionality

$$P \propto \frac{1}{x_0^{10}} \propto \rho^{10/3}. \tag{11.61}$$

11.2 Novae

Nova outbursts occur in white dwarfs that are binary stars. It is clear that mass transfer between the companion star and white dwarf must occur before the outburst. Most binary stars in which novae have occurred are close binary systems with orbital periods as low as $\simeq 4$ hours. In such binary stars mass transfer is the result of Roche lobe overflow from the larger but often less massive companion. The inflowing mass has a high angular momentum to mass ratio. Therefore, an accretion disk forms about the white dwarf. Mass accretion onto the white dwarf occurs as viscous forces cause mass within the accretion disk to spiral onto the surface of the white dwarf.

The light curves of novae show an initial rise that is quite rapid (~ 1 day) and more gradual decline in brightness. Some nova light curves show a pre-maximum halt that is followed by a $\simeq 1$–2 magnitude final rise to maximum light. The early decline, which is defined as the time interval from maximum light to $\simeq 3$–4 magnitudes below maximum light, occurs over a period of weeks. Early decline is followed by a much more gradual state of final decline that lasts for $\simeq 1$–10 years. Some nova light curves show large variations in brightness after the stage of early decline while others decline in brightness without appreciable variability.

The pre-maximum nova spectrum is characterized by broad absorption lines that change to a later spectral type (A or F supergiant) as the nova passes through maximum light. Strong, symmetrically broadened emission lines and blue shifted absorption lines appear shortly after light maximum. As decline from maximum light proceeds emission lines appear that require high excitation temperatures. These emission lines are caused by ultraviolet radiation from the central white dwarf, which is still quite luminous. Intense infrared continuum radiation is observed from some novae. The phase of maximum infrared brightness arises after maximum light. The observed infrared continuum radiation shows that physical conditions within the ejected mass shell are such that dust formation has occurred.

Approximately 50 novae occur each year within a massive spiral galaxy such as the Milky Way, and consequently the frequency of nova outbursts is appreciably greater than the white-dwarf formation rate. The spatial distribution of novae within the Galaxy shows concentration toward the Galactic plane and Galactic center. The distances to relatively nearby Galactic novae can be estimated by measuring Doppler shifts of spectral lines and the angular rate of expansion of the ejected mass shell.

The observed expansion velocities of nova shells are typically 500–2000 km s^{-1}. Since their masses are approximately 10^{-4}–$10^{-6} M_\odot$ the kinetic energies of nova shells are $\simeq 10^{43}$–10^{44} erg. Because the masses of nova shells are small many outbursts ($\sim 10^4$–10^5) can arise as the result of mass accretion in a single binary system. It follows that only about 10^7–10^8 close binary stars are required to explain the approximately 10^{11}–10^{12} nova outbursts that have probably occurred over the $\simeq 12$–15×10^9 year lifetime of the Milky Way.

The luminosities of novae at maximum light are approximately 10^4–$3 \times 10^5 L_\odot$. Novae with the highest luminosities at maximum light tend to decline in brightness more rapidly than those with lower maximum luminosities. This circumstance implies that novae can be used as distance indicators even when their distances are too large for changes in angular diameter to be measured. The total radiated energy during a nova outburst is $\simeq 10^{45}$–10^{46} erg. Therefore, energy in the form of radiation exceeds kinetic energy by a factor of $\sim 10^2$. On the other hand, the kinetic energies of supernova shells ($\sim 10^{50}$–10^{51} erg) are much greater than radiated photon energy ($\simeq 10^{48}$–10^{49} erg) during a supernova outburst.

Low-mass stars burn hydrogen quiescently as they evolve into white dwarfs. However, a thermonuclear runaway will occur if a star evolves into the white-dwarf end state and then accretes hydrogen-rich gas from a binary companion. Consider a white-dwarf with a thin ($\lesssim 10^{-4} M_\odot$) outer envelope of hydrogen-rich gas. The temperature and density within the hydrogen-rich envelope are too low for hydrogen burning to occur. If the white dwarf accretes mass gradually from a binary companion the envelope of the white dwarf will contract and the gas temperature increase. Mass-accretion rates of $\simeq 10^{-9}$– $10^{-10} M_\odot$/yr are probably typical of those present in the low-mass binary stars in which nova outbursts occur.

As mass accretion continues the temperature and/or density at the base of the hydrogen-rich envelope eventually becomes sufficiently high for hydrogen burning to begin. If gas densities are relatively high ($\simeq 10^4 \, \mathrm{g \, cm^{-3}}$) energy generation via the proton–proton chains may become significant before the CNO cycle becomes the dominant energy source. The CNO cycle is highly temperature sensitive and the hydrogen-burning shell becomes thermally unstable. The e-folding timescale τ_K for the resultant thermonuclear runaway is approximately equal to the local Kelvin timescale:

$$\tau_K \simeq \frac{\left(\frac{3P}{2\rho}\right)}{\varepsilon_{\mathrm{CNO}}}. \tag{11.62}$$

If τ_K in Equation (11.62) were to become less than the sound-travel time across the hydrogen-rich envelope of the white dwarf then a strong shock wave would result. This strong shock wave would propagate down a steep density gradient to the surface of the white dwarf and eject an appreciable fraction of the hydrogen-rich envelope. Since the relevant sound travel time is ~ 2 seconds and the temperature within the hydrogen-burning shell becomes $\simeq 10^8 \, \mathrm{K}$ during a thermonuclear runaway it follows from Equation (11.62) that $\varepsilon_{\mathrm{CNO}}$ must become $\gtrsim 10^{16} \, \mathrm{erg \, g^{-1} \, s^{-1}}$ for a strong shock wave to form. During quiescent hydrogen-burning stages the β-decay lifetimes of ^{13}N and ^{15}O are much shorter than thermonuclear reaction timescales required to complete the CNO cycle. However, at temperatures $\gtrsim 10^8 \, \mathrm{K}$ hydrogen-burning rates are sufficiently high that β-decays limit the rate at which the CNO cycle can go to completion.

The rate of thermonuclear energy generation $\varepsilon_{\mathrm{CNO}}$ cannot exceed about $10^{14} \, \mathrm{erg \, g^{-1} \, s^{-1}}$ if the CNO cycle is described by reactions (2.124) and C and N abundances are approximately equal to solar abundances. At temperature $T \gtrsim 10^8 \, \mathrm{K}$ the ^{13}N$(p, \gamma)^{14}$O reaction is faster than the ^{13}N decay time and consequently the maximum value of $\varepsilon_{\mathrm{CNO}}$ can be increased by a factor of $\simeq 8$ because the ^{14}O beta-decay to ^{14}N occurs approximately 8 times faster than the ^{13}N beta-decay to ^{13}C. It follows that ^{12}C and/or ^{14}N abundances $\simeq 12$ times greater than solar abundances could cause the rate of energy generation $\varepsilon_{\mathrm{CNO}}$ to become $\simeq 10^{16} \, \mathrm{erg \, g^{-1} \, s^{-1}}$. Observations of nova shells provide evidence that their ^{14}N abundances are appreciably greater than the solar abundance. High ^{14}N abundances are undoubtedly the result of the CNO cycle. The interiors of white dwarfs that produce nova outbursts are mostly ^{12}C and ^{16}O. Shear turbulence and rotational mixing are two physical mechanisms that might cause mixing between the core and hydrogen-rich outer layers of a pre-nova star. Measured CNO abundances of nova mass shells appear to demand that such mixing occur.

Other physical factors besides abundances limit the maximum thermonuclear burning rates attained during a thermonuclear runaway. Even if ^{12}C abundances are appreciably enhanced it is unlikely that the strong shock wave condition described above is exceeded during a nova outburst. Hydrodynamic calculations show that a nova shell ejected by a strong shock wave could have an outflow velocity of $\simeq 4000\,\mathrm{km\,s^{-1}}$. The initial outflow velocity of Nova Cygni 1992 was sufficiently high ($4500\,\mathrm{km\,s^{-1}}$) to indicate that mass ejection was caused by a strong shock wave. However, such high outflow velocities are rare. Rapid mass accretion causes the surface and outer envelope of a white dwarf to expand. Under conditions of rapid mass accretion the thermonuclear runaway in a hydrogen-burning shell will not lead to hydrodynamic mass ejection. On the other hand, if rates of mass accretion are sufficiently low ($\lesssim 10^{-9}\,M_{\odot}/\mathrm{yr}$) thermonuclear runaways can become sufficiently strong to cause the ejection of a nova shell.

A convection zone extends outwards from the hydrogen-burning shell as the thermonuclear runaway in a hydrogen-burning shell develops. The outward movement of this convective region continues until it extends nearly to the photosphere of the white dwarf. Luminosities in excess of $10^{6}L_{\odot}$ are reached within the hydrogen-burning shell and convective velocities become comparable to sonic velocities. The isotopes ^{13}N and ^{15}O, which are synthesized in the CNO cycle, mix throughout the convective envelope before undergoing beta-decay. Energy deposited throughout the convective envelope as the result of beta-decays plays an important role in causing the ejection of a nova mass shell. At high temperatures ($T \geq 10^{8}$ K) beta-decay of ^{14}O is also important.

As discussed in Section 10.4, dynamo magnetic fields can be generated in the presence of rotation and convection. Dynamo magnetic fields are likely to be produced prior to a nova outburst in the convective zone that extends outwards from the hydrogen-burning shell. Magnetic fields of $\simeq 10^{6}$–10^{8} gauss can be generated. Although the magnetic pressure created by such magnetic fields is small as compared to gas pressure at the base of the convective zone, steep density and pressure gradients exist, and consequently buoyancy of magnetic fields or diffusion of magnetic flux can cause magnetic field strengths in the outer layers of the convective zone to become sufficiently intense for magnetic pressure to become $\gtrsim$ gas pressure. Under such physical conditions convection is suppressed. DQ Her is an example of a post nova star that is known to have a strong magnetic field (i.e. $\sim 10^{6}$ gauss). The suppression of convection in white dwarfs undergoing thermonuclear runaways could lead to hydrodynamic expansion and resultant mass ejection because the sum of the radiative and convective luminosities exceeds the Eddington limit given in Equations (2.138) or (5.125).

There is observational evidence that high-velocity ($v \simeq 3000\,\mathrm{km\,s^{-1}}$) stellar winds are emitted from post-nova white dwarfs. Such outflow velocities are comparable to escape velocities. Physically similar stellar winds have been observed from the central stars of planetary nebulae. Stellar winds are probably produced because the luminosity of the post-nova star is comparable to the Eddington limit. Strong surface magnetic fields and rotation may play a significant role in generating post-nova stellar winds. The mass

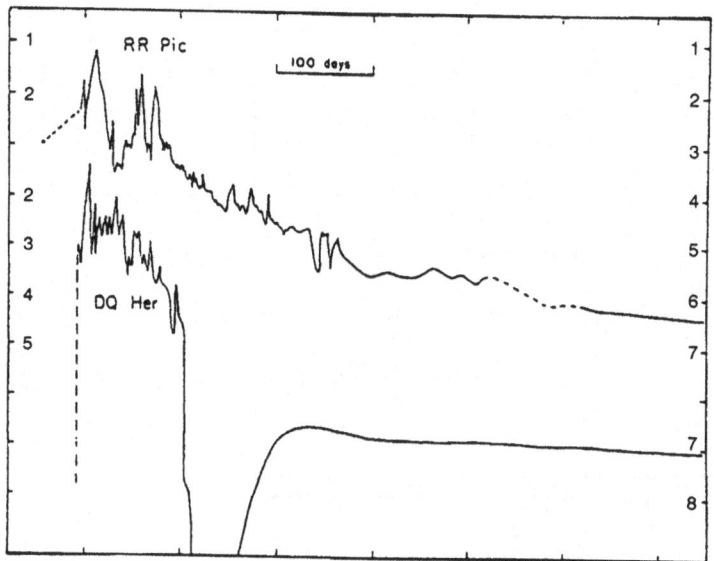

Figure 11.4. Light curves of two well-known novae, DQ Her 1934 and RR Pic 1925. The dip in optical radiation from DQ Her was probably not a reduction in power output, but due to dust formation causing re-radiation at infrared wavelengths. The ejecta of DQ Her has been shown to have a [N/He] ratio that is very high compared to the solar value and also high [O/He] and [C/He] ratios. RR Pic is known to be overabundant in N and He.

accreted onto a white dwarf has a high angular momentum to mass ratio, and consequently will cause the white dwarf to spin-up. As discussed in Section 10.1 stellar winds in the presence of strong magnetic fields can cause appreciable angular momentum loss because the corotation radius can be much greater than the stellar radius.

The light curves of two well-known novae, namely DQ Her 1934 and RR Pic 1925, are shown in Figure 11.4. Novae can be used as distance indicators even when they are too distant for rates of angular expansion of ejected mass shells to be measured because on average brighter novae decline from maximum light more rapidly than intrinsically fainter novae. This effect is shown in Figure 11.5 for novae in Milky Way and M31. Figure 11.6 shows how the $^{13}N(p, \gamma)^{14}O$ reaction bypasses ^{13}C at high temperatures and makes higher rates of nuclear energy via the CNO cycle possible.

11.3 Supernovae

Supernovae are exploding stars with luminosities at maximum light that are $\simeq 10^9$–$10^{10} L_\odot$. The rise to maximum light takes several weeks. Supernova shells have outflow velocities that are $\simeq 5$–$10 \times 10^3 \, \mathrm{km \, s^{-1}}$. Their masses are typically 1–$10 \, M_\odot$. The two major types of supernovae are known as Type I and Type II supernovae. In massive spiral galaxies the formation rate of supernovae is $\simeq 1$ Type I supernovae every 100 years

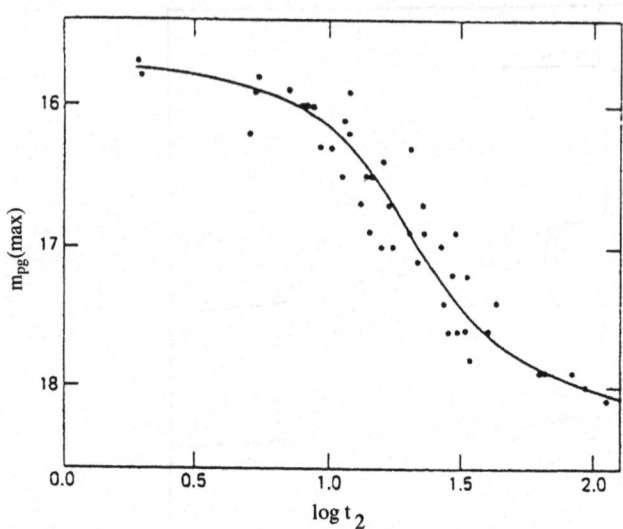

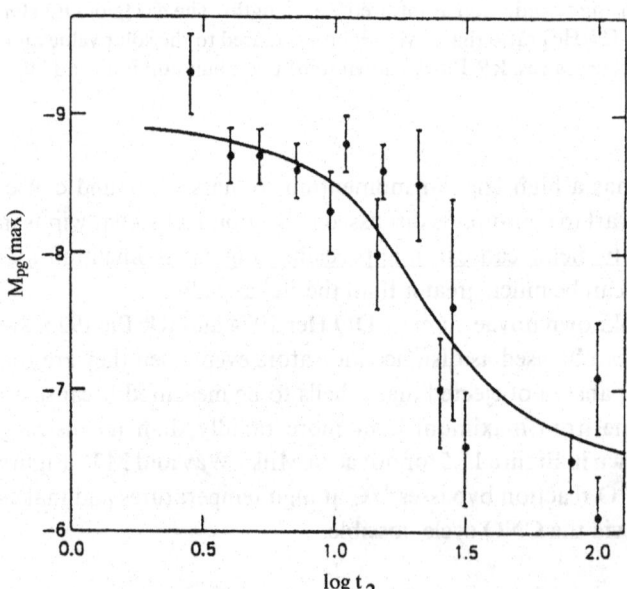

Figure 11.5. (Top) The photographic magnitude at maximum light m_{pq}(max) versus rate of decline $\log t_2$ for novae in M31. t_2 is the time in days for brightness to decline by 2 magnitudes. (Bottom) Absolute photographic magnitude at maximum light M_{pq}(max) and rate of decline $\log t_2$ for Galactic novae. The smooth curve is from the M31 observations shown above. From van den Bergh (1975).

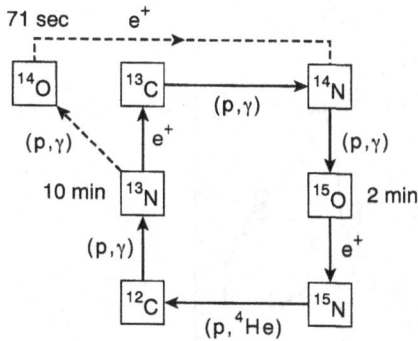

Figure 11.6. At high temperatures (i.e. $T > 10^8$ K) the isotopic ^{13}C is bypassed by the ^{13}N(p,γ)^{14}O reaction. Half-lives of ^{13}N, ^{14}O and ^{15}O are given.

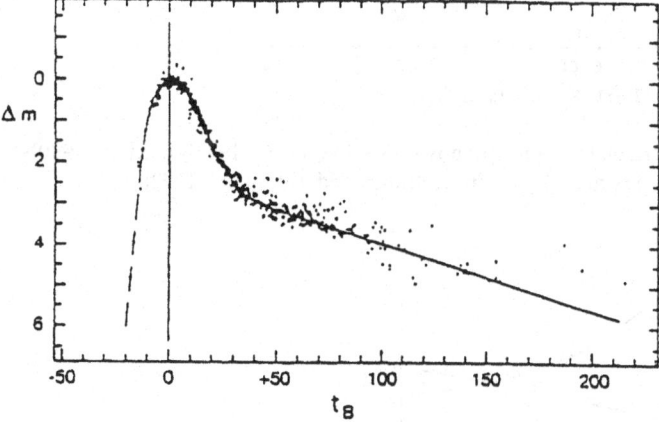

Figure 11.7. The *B* (blue filter) magnitude light curve for 22 Type Ia supernovae. From Branch and Tammann (1992).

and about one Type II supernova every 30 years. Type II supernovae occur in massive stars and are observed in the spiral arms of spiral galaxies and in irregular galaxies. As already noted in Chapter 7 a Type II supernova has recently (Feb 1987) been observed in the Large Magellanic Cloud.

Light curves as measured through the *B* (i.e. blue) filter are shown in Figure 11.7 for 22 Type Ia supernovae. Figure 11.8 shows the spectra of three Type Ia supernovae as they appeared about one week after maximum light. Because they are fairly similar and very bright Type Ia supernovae are important indicators of distances to galaxies. The early light curves of SN 1993J and SN 1987A, which occurred in the galaxies M81 and Large Magellanic Cloud respectively, are given in Figure 11.9. The initial spectrum of SN 1993J had hydrogen lines and therefore was that of a Type II supernova. However, as shown in Figure 11.10 its hydrogen lines disappeared after about two months. The infrared spectrum of SN 1987A shown in Figure 11.11 provides evidence for heavy element synthesis. Figure 11.12 gives measured γ-ray line fluxes from the SN 1987A ejecta.

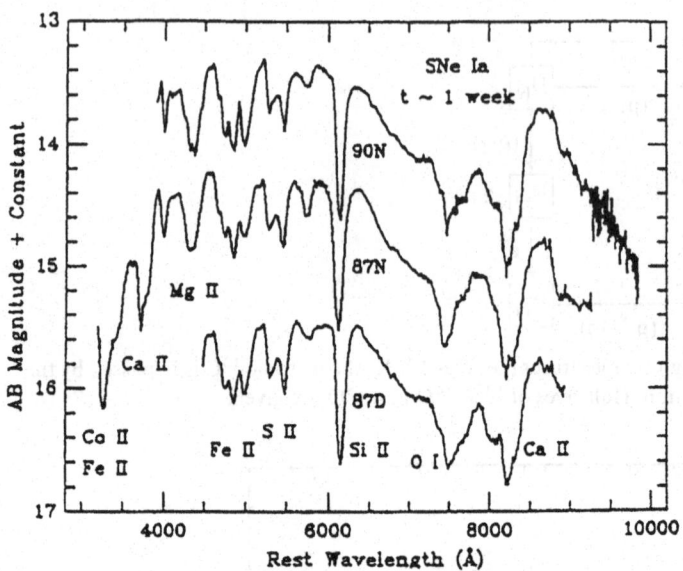

Figure 11.8. The spectra of Type Ia supernovae (SN 1987D, SN 1987N, and SN 1990N) about one week after maximum light. From Branch and Tammann (1992).

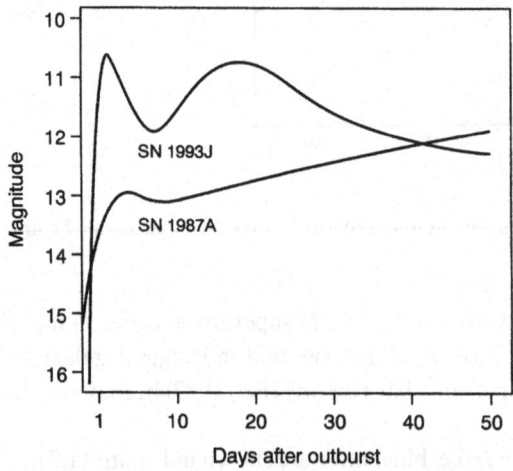

Figure 11.9. The light curves of SN 1993J and SN 1987A. SN 1993J occurred in the spiral galaxy M81 whereas SN 1987A exploded in the Large Magellanic Cloud. From Marschall (1994).

Type II supernova outbursts occur when the core of a massive star implodes causing the envelope of the star to be ejected at high velocities. Type II supernovae are less luminous at maximum light than Type I supernovae. Highly broadened spectral lines of atomic hydrogen are prominent in the spectra of Type II supernovae. Such spectral lines are by definition absent from the spectra of Type I supernovae.

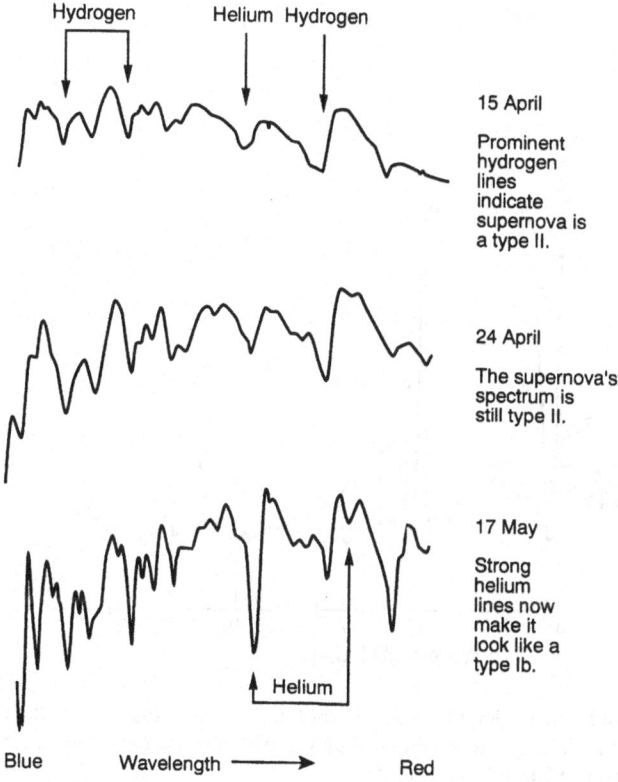

Figure 11.10. The spectrum of SN 1993J on three dates. Its initial spectrum was that of a Type II supernova because it had hydrogen lines. However, after about two months its hydrogen lines had dissappeared and its strong helium lines made it appear as a Type Ib supernova. Apparently, the progenitor star was a massive star that lost most but not all of its hydrogen-rich envelope before exploding as a supernova. From Marschall (1994).

Type I supernovae are known to occur in both spiral and elliptical galaxies. In spiral galaxies they are not generally observed in spiral arms. Recently a number of Type I supernovae have been discovered in the vicinity of HII regions and spiral arms. Such supernovae are referred to as Type Ib supernovae whereas classical Type I supernovae are referred to as Type Ia supernovae.

Type Ib supernovae are caused either by explosive carbon burning or core collapse in hydrogen-deficient stars such as Wolf–Rayet stars. Mass transfer in a binary system containing a white dwarf can cause a Type Ia supernova if mass transfer is sufficiently rapid that nova outbursts do not eject accreted mass, and consequently the white dwarf mass limit is exceeded. Mass accretion as a triggering mechanism for Type Ia supernovae explains their occurrence in elliptical galaxies, which do not contain massive or inter-mediate-mass stars.

Supernova explosions (Type I or Type II) cause a strong shock wave to form and propagate to the surface of the star. As the shock front propagates outwards it heats the

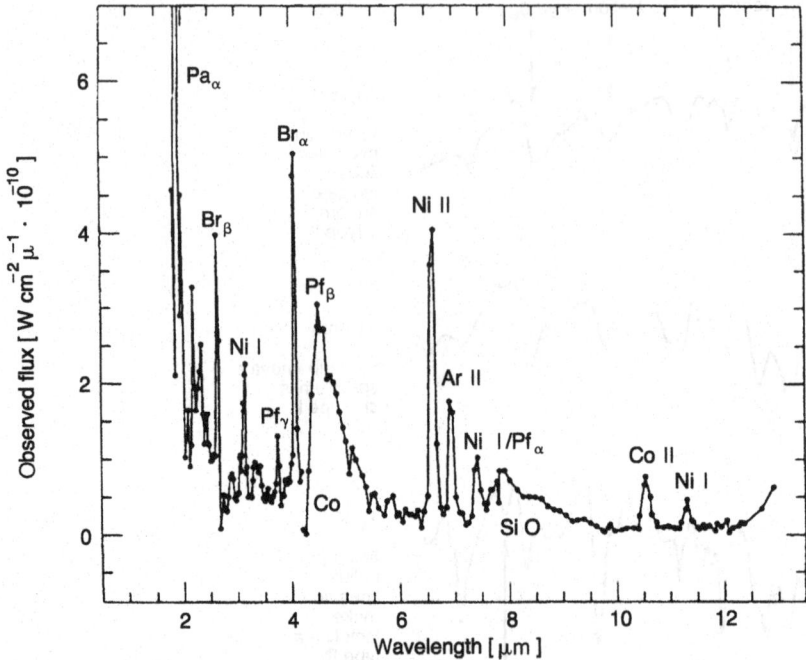

Figure 11.11. The mid-April 1988 infrared spectrum of SN 1987A taken from the Kuiper Airborne Observatory. The spectrum shows that many heavy elements were synthesized in the explosion. From Witteborn *et al.* (1989).

surrounding gas causing the stellar envelope (Type II supernova) or entire star (Type Ia supernova) to undergo rapid hydrodynamic expansion. The rise to maximum light occurs as the photosphere of the supernova shell expands and cools. Maximum light occurs when the photon diffusion timescale, $R\tau/c$ with τ equal to the optical depth and R equal to the radius of the shell, becomes approximately equal to the expansion timescale R/v.

A supernova outburst occurs because a large amount of energy E has been rapidly released within the core of a star. Insight into the nature of supernova outbursts is provided by self-similar blast wave solutions. We assume that a large energy release E has occurred at the center of a spherically symmetric star with uniform density ρ_0. The only dimensionless combination of E and ρ_0 is

$$\xi = r\left(\frac{\rho_0}{Et^2}\right)^{1/5}. \tag{11.63}$$

From Equation (11.63) the propagation distance of the strong shock front becomes

$$R(t) = \xi_0\left(\frac{E}{\rho_0}\right)^{1/5} t^{2/5} \tag{11.64}$$

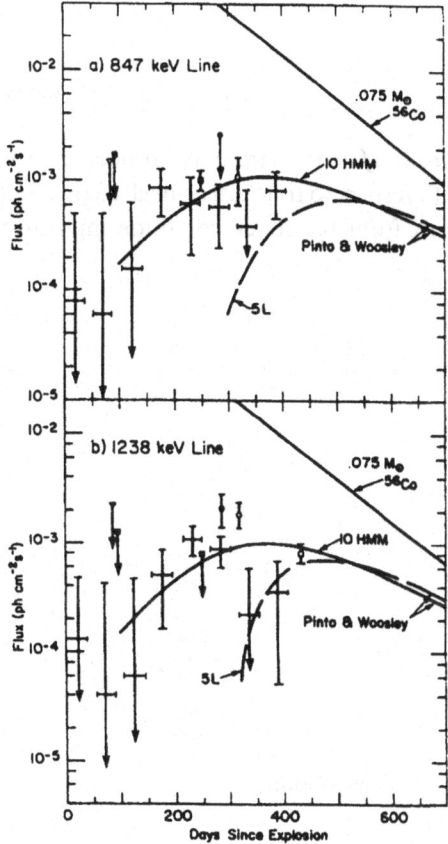

Figure 11.12. Measurements of 847 and 1238 keV gamma-ray line fluxes from ^{56}Co decay within the SN 1987A ejecta. From Gehrels, Leventhal and MacCallum (1988).

with ξ_0 equal to a constant. From Equation (11.64) the propagation velocity of the shock front is

$$\frac{dR(t)}{dt} = \frac{2}{5}\frac{R}{t} = \xi_0\frac{2}{5}\left(\frac{E}{\rho_0}\right)^{1/5}t^{-3/5}. \tag{11.65}$$

The values of physical variables on the upstream and downstream boundaries of a shock front are determined by the equations of motion, continuity and energy. Using the equation of continuity given by Equation (2.41) and setting the partial derivative of ρ with respect to time equal to zero we obtain the condition

$$\rho_0 v_0 = \rho_1 v_1 \tag{11.66}$$

where ρ_0 and v_0 (ρ_1 and v_1) are the upstream (downstream) values of gas density and velocity respectively.

The equations of motion can be expressed as

$$\rho\frac{d\boldsymbol{v}}{dt} = \rho\left(\frac{\partial\boldsymbol{v}}{\partial t} + \boldsymbol{v}\cdot\nabla\boldsymbol{v}\right) = -\nabla P + f \tag{11.67}$$

where P includes both gas and radiation pressure and f denotes the gravitational force per unit volume. Since the partial derivative of $\boldsymbol{v}$ with respect to time can be neglected and the gravitational force is continuous across a shock front the shock condition implied by Equation (11.67) becomes

$$P_0 + \rho_0 v_0^2 = P_1 + \rho_1 v_1^2. \tag{11.68}$$

The energy equation of gas and radiation is

$$\rho\left(\frac{dU}{dt} + P\frac{d\left(\frac{1}{\rho}\right)}{dt}\right) = -\nabla\cdot\boldsymbol{F} \tag{11.69}$$

with the radiative flux

$$F = -\frac{4acT^3}{3\kappa\rho}\frac{dT}{dr},$$

$$U = \frac{3P}{2\rho} + \frac{aT^4}{\rho} \text{ and } P = \frac{\rho kT}{\mu m_p} + \frac{aT^4}{3}.$$

Taking the scalar product of $\boldsymbol{v}$ and Equation (11.67) we obtain

$$\rho\left[\frac{d\left(\frac{1}{2}v^2\right)}{dt} + \boldsymbol{v}\cdot\nabla P\right] = \boldsymbol{v}\cdot f. \tag{11.70}$$

Adding Equations (11.69) and (11.70) gives

$$\rho\frac{d\left(\frac{1}{2}v^2 + U\right)}{dt} + \nabla\cdot(P\boldsymbol{v}) = \boldsymbol{v}\cdot f - \nabla\cdot\boldsymbol{F}. \tag{11.71}$$

Equation (11.71) can be rewritten as

$$\frac{\partial}{\partial t}\left(\frac{1}{2}\rho v^2 + \rho U\right) + \nabla\cdot\left[\left(\frac{1}{2}\rho v^2 + \rho U + P\right)\boldsymbol{v} + \boldsymbol{F}\right] = \boldsymbol{v}\cdot f. \tag{11.72}$$

From Equations (11.66) and (11.72) the third shock condition becomes

$$\frac{v_0^2}{2} + U_0 + \frac{P_0}{\rho_0} + \frac{F_0}{\rho_0 v_0} = \frac{v_1^2}{2} + U_1 + \frac{P_1}{\rho_1} + \frac{F_1}{\rho_0 v_0}. \tag{11.73}$$

In the reference frame moving with the shock the upstream gas is supersonic. Equations (11.66), (11.68) and (11.73) give the required relations between upstream and downstream physical variables across a shock front.

Since the gravitational binding energy of a stellar envelope is small as compared to the total energy release, the equations describing the spherical outflow are

$$\rho \frac{d^2r}{dt^2} = -\frac{dP}{dr} \tag{11.74}$$

$$\frac{\partial \rho}{\partial t} + \frac{1}{r^2}\frac{\partial}{\partial r}\left(r^2\rho\frac{dr}{dt}\right) = 0 \tag{11.75}$$

$$\frac{\partial s}{\partial t} + v\frac{\partial s}{\partial r} = 0 \tag{11.76}$$

$$P = \frac{\rho kT}{\mu m_n} + \frac{1}{3}aT^4$$

with s equal to the entropy. Self-similar solutions to equations can be found by making the substitutions $r = R(t)\xi/\xi_0$, $\rho = a(t)\rho_1\left(\frac{r}{R}\right)$, $P = b(t)P_1\left(\frac{r}{R}\right)$ and $T = c(t)T_1\left(\frac{r}{R}\right)$ into Equations (11.74–11.76) where $R(t)$ is given by Equation (11.64), using the shock conditions and the conditions that the total energy

$$E = \int_0^R \left(U + \frac{\rho v^2}{2}\right)4\pi r^2 dr \tag{11.77}$$

and energy interior to any sphere of smaller radius with ξ = constant remain constant. It is obviously easier to solve Equations (11.74–11.76) assuming either the gas pressure or radiation pressure dominant. The relations between $R(t)$, $a(t)$, $b(t)$ and $c(t)$ reduce the partial differential Equations (11.74–11.76) to first-order ordinary differential equations that can be integrated to determine the solutions $\rho_1(r/R)$, $P_1(r/R)$ and $T_1(r/R)$. Because we have assumed that solutions are self-similar the physical variables density, temperature, pressure and velocity can be determined as products of time-dependent functions multiplied by functions of $\xi = \xi_0 r/R$. It is clear that self-similar solutions are not relevant after the shock front has reached the surface of the star. Solutions to the hydrodynamic equations for realistic stellar models with density gradients can be obtained by the method of solution described in Chapter 8.

We can consider the kinetic energy of the ejected supernova mass shell as an explosive energy input to the interstellar medium. If the density ρ_0 of the interstellar gas is uniform self-similar solutions are relevant until radiative losses become appreciable. In such solutions the density is high just inside the shock front but decreases rapidly inward. The temperature increases toward the center of the expanding gas.

Explosive carbon-burning supernovae

As already discussed in Chapter 7, neutrino emission processes in the degenerate cores of low- and intermediate-mass stars ($M \lesssim 9\,M_\odot$) remove sufficient thermal energy to prevent carbon burning until the star loses sufficient mass during red-giant evolution to become a white dwarf or until the core mass becomes approximately equal to the white-dwarf mass limit. It follows that white dwarfs that have evolved from intermediate-mass main-sequence stars are predicted to have carbon-rich cores. Therefore, if such white dwarfs accrete mass from binary companions at rates sufficiently rapid that nova outbursts do not occur then the ignition of carbon burning takes place in a very strongly degenerate gas ($\rho \simeq 2 - 4 \times 10^9\,\mathrm{g\,cm^{-3}}$, $T \simeq 5 \times 10^8\,\mathrm{K}$) when the white-dwarf mass limit is approached. Under such physical conditions the pressure within the core is nearly independent of temperature and consequently heating of the core does not cause the core to expand and cool. Although temperature sensitive neutrino emission processes tend to cool the core, carbon burning is highly temperature sensitive and the core is also heated by gravitational contraction. Therefore, carbon-burning rates increase until a thermonuclear explosion takes place. As a consequence of explosive carbon-burning temperatures become sufficiently high for oxygen and silicon burning to occur almost simultaneously. The available thermonuclear energy release ($\simeq 2 \times 10^{51}\,\mathrm{erg}$) is significantly greater than the binding energy of the degenerate core ($\simeq 3 \times 10^{50}\,\mathrm{erg}$), and therefore the entire star is dispersed into space at high velocity. Silicon burning synthesizes large amounts of ^{56}Ni and ^{56}Co, which as described in Chapter 6, are transformed into ^{56}Fe. After two months the light curves of Type Ia supernovae (as measured in magnitudes) decline linearly with time. Gamma-rays emitted as the result of decay of ^{56}Ni and ^{56}Co heat the supernova shell and thereby provide a plausible explanation for the observed linear decline in brightness that characterizes Type I supernova light curves.

Gravitational collapse supernovae

Massive stars ($M \gtrsim 9$–$10\,M_\odot$) evolve through quiescent stages of carbon, oxygen and silicon burning and consequently their cores are primarily iron-group elements at the onset of gravitational collapse. The masses of their cores are $\simeq 1.5\,M_\odot$ independent of the total stellar mass when collapse begins. Convective zones caused by quiescent carbon, oxygen and silicon burning stages prior to collapse cause the core to be of homogenous chemical composition prior to collapse. This is an important result from the point of view of the formation of neutron stars because as will be discussed in Chapter 13 the limiting mass of a neutron star is $\simeq 2\,M_\odot$. Calculations have shown that core collapse of a $15\,M_\odot$ star begins when the central density and temperature are approximately $4 \times 10^9\,\mathrm{g\,cm^{-3}}$ and $8 \times 10^9\,\mathrm{K}$ respectively.

At the onset of gravitational collapse the gas pressure in the core is caused mostly by the relativistic, degenerate electron gas and therefore $P \simeq K\rho^\Gamma$ with $\Gamma \simeq \frac{4}{3}$. Initially the star is dynamically stable because Γ is slightly greater than $\frac{4}{3}$. As core contraction continues partial photodissociation of iron group elements triggers collapse by reducing Γ below $\frac{4}{3}$. The physical explanation for dynamical instability is that dissociation of iron-group elements removes thermal energy from the gas, and consequently gas pressure is reduced. As the core density increases during collapse the electron Fermi energies become higher and electrons are captured by protons inside nuclei. By removing degenerate electrons, which are the principal source of gas pressure, electron capture tends to reduce gas pressure. Both dissociation and electron capture cause the adiabatic index Γ to be less than $\frac{4}{3}$.

Photodissociation of iron-group elements into lighter elements involves many reactions. Examples of such reactions are the following:

$$\gamma + {}^{56}\text{Fe} \leftrightarrow 13\alpha + 4\text{n}$$
$$\gamma + \alpha \leftrightarrow 2\text{p} + 2\text{n}$$
$$\gamma + (Z, A) \leftrightarrow Z\text{p} + (A - Z)\text{n}$$
$$\gamma + (Z, A + 1) \leftrightarrow \text{n} + (Z, A)$$
$$\gamma + (Z + 1, A + 1) \leftrightarrow \text{p} + (Z, A). \tag{11.78}$$

Since under the physical conditions we consider all of the above reactions are in equilibrium (and the chemical potential of the photon is zero) Equations (3.256) and (11.78) imply

$$\mu_{\text{Fe}} = 13\mu_\alpha + 4\mu_\text{n}$$
$$\mu_\alpha = 2\mu_\text{p} + 2\mu_\text{n}$$
$$\mu(Z, A) = Z\mu_\text{p} + (A - Z)\mu_\text{n}$$
$$\mu(Z, A + 1) = \mu_\text{n} + \mu(Z, A)$$
$$\mu(Z + 1, A + 1) = \mu_\text{p} + \mu(Z, A) \tag{11.79}$$

Nucleons and nuclei are nonrelativistic and therefore from Equation (3.107) the chemical potential μ_i is

$$\frac{\mu_i}{kT} = \ln\left[\frac{n_i}{g_i}\left(\frac{h^2}{2\pi m_i kT}\right)^{3/2}\right]. \tag{11.80}$$

From Equations (11.79) and (11.80) we obtain

$$\frac{n_\alpha^{13} n_\text{n}^4}{n_{\text{Fe}}} = \left(\frac{2\pi kT}{h^2}\right)^{24}\left(\frac{m_\alpha^{13} m_\text{n}^4}{m_{\text{Fe}}}\right)^{3/2} e^{-Q_1/kT} \tag{11.81}$$

$$\frac{n_\text{p}^2 n_\text{n}^2}{n_\alpha} = \left(\frac{2\pi kT}{h^2}\right)^{9/2}\left(\frac{m_\text{p}^2 m_\text{n}^2}{m_\alpha}\right)^{3/2} e^{-Q_2/kT} \tag{11.82}$$

with $Q_1 = (13m_\alpha + 4m_\text{n} - m_{\text{Fe}})c^2$ and $Q_2 = (2m_\text{p} + 2m_\text{n} - m_\alpha)c^2$. Similarly Equations (11.79) and (11.80) imply

$$n(Z, A) = \frac{g(Z, A)A^{3/2}}{2^A}\left(\frac{h^2}{2\pi mkT}\right)^{A-1} n_\text{p}^Z n_\text{n}^{A-Z} e^{[Q(Z,A)/kT]} \tag{11.83}$$

with $Q(Z, A) = [Zm_\text{p} + (A - Z)m_\text{n} - M(Z, A)]$ equal to the nuclear binding energy.

The equilibrium number densities $n(Z, A)$, n_p and n_n are determined by equations of the form (11.79) and (11.80), and charge and baryon conservation, which are given by the conditions

$$\sum_i n_i A_i = \frac{\rho}{m_n} = n \tag{11.84}$$

and

$$\sum_i n_i Z_i = n\bar{Z}_e \qquad (11.85)$$

with $\bar{Z}_e$ equal to the mean number of electrons per baryon. The quantity $\bar{Z}_e$ must be specified if the number densities $n(Z, A)$ of the various isotopes are to be determined. The quantity $\bar{Z}_e$ depends on rates of electron capture and beta-decay. Until the latter stages of gravitational collapse these processes are not in thermodynamic equilibrium and therefore $\bar{Z}_e$ must be determined by solving the electron-capture and β-decay rate equations as functions of temperature, density and chemical composition of the gas during collapse.

Gravitational collapse is accelerated because as the Fermi energies of electrons increase the resultant electron captures reduce the mean number of electrons per baryon $\bar{Z}_e$. During the early stages of collapse neutrinos emitted by electron-capture processes escape directly from the star. Neutrinos are also emitted as the result of $e^- e^+$ annihilation and the Urca reactions. As the density of the core increases neutrino opacity increases until absorption and scattering of neutrinos is sufficiently large that the neutrino mean free path becomes less than the radius of the core. Relevant neutrino processes and cross sections have already been discussed in Chapter 7.

The existence of weak neutral currents that are created by the Z boson enhance neutrino scattering cross sections from nucleons, nuclei and electrons. Such scattering processes increase the neutrino opacity appreciably. Because neutrino energies are typically much greater than the rest mass of the electron the scattering of neutrinos from electrons can cause neutrino energies to change by a large amount. The trapping of neutrinos in the collapsing core of a star occurs at densities $\gtrsim 3 \times 10^{11}\,\mathrm{g\,cm^{-3}}$. After such densities have been reached thermalization of neutrinos occurs as the result of scattering with electrons and neutrino absorption and emission processes. Because heavy nuclei survive until core densities become comparable to nuclear densities (i.e. $\rho = 2.8 \times 10^{14}\,\mathrm{g\,cm^{-3}}$) the diffusion of neutrinos from the collapsed core is limited by coherent scattering from heavy nuclei.

The total neutrino–free-neutron scattering cross section is

$$\sigma_n \simeq \frac{1}{4}\sigma_0 \left(\frac{E_\nu}{mc^2}\right)^2 \qquad (11.86)$$

with $\sigma_0 = 1.76 \times 10^{-44}\,\mathrm{cm^2}$. Equation (11.86) is valid so long as the energy of the neutrino satisfies the condition $E_\nu \ll m_n c^2$ with m_n equal to the mass of the neutron. Because neutrinos can scatter coherently from heavy nuclei the coherent neutrino–nuclei cross section varies as A^2 with A equal to the number of nucleons in the nucleus. The precise coherent neutrino-scattering cross section depends on the details of the WSG (Weinberg–Salam–Glashow) theory of weak interactions. For nuclei with $Z \simeq A/2$ the coherent scattering cross section σ_A^c is approximately

$$\sigma_A^c \simeq \frac{A^2}{64}\sigma_0 \left(\frac{E_\nu}{mc^2}\right)^2 \qquad (11.87)$$

for neutrino energies $E_\nu \ll 300\,A^{-1.3}$ MeV. If we make the approximation that at a given density and temperature heavy nuclei all have the same number of nucleons (i.e. same A value) then the neutrino mean free path λ is

$$\lambda = \frac{1}{n_A \sigma_A^c} \tag{11.88}$$

with n_A equal to the number densities of heavy nuclei and σ_A^c equal to the coherent scattering cross section given in Equation (11.87). The corresponding neutrino diffusion timescale becomes

$$\tau_D \sim \frac{R^2}{\lambda c} \tag{11.89}$$

with R equal to the radius of the core.

During the final collapse stage, which takes $\simeq 1$–2 milliseconds, the corresponding neutrino diffusion timescale from the core is approximately 10^3 times longer than the collapse timescale ($\sim 1/\sqrt{G\rho}$). The binding energy of a neutron star of mass M is approximately $0.1\,Mc^2 \simeq 10^{53}\,(M/M_\odot)$ erg. In order for a neutron star to be formed the above amount of energy must escape. Most of the energy that leave the collapsed core of a star is in the form of neutrinos. If neutrinos escape from the core primarily by means of radiative diffusion then because the neutrino diffusion timescale τ_D is $\simeq 1$ second the maximum neutrino luminosity becomes

$$L_{\nu_{max}} \simeq \frac{10^{53}\left(\dfrac{M}{M_\odot}\right)}{\tau_D} \simeq 10^{53}\ \mathrm{erg\,s}^{-1}. \tag{11.90}$$

The condition for neutrinos to cause hydrodynamic outflow from the collapsed stellar core (pre-neutron star) is that the neutrino luminosity exceed the critical (or Eddington) luminosity given in Equation (5.125) with the neutrino opacity κ primarily caused by coherent scattering from heavy nuclei. When core collapse has been halted a shock wave is generated. At this epoch the interior temperature of the neutron star is $\simeq 5$–10×10^{10} K and neutrinos are trapped behind the outwards moving shock whose initial propagation velocity is $\simeq 0.1$–$0.2c$.

If we assume $A = 56$, $M \simeq 1$–$1.2\,M_\odot$ and neutrino energy $\simeq 15$ MeV then from Equations (5.125) and (11.87) the Eddington luminosity L_c becomes

$$L_c \simeq 10^{54}\ \mathrm{erg\,s}^{-1} \tag{11.91}$$

where we have neglected general relativistic effects and the outward acceleration of mass interior to the shock caused by the bounce of the core. A comparison between Equations (11.90) and (11.91) shows that the maximum luminosity is approximately a factor of 10 less than the Eddington luminosity given by Equation (5.125). Therefore, if neutrinos are to impulsively eject mass from the outer layers of the pre-neutron star neutrinos must be transported from the interior to the outer layers in a timescale approximately 10 times faster than by radiative diffusion.

Massive stars rotate rapidly. If angular momentum is conserved within the core of the star then contraction of the core prior to gravitational collapse will cause the pre-supernova core to rotate rapidly even if there is no differential rotation on the main sequence. The meridional circulation currents in such a rapidly rotating stellar core are fairly rapid. Because vorticity ($\nabla \times \boldsymbol{v}$) is conserved during gravitational collapse such circulation velocities are greatly enhanced (by a factor of $\simeq 10^3$). In addition, thermal convective velocities caused by steep temperature gradients in the pre-neutron star might also become sufficiently high to reduce the neutrino transport timescale sufficiently to cause the critical Eddington limit to be exceeded.

To explain the observed properties of Type II supernovae it is necessary to account for the rapid transfer of energy from the pre-neutron star to the surrounding envelope of the supergiant. As described above, impulsive neutrino energy transport to the outer layers of the pre-neutron star and the resultant ejection of these outer layers is one mechanism that could explain the observed Type II supernova outburst. Plausible physical parameters for an outer region of a pre-neutron star are $\rho = 3 \times 10^{11} \, \text{g cm}^{-3}$, radial distance $\simeq 5 \times 10^6 \, \text{cm} = 50 \, \text{km}$ and thickness $\simeq 2 \times 10^6 \, \text{cm}$. The mass of such an outer region is

$$M_s \simeq 0.03 \, M_\odot. \tag{11.92}$$

The impulse condition requires that momentum transport to the outer layer of the pre-neutron star occur in less than a sound-travel timescale. This condition is satisfied if

$$\frac{L_v}{c} = \frac{\Delta p}{\Delta t} = \frac{M_s v}{\Delta t} \tag{11.93}$$

with M_s given by Equation (11.92) $v \simeq 0.1c = 3 \times 10^9 \, \text{cm s}^{-1}$ and $\Delta t \simeq 5 \times 10^{-3}$ s. Substituting the above values for M_s, v and Δt into Equation (11.88) implies that the required value of L_v is $10^{54} \, \text{erg s}^{-1}$, which is the same as the critical Eddington luminosity given in Equation (11.91). For the physical parameters assumed above the kinetic energy of the ejected outer layer of the pre-neutron star is $\simeq 2.7 \times 10^{50} \, \text{erg}$. We note that from the point of view of required energy input as many as 10 such mass ejections could occur and therefore as much as $0.3 \, M_\odot$ could be ejected from the pre-neutron star.

When a stellar core of mass $M \simeq 1.4$–$1.5 \, M_\odot$ collapses approximately 1/2 of the core (i.e. 0.6–0.8 $M_\odot$) collapses homologously (i.e. infall velocity varies as r/R with R the radius of the core), and therefore a pre-neutron star of mass $M \simeq 0.6$–$0.8 \, M_\odot$ is formed when the core density ρ exceeds nuclear densities (i.e. $\rho \simeq 2.8 \times 10^{14} \, \text{g cm}^{-3}$). Because the neutrino diffusion timescale is much longer than the collapse timescale only a small amount of energy escapes from the core during collapse, and consequently the core will bounce. If general relativistic effects are neglected then the kinetic energy E of the core will be $E \simeq 2$–$3 \times 10^{52} \, \text{erg}$ and the initial bounce velocity v_b becomes

$$v_b \simeq \sqrt{\frac{2E}{M}} \simeq 4 \times 10^9 \, \text{cm s}^{-1}. \tag{11.94}$$

Mass infall continues after the homologous core has formed. Therefore, since the

Table 11.1. Measured properties of neutrino events from SN 1987A observed in two water Cerenkov detectors. Duration of neutrino detections is about 12 seconds. Some evidence for bunching of the 20 events is suggested by data. From Arnett *et al.* 1989.

Event	Event time (s)	Electron energy (MeV)	Electron angle (degrees)
KamiokandeII:			
1	0.0	20.0 ± 2.9	18 ± 18
2	0.107	13.5 ± 3.2	40 ± 27
3	0.303	7.5 ± 2.0	108 ± 32
4	0.324	9.2 ± 2.7	70 ± 30
5	0.507	12.8 ± 2.9	135 ± 23
6	0.686	6.3 ± 1.7	68 ± 77
7	1.541	35.4 ± 8.0	32 ± 16
8	1.728	21.0 ± 4.2	30 ± 18
9	1.915	19.8 ± 3.2	38 ± 22
10	9.219	8.6 ± 2.7	122 ± 30
11	10.433	13.0 ± 2.6	49 ± 26
12	12.439	8.9 ± 1.9	91 ± 39
1	0.0	38 ± 7	80 ± 10
2	0.41	$37 \pm$	44 ± 15
3	0.65	28 ± 6	56 ± 20
4	1.14	39 ± 7	65 ± 20
5	1.56	36 ± 9	33 ± 15
6	2.68	36 ± 6	52 ± 10
7	5.01	19 ± 5	42 ± 20
8	5.58	22 ± 5	104 ± 20

relative velocity v_r between outwards moving gas and infalling gas (i.e. $v_r \simeq 2v_b \simeq 8 \times 10^9 \, \mathrm{cm \, s^{-1}}$) is highly supersonic a shock wave propagates outwards. The initial velocity of this shock front is $\simeq 8 \times 10^9 \, \mathrm{cm \, s^{-1}}$ as measured in the reference frame of the infalling gas. Neutrinos are trapped behind the shock front because the neutrino diffusion timescale is much longer than the collapse timescale.

The arrival times and energies of neutrinos measured from SN 1987A by two water Cerenkov detectors are given in Table 11.1

Stellar model calculations have shown that stars whose initial main-sequence mass is approximately $10 \, M_\odot$ should evolve through a pre-supernova evolutionary state that differs significantly from the pre-supernova state of more massive stars. Stars of $\simeq 10 \, M_\odot$ are predicted to develop a degenerate oxygen–neon core following nondegenerate carbon burning. Oxygen and neon are retained until the central density becomes about $2.5 \times 10^{10} \, \mathrm{g \, cm^{-3}}$. Under such physical conditions thermonuclear reactions synthesize iron-group elements; however, energy release is not sufficient to cause a supernova. Instead, rapid electron capture on iron-group elements initiates a sudden decrease in pressure that causes the core to undergo nearly free-fall collapse. In more massive stars a stable iron-group element core is formed before collapse. It is plausible to assume that the Crab pulsar was formed from a star in which a degenerate oxygen–neon core developed prior to gravitational collapse. As a result of such special physical conditions the Crab nebula is an anomalous Type II supernova remnant from the point of view of relatively low outflow velocity, mass and amount of heavy element nucleosynthesis.

Evolutionary models of stars initially more massive than about $11 M_\odot$ predict the presence of pre-collapse iron-group element cores whose masses increase with initial main-sequence mass. Calculations also predict that neutron-star masses and the amount of heavy-element nucleosynthesis increase with initial main-sequence mass. More massive stars end up with massive iron-group cores whereas less massive stars (but with $M \geq 11 M_\odot$) end up with iron cores close to the white-dwarf mass limit and therefore eject smaller amounts of synthesized heavy elements into interstellar space. Because the number of stars formed decreases with mass, stars in the mass range 15–$25 M_\odot$ are predicted to produce most of the Type II supernova heavy element enrichment of interstellar gas and dust. In pre-supernova stars less massive than about $15 M_\odot$ electron capture on iron-group elements provides the pressure decrement that initiates core collapse. In more massive stars photodisintegration of nuclei leads to dynamical instability ($\Gamma < \frac{4}{3}$) and subsequent gravitational collapse.

The basic idea of a neutrino transport supernova was suggested during the 1960s but was for many years somewhat controversial. The discovery of neutral weak currents and the previously discussed suggestion that neutrino cross sections on heavy elements vary as A^2 (Equation (11.87)) made the neutrino transport supernova model more widely accepted. However, some calculations indicated that neutronization of heavy elements by neutrinos weakened the initial shock wave sufficiently that black holes rather than neutron stars would form. Black hole formation would contradict the fact that neutron stars rather than black holes are the most common stellar remnants of Type II supernova explosions. Considerable effort has been made to improve input physics to the point where numerical simulations are realistic representations of Type II supernova explosions. However, successful modeling of supernova explosions depends critically on the precise time development of neutrino energy transport from stellar core to stellar envelope and therefore the neutrino transport model of Type II supernovae remained in some doubt until predicted neutrino fluxes and energies were observed from Supernova 1987A.

Further comments on supernovae

Although there are two basic mechanisms for supernova explosions (gravitational collapse and thermonuclear explosion) and two common types of supernovae (Type II and Type I) the correspondence between physical mechanism and observational classification is not one-to-one. The reason for this is somewhat accidental. Although the energy release in a thermonuclear supernova is $\simeq 2 \times 10^{51}$ erg and that of a gravitational collapse supernova $\simeq 2 \times 10^{53}$ erg the efficiency with which neutrino energy is converted into kinetic energy is only about 1% whereas explosive thermonuclear energy release has an efficiency of almost unity. The physical consequences of an explosive process depend primarily on effective energy release, matter density distribution and composition. Type Ib supernovae were not identified until the 1980s. As already discussed it is not known whether they result from thermonuclear explosion or gravitational collapse. It is known that like Type II supernovae they occur in spiral arms and at maximum light their luminosities are about 1.5 magnitudes lower in the blue than Type Ia supernovae. However, like Type Ia supernovae their spectra lack hydrogen and their light curves show linear decline. Mass loss is very significant during certain stages of stellar evolution. If extensive mass loss did not occur on the red-giant branch then the rate of formation of Type I supernovae from intermediate-mass stars would be about one every several years. It appears that stars which have initial main-sequence mass in the 1.5–$9\,M_\odot$ range lose sufficient mass as red giants to become white dwarfs. However, binary-star evolution probably leads to exceptions. If Type Ib supernovae are the result of either gravitational collapse or thermonuclear explosion in a hydrogen-deficient Wolf–Rayet star then the initial main-sequence mass could have been greater than that of a typical Type II supernova progenitor star. Models for light curves of Type Ib supernovae indicate that at time of explosion the mass of the pre-supernova star was between 4–$8\,M_\odot$. It is possible that a star initially more massive than $20\,M_\odot$ could undergo extensive mass loss during early evolutionary stages, evolve into a hydrogen deficient Wolf–Rayet star and then form a Type Ib supernova outburst.

Radio emission caused by synchrotron radiation from relativistic electrons moving in weak magnetic fields is a well-known general characteristic of supernova nebular remnants. This radiation can be explained as resulting from the interaction of a supernova shock wave with ambient interstellar gas and magnetic field. In recent years radio emission has been observed shortly after several Type II supernova outbursts. Moreover, the Type II supernovae SN 1981K and SN 1986J were discovered because of their radio emission. However, similar radio emission has not been observed after several Type Ia supernovae. The existence of appreciable radio emission shortly after an outburst is probably the result of the supernova shock interacting with gas ejected as a stellar wind from the pre-supernova star prior to outburst. Radio flux from Supernova 1987A declined very rapidly (within weeks) after outburst presumably because the stellar wind from a B3 supergiant is low. Most Type II supernovae occur in red supergiants with strong stellar winds and consequently more radio emission is expected. If the mass loss rate from a particular presupernova star is very high then the occurrence of radio emission after an outburst will be delayed by free–free absorption. Radio observations of the Type II supernova SN 1979C in the galaxy NGC 4321 showed no radio emission a few weeks after its discovery in April 1979. However, a bright unresolved radio source was observed a year later. Multi-wavelength observations of SN 1979C showed rapid radio flux increases at progressively lower frequencies indicating the presence of free–free absorption in circumstellar gas. The observed change in angular size of the radio source together with measured optical spectral-line expansion velocities provide a direct measurement of the distance to NGC 4321 and therefore a direct determination of the Hubble constant. Wolf–Rayet stars are known to have strong stellar winds with inferred mass loss rates of 10^{-5}–$10^{-4}\,M_\odot/\mathrm{yr}$ and wind velocities of 1–$2\times10^3\,\mathrm{km\,s^{-1}}$. Therefore, if a Type Ib supernova occurs in a Wolf–Rayet star the supernova shock will propagate into circumstellar gas from the wind.

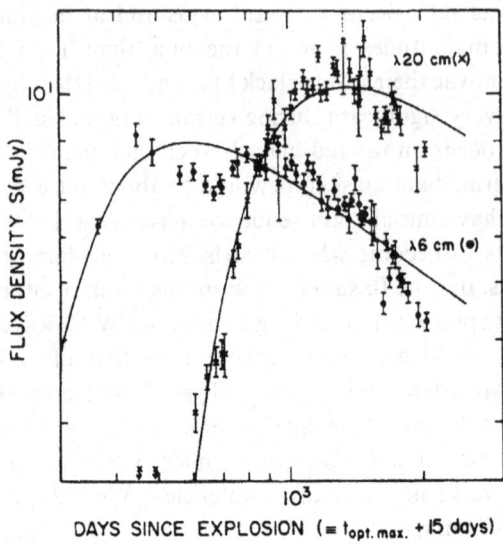

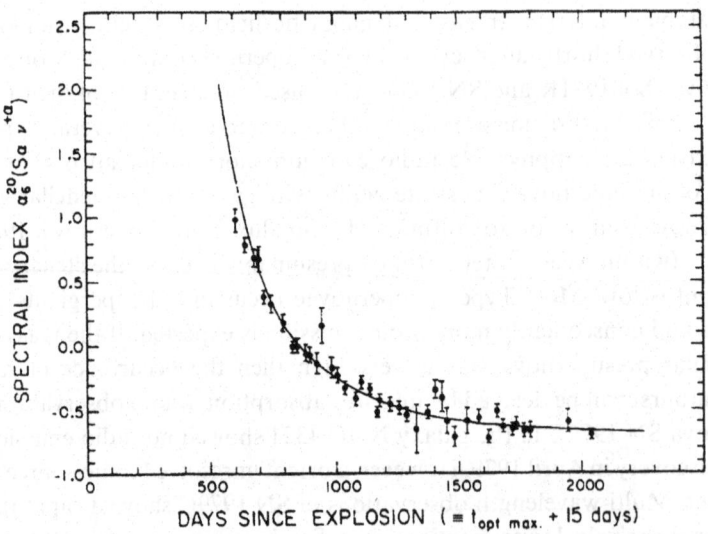

Figure 11.13. (Top) Radio light curves at 6 cm and 20 cm wavelengths for the Type II supernova SN 1979C in NGC 4321 (M100). Free–free absorption is responsible for the delay in emission of 20 cm radiation. (Bottom) Evolution of the spectral index $\alpha(S \propto \nu^{+\alpha})$ between 20 and 6 cm. From Weiler and Sramek (1988).

The propagation of a supernova shock wave into circumstellar gas can also lead to detectable x-ray emission. SN 1980K was the first extragalactic supernova detected at x-ray wavelengths. Its x-ray luminosity was about $2 \times 10^{39}\,\mathrm{ergs^{-1}}$ 40 days after discovery and then declined by about a factor of two during the next 50 days. If the observed x-ray emission was caused by thermal emission from shocked circumstellar gas then the inferred pre-supernova mass loss rate and wind velocity are $\simeq 2 \times 10^{-5}\,M_\odot/\mathrm{yr}$ and $10\,\mathrm{km\,s^{-1}}$, respectively. Circumstellar winds from red supergiants also contain dust and infrared emission from radiatively heated dust has been detected soon after supernova explosions.

Even if a particular pre-supernova star did not produce an appreciable stellar wind supernova ejecta eventually (after about 10^2 years) interact with significant amounts of ambient interstellar gas. The formation of a mass shell results from this interaction and previously discussed self-similar blast wave solutions approximately describe the evolution of a relatively low-temperature ($\simeq 10$–$20 \times 10^3\,\mathrm{K}$) and high-density mass shell outside a much hotter, x-ray emitting low-density gas. Cas A is a well-known example of a radio- and x-ray emitting supernova remnant that is about 300 years old. It is widely believed that cosmic rays as well as relativistic electrons are formed as the result of a supernova shock wave interacting with interstellar gas and magnetic fields.

Radio light curves at 6 cm and 20 cm as well as the evolution of the radio spectral index are shown in Figure 11.13 for SN 1979C, which exploded in the galaxy M100.

Problems

1. What is the white-dwarf mass limit of a ^{56}Fe white dwarf? Assume that the gas is fully ionized and at zero temperature. Estimate the density required for complete pressure ionization of Fe.

2. Determine how temperature T and density ρ vary as a function of r close to the surface of a white dwarf. The opacity is $\kappa = \kappa_0 \rho T^{-3.5}$. It can be assumed that $1 - r/R$ is $\ll 1$, with R equal to the radius of the white dwarf and $T = 0$ at $r = R$. Estimate the thickness of the outer layer in which electrons are nondegenerate.

3. Ions of charge Ze are separated by a mean distance r_i in the interior of a white dwarf and surrounded by free electrons. The ions are in the lowest energy level of a three-dimensional simple harmonic oscillator and therefore their zero-point energy is $\frac{1}{2}\hbar\omega_0$, with ω_0 given by Equation (11.26). Calculate the probability per unit time that an ion will penetrate a distance r_i through the surrounding potential barrier. What is the approximate reaction rate per ion pair if the nuclear reaction cross section is σ_n?

4. Estimate the time intervals for $0.5\,M_\odot$ and $1\,M_\odot$ white dwarfs to cool from $T_c = 3 \times 10^7\,\mathrm{K}$ to $T_c = 10^6\,\mathrm{K}$. Make an approximate correction for crystallization.

5. Consider a ^{12}C white dwarf of central density $\rho_c = 10^5\,\mathrm{g\,cm^{-3}}$ and interior temperature $T = 10^6\,\mathrm{K}$. Assume that the white-dwarf structure is supported by a nonrelativistic electron-degenerate gas and is therefore a polytrope of index

$n = 3/2$. What are the mass and radius of the white dwarf? Calculate the heat capacity per ion at densities of 10^5, 10^4 and $10^3 \, \mathrm{g \, cm^{-3}}$.

6. Solve for the structures of white dwarfs with $\mu_e = 2$ by numerically integrating the equations of hydrostatic equilibrium and mass conservation assuming a zero-temperature electron-degenerate gas. Assume some central pressure P_c and integrate until the surface boundary condition $P = 0$ is satisfied.

7. Assume that the equation of state of a ^{12}C white dwarf is described by a zero-temperature, ultrarelativistic electron-degenerate gas ($P \propto \rho^{4/3}$) and is therefore a polytrope of index $n = 3$. Show that the white-dwarf mass is independent of ρ_c and is therefore equal to the Chandrasekhar mass limit.

8. Solve Equation (11.54) numerically for $\phi(x_0)$. Compute the Coulomb correction to the pressure of a zero-temperature electron-degenerate gas at $\rho = 10^7$, 10^5 and $10^3 \, \mathrm{g \, cm^{-3}}$.

12 General relativity

12.1 Tensor analysis

Consider two continuous systems of coordinates x^i and $\bar{x}^j$ in an n-dimensional space. The coordinate systems are related by the transformation equations $x^i = x^i(\bar{x}^j)$ and $\bar{x}^j = \bar{x}^j(x^i)$. A contravariant vector A^i is a set of n quantities that transforms according to the rule

$$A^k = \frac{\partial x^k}{\partial \bar{x}^i} \bar{A}^i. \tag{12.1}$$

In Equation (12.1) a summation over the index i is understood. A covariant vector is a set of n quantities whose transformation equation is

$$A_k = \frac{\partial \bar{x}^i}{\partial x^k} \bar{A}_i. \tag{12.2}$$

A second-rank contravariant tensor is a set of n^2 quantities that transforms according to the equation

$$T^{kl} = \frac{\partial x^k}{\partial \bar{x}^i} \frac{\partial x^l}{\partial \bar{x}^j} \bar{T}^{ij}. \tag{12.3}$$

The corresponding covariant tensor is

$$T_{kl} = \frac{\partial \bar{x}^i}{\partial x^k} \frac{\partial \bar{x}^j}{\partial x^l} \bar{T}_{ij}. \tag{12.4}$$

Tensors with contravariant and covariant indices can also be defined. Such tensors are known as mixed tensors. A mixed second-rank tensor transforms by the rule

$$T_l^k = \frac{\partial x^k}{\partial \bar{x}^i} \frac{\partial \bar{x}^j}{\partial x^l} \bar{T}_j^i. \tag{12.5}$$

The rank of a tensor is determined by the number of indices. A second-rank tensor in an n-dimensional space has n^2 components. Vectors and scalars can be considered tensors of first and zeroth rank respectively.

Scalars are invariant under coordinate transformations. The scalar product of two vectors is

$$A_k B^k = A^k B_k. \tag{12.6}$$

The contraction of a second-rank tensor with mixed indices (i.e. T^k_k) is also a scalar. The coordinate differential dx^i is a contravariant vector because its transformation equation is

$$dx^k = \frac{\partial x^k}{\partial \bar{x}^i} d\bar{x}^i. \qquad (12.7)$$

It can readily be shown that the gradient of a scalar is a covariant vector.

12.2   Riemannian geometry

In special relativity the proper time interval $d\tau$ between two events defined by the coordinates $(t + dt, x + dx, y + dy, z + dz)$ and (t, x, y, z) respectively is determined by the equation

$$- ds^2 = - c^2 d\tau^2 = - c^2 dt^2 + dx^2 + dy^2 + dz^2 \equiv \eta_{ij} dx^j. \qquad (12.8)$$

In a strong gravitational field Riemannian geometry must be used to describe the properties of space because space has significant curvature. The Riemannian invariant interval ds between two neighboring points x^i, $x^i + dx^i$ is given by the equation

$$- ds^2 = g_{ij} dx^i dx^j, \qquad (12.9)$$

where g_{ij} is a symmetric tensor known as the metric tensor. A postulate of Riemannian geometry is that about any nonsingular point it is possible to find a coordinate system in which space is locally inertial. In general relativity this postulate is known as the principle of equivalence and it implies that in the neighborhood of an arbitrary, nonsingular point a gravitational field is equivalent to uniform acceleration. The Minkowski metric η_{ij} defined in Equation (12.8) is in Cartesian coordinates the metric of an inertial reference frame.

Consider a particle moving freely in a gravitational field. The principle of equivalence implies that there exists a locally inertial coordinate system ξ^i such that

$$\frac{d^2 \xi^i}{ds^2} = 0, \qquad (12.10)$$

with ds determined by the equation

$$- ds^2 = \eta_{ij} d\xi^i d\xi^j. \qquad (12.11)$$

If another arbitrary coordinate system x^k is defined then we may regard the free-falling coordinates ξ^i as functions of x^k and Equation (12.10) becomes

$$0 = \frac{d}{ds} \left(\frac{\partial \xi^i}{\partial x^k} \frac{dx^k}{ds} \right)$$

$$= \frac{\partial \xi^i}{\partial x^k} \frac{d^2 x^k}{ds^2} + \frac{\partial^2 \xi^i}{\partial x^k \partial x^l} \frac{dx^k}{ds} \frac{dx^l}{ds}. \qquad (12.12)$$

Multiplying Equation (12.12) by $\partial x^m/\partial \xi^i$ and using the identity

$$\frac{\partial \xi^i}{\partial x^k} \frac{\partial x^m}{\partial \xi^i} = \delta_k^m,$$

(12.13)

we obtain the equation of motion

$$\frac{d^2 x^m}{ds^2} + \Gamma_{kl}^m \frac{dx^k}{ds} \frac{dx^l}{ds} = \frac{du^m}{ds} + \Gamma_{kl}^m u^k u^l = 0$$

(12.14)

with $u^l = \dfrac{dx^l}{ds}$. From Equations (12.12–12.14) the affine connection Γ_{kl}^m (also called the Christoffel symbol of the second kind) is equal to

$$\Gamma_{kl}^m = \frac{\partial x^m}{\partial \xi^i} \frac{\partial^2 \xi^i}{\partial x^k \partial x^l}.$$

(12.15)

In curved space the trajectories of particles are geodesics described by Equation (12.14).

The action principle can also be used to obtain the equation of motion of a particle in a gravitational field. The action S is defined by the equation

$$S = - mc \int ds,$$

(12.16)

and the action principle states that

$$\delta S = - mc\delta \int ds = 0.$$

(12.17)

From the definition of the invariant interval ds given in Equation (12.9) we obtain

$$- \delta ds^2 = - 2ds\delta ds = \delta(g_{ik}dx^i dx^k)$$

$$= dx^i dx^k \frac{\partial g_{ik}}{\partial x^l} \delta x^l + 2g_{ik}dx^i d\delta x^k.$$

(12.18)

Equations (12.17) and (12.18) imply that

$$0 = \delta S = mc \int \left[\frac{1}{2} \frac{dx^i}{ds} \frac{dx^k}{ds} \frac{\partial g_{ik}}{\partial x^l} \delta x^l + g_{il} \frac{dx^i}{ds} \frac{d\delta x^l}{ds} \right] ds$$

$$= mc \int \left[\frac{1}{2} \frac{dx^i}{ds} \frac{dx^k}{ds} \frac{\partial g_{ik}}{\partial x^l} \delta x^l - \frac{d}{ds}\left(g_{il} \frac{dx^i}{ds} \right) \delta x^l \right] ds.$$

(12.19)

In obtaining the second term inside the integral of Equation (12.19) we have performed an integration by parts and used the conditions $\delta x^l = 0$ at the limits of the integration.

Since the small variations δx^l are arbitrary Equation (12.19) implies that

$$\frac{1}{2}u^iu^k\frac{\partial g_{ik}}{\partial x^l} - \frac{d}{ds}(g_{il}u^i)$$

$$=\frac{1}{2}u^iu^k\frac{\partial g_{ik}}{\partial x^l} - g_{il}\frac{du^i}{ds} - u^iu^k\frac{\partial g_{il}}{\partial x^k}$$

$$= -g_{il}\frac{du^i}{ds} + \frac{1}{2}u^iu^k\frac{\partial g_{ik}}{\partial x^l} - \frac{1}{2}u^iu^k\left(\frac{\partial g_{il}}{\partial x^k} + \frac{\partial g_{kl}}{\partial x^i}\right) = 0, \tag{12.20}$$

where $u^i = dx^i/ds$ is the four velocity. The second equality in Equation (12.20) follows because u^iu^k is symmetric in the indices i and k. Using the relation

$$g^{lm}g_{il} = \delta_i^m, \tag{12.21}$$

relabeling indices that are summed over, and noting that the metric tensor is symmetric Equation (12.20) becomes

$$\frac{du^i}{ds} + \frac{1}{2}g^{im}\left[\frac{\partial g_{mk}}{\partial x^l} + \frac{\partial g_{ml}}{\partial x^k} - \frac{\partial g_{kl}}{\partial x^m}\right]u^ku^l = 0. \tag{12.22}$$

Because the geodesic Equations (12.14) and (12.22) must be identical it follows that the affine connection Γ_{kl}^i and the metric tensor g_{jl} are related by the equation

$$\Gamma_{kl}^i = \frac{1}{2}g^{im}\left[\frac{\partial g_{mk}}{\partial x^l} + \frac{\partial g_{ml}}{\partial x^k} - \frac{\partial g_{kl}}{\partial x^m}\right]. \tag{12.23}$$

From Equation (12.1) we have

$$dA^k = \frac{\partial x^k}{\partial \bar{x}^i}d\bar{A}^i + \bar{A}^id\left(\frac{\partial x^k}{\partial \bar{x}^i}\right)$$

$$= \frac{\partial x^k}{\partial \bar{x}^i}d\bar{A}^i + \bar{A}^i\frac{\partial^2 x^k}{\partial \bar{x}^j\partial \bar{x}^i}d\bar{x}^j. \tag{12.24}$$

Equation (12.24) shows that dA^k does not transform as a contravariant vector unless x^k are linear functions of $\bar{x}^j$. The differential dA^k in Equation (12.24) is not a vector in curved space or in curvilinear coordinates defined in Euclidean space and consequently its components change under a parallel displacement. Consider a vector A^k defined at x^k. We denote the parallel displaced vector as $A^k + \delta A^k$ at $x^k + dx^k$ where the field vector is $A^k + dA^k$. In curved space or curvilinear coordinates in Euclidean space the difference

$$DA^k = dA^k - \delta A^k \tag{12.25}$$

transforms as a vector because its components remain unchanged under a parallel displacement. The ordinary derivative of a vector is not generally a tensor. However, the covariant derivative $A_{;j}^k$ which is defined by the equation

$$DA^k = A_{;j}^k dx^j \tag{12.26}$$

does transform as a tensor. The covariant derivative reduces to the ordinary derivative in Cartesian coordinates because $\delta A^k = 0$ in Equation (12.25).

Equation (12.14) can be written

$$\frac{Du^m}{Ds} = \frac{du^m}{ds} + \Gamma^m_{kl}u^k u^l = 0, \tag{12.27}$$

because geodesic motion is the generalization of parallel transport in Riemannian geometry. The tangent of a geodesic represents parallel motion along the particle trajectory. It follows immediately from Equation (12.27) that Γ^m_{kl} does not transform as a tensor. Since the vectors u^m and A^k transform similarly Equations (12.25–12.27) imply that

$$\delta A^k = - \Gamma^k_{ij} A^i dx^j \tag{12.28}$$

and

$$A^k_{;j} = \frac{\partial A^k}{\partial x^j} + \Gamma^k_{ij}A^i. \tag{12.29}$$

From Equations (12.25) and (12.28) it follows that the change in a vector after parallel transport over any infinitesimal closed contour is

$$\Delta A^k = \int \Gamma^k_{lm} A^l dx^m. \tag{12.30}$$

Consider a closed curve Γ bounding the two dimensional surface S in an n-dimensional space. Stokes' theorem implies that for any function $f_i(x^j)$ we have

$$\int_\Gamma f_i(x^j)dx^i = \int_S \int \frac{\partial f_j}{\partial x^i} dS^{ij}. \tag{12.31}$$

The surface element dS^{ij} can be expressed as

$$dS^{ij} = \left(\frac{\partial x^i}{\partial u}\frac{\partial x^j}{\partial v} - \frac{\partial x^j}{\partial u}\frac{\partial x^i}{\partial v}\right)dudv, \tag{12.32}$$

with the terms within the bracket the Jacobian of the coordinate transformation. The curves $u = $ constant and $v = $ constant divide the surface S into a grid.

Equation (12.31) can be written as

$$\int f_i(x^j)dx^i = \frac{1}{2}\int\int\left(\frac{\partial f_j}{\partial x^i} - \frac{\partial f_i}{\partial x^j}\right)dS^{ij} + \frac{1}{2}\int\int\left(\frac{\partial f_j}{\partial x^i} + \frac{\partial f_i}{\partial x^j}\right)dS^{ij}$$

$$= \frac{1}{2}\int\int\left(\frac{\partial f_j}{\partial x^i} - \frac{\partial f_i}{\partial x^j}\right)dS^{ij}. \tag{12.33}$$

Since the surface element dS^{ij} is antisymmetric in the indices i and j the second integral in Equation (12.33) is equal to zero.

From Equations (12.25) and (12.28) the parallel transport of any vector causes it to change by an amount

$$\Gamma^l_{ij}A^i\mathrm{d}x^j. \tag{12.34}$$

Subtracting this amount from A^l in Equation (12.30) gives

$$\Delta A^k = \int (\Gamma^k_{lm}A^l - \Gamma^k_{lm}\Gamma^l_{ij}A^i\mathrm{d}x^j)\mathrm{d}x^m. \tag{12.35}$$

Because changing the label of an index that is summed over does not change an equation we have

$$\Gamma^k_{lm}\Gamma^l_{ij}A^i\mathrm{d}x^j = \Gamma^k_{im}\Gamma^i_{lj}A^l\mathrm{d}x^j. \tag{12.36}$$

Substituting the right-hand side of Equation (12.36) into Equation (12.35) and using the symmetry relation $\Gamma^k_{im} = \Gamma^k_{mi}$ we obtain

$$\Delta A^k = \int A^l(\Gamma^k_{lm} - \Gamma^k_{mi}\Gamma^i_{lj}\mathrm{d}x^j)\mathrm{d}x^m. \tag{12.37}$$

In deriving Equation (12.38) below we choose coordinates such that $\dfrac{\partial\mathrm{d}x^j}{\partial x^m} = \delta_{jm}$. However, because the Riemann tensor in Equation (12.39) is a tensor its form is independent of coordinate system.

Defining the integrand of Equation (12.37) as B^k_m and using Stokes' theorem given in Equation (12.33) leads to the equation

$$\Delta A^k = \frac{1}{2}\int\int\left(\frac{\partial B^k_n}{\partial x^m} - \frac{\partial B^k_m}{\partial x^n}\right)\mathrm{d}S^{mn}$$

$$= \int\int R^k_{lmn}A^l\mathrm{d}S^{mn}, \tag{12.38}$$

where

$$R^k_{lmn} = \frac{\partial\Gamma^k_{ln}}{\partial x^m} - \frac{\partial\Gamma^k_{lm}}{\partial x^n} - \Gamma^k_{nj}\Gamma^j_{lm} + \Gamma^k_{mj}\Gamma^j_{ln}. \tag{12.39}$$

The fourth-rank tensor R^k_{lmn} is known as the Riemann tensor. It can be shown that the necessary and sufficient condition for a metric $g_{ij}(x)$ to be equivalent to a Minkowski metric is that R^k_{lmn} vanish everywhere. The Riemann tensor has the symmetry property $R^k_{lmn} = -R^k_{lnm}$.

From the definition of the metric tensor g_{ij} it follows that the covariant metric tensor g_{ij} can be used to lower the index of a vector or an index of a tensor. Similarly the contravariant metric tensor g^{ij} can be used to raise the indices of vectors and tensors. Lowering the upper index of the Riemann tensor gives the fourth rank covariant tensor

$$R_{klmn} = g_{ki}R^i_{lmn}. \tag{12.40}$$

The second-rank symmetric Ricci tensor R_{ln} is obtained by the contraction

$$R_{ln} = R_{lkn}^k. \tag{12.41}$$

Multiplying R_{ln} by g^{ln} and contracting over the two indices l and n gives the scalar

$$R = g^{lm} R_{lm}. \tag{12.42}$$

The Einstein tensor G_{lm} is defined as

$$G_{lm} = R_{lm} - \frac{1}{2} R g_{lm}. \tag{12.43}$$

The covariant Riemann tensor R_{klmn} has the symmetry properties

$$R_{klmn} = R_{mnkl}, \tag{12.44}$$

$$R_{klmn} = - R_{lkmn} = - R_{klnm} = R_{lknm}, \tag{12.45}$$

$$R_{klmn} + R_{knlm} + R_{kmnl} = 0. \tag{12.46}$$

An important property of the metric tensor g_{ik} is that its covariant derivative is zero. Consider an arbitrary vector A_i. The covariant vector A_i and the contravariant vector A^i are related by the equation

$$A_i = g_{ij} A^j. \tag{12.47}$$

Taking the covariant differential D of both sides of Equation (12.47) gives

$$DA_i = (Dg_{ij})A^j + g_{ij} DA^j$$

$$= (Dg_{ij})A^j + DA_i. \tag{12.48}$$

Since Equation (12.48) implies $Dg_{ij} = 0$ it follows that

$$g_{ij;k} = 0. \tag{12.49}$$

We choose a locally inertial frame of reference in which Γ_{lm}^k vanishes locally but not its derivatives. From Equations (12.23), (12.39), (12.40) and (12.49) it can be shown that the covariant derivative of R_{klmn} is equal to

$$R_{klmn;i} = \frac{1}{2} \frac{\partial}{\partial x^i} \left(\frac{\partial^2 g_{km}}{\partial x^n \partial x^l} - \frac{\partial^2 g_{lm}}{\partial x^n \partial x^k} - \frac{\partial^2 g_{kn}}{\partial x^l \partial x^m} + \frac{\partial^2 g_{ln}}{\partial x^m \partial x^k} \right). \tag{12.50}$$

If the indices m, n and i are permuted cyclically then we have the Bianchi identities

$$R_{klmn;i} + R_{klim;n} + R_{klni;m} = 0. \tag{12.51}$$

Operating separately on the three terms in Equation (12.51) with g^{mk} and contracting indices gives

$$g^{mk} R_{klmn;i} = R_{lmn;i}^m = R_{ln;i}, \tag{12.52}$$

$$g^{mk} R_{klim;n} = - R_{li;n}, \tag{12.53}$$

$$g^{mk}R_{klni;m} = R^m_{lni;m}.$$ (12.54)

Adding Equations (12.52–12.54) and using Equation (12.51) we obtain

$$R_{ln;i} - R_{li;n} + R^m_{lni;m} = 0.$$ (12.55)

After raising the index l in Equations (12.52–12.55) and contracting on the indices l and n we find

$$0 = R_{;i} - R^l_{i;l} - R^m_{i;m} = \left(R^l_i - \frac{1}{2} \delta^l_i R \right)_{;l}.$$ (12.56)

Using the metric tensor to raise the index i and recalling that the covariant derivative of the metric tensor is zero Equation (12.56) becomes

$$G^{ik}_{;i} = \left(R^{ik} - \frac{1}{2} g^{ik} R \right)_{;i} = 0.$$ (12.57)

Equation (12.57) implies $G_{ik;i} = 0$ and therefore the covariant derivative of Equation (12.43) is also zero.

12.3   The Einstein equations

Equation (12.22) must reduce to the Newtonian equation of motion in a weak gravitational field. Consider the metric tensor given by the equation

$$-ds^2 = -\left(1 - \frac{2\Phi}{c^2}\right)dt^2 + \frac{1}{c^2}(dx^2 + dy^2 + dz^2).$$ (12.58)

with Φ equal to the Newtonian gravitational potential. If $2\Phi/c^2$ and v^2/c^2 are both $\ll 1$ then it can readily be shown that Equation (12.22) reduces to the Newtonian equation of motion if the metric tensor is given by Equation (12.58). In Newtonian gravitation the density of matter and the gravitational potential are related by the equation

$$\nabla^2 \Phi = 4\pi G\rho.$$ (12.59)

For nonrelativistic matter the energy density is the T_{00} component of the energy–momentum tensor T_{ik} and therefore

$$T_{00} \simeq \rho c^2.$$ (12.60)

The g_{00} component of the metric given in Equation (12.58) is

$$g_{00} = -\left(1 + \frac{2\Phi}{c^2}\right).$$ (12.61)

Equations (12.59–12.61) imply that

$$\nabla^2 g_{00} = -\frac{8\pi}{c^2} G T_{00}.$$ (12.62)

To obtain the Einstein field equations we must replace Equation (12.62) with a tensor equation that reduces to Equation (12.62) in the Newtonian limit.

The covariant form of the energy–momentum tensor is

$$T_{ik} = (P + \rho c^2)u_i u_k + P g_{ik}. \tag{12.63}$$

In the absence of a gravitational field we have

$$\frac{\partial T_{ik}}{\partial x^k} = 0. \tag{12.64}$$

In Riemannian geometry the generalization of Equation (12.64) is the equation

$$T_{ik;k} = 0. \tag{12.65}$$

In general relativity the geometry of space is determined by the presence of mass and other forms of energy. The Einstein field equations can be obtained by setting the covariant Einstein tensor G_{ik} given in Equation (12.43) equal to a constant times the energy–momentum tensor given in Equation (12.63). Setting the Einstein tensor given in Equation (12.43) equal to the energy–momentum tensor in Equation (12.63) times a constant we obtain the Einstein equations

$$G_{ik} = R_{ik} - \frac{1}{2}g_{ik}R = \frac{8\pi G}{c^4}T_{ik}. \tag{12.66}$$

The tensor equation (12.66) reduces to Poisson's equation in the weak-gravity and low-velocity limits. From Equations (12.57) and (12.65) it follows that the covariant derivatives of both sides of Equation (12.66) equal zero. Different forms of the Einstein field equations can be readily derived. Using the metric tensor to raise the index k in Equation (12.66) gives

$$R_i^k - \frac{1}{2}g_i^k R = \frac{8\pi G}{c^4}T_i^k. \tag{12.67}$$

Contracting on the indices i and k reduces Equation (12.67) to the equation

$$R = -\frac{8\pi G}{c^4}T, \tag{12.68}$$

with $T = T_i^i$. Equations (12.67) and (12.68) lead to the equation

$$R_{ik} = \frac{8\pi G}{c^4}\left(T_{ik} - \frac{1}{2}g_{ik}T\right). \tag{12.69}$$

Equation (12.69) implies that in empty space we have

$$R_{ik} = 0. \tag{12.70}$$

The covariant divergence of a contravariant vector is

$$A^k_{;k} = \frac{\partial A^k}{\partial x^k} + \Gamma^k_{kl} A^l. \tag{12.71}$$

The affine connection Γ^k_{kl} with summation over the index k understood is from Equation (12.23) equal to

$$\Gamma^k_{kl} = \frac{1}{2} g^{lm} \frac{\partial g_{ml}}{\partial x^l}. \tag{12.72}$$

Because the derivative of a determinant is found by differentiating each row separately and then making a summation it follows that

$$\frac{1}{g} \frac{\partial g}{\partial x^i} = g^{lm} \frac{\partial g_{ml}}{\partial x^i}, \tag{12.73}$$

where g is the determinant of the metric tensor. Equations (12.71–12.73) imply

$$A^k_{;k} = \frac{1}{\sqrt{-g}} \frac{\partial \sqrt{-g} A^k}{\partial x^k}. \tag{12.74}$$

The product of two contravariant vectors $A^i B^k$ is a tensor. From Equation (12.28) it follows that under a parallel displacement we have

$$\begin{aligned} \delta(A^i B^k) &= A^i \delta B^k + B^k \delta A^i \\ &= - A^i \Gamma^k_{ml} B^m dx^l - B^k \Gamma^i_{ml} A^m dx^l. \end{aligned} \tag{12.75}$$

Since tensors T^{ik} transform similarly Equation (12.75) implies that

$$\delta T^{ik} = - (T^{im} \Gamma^k_{ml} + T^{mk} \Gamma^i_{ml}) dx^l. \tag{12.76}$$

From Equations (12.28) and (12.76) we obtain

$$DT^{ik} = dT^{ik} - \delta T^{ik} = T^{ik}_{;l} dx^l \tag{12.77}$$

with

$$T^{ik}_{;l} = \frac{\partial T^{ik}}{\partial x^l} + \Gamma^i_{ml} T^{mk} + \Gamma^k_{ml} T^{im}. \tag{12.78}$$

Using equation (12.78) and the symmetry relation $\Gamma^i_{ml} = \Gamma^i_{lm}$ the covariant divergence of a contravariant tensor becomes

$$T^{ik}_{;i} = \frac{\partial T^{ik}}{\partial x^i} + \Gamma^i_{im} T^{ml} + \Gamma^k_{im} T^{im}. \tag{12.79}$$

Equations (12.71–12.73) imply that Equation (12.79) reduces to the equation

$$T^{ik}_{;i} = \frac{1}{\sqrt{-g}} \frac{\partial}{\partial x^i} (\sqrt{-g} T^{ik}) + \Gamma^k_{im} T^{im}. \tag{12.80}$$

The contravariant energy–momentum tensor T^{ik} can be readily found by raising the indices i and k in Equation (12.63). From Equations (12.63), (12.65), (12.80) and the relation $T^{ik} = T^{ki}$ we obtain

$$T^{ki}_{;i} = \frac{\partial P}{\partial x^i} g^{ki} + (-g)^{-1/2} \frac{\partial}{\partial x^i} [(-g)^{1/2} (P + \rho c^2) u^k u^i] = 0. \tag{12.81}$$

In hydrostatic equilibrium we have

$$u^0 = (-g_{00})^{-1/2}, \tag{12.82}$$

$u^i = 0$ for $i = 1, 2, 3$, and

$$\frac{\partial}{\partial x^i} [(P + \rho c^2) u^k u^i] = 0. \tag{12.83}$$

Multiplying Equation (12.81) by g_{ki} and using Equations (12.73) and (12.81–12.83) gives

$$-\frac{\partial P}{\partial x^i} = [P + \rho c^2] \frac{\partial}{\partial x^i} \ln(-g_{00})^{1/2}. \tag{12.84}$$

12.4   The spherically symmetric gravitational field

The most general spherically symmetric expression for $-ds^2$ is

$$-ds^2 = A(r', t) dt^2 + B(r', t) dr' dt + C(r', t) dr'^2 + D(r', t)(d\theta^2 + \sin^2 \theta d\phi^2). \tag{12.85}$$

Because the particular choice of a coordinate system is arbitrary we can choose different coordinates according to the transformations:

$$r = f(r', t'), \tag{12.86}$$

$$t = g(r', t'), \tag{12.87}$$

where f and g are arbitrary functions of coordinates r' and t'. By means of coordinate transformations such as those given by Equations (12.86) and (12.87) we can find coordinates (r, θ, ϕ, t) such that the coefficient of $drdt$ is zero and the coefficient of $(d\theta^2 + \sin^2 \theta d\phi^2)$ is equal to r^2. In the new coordinate system it is convenient to write the coefficient of dr^2 and dt^2 as $-e^\lambda$ and $c^2 e^\nu$ respectively with $\lambda = \lambda(r, t)$ and $\nu = \nu(r, t)$. It follows that $-ds^2$ can be written

$$-ds^2 = -e^\nu c^2 dt^2 + e^\lambda dr^2 + r^2 (d\theta^2 + \sin^2 \theta d\phi^2). \tag{12.88}$$

The nonzero covariant and contravariant components of the metric tensor are

$$g_{00} = -e^\nu, \qquad g_{11} = e^\lambda, g_{22} = r^2, \qquad g_{33} = r^2 \sin^2 \theta \tag{12.89}$$

and

$$g^{00} = -e^{-\nu}, \qquad g^{11} = e^{-\lambda}, \qquad g^{22} = r^{-2}, \qquad g^{33} = r^{-2} \sin^{-2} \theta, \tag{12.90}$$

respectively.

In the space surrounding a star the energy–momentum tensor T^k_i is equal to zero. Evaluating the tensor $R^i_k = 0$ in free space we obtain the components

$$e^{-\lambda}\left(\frac{dv/dr}{r} + \frac{1}{r^2}\right) - \frac{1}{r^2} = 0, \tag{12.91}$$

$$e^{-\lambda}\left(\frac{d\lambda/dr}{r} - \frac{1}{r^2}\right) + \frac{1}{r^2} = 0, \tag{12.92}$$

$$\frac{d\lambda}{dt} = 0. \tag{12.93}$$

Equations (12.91) and (12.92) imply that

$$\frac{d\lambda}{dr} + \frac{dv}{dr} = 0. \tag{12.94}$$

From Equation (12.94) it follows that we can set $\lambda = -v$. Integrating Equation (12.94) we obtain

$$e^{-\lambda} = e^{v} = 1 + \frac{\text{constant}}{r}. \tag{12.95}$$

Equation (12.95) implies $e^{-\lambda} = e^{v} = 1$ as $r \to \infty$. To determine the constant in Equation (12.95) we note that in the weak gravitational field limit

$$g_{00} = e^{-\lambda} = -1 - \frac{2\Phi}{c^2}, \tag{12.96}$$

where $\Phi = -\dfrac{GM}{r}$. Equations (12.95) and (12.96) imply that

$$-c^2 d\tau^2 = -ds^2$$
$$= -c^2\left(1 - \frac{2GM}{c^2 r}\right)dt^2 + \frac{dr^2}{\left(1 - \dfrac{2GM}{c^2 r}\right)} + r^2(\sin^2\theta\, d\phi^2 + d\theta^2). \tag{12.97}$$

The metric given in Equation (12.97) is known as the Schwarzschild metric and has an event horizon at the radius $r_g = 2GM/c^2$. The Schwarzschild radius r_g is not a true singularity because it can be removed by a coordinate transformation. A singularity in the metric tensor surrounding a point mass that cannot be removed by a coordinate transformation does, however, exist at $r = 0$. Equation (12.97) implies that an interval dt of coordinate time and an interval $d\tau$ of proper time are related by the equation

$$d\tau = \left(1 - \frac{2GM}{c^2 r}\right)^{1/2} dt. \tag{12.98}$$

The proper time interval $d\tau$ represents an interval of time as measured in a rest frame at coordinate distance r. Consider a signal or spectral line of frequency $v_1 = 1/d\tau_1$ emitted at a radial distance r_1 and received at some radial distance $r_2 > r_1$. From Equation (12.98) the frequencies v_1 and v_2 of emitted and received photons are related by the equation

$$\frac{v_2}{v_1} = \frac{d\tau_1}{d\tau_2} = \frac{\left(1 - \dfrac{2GM}{c^2 r_1}\right)^{1/2}}{\left(1 - \dfrac{2GM}{c^2 r_2}\right)^{1/2}}.$$
(12.99)

As shown by Equation (12.99) a photon emitted outwards in a spherically symmetric gravitational field experiences a gravitational redshift. In the weak gravitational field limit (i.e. $2GM/rc^2 \ll 1$) Equation (12.99) implies

$$hv_2 = hv_1 - \frac{hv_1}{c^2}\frac{GM}{r_1},$$
(12.100)

if r_2 satisfies the condition $r_2 \gg r_1$.

The motions of particles in a spherical gravitational field are determined by Equations (12.22). In addition, particle orbits must satisfy the condition

$$g_{ij}\frac{dx^i}{ds}\frac{dx^j}{ds} = -1,$$
(12.101)

which can be used instead of one of the geodesic Equations (12.22). Equation (12.101) and the θ, ϕ and t components of Equation (12.22) are

$$\left(1 - \frac{2GM}{c^2 r}\right)\left(\frac{dt}{d\tau}\right)^2 - \frac{1}{c^2\left(1 - \dfrac{2GM}{c^2 r}\right)}\left(\frac{dr}{d\tau}\right)^2 - \frac{r^2}{c^2}\left[\sin^2\theta\left(\frac{d\phi}{d\tau}\right)^2 + \left(\frac{d\theta}{d\tau}\right)^2\right] = 1, \quad (12.102)$$

$$\frac{d}{d\tau}\left(r^2\frac{d\theta}{d\tau}\right) - r^2\sin\theta\cos\theta\left(\frac{d\phi}{d\tau}\right)^2 = 0,$$
(12.103)

$$\frac{d}{d\tau}\left(r^2\sin^2\theta\frac{d\phi}{d\tau}\right) = 0,$$
(12.104)

$$\frac{d}{d\tau}\left[\left(1 - \frac{2GM}{rc^2}\right)\frac{dt}{d\tau}\right] = 0.$$
(12.105)

We choose the spherical polar coordinates so that $\theta = \pi/2$. From Equation (12.104) it follows that if $\theta = \pi/2$ initially then θ remains equal to $\pi/2$. Integrating Equations (12.104) and (12.105) with $\theta = \pi/2$ gives

$$\frac{d\phi}{d\tau} = \frac{h}{r^2},$$
(12.106)

$$\frac{dt}{d\tau} = \frac{k}{1 - \frac{2GM}{rc^2}},$$ (12.107)

with h and k equal to constants. Substituting $dt/d\tau$ from Equation (12.107) into Equation (12.102) with $\theta = \pi/2$ leads to

$$\left(\frac{dr}{d\tau}\right)^2 + \frac{h^2}{r^2}\left(1 - \frac{2GM}{rc^2}\right) = c^2(k^2 - 1) + \frac{2GM}{r}.$$ (12.108)

Using Equation (12.106) to eliminate $d\tau$ from Equation (12.108) we find

$$\left(\frac{h}{r^2}\frac{dr}{d\phi}\right)^2 + \frac{h^2}{r^2} = c^2(k^2 - 1) + \frac{2GM}{r} + \frac{2GM}{c^2}\frac{h^2}{r^3}.$$ (12.109)

Substituting $u = 1/r$ as in Newtonian mechanics gives

$$\left(\frac{du}{d\phi}\right)^2 + u^2 = \frac{c^2}{h^2}(k^2 - 1) + \frac{2GMu}{h^2} + \frac{2GMu^3}{c^2}.$$ (12.110)

Differentiating Equation (12.110) with respect to ϕ we obtain

$$\frac{d^2u}{d\phi^2} + u = \frac{GM}{h^2} + \frac{3GMu^2}{c^2}.$$ (12.111)

The corresponding Newtonian equation of motion is

$$\frac{d^2u}{d\phi^2} + u = \frac{GM}{h^2},$$ (12.112)

with $h = r^2\dfrac{d\phi}{dt}$.

Deflection of electromagnetic waves

If we let $h \to \infty$ in Equation (12.111) then the resultant equation,

$$\frac{d^2u}{d\phi^2} + u = \frac{3GM}{c^2}u^2,$$ (12.113)

describes the trajectory of a photon (or neutrino). To the first approximation (i.e. neglecting the right-hand side of the equation) the solution of Equation (12.113) is

$$u_0 = \frac{\cos\phi}{b},$$ (12.114)

with b equal to the distance of closest approach. The approximate solution of Equation (12.113) given by Equation (12.114) represents a straight line with $\phi = 0$ at the distance of closest approach. Substituting the values for u given by Equation (12.114) on the right-hand side of Equation (12.113) gives

$$\frac{d^2u}{d\phi^2} + u = \frac{3GM}{b^2c^2}\cos^2\phi = \frac{3}{2}\frac{GM}{b^2c^2}(1 + \cos 2\phi). \tag{12.115}$$

Substituting

$$u_1 = A + B\cos 2\phi \tag{12.116}$$

into Equation (12.115) we obtain

$$-3B\cos 2\phi + A = \frac{3GM}{2b^2c^2}(1 + \cos 2\phi). \tag{12.117}$$

Equation (12.117) implies that u_1 given in Equation (12.116) is a solution of Equation (12.115) if

$$A = \frac{3GM}{2b^2c^2} = -3B, \tag{12.118}$$

the general solution of Equation (12.115) is the sum of the solution of the homogeneous equation given by Equation (12.114) and the solution given by Equations (12.116) and (12.118). Therefore we have

$$u = u_0 + u_1 = \frac{\cos\phi}{b} + \frac{GM}{2b^2c^2}(3 - \cos 2\phi). \tag{12.119}$$

Since u is equal to zero at the beginning and end of a trajectory the angle ϕ at those positions satisfies the equation

$$0 = \frac{\cos\phi}{b} + 3A - A(2\cos^2\phi - 1)$$

$$= -2A\cos^2\phi + \frac{\cos\phi}{b} + 4A = 0, \tag{12.120}$$

with A given by Equation (12.118). Solving Equation (12.120) for $\cos\phi$ and assuming that the deflection angle is small we have

$$\cos\phi = \frac{1}{4Ab} \pm \frac{1}{4A}\left(\left(\frac{1}{b}\right)^2 + 32A^2\right)^{1/2}$$

$$\simeq \frac{1}{4Ab} \pm \frac{1}{4Ab}(1 + 16A^2b^2). \tag{12.121}$$

Equation (12.121) can be written as

$$\cos\phi = \cos\left(\frac{\pi}{2} + \varepsilon\right) = -4Ab. \tag{12.122}$$

From Equations (12.118) and (12.122) we have

$$\cos\left(\frac{\pi}{2} + \varepsilon\right) \simeq \varepsilon \simeq -\frac{2GM}{bc^2}. \tag{12.123a}$$

Therefore, the total deflection $|2\varepsilon|$ becomes

$$|2\varepsilon| \simeq \frac{4GM}{bc^2} \qquad (|2\varepsilon| \ll 1). \tag{12.123b}$$

Advance of periastron

According to Newtonian gravitation the periastron of a binary star or the perihelion of a planet remains fixed in space. General relativity implies that the periastron (perihelion) moves in the same direction as the orbital motion. The solution of the Newtonian equation of motion given by Equation (12.112) can be expressed as

$$u_0 = \frac{GM}{h^2}(1 + e\cos(\phi - \phi_0)), \tag{12.124}$$

with e equal to the eccentricity of the orbit and ϕ_0 equal to the phase of the periastron (perihelion).

Substituting $u = u_0$ on the right-hand side of Equation (12.111) gives

$$\frac{d^2u}{d\phi^2} + u = \frac{GM}{h^2} + K[1 + e\cos(\phi - \phi_0)]^2, \tag{12.125}$$

with $K = \dfrac{3GM}{c^2}\left(\dfrac{GM}{h^2}\right)^2$.

We let

$$u_1 = A + B\phi\sin(\phi - \phi_0) + C\cos^2(\phi - \phi_0). \tag{12.126}$$

Equation (12.126) implies that

$$\frac{d^2u_1}{d\phi^2} = 2B\cos(\phi - \phi_0) - B\phi\sin(\phi - \phi_0) - 4C\cos 2(\phi - \phi_0). \tag{12.127}$$

Substituting Equations (12.126) and (12.127) into Equation (12.125) and using the equality

$$\cos^2(\phi - \phi_0) = \frac{1}{2}(\cos 2(\phi - \phi_0) + 1), \tag{12.128}$$

we obtain

$$2B\cos(\phi - \phi_0) + A - 4C\cos(\phi - \phi_0)$$

$$= \frac{GM}{h^2} + K[1 + 2e\cos(\phi - \phi_0) + e^2\cos^2(\phi - \phi_0)]$$

$$= \frac{GM}{h^2} + K + \frac{Ke^2}{2} + 2eK\cos(\phi - \phi_0) + \frac{Ke^2}{2}\cos 2(\phi - \phi_0). \tag{12.129}$$

From Equation (12.129) it follows that u_1 in Equation (12.126) is a solution of Equation

(12.125) if

$$A = \frac{GM}{h^2} + K + \frac{Ke^2}{2}, \tag{12.130a}$$

$$B = eK, \tag{12.130b}$$

$$C = -\frac{K}{8}e^2. \tag{12.130c}$$

From Equations (12.124), (12.126) and (12.130a,b,c) we have

$$u = u_0 + u_1 = \frac{GM}{h^2}(1 + e\cos(\phi - \phi_0)) + A + B\phi\sin(\phi - \phi_0) + C\cos 2(\phi - \phi_0). \tag{12.131}$$

Equation (12.131) can be rewritten as

$$u = \frac{GM}{h^2} + A + \frac{GMe}{h^2}(\cos(\phi - \phi_0) + K'\phi\sin(\phi - \phi_0))$$
$$+ C\cos 2(\phi - \phi_0), \tag{12.132}$$

with $K' = K/(GM/h^2)$. Since

$$\cos(\phi - \phi_0 - K'\phi) = \cos(\phi - \phi_0)\cos K'\phi + \sin(\phi - \phi_0)\sin K'\phi$$
$$\simeq \cos(\phi - \phi_0) + K'\phi\sin(\phi - \phi_0), \tag{12.133}$$

Equation (12.132) becomes

$$u = \frac{1}{r} \simeq \frac{GM}{h^2} + A + \frac{GMe}{h^2}\cos(\phi - \phi_0 - K'\phi) + C\cos 2(\phi - \phi_0). \tag{12.134}$$

The final term in Equation (12.134) is small. Because ϕ accumulates with each orbital period the periastron (or perihelion) motion implied by the $\cos(\phi - \phi_0 - K'\phi)$ term becomes appreciable. The perihelion of Mercury advances by about 43 seconds of arc per century due to general relativity.

12.5   Gravitational radiation

Suppose that the metric tensor g_{kl} can be approximated as

$$g_{kl} = \eta_{kl} + h_{kl}, \tag{12.135}$$

with η_{kl} equal to the Minkowski metric and $|h_{kl}| \ll 1$. To first order in h the Ricci tensor given by Equation (12.41) is

$$R_{kl} \simeq \frac{\partial}{\partial x^l}\Gamma^m_{mk} - \frac{\partial}{\partial x^m}\Gamma^m_{kl} + O(h^2), \tag{12.136}$$

and the affine connection given by Equations (12.23) and (12.135) is

$$\Gamma^m_{kl} = \frac{1}{2}\eta^{mn}\left[\frac{\partial h_{nl}}{\partial x^k} + \frac{\partial h_{nk}}{\partial x^l} - \frac{\partial h_{kl}}{\partial x^n}\right] + O(h^2).\tag{12.137}$$

If we restrict ourselves to first order in h we can raise and lower all indices using η^{kl} rather than g^{kl}. Therefore, we have

$$\eta^{km}h_{ml} = h^k_l,\tag{12.138}$$

$$\eta^{km}\frac{\partial}{\partial x^m} = \frac{\partial}{\partial x_k}.\tag{12.139}$$

Equations (12.136) and (12.137) lead to the first-order Ricci tensor

$$R_{kl} \simeq R^{(1)}_{kl} = \frac{1}{2}\left[\Box^2 h_{kl} - \frac{\partial\gamma_l}{\partial x^k} - \frac{\partial\gamma_k}{\partial x^l}\right],\tag{12.140}$$

where $\gamma_l = \dfrac{\partial h^m_l}{\partial x^m} - \dfrac{1}{2}\dfrac{\partial h^m_m}{\partial x^l}$,

$$\gamma_k = \frac{\partial h^m_k}{\partial x^m} - \frac{1}{2}\frac{\partial h^m_m}{\partial x^k}$$

and

$$\Box^2 = \frac{\partial^2}{c^2\partial t^2} - \frac{\partial^2}{\partial x^2} - \frac{\partial^2}{\partial y^2} - \frac{\partial^2}{\partial z^2}.$$

By transforming coordinates the variables γ_l (or γ_k) can be made to vanish. We choose coordinates by the transformation

$$x'^k = x^k + f^k(x^l),\tag{12.141}$$

where f^k satisfies the equation

$$\Box^2 f^k = \gamma_k.\tag{12.142}$$

Equation (12.142) and the definition of γ_k in Equation (12.140) imply that f^k is the same order of magnitude as h_{kl}. In the new coordinates the function h'_{kl} is defined by the equation

$$h'_{kl} = g'_{kl} - \eta_{kl}.\tag{12.143}$$

From the tensor transformation we have

$$h'_{kl} = h_{kl} - \frac{\partial f^k}{\partial x^l} - \frac{\partial f^l}{\partial x^k}.\tag{12.144}$$

Equation (12.144) implies that

$$\gamma'_k = \gamma_k - \Box^2 f^k.\tag{12.145}$$

Therefore, from Equations (12.142) and (12.145) we obtain

$$\gamma'_k = 0. \tag{12.146}$$

From Equations (12.140) and (12.146) it follows that a coordinate system can be found such that

$$R_{kl} = \frac{1}{2}\Box^2 h_{kl}. \tag{12.147}$$

In this coordinate system h_{kl} satisfy the equation

$$\frac{\partial h_{kl}}{\partial x^l} - \frac{1}{2}\frac{\partial h_l^l}{\partial x^k} = 0. \tag{12.148}$$

Summing over the indices kl of the Ricci tensor R_{kl} in Equation (12.147) we obtain

$$R = R_{kk} = \frac{1}{2}\Box^2 h_{kk}. \tag{12.149}$$

It follows that to first order in h_{kl} the Einstein tensor is

$$R_{kl} - \frac{1}{2}g_{kl}R = \frac{1}{2}\Box^2 h_{kl} - \frac{1}{4}\eta_{kl}\Box^2 h_{mm} = \frac{1}{2}\Box^2 h'_{kl}, \tag{12.150}$$

where $h'_{kl} = h_{kl} - \frac{1}{2}\eta_{kl}h_{mm}$.

Equation (12.150) implies that the Einstein Equation (12.66) becomes

$$\Box^2 h'_{kl} = 16\pi\frac{G}{c^4}T_{kl}. \tag{12.151}$$

Equation (12.151) which is similar to the corresponding equation in electromagnetic theory, has the well-known solution

$$h'_{kl}(x, t) = \frac{4G}{c^4}\int d^3x \frac{T_{kl}(x, ct - |x - x'|)}{|x - x'|}. \tag{12.152}$$

The divergence $T_{kl,l}$ of the energy–momentum tensor T_{kl} is equal to zero and therefore we have

$$T_{\alpha 0,0} + T_{\alpha\gamma,\gamma} = 0 \qquad (\alpha = 1, 2, 3; \gamma = 1, 2, 3). \tag{12.153}$$

Using the identity

$$(T_{\alpha\gamma}x^\beta)_{,\gamma} = T_{\alpha\beta} - T_{\alpha 0,0}x^\beta, \tag{12.154}$$

and integrating over space we obtain

$$0 = \int T_{\alpha\beta}d^3x - \int T_{\alpha 0,0}x^\beta d^3x. \tag{12.155}$$

The divergence theorem has been used to establish that the left-hand side of Equation

(12.155) is equal to zero. After exchanging the indices α and β and recalling that $T_{\alpha\beta} = T_{\beta\alpha}$ Equation (12.155) becomes

$$\int T_{\alpha\beta}d^3x = \frac{1}{2c}\frac{d}{dt}\int (T_{\alpha 0}x^\beta + T_{\beta 0}x^\alpha)d^3x. \tag{12.156}$$

Integrating the identity

$$(T_{\gamma 0}x^\alpha x^\beta)_{,\gamma} = T_{0\gamma,\gamma}x^\alpha x^\beta + T_{\alpha 0}x^\beta + T_{\beta 0}x^\alpha \tag{12.157}$$

over space and using the divergence theorem implies that

$$\int (T_{\alpha 0}x^\beta + T_{\beta 0}x^\alpha)d^3x = -\int T_{0\gamma,\gamma}x^\alpha x^\beta d^3x. \tag{12.158}$$

Equations (12.153), (12.156) and (12.158) imply that

$$\int T_{\alpha\beta}d^3x = \frac{1}{2c^2}\frac{d^2}{dt^2}\int T_{00}x^\alpha x^\beta d^3x. \tag{12.159}$$

From Equations (12.152) and (12.159) we find that

$$h'_{\alpha\beta} = \frac{2G}{c^4 r}\frac{d^2}{dt^2}\int \rho c^2 x^\alpha x^\beta d^3x. \tag{12.160}$$

The components h'_{0k} are time independent and therefore do not contribute to the gravitational wave field. If we further assume that gravitational radiation occurs at a single frequency Ω then Equation (12.160) can be expressed as

$$h'_{\alpha\beta} = \frac{2G}{c^4}\Omega^2 I_{\alpha\beta}\frac{e^{i\frac{\Omega}{c}(r-ct)}}{r}. \tag{12.161}$$

The solution for $h'_{\alpha\beta}$ given in Equation (12.161) satisfies the gauge condition $h'_{\alpha\beta;\beta} = 0$. We are interested in plane-wave solutions and we can assume that the plane wave travels along the z-axis. The gauge condition on the solutions given in Equation (12.161) can be further restricted. We assume that the solutions of Equation (12.161) satisfy the transverse gauge conditions $h^{TT}_{\alpha\alpha} = 0$ and $h^{TT}_{\alpha\beta}n^\beta = 0$ with $\hat{n}$ a unit vector in the direction of the wave. In the transverse gauge plane-wave solutions become

$$h^{TT}_{z\alpha} = 0, \tag{12.162a}$$

$$h^{TT}_{xx} = -h^{TT}_{yy} = \frac{G\Omega^2}{c^4}(\mathcal{I}_{xx} - \mathcal{I}_{yy})\frac{e^{i\Omega r/c}}{r}, \tag{12.162b}$$

$$h^{TT}_{xy} = \frac{2G\Omega^2}{c^4}\mathcal{I}_{xy}\frac{e^{i\Omega r/c}}{r}, \tag{12.162c}$$

where $\mathcal{I}_{\alpha\beta} \equiv I_{\alpha\beta} - \frac{1}{3}\delta_{\alpha\beta}I_{\gamma\gamma}.$

Gravitational radiation is quadrupole radiation. In the transverse gauge it is clear that dipole radiation does not contribute to the radiation field.

In order to determine the rate of energy loss caused by gravitational radiation we must find an expression for the energy flux in a gravitational wave. We assume that a gravitational wave of the form

$$h_{xx}^{TT} = A \cos[\Omega(z - ct)/c], \tag{12.163a}$$

$$h_{yy}^{TT} = - h_{xx}^{TT} \tag{12.163b}$$

propagates along the z-axis and interacts with oscillators in a plane perpendicular to the z-axis. There are N oscillators per unit area and under steady-state conditions the rate of energy loss of an oscillator excited by the gravitational wave is

$$\frac{dE}{dt} = m\gamma \left(\frac{d\xi}{dt}\right)^2 = m\gamma\Omega^2\xi_0^2 \sin^2(\Omega t + \phi), \tag{12.164}$$

with $\xi = \xi_0 \cos(\Omega t + \phi)$ equal to the displacement of an oscillator. Averaging over a period of oscillation the energy per unit time supplied to an oscillator becomes

$$\frac{\Omega}{2\pi}\int_0^{2\pi/\Omega} m\gamma\Omega^2\xi_0^2 \sin^2(\Omega t + \phi)dt = \frac{1}{2}m\gamma\Omega^2\xi_0^2. \tag{12.165}$$

Equation (12.165) implies that the flux F of the gravitational wave decreases by the amount

$$\delta F = -\frac{N}{2}m\gamma\Omega^2\xi_0^2 \tag{12.166}$$

with N equal to the number of oscillators per unit area. Each oscillator has the moment of inertia

$$I_{xx} = m(x_1^2 + x_2^2) = m\left(-\frac{l_0}{2} - \xi_0 \cos\Omega t\right)^2 + m\left(\frac{l_0}{2} + \xi_0 \cos\Omega t\right)^2$$

$$= \text{constant} + m\xi_0^2 \cos 2\Omega t - ml_0\xi_0 \cos\Omega t \tag{12.167}$$

where l_0 is the undisplaced length. For $\xi_0 \ll l_0$ Equation (12.167) becomes

$$I_{xx} = - ml_0\xi_0 \cos(\Omega t + \phi). \tag{12.168}$$

Equations (12.160), (12.162b) and (12.168) imply that the wave amplitude of each oscillator is

$$\delta h_{xx}^{TT} = - \frac{G\Omega^2}{c^4 r} ml_0\xi_0 \cos[\Omega(r - ct)/c + \phi]. \tag{12.169}$$

Summing over all oscillators in the plane perpendicular to the z-direction we obtain

$$\delta h_{xx}^{TT} = \frac{- 2\pi G}{c^4} ml_0\Omega^2\xi_0 \int_0^\infty N \cos[\Omega(r - ct)/c + \phi]\frac{RdR}{r}$$

$$= \frac{-2\pi G}{c^4} m l_0 \Omega^2 \xi_0 \int_z^\infty N \cos[\Omega(r - ct)/c + \phi]dr \qquad (12.170)$$

with $r = (R^2 + z^2)^{1/2}$ and N equal to the number of oscillators per unit area. The final integral in Equation (12.170) appears to have no well-defined value at the limit $r = \infty$. However, this integral can be evaluated by the following mathematical trick. Multiply the integrand by $e^{-\varepsilon r}$, evaluate the integral and then let $\varepsilon \to 0$. We obtain

$$\delta h_{xx}^{TT} = - \delta h_{yy}^{TT} = - 2\pi \frac{G}{c^3} N m \Omega l_0 \xi_0 \sin[\Omega(z - ct)/c + \phi]. \qquad (12.171)$$

Adding δh_{xx}^{TT} to h_{xx}^{TT} in Equation (12.163a) we have

$$h_{xx}^{TT} + \delta h_{xx}^{TT} = \left(A - 2\pi \frac{G}{c^3} N m \Omega l_0 \xi_0 \sin \phi \right) \cos[(\Omega(z - ct)/c - \psi] \qquad (12.172)$$

with ψ equal to a small phase shift. From Equation (12.172) the change in amplitude of the wave caused by the oscillators is

$$\delta A = - 2\pi \frac{NG}{c^3} m \Omega l_0 \xi_0 \sin \phi. \qquad (12.173)$$

In order to determine δF, the change in flux of the gravitational wave, we must relate the damping constant γ given in Equation (12.164) to the amplitudes A and ξ (i.e. the amplitude of the incident gravitational wave and excited oscillator). The proper length $l(t)$ of the spring, which is assumed to contain point masses of mass m at positions x_1 and x_2, is

$$l(t) = \int_{x_1(t)}^{x_2(t)} [1 + h_{xx}^{TT}(t)]^{1/2}dt. \qquad (12.174)$$

The corresponding force equations for the spring are

$$m \frac{d^2 x_1}{dt^2} = - k(l_0 - l) - \gamma m \frac{d(l_0 - l)}{dt} \qquad (12.175)$$

and

$$m \frac{d^2 x_2}{dt^2} = - k(l - l_0) - \gamma m \frac{d(l - l_0)}{dt}. \qquad (12.176)$$

To first order in h_{xx}^{TT} Equation (12.174) implies that

$$\xi = l - l_0 = x_2 - x_1 - l_0 + \frac{1}{2} h_{xx}^{TT}(x_2 - x_1). \qquad (12.177)$$

Solving Equation (12.177) to first order in h_{xx}^{TT} gives

$$x_2 - x_1 = l_0 + \xi - \frac{1}{2} h_{xx}^{TT} l_0. \qquad (12.178)$$

Equations (12.175), (12.176) and (12.178) imply that

$$\frac{d^2\xi}{dt^2} + 2\gamma\frac{d\xi}{dt} + \omega_0^2\xi = \frac{1}{2}l_0\frac{d^2h_{xx}^{TT}}{dt^2} \tag{12.179}$$

where $\omega_0^2 = k/m$. The incident gravitational wave has the solution $h_{xx}^{TT} = A\cos\Omega t$, and therefore the steady-state solution of Equation (12.179) is

$$\xi = \xi_0\cos(\Omega t + \phi) \tag{12.180}$$

with

$$\xi_0 = \frac{1}{2}l_0\Omega^2 A/[(\omega_0 - \Omega)^2 + 4\Omega^2\gamma^2]^{1/2}$$

$$\tan\phi = 2\gamma\Omega/(\omega_0^2 - \Omega^2).$$

From Equations (12.166), (12.173) and (12.180) we find that the change in flux δF of the gravitational wave divided by the change in amplitude δA is

$$\frac{\delta F}{\delta A} = \frac{c^3}{16\pi}\frac{\Omega^2 A}{G} \tag{12.181}$$

at the resonant frequency $\Omega = \omega_0$. Equation (12.163a) implies that the average square of the wave amplitude is

$$\langle(h_{xx}^{TT})^2\rangle = \frac{1}{2}A^2. \tag{12.182}$$

The integral of Equation (12.181) and Equation (12.182) imply that the gravitational energy flux is

$$F = \frac{c^3}{32\pi}\frac{\Omega^2}{G}\langle h_{\alpha\beta}^{TT}h^{TT\alpha\beta}\rangle. \tag{12.183}$$

Using Equations (12.162b, c) and (12.183) we obtain

$$F = \frac{G\Omega^6}{32\pi c^5 r^2}\langle 2(\mathcal{I}_{xx} - \mathcal{I}_{yy})^2 + 8\mathcal{I}_{xy}^2\rangle. \tag{12.184}$$

We can assume that the source of gravitational radiation is at a distance r along the z-axis. Moreover, $\mathcal{I}_{\alpha\alpha}$ satisfies the identity

$$\mathcal{I}_{\alpha\alpha} = \mathcal{I}_{xx} + \mathcal{I}_{yy} + \mathcal{I}_{zz} = 0. \tag{12.185}$$

Expanding the terms inside the brackets of Equation (12.184) gives

$$2\mathcal{I}_{xx}^2 + 2\mathcal{I}_{yy}^2 - 4\mathcal{I}_{xx}\mathcal{I}_{yy} + 8\mathcal{I}_{xy}^2. \tag{12.186}$$

Equations (12.184–12.186) imply that

$$F = \frac{G\Omega^6}{16\pi c^5 r^2}\langle 2\mathcal{I}_{\alpha\beta}\mathcal{I}_{\alpha\beta} - 4\mathcal{I}_{z\alpha}\mathcal{I}_{z\alpha} + \mathcal{I}_{zz}^2\rangle. \tag{12.187}$$

In order to calculate the rate of gravitation energy loss we must first express Equation (12.187) in a covariant form and then integrate over the solid angle. Let $n^j = x^j/r$ be the unit vector defined on a sphere that surrounds the radiating source. The covariant form of Equation (12.187) then becomes

$$F = \frac{G\Omega^6}{16\pi c^5 r^2} \langle 2\mathcal{I}_{\alpha\beta}\mathcal{I}_{\alpha\beta} - 4n^\alpha n^\beta \mathcal{I}_{\alpha\gamma}\mathcal{I}_{\beta\gamma} + n^\alpha n^\beta n^\gamma n^\delta \mathcal{I}_{\alpha\beta}\mathcal{I}_{\gamma\delta} \rangle. \tag{12.188}$$

By choosing $\hat{n}$ along the z-axis it can readily be shown that Equation (12.188) reduces to Equation (12.187). From Equation (12.188) and the integrals

$$\int n^\alpha n^\beta \sin\theta d\theta d\phi = \frac{4\pi}{3}\delta^{\alpha\beta} \tag{12.189}$$

$$\int n^\alpha n^\beta n^\gamma n^\delta \sin\theta d\theta d\phi = \frac{4\pi}{15}(\delta^{\alpha\beta}\delta^{\gamma\delta} + \delta^{\alpha\gamma}\delta^{\beta\delta} + \delta^{\alpha\delta}\delta^{\beta\gamma}) \tag{12.190}$$

the rate of energy loss becomes

$$\frac{dE}{dt} = \int Fr^2 \sin\theta d\theta d\phi$$

$$= \frac{G\Omega^6}{4c^5}\langle (2 - 4/3)\mathcal{I}_{\alpha\beta}\mathcal{I}_{\alpha\beta} + \frac{1}{15}(\mathcal{I}_{\alpha\alpha}\mathcal{I}_{\beta\beta} + 2\mathcal{I}_{\alpha\beta}\mathcal{I}_{\alpha\beta})\rangle = \frac{G\Omega^6}{5c^5}\langle \mathcal{I}_{\alpha\beta}\mathcal{I}_{\alpha\beta}\rangle. \tag{12.191}$$

In evaluating Equation (12.191) we have used the identities $\mathcal{I}_{\alpha\alpha} = 0$ and $\mathcal{I}_{\alpha\beta} = \mathcal{I}^{\alpha\beta}$. A more general expression for the gravitational energy loss rate is

$$\frac{dE}{dt} = \frac{G}{5c^5}\langle \dddot{\mathcal{I}}_{\alpha\beta}\dddot{\mathcal{I}}_{\alpha\beta}\rangle. \tag{12.192}$$

Binary stars in circular orbit radiate gravitational radiation because the mass distribution is time varying and non-axially symmetric. If the axis of rotation is chosen to be the z-axis then I_{xx} becomes

$$I_{xx} = (M_1 a_1^2 + M_2 a_2^2)\cos^2\Omega t + \text{constant} \tag{12.193}$$

where $M_1 a_1 = M_2 a_2$. Using Equation (12.193) and similar equations for I_{yy}, I_{xy} and I_{yx} we obtain

$$I_{xx} = \frac{1}{2}\mu a^2 \cos 2\Omega + \text{constant}$$

$$I_{yy} = -\frac{1}{2}\mu a^2 \cos 2\Omega + \text{constant}$$

$$I_{xy} = I_{yx} = \frac{1}{2}\mu a^2 \sin 2\Omega + \text{constant} \tag{12.194}$$

where $a = a_1 + a_2$ and $\mu = \dfrac{M_1 M_2}{M_1 + M_2}$.

Equations (12.192) and (12.194) imply that

$$\frac{dE}{dt} = \frac{2}{5}\frac{G}{c^5}(2\Omega)^6\left(\frac{\mu a^2}{2}\right)^2\langle\sin^2 2\Omega t + \cos^2 2\Omega t\rangle$$

$$= \frac{32}{5}\frac{G^4}{c^5}\frac{M^3\mu^2}{a^5} \tag{12.195}$$

where $\Omega^2 = GM/a^3$ and $M = M_1 + M_2$.

Problems

1.   a. Prove that any tensor T^{ik} can be expressed as the sum of a symmetric and antisymmetric tensor.
   b. A_{ik} and S^{ik} are antisymmetric and symmetric tensors, respectively. Prove the relation $A_{ik}S^{ik} = 0$.
   c. For any tensor T^{ik} prove that T^i_i is invariant under coordinate transformations but T^{ii} and T_{ii} are not invariant. (Summations over indices are understood.)

2.   a. Prove that the Kronecker delta δ^i_k is a tensor.
   b. Prove that the affine connection Γ^i_{jk} is not a tensor.

3.   Prove that the invariant proper volume element in four-dimensional space is

$$dV = (-g)^{1/2}cdxdydzdt,$$

with g equal to the determinant of the metric tensor g_{ik}. *Hint:* One can always find a coordinate system in which the metric is locally a Minkowski metric (η_{ij}).

4.   a. Prove the symmetry relations $\Gamma^k_{mi} = \Gamma^k_{im}$ and $R^k_{lmn} = -R^k_{lnm}$.
   b. Prove the symmetry properties given by Equations (12.44–12.46).
   c. Verify Equation (12.52).

5.   Verify Equation (12.50).

6.   Prove Equations (12.102–12.105) explicitly.

7.   Binary stars of mass M_1 and M_2 are in circular orbit. The distances of the stars from the center of mass are a_1 and a_2 respectively, and therefore their initial separation is $a = a_1 + a_2$. Gravitational radiation causes the orbital energy to decrease and the binary stars move closer together. Let t_0 equal the time at which the two stars coalesce and determine how a decreases as a function of time.

8.   In Newtonian mechanics particle trajectories in a gravitational field satisfy the equation

$$\frac{d^2x^\alpha}{dt^2} = -\frac{\partial\Phi}{\partial x^\alpha}, \quad \alpha = 1, 2, 3.$$

Consider the above equation a geodesic equation and prove that the nonzero components of the Riemann tensor are

$$R^\alpha_{0\beta 0} = -R^\alpha_{00\beta} = \frac{\partial^2 \Phi}{\partial x^\alpha \partial x^\beta}.$$

9.   Obtain an equation of hydrostatic equilibrium from Equation (12.84) and the Schwarzschild metric given by Equation (12.88).

13 Neutron stars and black holes

13.1 Neutron stars

Hydrostatic equilibrium from the Einstein equations

We consider a spherically symmetric neutron star. For matter at rest the four-velocity u^i is $(u^0, 0, 0, 0)$, and therefore, since

$$- ds^2 = g_{00}(dx^0)^2 \tag{13.1}$$

and

$$u^0 = \frac{dx^0}{ds}, \tag{13.2}$$

we have

$$u_0 = g_{0i}u^i = g_{00}u^0 = \sqrt{-g_{00}}, \quad u_1 = u_2 = u_3 = 0. \tag{13.3}$$

Equations (12.88) and (13.3) imply that the energy–momentum tensor given in Equation (12.63) becomes

$$T_{ik} = \begin{bmatrix} \rho c^2 e^\nu & 0 & 0 & 0 \\ 0 & Pe^\lambda & 0 & 0 \\ 0 & 0 & Pr^2 & 0 \\ 0 & 0 & 0 & Pr^2 \sin^2 \theta \end{bmatrix}. \tag{13.4}$$

From The Schwarzschild metric given in Equation (12.88), and the equation for the affine connection given in Equation (12.23), the Einstein tensor defined in Equation (12.43) can be evaluated from Equations (12.39), (12.41) and (12.42). Calculating the G_{00}, G_{11} and G_{22} components of the Einstein equations (12.66) with the energy–momentum tensor equal to Equation (13.4), we obtain

$$\frac{8\pi G\rho}{c^2} = -e^{-\lambda}\left(\frac{1}{r^2} - \frac{\dfrac{d\lambda}{dr}}{r}\right) + \frac{1}{r^2}, \tag{13.5}$$

$$\frac{8\pi GP}{c^4} = -\frac{1}{r^2} + e^{-\lambda}\left(\frac{1}{r^2} + \frac{\dfrac{d\nu}{dr}}{r}\right), \tag{13.6}$$

$$\frac{8\pi GP}{c^4} = e^{-\lambda}\left[-\frac{\dfrac{dv}{dr}\dfrac{d\lambda}{dr}}{4} + \frac{1}{4}\left(\frac{dv}{dr}\right)^2 + \frac{1}{2}\frac{d^2v}{dr^2} + \frac{1}{2}\left(\frac{\dfrac{dv}{dr} - \dfrac{d\lambda}{dr}}{r}\right)\right].$$

(13.7)

Eliminating P from Equations (13.6) and (13.7) gives

$$\frac{e^\lambda}{r^2} = \frac{\dfrac{dv}{dr}\dfrac{d\lambda}{dr}}{4} - \frac{1}{4}\left(\frac{dv}{dr}\right)^2 - \frac{\dfrac{d^2v}{dr^2}}{2r} + \frac{\dfrac{dv}{dr} + \dfrac{d\lambda}{dr}}{2r} + \frac{1}{r^2}.$$

(13.8)

Adding Equations (13.5) and (13.6), we have

$$-\frac{8\pi G}{c^2}\left(\rho + \frac{P}{c^2}\right) = \frac{e^{-\lambda}}{r}\left(\frac{dv}{dr} + \frac{d\lambda}{dr}\right).$$

(13.9)

Differentiating Equation (13.6) with respect to r leads to the equation

$$\frac{8\pi G}{c^4}\frac{dP}{dr} = -\frac{2}{r^3} + e^{-\lambda}\left(\frac{\dfrac{d\lambda}{dr}}{r^2} + \frac{\dfrac{d\lambda}{dr}\dfrac{dv}{dr}}{r} + \frac{2}{r^3} + \frac{\dfrac{dv}{dr}}{r^2} - \frac{\dfrac{d^2v}{dr^2}}{r}\right).$$

(13.10)

Equations (13.8) and (13.10) imply that

$$\frac{8\pi G}{c^4}\frac{dP}{dr} = e^{-\lambda}\frac{1}{2r}\frac{dv}{dr}\left(\frac{d\lambda}{dr} + \frac{dv}{dr}\right).$$

(13.11)

From Equations (13.9) and (13.11) we obtain

$$\frac{1}{c^2}\frac{dP}{dr} = -\frac{1}{2}\frac{dv}{dr}\left(\rho(r) + \frac{P(r)}{c^2}\right).$$

(13.12)

Equation (13.5) can be re-expressed as

$$\frac{dre^{-\lambda}}{dr} = 1 - \frac{8\pi r^2 G\rho}{c^2}.$$

(13.13)

Integrating Equation (13.13) gives

$$e^{-\lambda} = 1 - \frac{8\pi G}{rc^2}\int_0^r \rho r^2\, dr$$

$$= 1 - \frac{2GM_r}{rc^2},$$

(13.14)

where M_r in Equation (13.14) is identical to the Newtonian expression for M_r. Solving for $\dfrac{dv}{dr}$ in Equation (13.12), we have

$$\frac{dv}{dr} = -2\frac{dP}{dr}\frac{1}{(\rho c^2 + P)}.$$

(13.15)

Equations (13.6) and (13.15) imply that

$$e^{-\lambda}\left(-\frac{2}{r}\frac{dP}{dr}\frac{1}{(\rho c^2 + P)} + \frac{1}{r^2}\right) - \frac{1}{r^2} = \frac{8\pi GP}{c^4}.$$ (13.16)

Finally, solving for dP/dr from Equations (13.14) and (13.16), we obtain

$$\frac{dP}{dr} = -\rho\frac{GM_r}{r^2}\frac{\left(1 + \frac{P}{\rho c^2}\right)\left(1 + \frac{4\pi r^3 P}{M_r c^2}\right)}{\left(1 - \frac{2GM_r}{rc^2}\right)}.$$ (13.17)

Equation (13.17) is the equation for hydrostatic equilibrium in general relativity and is known as the Oppenheimer–Volkoff equation. It reduces to the Newtonian equation of hydrostatic equilibrium in the nonrelativistic limit. The term in the denominator of Equation (13.17) equals zero when

$$r = r_g = \frac{2GM_r}{c^2}.$$ (13.18)

The radial distance r_g given in Equation (13.18) is known as the Schwarzschild radius. Equation (13.17) implies that r must exceed r_g everywhere throughout the interior of a neutron star.

Hydrostatic equilibrium from a variational principle

Equation (13.17) can also be derived from a variational principle. We require that the expression

$$M(r) = 4\pi\int_0^r \rho r'^2\,dr'$$ (13.19)

is an extremum with respect to adiabatic Eulerian variations in which the number of baryons,

$$N = \int_0^R 4\pi r^2 n(r)\left(1 - \frac{2GM_r}{c^2 r}\right)^{-1/2}\,dr,$$ (13.20)

remains unchanged. Using the Lagrangian-multiplier method of solution, we require that there exist a constant λ and that $M + \lambda N$ remains stationary with respect to arbitrary Eulerian variations. From Equations (13.19) and (13.20) we obtain

$$\delta M + \lambda\delta N = \int_0^\infty 4\pi r^2 \delta\rho(r)\,dr$$

$$+ \lambda\int_0^\infty 4\pi r^2\left(1 - \frac{2GM(r)}{c^2 r}\right)^{-1/2}\delta n(r)\,dr$$

$$+ \lambda\frac{G}{c^2}\int_0^\infty 4\pi r\left(1 - \frac{2GM(r)}{c^2 r}\right)^{-3/2} n(r)\,\delta M(r)\,dr = 0.$$ (13.21)

Using Equation (13.19) and changing the order of integration we find that the final term in Equation (13.21) becomes

$$\lambda \frac{G}{c^2} \int_0^\infty 4\pi r \left(1 - \frac{2GM(r)}{c^2 r}\right)^{-3/2} n(r)\,dr \int_0^r 4\pi r'^2 \delta\rho(r')\,dr'$$

$$= \lambda \frac{G}{c^2} \int_0^\infty \left[\int_0^r 4\pi r' \left(1 - \frac{2GM(r')}{c^2 r'}\right)^{-3/2} n(r')\,dr'\right] 4\pi r^2 \delta\rho(r)\,dr. \tag{13.22}$$

For adiabatic changes the first law of thermodynamics implies that

$$0 = \delta\left(\frac{\rho c^2}{n}\right) + P\delta\left(\frac{1}{n}\right). \tag{13.23}$$

From Equation (13.23) we have

$$\delta n = \frac{nc^2}{P + \rho c^2}\delta\rho. \tag{13.24}$$

Equations (13.22) and (13.24) imply that Equation (13.21) becomes

$$0 = \delta M + \lambda \delta N = \int_0^\infty 4\pi r^2 \left[1 + \frac{\lambda n(r)c^2}{P + \rho c^2}\left(1 - \frac{2GM(r)}{c^2 r}\right)^{-1/2}\right.$$

$$\left. + \frac{\lambda G}{c^2} \int_0^r 4\pi r' \left(1 - \frac{2GM(r')}{c^2 r'}\right)^{-3/2} n(r')\,dr'\right] \delta\rho(r)\,dr. \tag{13.25}$$

Because $\delta M + \lambda \delta N$ must equal zero for arbitrary small variations $\delta\rho$, Equation (13.25) leads directly to the equation

$$-\frac{1}{\lambda} = \frac{n}{P + \rho c^2}\left(1 - \frac{2GM(r)}{c^2 r}\right)^{-1/2} + \frac{G}{c^2}\int_0^r 4\pi r' n(r')\left(1 - \frac{2GM(r')}{c^2 r'}\right)^{-3/2} dr'. \tag{13.26}$$

Since λ is a constant and therefore independent of r, the derivative of Equation (13.26) with respect to r is

$$0 = \frac{d\left(-\frac{1}{\lambda}\right)}{dr} = c^2\left[\frac{\frac{dn}{dr}}{P + \rho c^2} - \frac{n}{(P + \rho c^2)^2}\left(\frac{dP}{dr} + c^2\frac{d\rho}{dr}\right)\right]\left(1 - \frac{2GM(r)}{c^2 r}\right)^{-1/2}$$

$$+ \frac{nc^2}{P + \rho c^2}\left(\frac{4\pi G\rho r}{c^2} - \frac{GM(r)}{c^2 r^2}\right)\left(1 - \frac{2GM(r)}{c^2 r}\right)^{-3/2} - \frac{4\pi Gnr}{c^2}\left(1 - \frac{2GM(r)}{c^2 r}\right)^{-3/2}.$$

$$\tag{13.27}$$

From Equation (13.23) we have

$$\frac{c^2\,d\rho}{dr} = \frac{P + \rho c^2}{n}\frac{dn}{dr}. \tag{13.28}$$

Equation (13.28) implies that the first term on the right-hand side of Equation (13.27) reduces to

$$\frac{-nc^2}{(P + \rho c^2)^2} \frac{dP}{dr} \left(1 - \frac{2GM(r)}{c^2 r}\right)^{-1/2}.$$

(13.29)

It follows that Equation (13.27) becomes

$$\frac{n}{(P + \rho c^2)^2} \frac{dP}{dr} = \frac{n}{P + \rho c^2} \left(\frac{4\pi\rho Gr}{c^2} - \frac{GM(r)}{c^2 r^2} - \frac{4\pi Gr(P + \rho c^2)}{c^4}\right) \left(1 - \frac{2GM(r)}{c^2 r}\right)^{-1}.$$

(13.30)

Equation (13.30) can be re-expressed as

$$r^2 \frac{dP}{dr} = -\frac{G}{c^2}(P + \rho c^2)\left(1 - \frac{2GM(r)}{c^2 r}\right)^{-1}\left(M(r) + \frac{4\pi r^3 P}{c^2}\right).$$

(13.31)

Equation (13.31) is identical to Equation (13.17).

In the interiors of neutron stars, as in white dwarfs, the equation of state has the form $P = P(\rho)$. It follows that the structure of a neutron star is determined by Equations (13.19), (13.17) (or 13.31) and the equation of state.

Interaction between neutrons

The interiors of neutron stars consist mostly of nonrelativistic, degenerate neutrons. The equation of state of a degenerate neutron gas in which interactions between neutrons is neglected is similar to that of a degenerate, noninteracting electron gas, described in Chapter 3, except that the neutron mass replaces the electron mass. As already described in Chapters 3 and 11, corrections to the equation of state of a degenerate electron gas caused by Coulomb interactions between electrons and other particles are small. Because neutrons interact with each other by means of the strong interaction, the equation of state of a degenerate neutron gas is changed appreciably by nuclear interactions. At neutron densities comparable to those inside the atomic nucleus (i.e. $\simeq 1.4 \times 10^{14}\,\mathrm{g\,cm^{-3}}$) the force between neutrons is attractive, and consequently the pressure of the interacting neutrons is reduced as compared to the pressure of a free neutron gas.

The strong interaction between two neutrons can be described approximately by an attractive Yukawa potential at densities comparable to those inside the atomic nucleus and by a repulsive Yukawa potential at higher densities. When the nuclear force is attractive, the interaction potential is

$$V_{\mathrm{INT}} = -g_1^2 \frac{e^{-\mu_1 r}}{r},$$

(13.32)

with $g_1^2/\hbar c \simeq 15$, $\mu_1^{-1} \simeq \dfrac{\hbar}{m_\pi c} = 1.4 \times 10^{-13}\,\mathrm{cm}$. When the interaction between neutrons is repulsive, we have

$$V_{\text{INT}} = g_2^2 \frac{e^{-\mu_2 r}}{r}, \tag{13.33}$$

with $g_2^2/\hbar c \simeq 15$, $\mu_2^{-1} = 0.4 \times 10^{-13}$ cm. Calculating the strong interaction energy E_{INT} in a volume V classically from Equation (13.32) or (13.33) gives

$$E_{\text{INT}} = \frac{1}{2} \sum_{i \neq j} V_{ij} = \mp \frac{1}{2} n^2 g^2 \int \int \frac{e^{-\mu r_{ij}}}{r_{ij}} \, dV_i dV_j, \tag{13.34}$$

where n is the neutron number density.

If we substitute $r = r_{ij}$ in Equation (13.34), we obtain

$$V \int \frac{e^{-\mu r}}{r} 4\pi r^2 \, dr = \frac{4\pi}{\mu^2} V. \tag{13.35}$$

Equations (13.34) and (13.35) imply that

$$E_{\text{INT}} = \mp \frac{1}{2} n^2 g^2 \frac{4\pi}{\mu^2} v. \tag{13.36}$$

From Equations (3.81–3.83) and (13.35) the total energy per unit volume becomes

$$\varepsilon = \varepsilon_k \mp \frac{2\pi n^2}{\mu^2} g^2, \tag{13.37}$$

where $\varepsilon_k = \frac{3}{10} (3\pi^2)^{2/3} \frac{\hbar^2}{m_n} n^{5/3}$.

We note that in neutron stars the density never becomes sufficiently high for the neutrons to be relativistic, and therefore the Fermi- (or kinetic-) energy density ε_k is always proportional to $n^{5/3}$.

From Equation (13.37) and the first law of thermodynamics the pressure becomes

$$P = \frac{-d(\varepsilon/n)}{d(1/n)} = n^2 \frac{d(\varepsilon/n)}{dn} = P_K \mp \frac{2\pi n^2 g^2}{\mu^2}, \tag{13.38}$$

with $P_K = 2\varepsilon_k/3$. At high densities the pressure caused by the kinetic-energy density ε_k is less than that due to the interaction between neutrons, and therefore the kinetic-energy density $\varepsilon = \rho c^2$ is proportional to n^γ with $\gamma = 2$. From Equation (13.38) we obtain

$$P = \frac{n \, d\rho \, c^2}{dn} - \rho c^2 = (\gamma - 1)\rho c^2 \underset{\rho \to \infty}{\rightarrow} \rho c^2. \tag{13.39}$$

Equation (13.32) implies that at high densities the velocity of sound is

$$v_s = \left(\frac{dP}{d\rho}\right)^{1/2} \rightarrow c. \tag{13.40}$$

It is of interest to compare the expression for the pressure given in Equation (13.39) with that of a relativistic gas of noninteracting particles. For an ideal relativistic gas the pressure is

$$P = \frac{1}{3}\rho c^2, \tag{13.41}$$

and therefore the speed of sound is

$$v_s = \frac{1}{\sqrt{3}}c \text{ as } \rho \to \infty. \tag{13.42}$$

The interaction energy of neutrons can also be calculated quantum mechanically. Because the strong interaction is short range, we can assume that the neutrons have plane-wave solutions. If a total of N neutrons are present in some volume V, then the wave function that describes their motion must be antisymmetric with respect to the interchange of particles, because neutrons are fermions. Therefore, the total wave function ψ becomes the Slater determinant

$$\psi = \frac{1}{(N!)^{1/2}} \begin{vmatrix} u_1(1) u_1(2) & \cdots & u_1(N) \\ \vdots & & \vdots \\ u_N(1) u_N(2) & \cdots & u_N(N) \end{vmatrix}, \tag{13.43}$$

where the wave functions $u_n(m)$ are individual particle wave functions. The index n in $u_n(m)$ labels the particle, and the index m in $u_n(m)$ denotes the position r_m of particles. Equation (13.43) implies that if the positions of any two particles are interchanged, then the sign of the total wavefunction changes and consequently ψ is antisymmetric. The individual particle wave functions $u_n(m)$ given in Equation (13.43) can be expressed as the products of position-dependent and spinor-dependent wave functions, i.e.

$$u_n(m) = u_n(r_m)\chi_n(\sigma), \tag{13.44}$$

where $u_n(r_m)$ are plane-wave solutions and $\chi_n(\sigma)$ are some linear combination of the spin wave functions $\begin{pmatrix} 1 \\ 0 \end{pmatrix}$ and $\begin{pmatrix} 0 \\ 1 \end{pmatrix}$. The wave functions $u_n(m)$ given in Equation (13.44) must satisfy the orthonormal conditions

$$\frac{1}{2}\sum_{\text{spin}} \int dV_m u_n^\star(m)u_l(m) = \delta_{nl}, \tag{13.45}$$

where the factor $\frac{1}{2}$ represents an average over initial spin states.

The Hamiltonian describing the interaction between neutrons is

$$H = -\frac{\hbar^2}{2m_n}\nabla^2 + V_{\text{INT}}, \tag{13.46}$$

with V_{INT} given by Equation (13.32) (or 13.33). From Equations (13.43) and (13.46) we obtain

$$\langle \psi | H | \psi \rangle = \sum \frac{p^2}{2m_n} + \sum_{m<n} [\langle mn | V_{mn} | mn \rangle - \langle mn | V_{mn} | nm \rangle]$$

$$= \sum \frac{p^2}{2m_n} + \frac{1}{2} \sum_{m,n} [\langle mn | V_{mn} | mn \rangle - \langle mn | V_{mn} | nm \rangle]. \tag{13.47}$$

The first two terms in Equation (13.47) normalized to a unit volume are equal to

$$\varepsilon_k \mp \frac{2\pi n^2 g^2}{\mu^2}, \tag{13.48}$$

which is identical to the interaction-energy density given by Equation (13.37). It follows that the quantum-mechanical energy density differs from the classical value only because of the exchange term that appears in Equation (13.47).

Using the orthonormal conditions given by Equation (13.45), we see that the exchange term in Equation (13.47) becomes

$$I = -\frac{1}{2} \sum_{m,n} \langle mn | V_{\text{INT}} | nm \rangle$$

$$= -\frac{1}{2} \int dV_1 dV_2 V_{\text{INT}} | \rho(r_1, r_2) |^2, \tag{13.49}$$

with $\rho(r_1, r_2) = \sum_{m=1}^{N} u_m^{\star}(r_2) u_m(r_1)$.

For plane-wave solutions normalized to unit volume, $\rho(r_1, r_2)$ in Equation (13.49) is equal to

$$\rho(r_1, r_2) = \frac{1}{(2\pi)^3} \sum_k e^{ik \cdot (r_1 - r_2)}$$

$$= \frac{1}{(2\pi)^3} \int_0^{k_F} e^{ik \cdot r_{12}} d^3k$$

$$= \frac{1}{(2\pi)^2} \int_0^{k_F} \int_{-1}^{+1} e^{ikr_{12} \cos \theta} d(\cos \theta) k^2 dk, \tag{13.50}$$

with the Fermi momentum $p_F = \hbar k_F$. Integrating Equation (13.50) gives

$$\rho(r_1, r_2) = \frac{1}{(2\pi)^2} \frac{1}{r_{12}^3} (\sin k_F r_{12} - k_F r_{12} \cos k_F r_{12}). \tag{13.51}$$

Making the change of variables

$$R = \frac{1}{2}(r_1 + r_2)$$

$$r = r_{12} = r_1 - r_2 \tag{13.52}$$

the exchange term given by Equation (13.51) and normalized to unit volume becomes

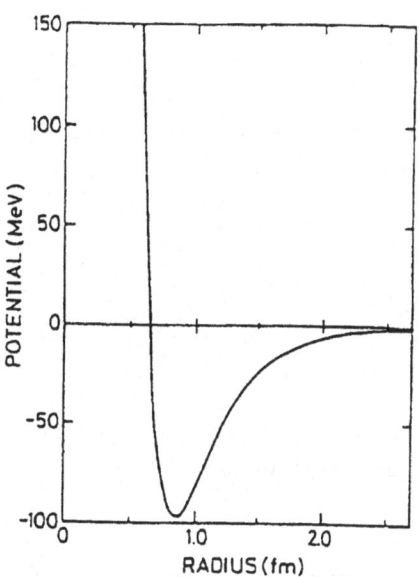

Figure 13.1. The approximate nucleon-nucleon potential shown as a function of the distance between two nucleons. The radius of the hard core nuclear potential is about 0.4 fm. The long-range (2 fm) attractive part of the potential is caused by π meson exchange. The most attractive portion of the nucleon potential which occurs at about 1 fm is more complicated. It is believed to result mainly from scalar meson exchange. The hard core (i.e. repulsive) part of the nucleon potential is believed to be caused by ρ and ω vector meson exchange between nucleons.

$$I = \mp \frac{g^2}{2} \int d^3R d^3r \rho^2(r) \frac{e^{-\mu r}}{r}$$

$$= \mp 2\pi g^2 \int \rho^2(r) \frac{e^{-\mu r}}{r} r^2 \, dr. \tag{13.53}$$

Figure 13.1 gives a schematic nucleon–nucleon interaction potential. Approximate neutron star mass–density relations are shown in Figure 13.2.

Superfluidity and superconductivity in neutron stars

In a diagram of pressure versus temperature normal substances possess a triple point. Small perturbations about a triple point change the phase to either gaseous, liquid or solid. Unlike other atoms and molecules helium has two triple points. The triple point separating normal liquid, gaseous and superfluid helium is referred to as the λ-point. In the vicinity of the λ-point, which occurs at $T \simeq 2$ K helium undergoes a phase transition and becomes a superfluid. Helium atoms pass through small holes without resistance when helium is a superfluid. The simplest explanation of this phenomenon is that

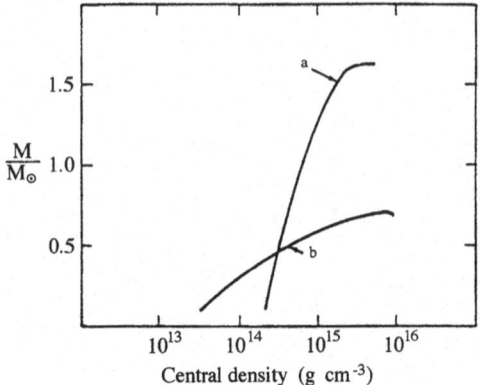

Figure 13.2. The neutron-star mass–density relation for (a) a degenerate neutron gas with strong interaction included and (b) a degenerate neutron gas with strong interaction neglected.

superfluid helium has no viscosity. This is not quite correct, however, because superfluid helium has some properties of a normal fluid that require finite viscosity. The two-fluid model of superfluid helium explains the above apparent paradox. In the two-fluid model it is assumed that the density ρ can be expressed as

$$\rho = \rho_s + \rho_n \tag{13.54}$$

where ρ_s is the density of the purely superfluid component and ρ_n the density of the normal component. The mass current density is

$$J = \rho u = \rho_s u_s + \rho_n u_n \tag{13.55}$$

where u_s and u_n are the velocities of the superfluid and normal components respectively. The superfluid velocity u_s satisfies the equation

$$\nabla \times u_s = 0. \tag{13.56}$$

Because the vorticity $\nabla \times u_s$ is equal to zero it can readily be shown that the viscosity of the superfluid component is also zero. The normal fluid component does, however, have a finite viscosity. A superfluid is a superconductor of heat and like a superconductor can have persistent current states of constant velocity.

Atomic helium obeys Bose–Einstein statistics because it has zero spin. The superfluid state of liquid helium is in some respects similar to Bose condensation, which occurs because many bosons can occupy the lowest energy level and consequently collisional excitation to higher energy levels does not occur at low temperatures. However, the superfluidity of liquid helium cannot be explained completely in terms of Bose condensation because liquid helium cannot be treated as a perfect gas.

Superconductivity is a physical phenomenon that is closely connected to superfluidity except that charged particles are involved. Electrons become superconducting in certain metals. Superconductivity occurs in metals because electrons have a tendency to form pairs (called Cooper pairs). The electrons in such pairs have opposite momenta and spin. Therefore, like helium atoms in liquid helium the pairs obey Bose–Einstein statistics. The attractive interaction between electrons in a superconductor results because the negatively charged electrons distort the lattice, which consists of positively charged ions. The distortion of the lattice is quantized in units of phonons. An electron moving through a lattice is surrounded by a cloud of phonons. Such phonons are referred to as virtual phonons because they have short lifetimes and consequently the uncertainty principle implies that energy need not be conserved in electron–phonon interactions. In superconductivity the fundamental interaction is between a pair of electrons. An electron of momentum $\hbar k'$ absorbs a phonon of momentum $\hbar q$ that was emitted by an electron of momentum $\hbar k$. Momentum is conserved and therefore the resultant momenta of the two electrons are $\hbar(k' + q)$ and $\hbar(k - q)$ respectively. If the phonon energy $\hbar\omega_q$ is greater than the energy exchange caused by the Coulomb interaction between electrons then the resultant electron–electron interaction is attractive. If the charge fluctuation of the lattice surrounds one electron charge then the second electron is attracted by a net positive charge. Superconductivity occurs when the attractive phonon exchange force between electrons exceeds the repulsive Coulomb interaction.

In a neutron star the electrons are relativistic and do not become superconducting. However, nuclear pairing forces can cause neutrons to become superfluid and protons to become a superconductor. Although two neutrons (or two protons) do not have a bound state in a vacuum they can become bound when they are in the field of other nucleons. In degenerate fermion systems pairing occurs between states close to the Fermi surface, which in momentum space is defined by the condition that particle energy equals E_F.

At nuclear densities ($\rho \simeq 2.8 \times 10^{14}\,\mathrm{g\,cm^{-3}}$) neutrons and protons are known to undergo a pairing interaction inside the atomic nucleus. The pairing energy Δ is $\Delta \sim 1$–$2\,\mathrm{MeV}$. Since the temperature in the interiors of neutron stars is $T \lesssim 10\,\mathrm{KeV}$ neutron (and proton) superfluidity is predicted to exist. In the inner crust of a neutron star ($4 \times 10^{11} < \rho < 2 \times 10^{14}\,\mathrm{g\,cm^{-3}}$) heavy, neutron-rich nuclei are present and neutrons pair in the ^1S state to form a superfluid. In the cores of neutron stars where ρ is $\gtrsim 2 \times 10^{14}\,\mathrm{g\,cm^{-3}}$ nuclei have dissolved into a degenerate fluid of neutrons and protons. Under such physical conditions neutrons are predicted to pair in the ^3P state and protons in the ^1S state.

The heat capacity of degenerate fermions such as neutrons is caused by fermions close to the Fermi surface. Because of the tendency of such neutrons to form bound pairs the heat capacity varies as $\exp(-\Delta/kT)$ where Δ is the binding energy of the pair. It follows that the production of bound pairs shortens the cooling time of the interior and at temperatures just above the bound-pair transition temperature neutron-star cooling timescales are increased whereas at lower temperatures they are decreased. In addition, neutron-star cooling timescales are increased because neutrino emission processes discussed in Chapter 7 are inhibited in a superfluid.

Magnetic fields are expelled from superconductors. This effect is known as the Meissner effect. The free electrons in a neutron star are not superconducting and therefore magnetic fields can remain in neutron stars for an extended interval of time. It follows that the crust and core of a neutron star are tied together by magnetic lines of force and therefore corotate. The superfluid neutrons are only weakly coupled to the electron component within both the crust and core of the neutron star. As pulsars slow down because of relativistic charged particle emission from their polar regions and radiative losses the charged particle component of their crusts and interiors are continuously decelerated. Because superfluid neutrons are only weakly coupled to electrons they rotate slightly faster than the pulsar magnetosphere. Glitches in the rotation periods of the Crab and Vela pulsars indicate that relaxation of superfluid neutrons to rigid body rotation can take days or months.

Because pulsars and other neutron stars rotate superfluid neutrons must have angular momentum. A rotating superfluid has a discrete number of vortices that are aligned parallel to the rotation axis. Each vortex must satisfy the quantum condition

$$\int v \cdot dl = \frac{\hbar}{2m_n} \tag{13.57}$$

where $2m_n$ is the mass of a neutron pair. If n_v is the number of vortex lines per unit area and Ω the uniform angular velocity of the neutron star then the equation

$$\frac{n_v \hbar}{2m_n} = n_v \int v \cdot dl = n_v \int (\nabla \times v) \cdot dA = 2\Omega \tag{13.58}$$

must be satisfied. Equation (13.58) implies that the number of vortices per unit area is

$$n_v = \frac{4\Omega m_n}{\hbar}. \tag{13.59}$$

The periods of pulsars normally increase continuously. However, during a pulsar glitch the rotation period decreases abruptly. The pulsar period then increases and after a characteristic relaxation time the pulsar rate of slow down returns to the pre-glitch rate. The outer region (crust) solidifies after the formation of a neutron star. The oblate shape of the crust is determined by the angular rotational velocity at the time the crust solidifies. As the angular velocity decreases stresses build up within the crust. Eventually stresses reach a critical point and a starquake results. The sudden cracking of the crust of a neutron star decreases the moment of inertia and consequently the angular velocity increases. It follows that the spin-up of a neutron star during a glitch may be caused by a starquake.

The spin-up of a pulsar during a glitch is quickly (i.e. in the timescale for an Alfvén wave to traverse the neutron star) transmitted to the core. However, the charged electrons and heavy nuclei that are strongly coupled to the magnetic field are only weakly coupled to the neutron superfluid. Therefore, neutrons do not spin-up at the time of a starquake and corotation between superfluid and charged particles occurs only after a characteristic relaxation time.

The interaction between the normal and superfluid components of a neutron star is described by the equations

$$I_c\dot{\omega} = -\alpha - \frac{I_c(\omega - \omega_n)}{\tau}$$ (13.60)

$$I_n\dot{\omega}_n = \frac{I_c(\omega - \omega_n)}{\tau}$$ (13.61)

where ω and ω_n are the angular velocities of the normal and superfluid components respectively, I_c and I_n are the corresponding moments of inertia, τ is the relaxation timescale between normal and superfluid components, and α is the external braking torque caused by electromagnetic radiation and emission of the pulsar charged particle beam. The braking torque is assumed to act instantaneously on the nonsuperfluid component of the neutron star.

If the moments of inertia I_c and I_n are such that $I_n \gg I_c$ then Equations (13.60) and (13.61) can be combined to give the equation

$$\frac{dy}{dt} = -\frac{\alpha}{I_c} - \frac{y}{\tau}.$$ (13.62)

Multiplying Equation (13.62) by $e^{t/\tau}$ we obtain

$$d(ye^{t/\tau}) = -\frac{\alpha}{I_c}e^{t/\tau}\,dt$$ (13.63)

where $y = \omega - \omega_n$. Integrating Equation (13.63) gives

$$ye^{t/\tau} = \frac{-\alpha}{I_c}\tau e^{t/\tau} + \text{constant.}$$ (13.64)

Since $y = \omega - \omega_n = \Delta\omega$ at $t = 0$ it follows that the constant in Equation (13.64) is

$$\text{constant} = \Delta\omega + \frac{\alpha\tau}{I_c}.$$ (13.65)

Equations (13.64) and (13.65) lead to the equation

$$\omega - \omega_n = \frac{-\alpha\tau}{I_c}(1 - e^{-t/\tau}) + \Delta\omega e^{-t/\tau}.$$ (13.66)

Equation (13.66) implies that the difference $\omega - \omega_n$ approaches a value of $\omega - \omega_n = -\alpha\tau/I_c$.

In the discussion above we have assumed that pulsar glitches are caused by starquakes. The observed frequency of pulsar glitches is too high to be explained entirely on the basis of the starquake model. Another explanation for pulsar glitches is the following. Charged particles can accumulate in the magnetospheres of pulsars and then be released suddenly as the result of instability. The abrupt release of charged particles from the magnetosphere of a neutron star causes the moment of inertia of the non-superfluid component to change instantaneously. Therefore, heavy nuclei and electrons spin-up as in the starquake model and the time rate of change of $\omega - \omega_n$ given in Equation (13.66) is the same as in the starquake model.

13.2   Black holes

Spherical black holes

Consider an object free-falling radially in a gravitational field described by the Schwarzschild metric given in Equation (12.88). At some initial time $t = 0$ we assume that the object is at some radial distance $r = R > r_g = 2GM/c^2$ and that $dr/dt = 0$ at $t = 0$. Eliminating the proper time interval $d\tau$ from Equations (12.102) and (12.105) leads to the equation of motion

$$\left(\frac{dr}{dt}\right)^2 = 2GM\left(1 - \frac{r_g}{R}\right)^{-1}\left(1 - \frac{r_g}{r}\right)^2\left(\frac{1}{r} - \frac{1}{R}\right). \tag{13.67}$$

Integrating Equation (13.67) we obtain

$$ct = \left(\frac{R}{r_g} - 1\right)^{1/2}\int_r^R \frac{r^{3/2}\,dr}{(r - r_g)(R - r)^{1/2}} \tag{13.68}$$

with $r_g = \dfrac{2GM}{c^2}$. The integral given in Equation (13.68) diverges as $r \to r_g$. It follows that as measured in Schwarzschild coordinates (i.e. as measured by an observer at large distances from the black hole) it takes an infinite amount of time t for the infalling object to reach the Schwarzschild radius r_g. The above discussion is illustrated in Figure 13.3.

If the coordinate time is eliminated from Equations (12.102) and (12.105) we obtain

$$\left(\frac{dr}{d\tau}\right)^2 = 2GM\left(\frac{1}{r} - \frac{1}{R}\right). \tag{13.69}$$

Equation (13.69) describes how r depends on the proper time τ, which is the time as measured by a clock attached to the infalling object. Integrating Equation (13.69) with the initial boundary conditions $r = R$ and $dr/d\tau = 0$ at $\tau = 0$ gives

$$c\tau = (r^3/2r_g)^{1/2}\left[\left(\frac{r}{R} - \left(\frac{r}{R}\right)^2\right)^{1/2} + \frac{1}{2}\cos^{-1}\left(2\frac{r}{R} - 1\right)\right]. \tag{13.70}$$

Equation (13.70) shows that the proper time τ remains finite as r becomes $\le r_g$.

Although the Schwarzschild coordinates have a singularity at $r = r_g$ they provide a valid solution of the Einstein equations over the region $0 < r < r_g$. Consider a radially free-falling object in the region $0 < r < r_g$. The Schwarzschild metric can be expressed as

$$c^2 d\tau^2 = \frac{1}{\left(\frac{r_g}{r} - 1\right)}\left[dr^2 - c^2\left(\frac{r_g}{r} - 1\right)^2 dt^2\right]. \tag{13.71}$$

Since the proper time interval $d\tau$ must be real, Equation (13.71) implies either

$$\frac{dr}{dt} > c\left(\frac{r_g}{r} - 1\right) \text{ or } \frac{dr}{dt} < c\left(\frac{r_g}{r} - 1\right). \tag{13.72}$$

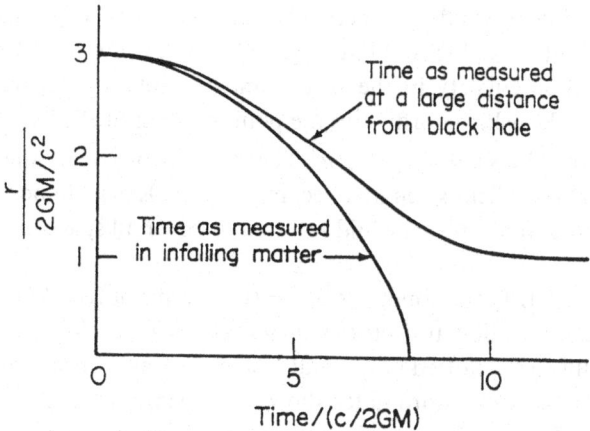

Figure 13.3. Radial distance of infalling object versus time as measured at large distances from black hole and as measured from infalling clock.

The inequalities (13.72) show that an object inside the Schwarzschild radius $r = r_g$ cannot be stationary with respect to the coordinate r at radial distances $r < r_g$. Although inequalities (13.72) allow us to infer that static solutions cannot exist inside the Schwarzschild radius it should be emphasized that as shown above objects inside the Schwarzschild radius r_g cannot be observed from radial distances $r > r_g$. The Schwarzschild radius is referred to as an event horizon.

Because the g_{11} component of the Schwarzschild metric given in Equation (12.97) becomes infinite as $r \to r_g$ it does not describe the physical properties of space on the sphere $r = r_g$. The Schwarzschild radius r_g is not a physical singularity but only a coordinate singularity that results because of the particular form of the Schwarzschild metric. This circumstance can be shown by means of the following transformation of coordinates:

$$u = \left(\frac{r}{r_g} - 1\right)^{1/2} e^{r/2r_g} \cosh\left(\frac{ct}{2r_g}\right) \tag{13.73}$$

$$v = \left(\frac{r}{r_g} - 1\right)^{1/2} e^{r/2r_g} \sinh\left(\frac{ct}{2r_g}\right). \tag{13.74}$$

The inverse transformations are

$$\left(\frac{r}{r_g} - 1\right) e^{r/r_g} = u^2 - v^2 \tag{13.75}$$

$$\tanh\frac{ct}{2r_g} = \frac{v}{u}. \tag{13.76}$$

Substituting the coordinates u and v for t and r in Equation (12.97) the metric becomes

$$ds^2 = \frac{4r_g^3}{r} e^{-r/r_g}(du^2 - dv^2) + r^2(d\theta^2 + \sin^2\theta d\phi^2). \tag{13.77}$$

In terms of the Kruskal coordinates u and v the metric describing a spherically symmetric gravitational field about a point mass does not have a singularity at $r = r_g$. However, the singularity at $r = 0$ is a real singularity in the metric and cannot be removed by a coordinate transformation. In Kruskal coordinates the point $r = 0$ in the Schwarzschild metric transforms into the two curves $v = \pm \sqrt{1 + u^2}$. The existence of a (real) singularity at $r = 0$ implies that gravitational field strengths become infinite. Kruskal coordinates give an analytic continuation of the Schwarzschild solution to cover all space except for the point $r = 0$.

The trajectories of photons satisfy the condition $ds^2 = 0$. In terms of Kruskal coordinates the radial paths of photons follow trajectories such that $u = v$. It follows that in a diagram of v plotted vertically and u plotted horizontally photons travel on straight lines that subtend an angle of 45°. Kruskal coordinates show quite clearly that any timelike trajectory ($ds^2 > 0$) inside the Schwarzschild radius r_g must reach the singularity at $r = 0$.

Intense x-ray emission comes from the inner regions of accretion disks that surround black holes. Fluid elements in an accretion disk satisfy Equation (12.108) because pressure gradients are small in the radial direction. Equation (12.108) can be expressed as

$$\left(\frac{dr}{d\tau}\right)^2 + V(r) = \varepsilon \tag{13.78}$$

with $\varepsilon = c^2(k^2 - 1)$

$$V(r) = \frac{-2GM}{r} + \frac{h^2}{r^2} - \frac{h^2 r_g}{r^3}. \tag{13.79}$$

For circular orbits we have $dr/d\tau = 0$.

In the Newtonian limit $V(r)$ becomes

$$V(r) = \frac{-2GM}{r} + \frac{h^2}{r^2} \tag{13.80}$$

with h equal to the angular momentum per unit mass. The equilibrium condition is

$$0 = \frac{dV(r)}{dr} = \frac{2GM}{r^2} - \frac{2h^2}{r^3}$$

or

$$r = \frac{h^2}{GM}. \tag{13.81}$$

The second derivative of $V(r)$ with respect to r determines the stability of circular orbits. From Equation (13.81) we find

$$\frac{d^2 V(r)}{dr^2} = \frac{1}{r^3}\left(-2GM + \frac{3h^2}{r}\right). \tag{13.82}$$

Since for circular orbits $h = vr$ and $v = \sqrt{\dfrac{GM}{r}}$ Equation (13.82) implies that

$$\frac{d^2 V(r)}{dr^2} = \frac{1}{r^3}(-2GM + 3GM) > 0. \tag{13.83}$$

Inequality (13.83) implies that in Newtonian gravitation circular orbits about a point mass are stable for all $r > 0$.

Equation (13.79) implies that the equilibrium condition for circular orbits in a Schwarzschild metric is

$$0 = \frac{dV}{dr} = \frac{2GM}{r^2} - \frac{2h^2}{r^3} + \frac{3r_g h^2}{r^4}. \tag{13.84}$$

From Equation (13.84) we obtain

$$h^2 = \frac{2GMr^2}{2r - 3r_g}. \tag{13.85}$$

Equation (13.85) implies that circular orbits exist for $r > 1.5r_g$.

The condition

$$\frac{d^2 V(r)}{dr^2} = 0 \tag{13.86}$$

separates stable and unstable circular orbits. From Equation (13.84) we have

$$\frac{d^2 V(r)}{dr^2} = \frac{-4GM}{r^3} - \frac{6h^2}{r^4} - \frac{12r_g h^2}{r^5}. \tag{13.87}$$

Equations (13.85), (13.86) and (13.87) imply that the radial distance r at which orbits becomes unstable is determined by the equation

$$0 = -1 + \frac{3r_g}{r}. \tag{13.88}$$

Equation (13.88) implies that stable circular orbits exist for $r > 3r_g$.

Rotating black holes

In our discussion of rotating black holes we will set the physical constants G and c equal to unity. It can be shown that the metric surrounding a black hole can be expressed as

$$-ds^2 = -\left(1 - \frac{2Mr}{\Sigma}\right)dt^2 - \frac{4aMr\sin^2\theta}{\Sigma}\,dt d\phi$$

$$+ \frac{\Sigma}{\Delta}dr^2 + \Sigma d\theta^2 + \left(r^2 + a^2 \frac{2Mra^2\sin^2\theta}{\Sigma}\right)\sin^2\theta d\phi^2 \tag{13.89}$$

where

$$\Delta = r^2 - 2Mr + a^2,$$
$$\Sigma = r^2 + a^2 \cos^2 \theta,$$

$$a = \frac{J}{M}.$$

In Equation (13.89) the constant J equals the angular momentum of the black hole. If we set $a = 0$ then the metric given by Equation (13.89), which is known as the Boyer–Lindquist form of the Kerr metric, reduces to the Schwarzschild metric. Event horizons exist in the Kerr metric when the function Δ becomes equal to zero. Solutions to the quadratic function Δ give two event horizons. The larger of the two event horizons is

$$r_+ = M + (M^2 - a^2)^{1/2}. \tag{13.90}$$

As $a \to 0$ the radial distance r_+ in Equation (13.90) approaches r_g. From the definition of Δ given in Equation (13.89) it follows that a and J attain maximum values when $a = M$. If $a = M$ then r_+ in Equation (13.90) equals $0.5 r_g$. A black hole in which $a = M$ is called a maximally rotating black hole because for a fixed mass M its angular momentum has the maximum allowed value.

As will be shown below a rotating black hole possesses a region of space surrounding its (outer) event horizon in which static objects cannot exist. This region of space is referred to as the ergosphere. We recall that in our discussion of the Schwarzschild metric we proved that static objects cannot exist inside the Schwarzschild radius of a spherical (i.e. nonrotating) black hole.

We consider at fixed r and θ an observer rotating with uniform angular velocity

$$\Omega = \frac{d\phi}{dt} = \frac{u^\phi}{u^t}. \tag{13.91}$$

The condition that this observer follows a time like world line is

$$u^i u_i = -1. \tag{13.92}$$

Equations (13.91) and (13.92) imply

$$-1 = (u^t)^2 (g_{tt} + 2\Omega g_{t\phi} + \Omega^2 g_{\phi\phi}). \tag{13.93}$$

The quadratic function for Ω^2 in Equation (13.93) must be negative. Since $g_{\phi\phi}$ in Equation (13.89) is positive the right-hand side of Equation (13.93) is negative if Ω satisfies the inequality

$$\Omega_{\min} < \Omega < \Omega_{\max}$$

where

$$\Omega_{\max} = \frac{-g_{t\phi} + (g_{t\phi}^2 - g_{tt} g_{\phi\phi})^{1/2}}{g_{\phi\phi}} \tag{13.94}$$

$$\Omega_{min} = \frac{-g_{t\phi} - (g_{t\phi}^2 - g_{tt}g_{\phi\phi})^{1/2}}{g_{\phi\phi}}. \tag{13.95}$$

Equation (13.94) implies that $\Omega_{max} = 0$ when $g_{tt} = 0$. From Equation (13.89) it follows that the conditon $g_{tt} = 0$ leads to the equation

$$r^2 - 2Mr + a^2 \cos^2 \theta = 0. \tag{13.96}$$

Solving Equation (13.96) for r we obtain

$$r_0 = M + (M^2 - a^2 \cos^2 \theta)^{1/2}. \tag{13.97}$$

Objects between the event horizon r_+ given by Equation (13.90) and r_0 given by Equation (13.97) must have $\Omega \neq 0$ (i.e. no static objects can exist inside the ergosphere). The outer surface r_0 of the ergosphere is called the static limit. The Kerr metric unlike the Schwarzschild metric is not the exterior metric during gravitational collapse. It is the asymptotic form of the exterior metric after collapse.

The Euler–Lagrange equations describing the motion of particles in a Kerr metric can be obtained from the variational principle

$$\delta \int ds = \delta \int \left[\left(1 - \frac{2Mr}{\Sigma}\right)\left(\frac{dt}{ds}\right)^2 + \frac{4aMr\sin^2\theta}{\Sigma}\frac{dt}{ds}\frac{d\phi}{ds} - \frac{\Sigma}{\Delta}\left(\frac{dr}{ds}\right)^2 - \Sigma\left(\frac{d\theta}{ds}\right)^2 \right.$$

$$\left. - \left(r^2 + a^2 + \frac{2Mra^2\sin^2\theta}{\Sigma}\right)\sin^2\theta\left(\frac{d\phi}{ds}\right)^2 \right]^{1/2} ds \tag{13.98}$$

where the integrand is the Kerr metric given in Equation (13.89). If F equals the square of the integrand of Equation (13.98) then since $\delta F^{1/2} = \frac{1}{2}F^{-1/2}\delta F$ the Euler–Lagrange equations become

$$\frac{d}{ds}\left(\frac{\partial F}{\partial x^i}\right) = \frac{\partial F}{\partial x^i}. \tag{13.99}$$

Because the function F does not depend explicitly on t or ϕ we have

$$\frac{\partial F}{\partial \dot{t}} = 2E \tag{13.100}$$

$$\frac{\partial F}{\partial \dot{\phi}} = -2l \tag{13.101}$$

with E and l equal to constants.

Equations (13.98), (13.100) and (13.101) with θ set equal to $\pi/2$ in Equation (13.98) imply

$$A\dot{t} + B\dot{\phi} = 2E \tag{13.102}$$

$$B\dot{t} + C\dot{\phi} = -2l \tag{13.103}$$

with $A = 2\left(1 - \dfrac{2Mr}{\Sigma}\right)$, $B = \dfrac{4aMr}{\Sigma}$ and $C = -\left(r^2 + a^2 + \dfrac{2Mra^2}{\Sigma}\right)$.

Solving Equations (13.102) and (13.103) for $\dot\phi$ and $\dot t$ we obtain

$$\dot\phi = \frac{BE + Al}{B^2 - CA} = \frac{[(r - 2M)l + 2aME]}{r(r^2 - 2Mr + a^2)} \tag{13.104}$$

and

$$\dot t = \frac{CE + Bl}{CA - B^2} = \frac{[(r^3 + ra^2 + 2Ma^2)E - 2aMl]}{r(r^2 - 2Mr + a^2)}. \tag{13.105}$$

Using Equation (13.89) with $\theta = \pi/2$ gives

$$g_{ij}\frac{dx^i}{ds}\frac{dx^j}{ds} = -\left(1 - \frac{2M}{r}\right)\left(\frac{dt}{ds}\right)^2 - \frac{4aM}{r}\frac{dt}{ds}\frac{d\phi}{ds} + \frac{r^2}{\Delta}\left(\frac{dr}{ds}\right)^2$$

$$+ \left(r^2 + a^2 + \frac{2Ma^2}{r}\right)\left(\frac{d\phi}{ds}\right)^2 = -m^2 \tag{13.106}$$

with $\Delta = r^2 - 2Mr + a^2$ and m^2 equal to a positive constant. Substituting Equations (13.104) and (13.105) into Equation (13.106) leads to the equation

$$r^3\frac{dr}{ds} = (r^3 + a^2r + 2Ma^2)E^2 - 4aMEl - (r - 2M)l^2 - m^2r\Delta. \tag{13.107}$$

For circular orbits the derivative dr/ds in Equation (13.107) is equal to zero. Regarding the right-hand side of Equation (13.107) as an effective potential $V(r)$ it follows that E and l are determined as a function of r and a by the equations

$$V(r) = (r^3 + a^2r + 2Ma^2)E^2 - 4aMEl - (r - 2M)l^2 - m^2r\Delta = 0 \tag{13.108}$$

and

$$\frac{dV(r)}{dr} = (3r^2 + a^2)E^2 - l^2 - m^2(3r^2 - 4Mr) = 0. \tag{13.109}$$

Photon orbits are also determined by Equations (13.108) and (13.109). If we set $m = 0$ in Equation (13.109) we obtain

$$l^2 = (3r^2 + a^2)E^2. \tag{13.110}$$

Substituting $l = \pm(3r^2 + a^2)^{1/2}E$ into Equation (13.108) with $m = 0$ gives

$$E^2[(r^3 + a^2r + 2Ma^2) \mp 4aM(3r^2 + a^2)^{1/2} - (r - 2M)(3r^2 + a^2)] = 0. \tag{13.111}$$

For $E \neq 0$ Equation (13.111) reduces to the equation

$$-r^3 + M(3r^2 + 2a^2) = \pm 2aM(3r^2 + a^2)^{1/2}. \tag{13.112}$$

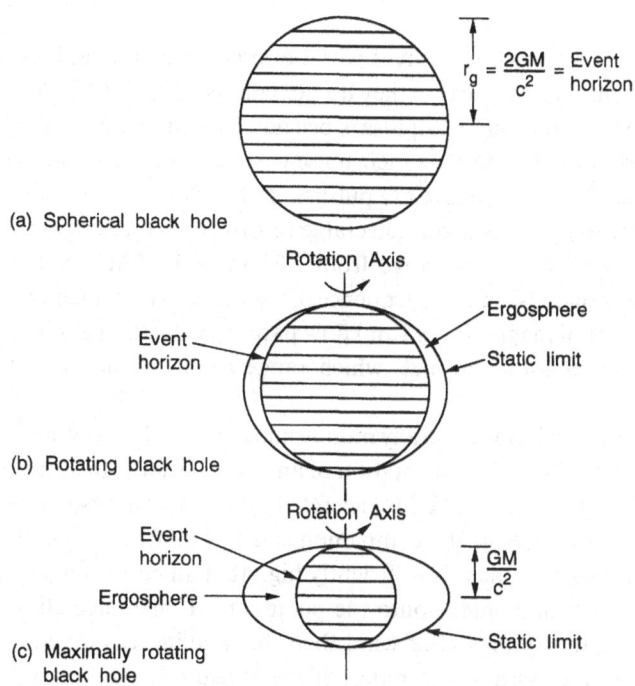

$$r_g = \frac{2GM}{c^2} = \text{Event horizon}$$

(a) Spherical black hole

Rotation Axis

Ergosphere

Event horizon

Static limit

(b) Rotating black hole

Rotation Axis

Event horizon

Ergosphere

$$\frac{GM}{c^2}$$

Static limit

(c) Maximally rotating black hole

Figure 13.4. Regions of space surrounding spherical and rotating black holes.

In a maximally rotating black hole $a = M$ and Equation (13.112) has the solutions $r = M$ and $r = 4M$. The solution $r = 4M$ corresponds to a retrograde orbit. Figure 13.4 compares spherical and rotating black holes.

13.3   Compact x-ray sources

Most compact Galactic x-ray sources are the result of mass accretion from a binary companion onto a neutron star or black hole. Significant x-ray emission can also be caused by mass accretion onto a white dwarf. Most (radio) pulsars do not emit detectable amounts of x-rays. However, strong x-ray pulses are emitted from the Crab pulsar. The x-ray (optical-γ-ray) pulses from the Crab pulsar are the result of synchrotron radiation from a relativistic electron–positron beam. As relativistic electrons and positrons stream away from the polar caps of the Crab pulsar the particle energy density becomes greater than the energy density of the magnetic field and consequently electrons and positrons are deflected to large pitch angles. The major identified classes of compact stellar mass x-ray sources are x-ray pulsars and x-ray burst sources (bursters). Both of these types of x-ray sources are caused by mass accretion onto a neutron star. Cygnus X-1 and SS 433 are compact stellar mass x-ray sources that are probably the result of mass accretion onto a black hole.

X-ray pulsars

X-ray pulsars are compact binary x-ray sources with periodic variations in brightness. These are distinguished from pulsars because they do not emit coherent radio pulses and are always binary stars. They are similar to pulsars because they are neutron stars with very strong magnetic fields and also because their pulse periods are the rotation periods of neutron stars. Examples of well-studied x-ray pulsars are Her X-1, Cen X-3 and SMC X-1. The pulse profiles of x-ray pulsars show detectable flux over a large fraction of their periods. Measured x-ray pulsar periods range from 0.71 seconds (SMC X-1) to 836 seconds (X-Per). These periodicities are a consequence of magnetic axes not being aligned with rotation axes. The external magnetic field and flow pattern of infalling gas determine the shapes of x-ray pulsar emission profiles, which range from symmetric to highly asymmetric.

Although x-ray pulsar spin-up is not generally monotonic the periods of pulsars tend to decrease as a function of time. The sense of rotation of an x-ray pulsar is the same as the direction of orbital motion. Torques exerted by accreted mass are responsible for variations in neutron star rotation rates. If the companion star fills its Roche lobe then the angular momentum of the accreted mass is sufficiently high that an accretion disk forms about the neutron star before mass infalls onto the polar caps. If mass accretion is the result of gravitational capture from a stellar wind then the infalling gas may not have sufficient angular momentum to form an accretion disk and neutron star spin-up is not likely to occur.

Accurate measurements of x-ray pulse arrival times can be used to determine the binary period of an x-ray pulsar. The masses of binary companions of x-ray pulsars are usually $\gtrsim 15 M_\odot$, and therefore their ages must be less than $\simeq 10^7$ years. An exception is Her X-1 whose binary companion HZ Her has a mass of $\simeq 2 M_\odot$. Her X-1 has three measured periodicities namely a pulse period of 1.24 s, an orbital period of 1.7 days and a 35-day periodicity, which is the result of precession of the accretion disk caused by torques exerted by the massive binary companion. Infalling mass passes through an accretion disk surrounding the magnetosphere of the neutron star before generating x-rays by infall onto the polar caps. Because the accretion disk precesses x-ray emission is periodically shielded. Periodic variations in optical brightness at the 1.7 day orbital period are also observed. These intensity variations are caused by absorption of x-rays by the binary companion and re-emission at optical wavelengths. The luminosity and distance of Her X-1 are $\simeq 10^{37} \, \mathrm{erg \, s}^{-1}$ and $\simeq 4 \, \mathrm{kpc}$. Unlike most x-ray pulsars Her X-1 is located well above the galactic plane.

From Equation (9.7) and the assumption that the orbit is circular we have

$$f = \frac{P v_1^3}{2\pi G} = \frac{(M_2 \sin i)^3}{(M_1 + M_2)^2} \qquad (13.113)$$

where $v_1 = \omega a_1 \sin i$ is the projected velocity, and M_2 and M_1 are the masses of the neutron star and binary companion respectively. The orbital inclination angle i can be calculated approximately if it is assumed that the companion star, which fills its Roche

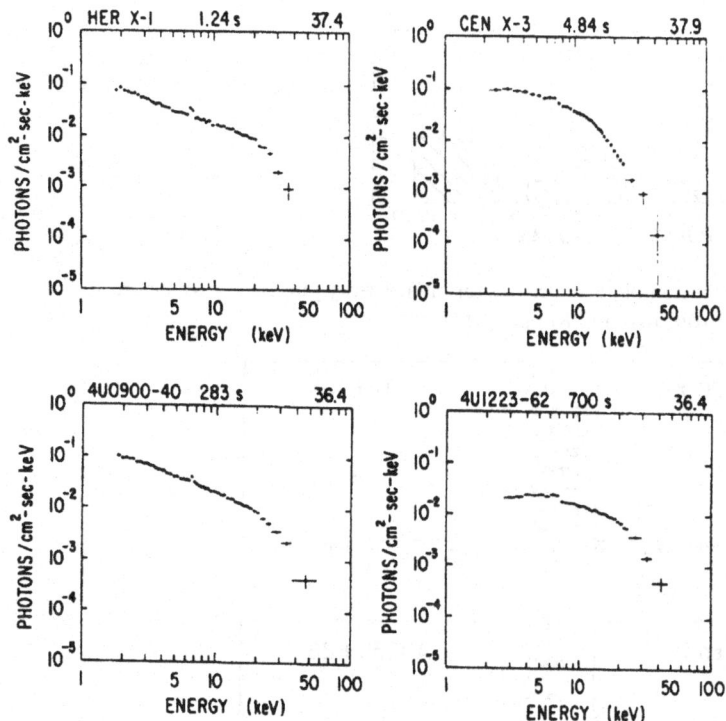

Figure 13.5. X-ray spectra of four pulsars averaged over pulse phase. Iron K-shell emission near 7 keV is often present in x-ray pulsar spectra which generally show a rapid decline in emission above ~ 20 keV. From White, Swank and Holt (1983).

lobe, is a sphere whose volume $4\pi R^3/3$ equals the volume of the Roche lobe. The inclination angle i and eclipse half-angle θ_e are then determined by the relation

$$R = a(\cos^2 i + \sin^2 i \sin^2 \theta_e)^{1/2} \tag{13.114}$$

where a is the separation of the binary stars. It can be shown that for the mass ratios of interest the Roche lobe radius can be approximated as

$$R_L = a\left[0.38 + 0.2\log\frac{M_1}{M_2}\right]. \tag{13.115}$$

If β is defined by the relation $R = \beta R_L$ then solving for $\sin i$ from Equations (13.114) and (13.115) we obtain

$$\sin i = \left[1 - \beta^2\left(0.38 + 0.2\log\frac{M_1}{M_2}\right)^2\right]^{1/2}/\cos\theta_e. \tag{13.116}$$

It can be shown that β is $\simeq 0.9$ and therefore if θ_e is measured the binary orbital parameters are determined by Equations (13.113) and (13.116).

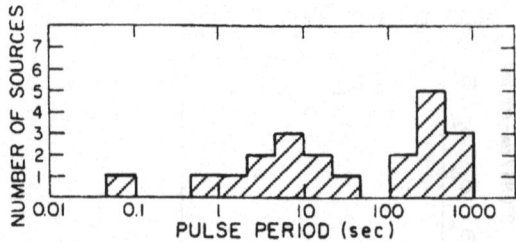

Figure 13.6. Distribution of pulse periods for binary x-ray pulsars known before September 1983. From Joss and Rappaport (1984).

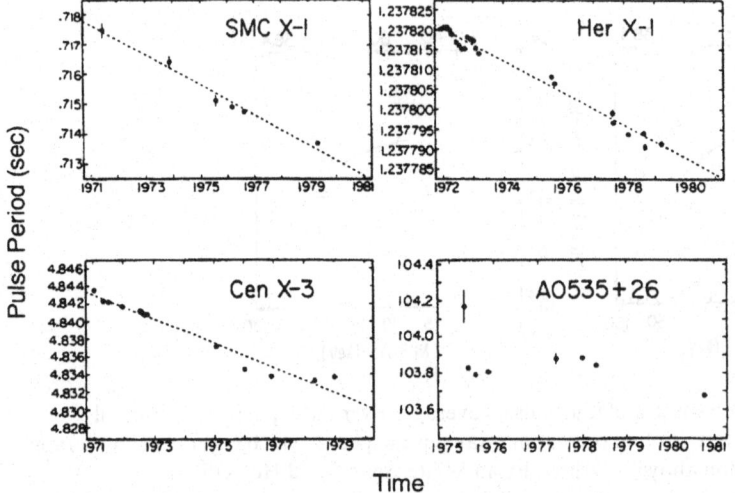

Figure 13.7. Time dependences of x-ray pulsar pulse periods. From Joss and Rappaport (1984).

The binary star systems containing the x-ray pulsars Her X-1, Cen X-3 and SMC X-1 are known to have nearly circular orbits. Estimated masses of these latter x-ray pulsars and several other similar x-ray pulsars give neutron-star masses in the range $\simeq 1.2$–$1.6\,M_\odot$. These measured neutron-star masses are consistent with formation by supernova implosion and also consistent with a neutron-star upper mass limit of $\leq 2\,M_\odot$.

Figure 13.5 gives the spectra of four x-ray pulsars. These spectra cannot be fit to blackbody radiation curves. As shown in Figure 13.6 the periods of x-ray pulsars are often longer than those of radio pulsars. The observed rotation periods of radio pulsars are less than about 3.5 seconds presumably because their pulses cease as they slow down. Decay of neutron-star magnetic fields probably contributes to the turnoff of coherent emission. Figure 13.7 shows that the periods of x-ray pulsars unlike those of most radio pulsars tend to spin-up rather than spin-down. Mass accretion from a binary companion is the cause of spin-up. Changes in x-ray pulsar periodicities are not entirely linear with respect to time as illustrated in the above figure. The masses of x-ray pulsars can be estimated because they are in binary systems. Mass determinations of six x-ray pulsars and the

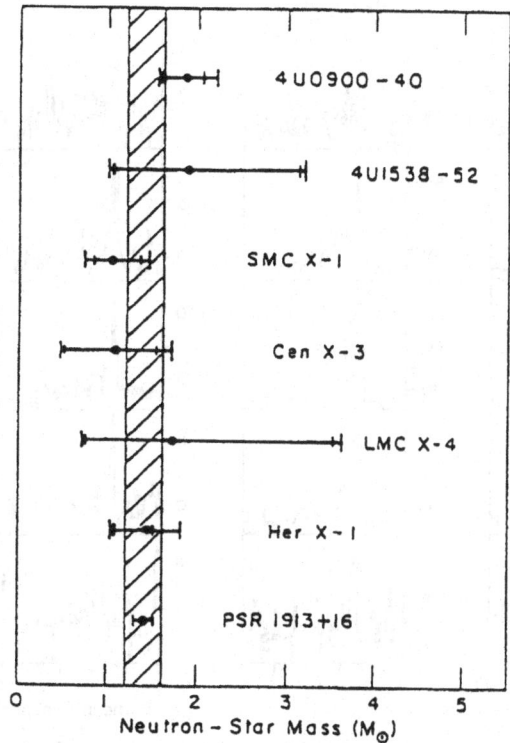

Figure 13.8. Estimate of neutron-star masses. Six of the inferred masses are from binary x-ray pulsar observations and one from measurements of the binary radio pulsar PSR 1913 + 16. From Joss and Rappaport (1984).

binary radio pulsar PSR 1913 + 16 are given in Figure 13.8. More recently the masses of several additional binary radio pulsars have been measured to be $\simeq 1.4\,M_\odot$. The results indicate that the actual neutron-star mass limit is close to the Chandrasekhar mass limit. They have important consequences from the point of view of predicting whether a particular object such as Cygnus X-1 or SS 433 is a neutron star or black hole. Figure 13.9 shows in different energy channels bursts from x-ray bursters, which are discussed below. Binary evolution plays an essential role in the formation of x-ray pulsars. An example of a scenario for x-ray pulsar formation is shown in Figure 13.10.

The luminosities of x-ray pulsars are measured to be $L_x \simeq 10^{36}$–$10^{38}\,\mathrm{erg\,s^{-1}}$. Their spectra are broadband and not dominated by sharp spectral features. Most of their emitted power is in the 2–20 keV photon energy range. Rapid decline in radiated flux occurs at photon energies above 20 keV. An iron K-shell emission feature exists at wavelengths $\simeq 6.7$–7.0 keV in several x-ray pulsars. The continuum spectra of x-ray pulsars cannot be described as blackbody radiation, thermal bremsstrahlung or by a simple power-law spectrum.

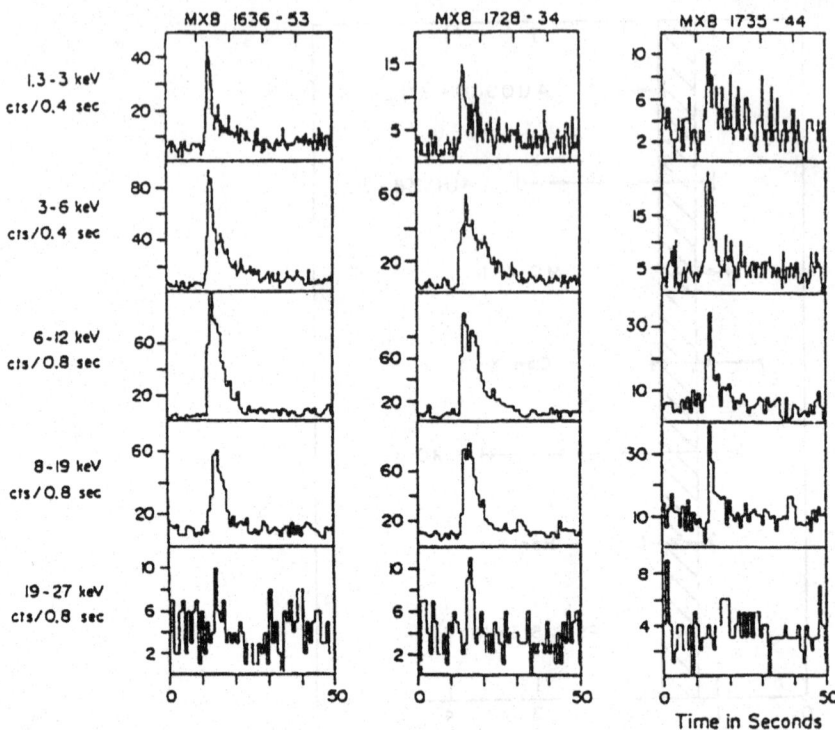

Figure 13.9. Profiles in five energy channels of typical bursts from three x-ray bursters. From Lewin and Joss (1977).

X-ray bursters

X-ray bursters are concentrated toward the center of the Galaxy and some are found in directions close to the centers of globular clusters. They are similar to x-ray pulsars because they are neutron stars that accrete mass. However, x-ray bursters unlike x-ray pulsars have weak magnetic fields. Because they are members of stellar systems with ages $> 10^9$ years their initial magnetic fields have decayed. It follows that accreted mass is not funneled toward their polar regions and consequently periodic x-ray pulses are not observed.

The time intervals between x-ray bursts are sometimes quasi-periodic but never strictly periodic. The rise time of an x-ray burst is $\sim 1\text{--}5\,\text{s}$ and the decay timescale $\simeq 3\text{--}100\,\text{s}$. The total amount of energy radiated during an x-ray burst is $\sim 10^{39}$ erg. Peak x-ray luminosities are $\sim 10^{38}\,\text{erg}\,\text{s}^{-1}$ and therefore they are close to the Eddington limit. The spectra of emitted radiation during an x-ray burst is approximately that of a blackbody of temperature $T \simeq 3 \times 10^7\,\text{K}$ and radius $R \sim 10\,\text{km}$. Weak x-ray emission is observed from x-ray bursters between bursts. The time averaged x-ray luminosity between x-ray bursts is approximately 10^2 times greater than the time averaged luminosity in the form of x-ray bursts. This circumstance indicates that helium burning (and/or carbon burning) is the energy source for x-ray bursts because the gravitational energy release due to mass accretion onto the surface of a neutron star is approximately 10^2 times greater than that

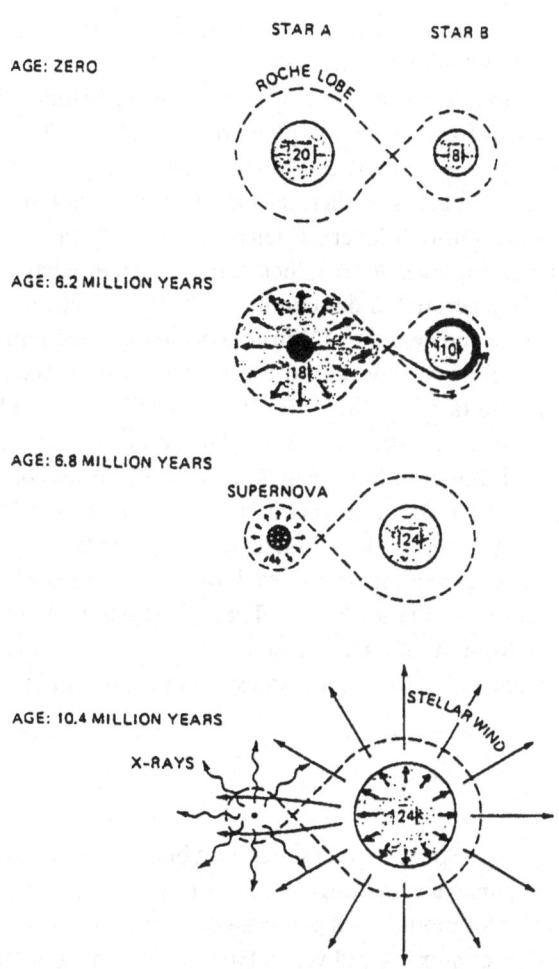

Figure 13.10. Schematic diagram showing scenario for formation of an x-ray pulsar via binary evolution of two stars with initial masses of 20 and 8 solar masses. A $4M_\odot$ hydrogen-deficient star produces the supernova outburst. From Gursky and Ruffini (1975).

available from helium burning. When they can be observed the optical counterparts of x-ray sources are found to be faint, hot objects, with emission line spectra. Most of the optical radiation observed from binary companions of x-ray bursters is probably caused by re-radiation of x-rays emitted by the neutron star.

X-ray bursters appear to be the result of thermonuclear runaways in the surface layers of neutron stars that are accreting mass at low rate from a low-mass companion star. The masses of binary companion stars of x-ray bursters are estimated to be $\lesssim 0.5\,M_\odot$. Binary star separations and orbital periods are $\simeq 10^{11}$ cm and $\lesssim 0.3$ days respectively. Gravitational radiation results in loss of orbital angular momentum, which causes the binary stars to move closer together over a timescale of $\simeq 10^{10}$ years. Therefore, low rates of mass accretion can be sustained for a long time interval.

Accreted mass onto the surface of a neutron star is primarily hydrogen and helium. A hydrogen-rich layer extends inwards from the photosphere to a depth of $\sim 10^2$–10^3 cm. At the bottom of the hydrogen-rich layer a hydrogen-burning shell exists at a density of $\rho \sim 10^{5-6}$ g cm^{-3}. Hydrogen burning occurs in thermal pulses that are not sufficiently strong to cause x-ray bursts. A helium-rich layer of density $\rho \simeq 10^{6-8}$ g cm^{-3} extends downwards from the hydrogen-burning shell. After sufficient mass accretion has occurred a very strong helium-shell flash is generated under conditions of high electron degeneracy. Because of the high temperature and density ^{12}C burning occurs simultaneously with ^{4}He burning. Thermal energy is rapidly transported to the surface and radiated away as an x-ray burst. The observed spectra of x-ray bursts are similar to Comptonized blackbody radiation at approximately the predicted temperature. The observed peak luminosities of x-ray bursts approach the Eddington limit, which must be corrected for general relativistic effects. The maximum observed luminosity is reduced to $L_c/(1 + z)^3$ with L_c equal to the uncorrected luminosity and $z = 1 + 2GM/c^2R$. The three factors of $1 + z$ that make up the general relativistic correction are the result of the gravitational redshift, curvature of space correction and time-dilation factor. The time-dilation factor occurs because if N photons are emitted from the surface of a neutron star in a time interval dt_1 then they are received at a distant point from the surface of the star within the time interval $dt_2 = (1 + z)dt_1$.

Cygnus X-1

Cygnus X-1 is a compact galactic x-ray source with a supergiant binary companion (HDE 226868). Although Cygnus X-1 exhibits a wide range of time variability the variations in x-ray emission are aperiodic except for modulation at the 5.6 day orbital period. Variability has been observed over periods of months and years. However, the most remarkable type of variation is the occurrence of x-ray bursts that last for approximately 1 ms. Such short bursts indicate that x-ray emission comes form a region of space whose dimension is ~ 300 km. The short timescale variability and lack of periodic variations except at the orbital period suggests that Cygnus X-1 is a black hole surrounded by an accretion disk.

The binary companion of Cygnus X-1 is the OB supergiant star HDE 226868 whose absolute visual magnitude is $M_v \simeq -6$. The distance to HDE 226868 is somewhat uncertain because of corrections for interstellar extinction. Distance determinations range from 2 kpc to 2.5 kpc. OB supergiants with spectra similar to HDE 226868 have masses $\gtrsim 20 M_\odot$. Because Cygnus X-1 does not have periodic x-ray pulses only the mass function given by Equation (9.8) can be determined. Measurements of spectral line Doppler shifts of HDE 226868 give an orbital period of 5.60 days, $a_1 \sin i = 5.82 \times 10^6$ km and $f = 0.252 M_\odot$. If it is assumed that the mass of HDE 226868 is $\gtrsim 20 M_\odot$ then from the measured mass function, $f = 0.252 M_\odot$, and the inequality $\sin i \le 1$ it follows that the mass of Cygnus X-1 is greater than $9 M_\odot$, which is appreciably greater than the neutron-star mass limit. Stellar model calculations show that two types

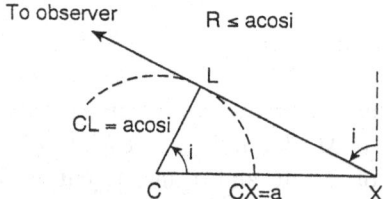

Figure 13.11. Geometry of Cygnus X-1 binary system. The angle i is the inclination angle between line of sight and orbital plane. X is the center of Cygnus X-1 and C is the center of the companion star. The absence of observed eclipses implies that the radius R of the binary companion must be less than the distance CL.

of stars can evolve to the position in the H–R diagram occupied by Cygnus X-1. These two types of stars are massive (i.e. $M \gtrsim 20\,M_\odot$) stars that are evolving off the main-sequence or low-mass stars ($M \lesssim 1.4\,M_\odot$) that are evolving from the red-giant branch to white-dwarf end states. It follows that measurement of the mass function f does not give conclusive evidence that Cygnus X-1 is a black hole because the binary companion might have mass $\lesssim 1.4\,M_\odot$. Fortunately additional determinations of a lower limit to the mass of Cygnus X-1 can be found.

The measured eccentricity of the orbit of Cygnus X-1 is small and therefore the orbit can be assumed to be circular. The absence of x-ray eclipses implies the inequality $\cos i > R/a$ with R equal to the radius of the OB supergiant and a equal to the separation of the two stars (see Figure 13.11). Using the above inequality and the relation

$$a = \frac{M_1 + M_2}{M_1} a_1 \tag{13.117}$$

we obtain

$$\cos i \geq \frac{R}{a_1 \sin i} \frac{M_2 \sin i}{M_1 + M_2}. \tag{13.118}$$

Equations (13.113) and (13.118) imply

$$M_2 \sin i \cos^2 i \geq \frac{f R^2}{(a_1 \sin i)^2} \tag{13.119}$$

with f and $a_1 \sin i$ known. The maximum value of $\sin i \cos^2 i$ is $2/(3\sqrt{3})$. Therefore, we have

$$M_2 \geq \frac{3\sqrt{3} f R^2}{2(a_1 \sin i)^2} \tag{13.120}$$

with R^2 unknown. The measured effective temperature of HDE 226868 is $T_{\text{eff}} = 30\,000$ K. Since $L = 4\pi R^2 \sigma T_{\text{eff}}^4$ it follows that R^2 is determined if the luminosity L can be determined. The apparent visual magnitude of a star with $L = L_\odot$ is 4.72 if it is at a distance of 10 pc. Therefore, R^2 can be expressed as

$$R^2 = \frac{L_\odot}{4\pi\sigma T_{\text{eff}}^4}\left(\frac{d}{10\text{pc}}\right)^2 10^{0.4(4.72-M_v-BC+A_v)} \tag{13.121}$$

where M_v is the visual magnitude, BC is the bolometric correction and A_v is the absorption in the V band. The measured values of M_v, BC and A_v are 8.87, -2.9 and 3.3 respectively. Substituting the measured values of M_v, BC and A_v into Equation (13.121) with $T_{\text{eff}} = 30\,000\,\text{K}$ we obtain

$$R^2 = (12.24 \times 10^6\,\text{km})^2 \left(\frac{d}{2\,\text{kpc}}\right)^2. \tag{13.122}$$

Equations (13.120) and (13.122) imply

$$M_2 \geq 3.4\,M_\odot \left(\frac{d}{2\,\text{kpc}}\right)^2. \tag{13.123}$$

Since the distance d to Cygnus X-1 is known to be greater than 2 kpc the inequality (13.123) implies that the mass of Cygnus X-1 is greater than the neutron-star mass limit. Stellar atmosphere calculations show that the surface gravity of the supergiant HDE 226868 is

$$g = \frac{GM}{R^2} = 1.6 \times 10^3\,\text{cm}\,\text{s}^{-2}. \tag{13.124}$$

Equations (13.122) and (13.124) imply that the mass of HDE 226868 is $M \gtrsim 15\,M_\odot$. Therefore, from the known value of the mass function f given by Equation (13.113) we can infer that the mass of Cygnus X-1 is greater than $6\,M_\odot$.

The observational evidence indicates that most supernova implosions lead to the formation of neutron stars. One scenario for the formation of Cygnus X-1 is that a supernova implosion in a binary system produced a neutron star. Subsequent mass accretion caused the neutron star to exceed the neutron-star mass limit and implosion to black-hole end state followed. Mass accretion through an accretion disk is limited because radiative flux cannot exceed the Eddington limit. Direct implosion to black-hole end state is also possible because the white-dwarf mass limit is close to the neutron-star mass limit.

SS 433

SS 433, which was first identified as the emission line object V134 Aquilae, is surrounded by the supernova remnant W50. Its spectrum shows broad Balmer and HeI emission lines. Radio and x-ray emission as well as emission lines are also observed from SS 433. The emission lines are Doppler-shifted Balmer and neutral helium lines. One set of these emission lines is strongly redshifted whereas the second set is strongly blueshifted. Changes in the wavelengths of the Doppler shifts have a period of approximately 164 days. The Doppler shifts are caused by two collimated jets that move with outflow velocities of $\simeq 0.26c$ in opposite directions. The jet axis precesses with a period of 164

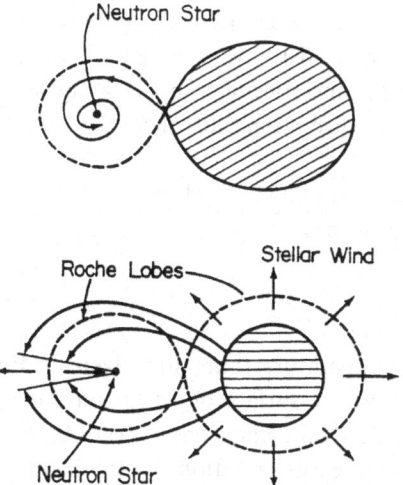

Figure 13.12. (Top) Mass transfer occurs because the binary companion of the neutron star has filled its Roche lobe. Infalling mass has enough angular momentum to form an accretion disk. (Bottom) Mass transfer occurs via stellar wind. The gravitational field causes some of the stellar wind to pass through shock waves and infall onto the neutron star.

days. The 13-day period associated with the nearly stationary Balmer and neutral helium lines is the orbital period of the massive, early type binary companion of SS 433.

The observations indicate that SS 433 is a compact star surrounded by an accretion disk whose principal precession period is 164 days. Some of the mass accreted onto the compact star, which is probably a black hole, is emitted as jets. The distance to SS 433 is estimated to be 5 kpc and the absolute optical magnitude more luminous than -7. It follows that SS 433 is intrinsically a very luminous object. The x-ray luminosity of SS 433 is $\sim 3 \times 10^{35}\,\mathrm{erg\,s^{-1}}$, and therefore appreciably less than the optical luminosity. A variable radio source is coincident with SS 433. The ejection of jets from SS 433 is probably caused by high radiative fluxes that are a consequence of rapid mass accretion ($\dot{M} = 5 - 10 \times 10^{-7}\,M_\odot/\mathrm{yr}$). Observations indicate that the mass of SS 433 satisfies the inequality $4\,M_\odot \le M \le 15\,M_\odot$. Therefore, SS 433 is too massive to be a neutron star. The estimated mass and high luminosity of SS 433 show that it is a stellar-mass black hole.

Figure 13.12 shows two types of mass transfer that occur in binary systems containing compact stars.

13.4 Accretion onto neutron stars and white dwarfs

The kinetic energy of infalling mass is transformed into thermal energy and radiated from the surface of a star. For spherical accretion the luminosity is

$$L = \frac{GM}{R}\frac{dM}{dt} = \frac{1}{2}\rho 4\pi R^2 v^3 \tag{13.125}$$

with $v \simeq \sqrt{2GM/R}$. The radii of neutron stars are $\simeq 10\,\mathrm{km}$ and therefore if the mass accretion rate onto the surface of a neutron star is $\simeq 10^{-9}\,M_\odot/\mathrm{yr}$ the luminosity implied by Equation (13.125) is $\simeq 10^{37}\,\mathrm{erg\ s^{-1}}$. The luminosities of compact x-ray sources are typically $\simeq 10^{36}$–$10^{38}\,\mathrm{erg\,s^{-1}}$ and therefore mass accretion rates of $\simeq 10^{-8}$–$10^{-9}\,M_\odot/\mathrm{yr}$ are most commonly relevant.

If infalling gas falls directly into the surface layer of a star and the emitted spectrum is that of a blackbody then the radiation temperature is

$$T = \left(\frac{L}{4\pi R^2 \sigma} \right)^{1/4}. \tag{13.126}$$

The radiation would have a much higher temperature than that given by Equation (13.126) if a standing shock front formed above the surface of a neutron star or white dwarf. Thermal bremsstrahlung emission would then come from a layer of gas between the shock front and underlying photosphere that emits radiation at a temperature lower than that of the gas above it. The overlying gas can be optically thin to free–free absorption but optically thick to electron scattering. The gas temperature behind such a shock front is approximately

$$T \lesssim \frac{\frac{1}{2}m_\mathrm{p}v^2}{k} \simeq \frac{Gm_\mathrm{p}M}{\sqrt{2kR}}. \tag{13.127}$$

For a collisional shock front to form the protons, electrons and ions in the infalling gas must be deflected through an angle of 90°. An approximate expression for the collisional cross section can be readily obtained from the Rutherford cross section given in Equation (3.332) but with the proton mass substituted for the electron mass. Because most of the incoming gas is protons the Rutherford scattering cross section must be corrected for particle symmetry and the Mott scattering cross section used to obtain an accurate stopping length. An approximate expression for the deflection cross section can be found by setting

$$\frac{e^2}{r_0} = \frac{1}{2}m_\mathrm{p}v^2 \tag{13.128}$$

with v equal to the infall velocity. From Equation (13.128) the deflection cross section becomes

$$\sigma_\mathrm{D} \simeq \pi r_0^2 \ln \Lambda \simeq \frac{4\pi e^4}{m_\mathrm{p}v^4} \ln \Lambda \tag{13.129}$$

where we have included the logarithmic factor $\ln \Lambda$ to take long-range collisions into account. Equation (13.129) implies that the mean free path for deflection through 90°, which is the shock thickness, is

$$\lambda = \frac{m_\mathrm{p}}{\rho \sigma_\mathrm{D}}. \tag{13.130}$$

In Equation (13.130) we have assumed that the gas is pure hydrogen.

Above the photosphere of the neutron star the gas density assuming spherical infall is

$$\rho = \frac{1}{4\pi R^2 v} \frac{dM}{dt}.$$

(13.131)

Therefore, the ratio of shock thickness λ implied by Equation (13.130) divided by stellar radius R becomes

$$\frac{\lambda}{R} = \frac{m_p}{R\rho\sigma_D} \simeq \frac{m_p^3 v^4}{R\rho 4\pi e^4 \ln \Lambda}.$$

(13.132)

It can readily be shown from Equations (13.131) and (13.132) that the required condition for shock formation (i.e. $\lambda/R \ll 1$) is not satisfied above the photosphere of a neutron star. Protons infalling onto the surface of a neutron star have energies of $\simeq 100$ MeV. The total nuclear and Coulomb scattering cross section of 100 MeV protons, which is about 3×10^{-26} cm^2, is appreciably greater than the Coulomb collisional cross section given by Equation (13.129). Protons infalling onto the surface of a neutron star will penetrate to optical depths greater than unity before being stopped by nuclear collisions, and consequently the emitted radiation is comptonized blackbody radiation.

The two stream instability between infalling and quasi-stationary electrons causes the excitation of plasma waves. Resonant plasma waves excited by the two stream instability couple to nonresonant plasma waves and ion acoustic waves. Heating of plasma electrons results. Because protons are initially unaffected by the two stream instability an electric field is generated by the charge separation between protons and electrons. The induced electric field tends to slow down protons. In this manner the infalling gas is decelerated. The collisionless excitation of plasma waves may decelerate mass infalling onto the surface of a neutron star in a path length appreciably less than the path length determined by the nuclear-scattering cross section. Therefore, a collisionless shock wave may form above the photosphere.

Neutron stars in binary systems are likely to be formed with strong magnetic fields. Pulsars and x-ray pulsars contain magnetic fields of $\simeq 10^{12}$ gauss and have ages $\leq 10^7$ years. X-ray bursters do not have strong magnetic fields but their ages ($\simeq 10^{10}$ years) are sufficiently great that their magnetic fields have decayed. Infalling gas is stopped by the magnetic field surrounding a star when the condition

$$\rho v^2 \simeq \frac{B^2}{4\pi}$$

(13.133)

is satisfied. Equation (13.133) is equivalent to the condition that the Alfvén velocity $V_A = \sqrt{B^2/4\pi\rho}$ equal the free-fall velocity.

If the magnetic field outside a neutron star is a dipole field then in spherical polar coordinates we have

$$\boldsymbol{B} = B_p R^3 \left(\frac{\cos\theta}{r^3} \hat{\boldsymbol{r}} + \frac{\sin\theta}{2r^3} \hat{\boldsymbol{\theta}} \right)$$

(13.134)

where R is the radius of the neutron star and B_p the field strength at the magnetic poles. In the equatorial plane $\theta = \pi/2$ and therefore Equation (13.134) becomes

$$B \simeq \frac{B_p R^3}{2r^3} \hat{\theta}. \tag{13.135}$$

From Equations (13.133) and (13.135) we obtain

$$\rho \simeq \frac{B_p^2 R^6}{8\pi GM r^5}. \tag{13.136}$$

Equations (13.131) and (13.136) imply that infalling mass close to the equatorial plane of a neutron star is stopped at a radial distance

$$r_A = B_p^{4/7} R^{12/7} \left(\frac{dM}{dt}\right)^{-2/7} (2GM)^{-1/7}. \tag{13.137}$$

The radial distance r_A given in Equation (13.137) is called the Alfvén radius. The thickness of the shock front formed at the Alfvén radius is approximately equal to the proton gyroradius.

After infalling gas has reached the Alfvén radius it is channeled along magnetic field lines into the magnetic polar regions of the neutron star. From Equation (10.121) the critical magnetic field line becomes

$$\sin^2 \theta_c = \frac{R}{r_A} \tag{13.138}$$

and therefore the area of the polar cap onto which the gas streams is

$$A \simeq \pi R^2 \sin^2 \theta_c. \tag{13.139}$$

The infalling gas is channeled into an area that is small as compared to the surface area of the neutron star. For sufficiently low mass-accretion rates the accreted mass will infall through the surface of the neutron star at the free-fall velocity (i.e. $v \simeq 0.4c$). Under such physical conditions radiation from the infalling gas is small and almost all emitted radiation comes from the polar regions.

13.5   Accretion disks surrounding black holes

Some of the physical properties of accretion disks have already been discussed in Section 9.2. As shown in Section 13.2 the inner radius of a black hole varies from $3r_g$ to $1.5r_g$ as the angular momentum varies from 0 to its maximum value. In terms of the surface density $\Sigma = \int \rho dz$ the mass-accretion rate is

$$\frac{dM}{dt} = -2\pi r \Sigma v_r \tag{13.140}$$

with v_r negative because the gas motion is inward. If r_1 is the inner boundary of an accretion disk then the luminosity L is approximately

$$L \simeq \frac{GM}{2r_1} \frac{dM}{dt}.$$
(13.141)

For an optically thick accretion disk the emitted flux F_ν perpendicular to the plane of symmetry is

$$F_\nu = \int_{r_1}^{R} B_\nu(T_s(r)) 2\pi r dr$$
(13.142)

with B_ν equal to the Planck function given in Equation (3.91), r_1 and R equal to the inner and outer radii of the accretion disk respectively and $T_s(r)$ given by Equation (9.86).

The principal uncertainty in determining the structures of accretion disks is the viscosity. In a turbulent gas the difference between the gas velocity at some particular position and the time averaged mean velocity varies irregularly as a function of time. The mean velocity u varies as a function of position. In a normal gas turbulent energy flow passes from larger to smaller eddies. Energy dissipation occurs in eddies with some characteristic dimension λ_0, which depends on particular physical conditions.

Suppose that a turbulent gas has some characteristic turbulent eddy size λ and Δu is equal to the variation in mean velocity over λ. The physical properties of the turbulence can be described by Δu and λ only, and therefore from dimensional analysis it follows that the mean rate of energy dissipation per unit mass is

$$\varepsilon \sim \frac{(\Delta u)^3}{\lambda} \text{erg g}^{-1}\text{s}^{-1}.$$
(13.143)

The turbulent kinetic viscosity v_{turb} is

$$v_{\text{turb}} \sim \lambda \Delta u.$$
(13.144)

The Reynolds number of a turbulent gas is equal to the ratio of the turbulent viscosity v_{turb} divided by the normal gas viscosity v discussed in Section 3.8. For the physical conditions we consider the Reynolds number R satisfies the inequality

$$R \equiv \frac{v_{\text{turb}}}{v} \gg 1$$
(13.145)

because v is small as compared to v_{turb} except at the smallest eddy sizes.

Equations (13.143) and (13.144) imply that the rate of energy dissipation ε can be expressed as

$$\varepsilon \sim v_{\text{turb}} \left(\frac{\Delta u}{\lambda}\right)^2.$$
(13.146)

The kinematic viscosity v and dynamic viscosity η discussed in Section 9.2 are related by the equation $\eta = \rho v$.

The variations in pressure ΔP and mean velocity Δu over a characteristic eddy size λ are related by the equation

$$\Delta P \sim \rho(\Delta u)^2. \tag{13.147}$$

Equations (13.144) and (13.147) imply that ΔP can be expressed as

$$\Delta P \sim \frac{1}{v_s^2}\left(\frac{v_{\text{turb}}}{\lambda}\right)^2 P \tag{13.148}$$

with $v_s^2 \simeq P/\rho$. Since the condition $\Delta P/P < 1$ must hold Equations (13.144), (13.147) and (13.148) imply that Δu must be less than the sound velocity v_s. The characteristic scale λ in Equations (13.144) and (13.148) must be less than or equal to the half-thickness of the accretion disk. Equations (9.47) and (9.74) imply

$$\frac{\eta_{\text{turb}}}{\rho} = v_{\text{turb}} = \frac{2}{3}\alpha\frac{P}{\rho}\frac{r}{v_\phi} = \frac{2}{3}\alpha\frac{v_s^2 r}{v_\phi} \tag{13.149}$$

where $v_\phi = (GM/r)^{1/2}$, $\alpha = \alpha(r) < 1$ and the turbulent viscosity v_{turb} satisfies the condition $v_{\text{turb}} \ll v_s h$ with h equal to the half-thickness of the accretion disk.

Computed disk models for Cygnus X-1 ($\dot{M} \simeq 2 \times 10^{-9} M_\odot/\text{yr}$) indicate that the accretion disk is convective. For a convective disk the kinematic viscosity is

$$v \sim v_c h \tag{13.150}$$

where v_c is the convective velocity and h the disk half-thickness.

The magnetic lines of force in an accretion disk are strongly bound to the gas because the gas is ionized and has a high electric conductivity. It follows that differential rotation causes a weak magnetic field within an accretion disk to be amplified. Under certain physical conditions magnetic field annihilation and reconnection can occur. The kinematic magnetic viscosity v_m is

$$v_m \sim v_A h \tag{13.151}$$

with $v_A = (B^2/4\pi\rho)^{1/2}$ and h equal to the half-thickness of the accretion disk.

Equilibrium models of accretion disks are subject to both thermal and viscous instabilities. Under equilibrium conditions the local heating rate Γ must equal the local cooling rate Λ. If T is the equilibrium temperature then the condition for thermal instability becomes

$$\frac{d\ln\Gamma}{d\ln T} > \frac{d\ln\Lambda}{d\ln T}. \tag{13.152}$$

When inequality (13.152) is satisfied small increases in T lead to a heating rate that is greater than the cooling rate.

The mass and angular momentum of an annulus of an accretion disk with inner radius r and outer radius $r + \Delta r$ are $2\pi r\Delta r\Sigma$ and $2\pi r\Delta r\Sigma r^2\omega$ respectively. The time rate of change of mass into and out of the annulus is

$$\frac{\partial(2\pi r\Delta r\Sigma)}{\partial t} = v_r(r,t)2\pi r\Sigma - v_r(r+\Delta r,t)2\pi(r+\Delta r)\Sigma(r+\Delta r,t) \tag{13.153}$$

with $\Sigma = \int\rho dz$ equal to the surface mass density. In the limit $\Delta r \to 0$ Equation (13.153) becomes

$$r\frac{\partial\Sigma}{\partial t} + \frac{\partial(r\Sigma v_r)}{\partial r} = 0. \tag{13.154}$$

Conservation of angular momentum leads to the equation

$$r\frac{\partial(\Sigma r^2\omega)}{\partial t} + \frac{\partial(r\Sigma v_r r^2\omega)}{\partial r} = G \tag{13.155}$$

where G is the torque per unit change in the circumference of the annulus. If $T(r,t)$ is the torque of an outer annulus acting on a neighboring inner annulus at radius r then $T(r,t)$ and $G(r,t)$ given in Equation (13.155) are related by the equation

$$G(r,t) = \frac{1}{2\pi}\frac{\partial T(r,t)}{\partial r}. \tag{13.156}$$

The viscous force per unit length along the circumference of the annulus is

$$\nu\Sigma\left(\frac{dv_\phi}{dr} - \frac{v_\phi}{r}\right) = \nu\Sigma r\frac{d\omega}{dr} \tag{13.157}$$

with ν equal to the kinematic viscosity. Equation (13.157) implies that the torque $T(r,t)$ on the annulus is

$$T(r,t) = 2\pi r\nu\Sigma r^2\frac{d\omega}{dr}. \tag{13.158}$$

Equations (13.155), (13.156) and (13.158) imply that

$$\frac{\partial}{\partial t}(\Sigma r^2\omega) + \frac{1}{r}\frac{\partial}{\partial r}(\Sigma r^3\omega v_r) = \frac{1}{r}\frac{\partial}{\partial r}\left(\nu\Sigma r^3\frac{d\omega}{dr}\right). \tag{13.159}$$

Using Equation (13.154) and substituting $\omega = (GM/r^3)^{1/2}$ we obtain

$$\frac{\partial\Sigma}{\partial t} = \frac{3}{r}\frac{\partial}{\partial r}\left[r^{1/2}\frac{\partial}{\partial r}[\nu\Sigma r^{1/2}]\right]. \tag{13.160}$$

Equation (13.160) is a time-dependent diffusion equation. It can be shown that if, at some initial time, mass is located at radial distance r then most of the mass will spiral inwards. However, because angular momentum must be conserved a small amount of mass will move outwards to large distances from the central star.

If we let $\mu = \nu\Sigma$ then the variables μ can be regarded as functions of r (or Σ). Varying Σ and μ by a small amount we have

$$\Sigma' = \Sigma + \delta\Sigma$$

$$\mu' = \mu + \frac{\partial \mu}{\partial \Sigma} \delta \Sigma. \tag{13.161}$$

Substituting the above perturbed values for Σ' and μ' into Equation (13.160) leads to the equation

$$\frac{\partial \delta \mu}{\partial t} = \frac{\partial \mu}{\partial \Sigma} \frac{3}{r} \frac{\partial}{\partial r} \frac{[r^{1/2} \partial (r^{1/2} \delta \mu)]}{\partial r} \tag{13.162}$$

because Σ and μ in Equation (13.161) are solutions to Equation (13.160). The solutions to Equation (13.162) are well behaved only if the diffusion coefficient is positive. Therefore, the condition for viscous instability of the accretion disk becomes

$$\frac{\partial \mu}{\partial \Sigma} \equiv \frac{\partial (v\Sigma)}{\partial \Sigma} < 0. \tag{13.163}$$

When an accretion disk is viscously unstable a region that is slightly overdense will become even more overdense, and conversely a region that is slightly underdense will become even more underdense.

The Kompaneets equation

Photons that undergo electron scattering exchange energy with electrons. However, if blackbody photons of temperature T_b are injected into an ionized gas of temperature $T > T_b$ and optical depth such that free–free absorption can be neglected then the photon distribution function will not approach the Planck function $B_v(T)$ because electron scattering conserves the number of photons. Absorption and emission of photons must occur for the photon distribution function to reach thermodynamic equilibrium.

We wish to calculate the photon number density $n(\omega)$ given that some initial distribution of photons interact only by means of electron scattering. The electron distribution function is assumed to be a nonrelativistic, Maxwell–Boltzmann distribution function. Photons of frequency ω moving in some particular direction interact with electrons of momentum p_1 and scatter with frequency ω_1. The inverse reaction also occurs. The time rate of change of $n(\omega)$ is determined by the Boltzmann equation

$$\frac{\partial n(\omega)}{\partial t} = c \int d^3 p \int \frac{d\sigma}{d\Omega} d\Omega [f(p_1)n(\omega_1)(1 + n(\omega)) - f(p)n(\omega)(1 + n(\omega_1))] \tag{13.164}$$

with $d\sigma/d\Omega$ and $f(p)$ given by Equations (2.89) and (3.11) respectively. The integral over solid angle $d\Omega$ and $d^3 p$ of the positive term within the brackets in Equation (13.164) gives the increase in $n(\omega)$ caused by photons of frequency ω_1 undergoing electron scattering. The corresponding negative term in Equation (13.164) represents the decrease in $n(\omega)$ caused by electron scattering from frequency ω to ω_1. The effect of stimulated emission is given by the terms $1 + n(\omega)$ and $1 + n(\omega_1)$ in Equation (13.164). We assume physical conditions such that changes in photon energy caused by electron scattering are small as compared to kT. Therefore, we have

$$\Delta = \frac{\hbar|\omega_1 - \omega|}{kT} \ll 1.$$

(13.165)

Expanding $n(\omega_1)$ and $f_1(E_1)$ to second order in $\omega_1 - \omega$ gives

$$n(\omega_1) = n(\omega) + (\omega_1 - \omega)\frac{\partial n(\omega)}{\partial \omega} + \frac{1}{2}(\omega_1 - \omega)^2\frac{\partial^2 n(\omega)}{\partial \omega^2}$$

(13.166)

and

$$f(E_1) = f(E) + \frac{\partial f}{\partial E}\Delta E + \frac{1}{2}\frac{\partial^2 f}{\partial E^2}(\Delta E)^2$$

(13.167)

where $\Delta E = \hbar(\omega_1 - \omega)$. Conservation of photon energy and momentum imply that

$$\hbar\omega + \frac{p^2}{2m} = \hbar\omega_1 + \frac{p_1^2}{2m}$$

(13.168)

and

$$\frac{\hbar\omega}{c}\hat{n} + p = \frac{\hbar\omega_1}{c}\hat{n}_1 + p_1.$$

(13.169)

From Equation (13.169) we have

$$p_1^2 = \left(\frac{\hbar\omega_1}{c}(\hat{n} - \hat{n}_1) + p\right)^2 \simeq p^2 + \frac{2\hbar\omega_1}{c}p \cdot (\hat{n} - \hat{n}_1).$$

(13.170)

Equations (13.169) and (13.170) imply that

$$\hbar(\omega_1 - \omega) = \frac{\hbar\omega_1}{mc}p \cdot (\hat{n}_1 - \hat{n}).$$

(13.171)

If we define the dimensionless variable $x = \hbar\omega/kT$ and substitute Equations (13.165–13.167) and (13.171) into Equation (13.164) we obtain the equation

$$\frac{\partial n}{\partial t} = c\left[\frac{dn}{dx} + n(1 + n)\right]\iint d^3p\frac{d\sigma}{d\Omega}d\Omega f\Delta$$

$$+ c\left[\frac{1}{2}\frac{d^2n}{dx^2} + \frac{dn}{dx}(1 + n) + \frac{1}{2}n(1 + n)\right]\iint d^3p\frac{d\sigma}{d\Omega}d\Omega f\Delta^2$$

(13.172)

with $\Delta = \hbar(\omega_1 - \omega)/kT$. The second integral in Equation (13.172) can be re-expressed as

$$\iint d^3p\frac{d\sigma}{d\Omega}d\Omega f\hbar^2(\omega_1 - \omega)^2 = \left(\frac{\hbar\omega}{mc}\right)^2\int d^3pp^2\cos^2\chi f\int|\hat{n}_1 - \hat{n}|^2\frac{d\sigma}{d\Omega}d\Omega$$

(13.173)

with χ equal to the angles between p and $\hat{n}_1 - \hat{n}$.
The integral over d^3p in Equation (13.173) is

$$-2\pi\int_0^\pi\int_0^\infty p^4f(p)dp\cos^2\chi d(\cos\chi) = n_e kTm.$$

(13.174)

From the relation

$$|\hat{n}_1 - \hat{n}|^2 = 2(1 - \cos\theta) \tag{13.175}$$

and Equation (2.89) the integral over θ in Equation (13.173) becomes

$$2\pi r_0^2 \int_{\pi/2}^{0} (1 - \cos\theta)(1 + \cos^2\theta)d(\cos\theta) = \frac{16\pi r_0^2}{3} \tag{13.176}$$

with $r_0 = \dfrac{e^2}{mc}$ and $\sigma_T = 8\pi r_0^2/3$. Equations (13.174) and (13.176) imply that the second integral in Equation (13.172) is

$$\int\int d^3p \frac{d\sigma}{d\Omega} d\Omega f \Delta^2 = 2x^2 n_e \sigma_T \frac{kT}{mc^2} \tag{13.177}$$

with $x = \hbar\omega/kT$.

Because the energy of a photon is $\hbar\omega$ it follows that frequency space ω is equivalent to energy space. The photon number density $n(\omega)$ must satisfy the equation

$$\frac{\partial n(\omega)}{\partial t} + \nabla \cdot J(\omega) = 0. \tag{13.178}$$

Because the current density $J(\omega)$ does not depend on direction we have

$$\frac{\partial n}{\partial t} = -\nabla \cdot J = \frac{-1}{\omega^2} \frac{\partial \omega^2 J}{\partial\omega} = \frac{-2J}{\omega} - \frac{\partial J}{\partial\omega}. \tag{13.179}$$

We assume that the current density J has the functional form

$$J = g(x)\left[\frac{dn}{dx} + h(n, x)\right] \tag{13.180}$$

with $x = \hbar\omega/kT$. If we require that the current density $J = 0$ for the Bose–Einstein distribution function

$$n = \frac{1}{e^{\alpha + x} - 1} \tag{13.181}$$

then the resultant function form of $h(n,x)$ is $n(1 + n)$. It follows from Equations (13.179) and (13.180) that the partial time derivative of $n(x, t)$ is

$$\frac{\partial n}{\partial t} = \frac{1}{x^2} \frac{\partial}{\partial x}\left(x^2 g(x)\left[\frac{dn}{dx} + n(1 + n)\right]\right)$$

$$= \frac{2g(x)}{x}\left[\frac{dn}{dx} + n(1 + n)\right] + \frac{dg}{dx}\left[\frac{dn}{dx} + n(1 + n)\right]$$

$$+ g(x)\left[\frac{d^2n}{dx^2} + \frac{dn}{dx}(1 + n) + n\frac{dn}{dx}\right]. \tag{13.182}$$

We substitute the expression for $\dfrac{\partial n}{\partial t}$ given by Equation (13.182) into Equation (13.172) and require that the coefficients of n, $\dfrac{dn}{dx}$ and $\dfrac{d^2 n}{dx^2}$ are equal on both sides of the resultant equation. The requirement that the coefficients of $\dfrac{d^2 n}{dx^2}$ are the same implies that

$$g(x) = -cx^2 n_e \sigma_T \frac{kT}{mc^2}. \tag{13.183}$$

Therefore, we have

$$\frac{dg(x)}{dx} = -2cxn_e \sigma_T \frac{kT}{mc^2}. \tag{13.184}$$

Similarly Equations (13.172), (13.177) and (13.182–13.184) imply that

$$\int\int d^3 p \frac{d\sigma}{d\Omega} d\Omega f \Delta = n_e \sigma_T x (4 - x) \frac{kT}{mc^2}. \tag{13.185}$$

Substituting Equations (13.177) and (13.185) into Equation (13.172) we obtain

$$\frac{\partial n}{\partial t_c} = \frac{kT}{mc^2} \frac{1}{x^2} \frac{\partial}{\partial x} \left[x^4 \left(\frac{dn}{dx} + n + n^2 \right) \right] \tag{13.186}$$

with $x = \hbar\omega/kT$ and $t_c = cn_e \sigma_T t$ equal to the time as measured in units of the Thomson-scattering timescale. Equation (13.186) is known as the Kompaneets equation. An equilibrium photon distribution exists if

$$\frac{dn}{dx} + n + n^2 = 0. \tag{13.187}$$

Substituting the Bose–Einstein distribution function given in Equation (13.181) into Equation (13.187) we obtain

$$\frac{-e^{\alpha + x}}{(e^{\alpha + x} - 1)^2} + \frac{1}{e^{\alpha + x} - 1} + \frac{1}{(e^{\alpha + x} - 1)^2} = 0. \tag{13.188}$$

Equation (13.188) shows that the Bose–Einstein distribution function is (as previously assumed) a time independent solution of Equation (13.186).

Suppose that input photon energies $\hbar\omega$ are much less than kT and therefore $x = \hbar\omega/kT$ satisfies the condition $x \ll 1$. If the photon occupation number n is small (i.e. α is large) then n is proportional to e^{-x} and we can neglect n and n^2 in Equation (13.186) as compared to dn/dx. Therefore the photon energy density E initially increases according to the equation

$$\frac{dE}{dt_c} = \frac{8\pi}{c^3h^3}(kT)^2\frac{d}{dt_c}\int_0^\infty nx^3dx = \frac{8\pi}{c^3h^3}\frac{(kT)^3}{mc^3}\int\frac{d\left(x^4\frac{dn}{dx}\right)}{dx}xdx$$

$$= \frac{-8\pi}{c^3h^3}\frac{(kT)^3}{mc^2}\int x^4\frac{dn}{dx}dx = \frac{4kTE}{mc^2} \tag{13.189}$$

where we have used Equation (13.186) and performed an integration. Equation (13.189) implies that the initial increase in photon energy is

$$\varepsilon(t) = \varepsilon(0)\exp\left(\frac{4kT}{mc^2}cn_e\sigma_T t\right). \tag{13.190}$$

Equation (13.185) implies that the sign of the energy transfer $\Delta\varepsilon$ to a photon in a single scattering is positive if ε is less than $4kT$ and negative if ε is greater than $4kT$. If ε satisfies the condition $\varepsilon \ll 4kT$ then the energy gain per scattering is

$$\frac{d\varepsilon}{dN} \simeq \varepsilon\frac{4kT}{mc^2}. \tag{13.191}$$

Therefore, after N scatterings the photon energy becomes

$$\frac{\varepsilon_N}{\varepsilon_i} \simeq e^{(4kT/mc^2)N} \quad (\varepsilon_N \ll 4kT) \tag{13.192}$$

with ε_i equal to the initial photon energy. If a corona above an accretion disk around a black hole has an optical depth $\tau_{es} \gg 1$ then a photon scatters with electrons $\simeq \tau_{es}^2$ times before escaping and Equation (13.192) can be expressed as

$$\frac{\varepsilon_f}{\varepsilon_i} \simeq e^y \quad (\varepsilon_f \ll 4kT) \tag{13.193}$$

with $y = \frac{4kT}{mc^2}\tau_{es}^2$.

After traversing some optical depth τ_c photons attain an energy $\varepsilon_f \simeq 4kT$. The critical optical depth τ_c can be estimated from Equation (13.193) by setting

$$\frac{4kT}{\varepsilon_i} \simeq e^{y_c} \tag{13.194}$$

with $y_c = \frac{4kT}{mc^2}\tau_c^2$.

Solving Equation (13.194) for τ_c we obtain

$$\tau_c = \left[\frac{mc^2}{4kT}\ln\left(\frac{4kT}{\varepsilon_i}\right)\right]^{1/2}. \tag{13.195}$$

Equation (13.195) implies that saturation occurs when τ_{es} equals τ_c.

The steady state Kompaneets equation given as Equation (13.186) with the radiative loss term and photon source term included is

$$0 = \frac{kT}{mc^2} \frac{1}{x^2} \frac{\partial}{\partial x}\left[x^4\left(\frac{dn}{dx} + n + n^2\right)\right] - \frac{n}{\max(\tau_{es}, \tau_{es}^2)} + Q(x).$$

(13.196)

The second term on the right-hand side of Equation (13.196) represents the change in n caused by photons escaping from the corona and $Q(x)$ gives the low-energy ($x = \hbar\omega/kT \ll 1$) source of photons. We assume that the n and n^2 terms on the right-hand side of Equation (13.196) can be neglected as compared to the dn/dx term. Substituting $n = Kx^m$ into Equation (13.196) with n and n^2 omitted, $Q(x) = 0$, and $\tau_{es}^2 \gg 1$ implies that

$$\frac{kT}{mc^2} \frac{Km(m + 3)x^{m + 2}}{x^2} - \frac{Kx^m}{\tau_{es}^2} = 0.$$

(13.197)

From Equation (13.197) we obtain

$$m^2 + 3m - \frac{4}{y} = 0$$

(13.198)

with $y = \dfrac{4kT}{mc^2}\tau_{es}^2$.

Solving the quadratic equation (13.198) we find that

$$m = -\frac{3}{2} \pm \left(\frac{9}{4} + \frac{4}{y}\right)^{1/2}.$$

(13.199)

If y in Equation (13.199) satisfies the condition $y \gg 1$ the positive root is relevant and $m = 0$. Since $n = Kx^m$ it follows that the intensity of radiation I_ν varies as $x^3 n \propto x^3$ with $x = \hbar\omega/kT$. Therefore, the $y \gg 1$ limit represents the low frequency limit. If y satisfies the condition $y \ll 1$ then the negative root in Equation (13.199) applies and we have

$$I_\nu \simeq \nu^{3 + m}.$$

(13.200)

For values of y not satisfying either of the above conditions both roots of Equation (13.199) are relevant and I_ν does not have a single power law spectrum.

Figure 13.13 shows a solution to the Kompaneets equation fit to the x-ray spectrum of Cygnus X-1.

13.6   X-ray pulsar spin-up

The periods of x-ray pulsars tend to decrease as a function of time. If mass accretion occurs via Roche lobe overflow then an accretion disk will be formed outside the magnetosphere of an x-ray pulsar. The inner region of the accretion disk exerts a torque on the neutron star by interacting with its magnetic field. In this section we describe the theory of x-ray pulsar spin-up.

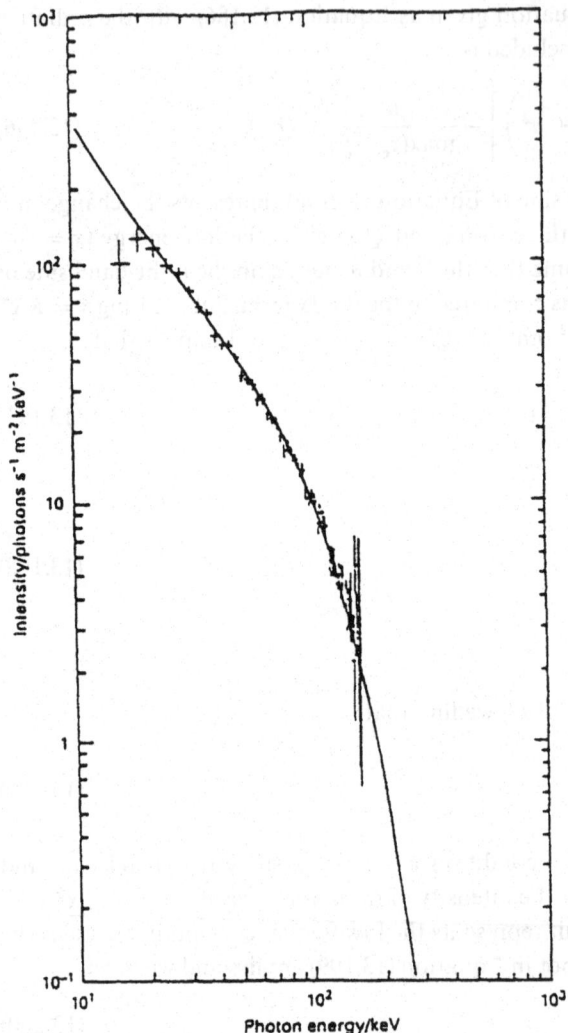

Figure 13.13. The hard-x-ray spectrum of Cygnus X-1 compared to a solution of the Kompaneets equation. From Sunyaev and Titarchuk (1980).

In cylindrical coordinates the equation of continuity for steady-state mass accretion is

$$\mathbf{V} \cdot (\rho \mathbf{v}) = \frac{1}{r} \frac{d(r\rho v_r)}{dr} = 0. \tag{13.201}$$

Equation (13.201) implies that the corresponding rate of mass accretion $\dot{M}$ is

$$\dot{M} = 4\pi r h v_r \rho \tag{13.202}$$

where h is the half-thickness of the accretion disk.

The steady-state force equation is

$$\rho(\boldsymbol{v} \cdot \nabla)\boldsymbol{v} = -\nabla P - \rho \nabla \phi + \frac{1}{4\pi}(\nabla \times \boldsymbol{B}) \times \boldsymbol{B}. \tag{13.203}$$

Magnetic fields of x-ray pulsars penetrate the inner regions (transition regions) of accretion disks and brake azimuthal gas motions. If the magnetic field is assumed to be a function of r and z only then the θ component of the magnetic force term in Equation (13.203) becomes

$$\frac{1}{4\pi}[(\nabla \times \boldsymbol{B}) \times \boldsymbol{B}]_\theta = \frac{1}{4\pi}\left[\frac{B_r}{r}\frac{\partial r B_\theta}{\partial r} + B_z \frac{\partial B_\theta}{\partial z}\right] \tag{13.204}$$

where the relevant components of $\nabla \times \boldsymbol{B}$ in cylindrical coordinates are

$$(\nabla \times \boldsymbol{B})_r = \frac{1}{r}\frac{\partial B_z}{\partial \theta} - \frac{\partial B_\theta}{\partial z} = -\frac{\partial B_\theta}{\partial z} \tag{13.205}$$

$$(\nabla \times \boldsymbol{B})_z = \frac{1}{r}\left[\frac{\partial r B_\theta}{\partial r} - \frac{\partial B_r}{\partial \theta}\right] = \frac{1}{r}\frac{\partial r B_\theta}{\partial r}. \tag{13.206}$$

Using the Maxwell equation

$$\nabla \cdot \boldsymbol{B} = \frac{1}{r}\frac{\mathrm{d}r B_r}{\mathrm{d}r} + \frac{\mathrm{d}B_z}{\mathrm{d}z} = 0 \tag{13.207}$$

Equation (13.204) can be written

$$\frac{1}{4\pi}[(\nabla \times \boldsymbol{B}) \times \boldsymbol{B}]_\theta = \frac{1}{4\pi}\left[\frac{1}{r^2}\frac{\mathrm{d}(r^2 B_r B_\theta)}{\mathrm{d}r} + \frac{\mathrm{d}(B_z B_\theta)}{\mathrm{d}z}\right]. \tag{13.208}$$

If we neglect the non-magnetic force terms on the right-hand side of Equation (13.203) then Equations (13.203) and (13.208) imply

$$\rho v_r r \frac{\mathrm{d}r v_\theta}{\mathrm{d}r} = \frac{1}{4\pi}\left[\frac{\mathrm{d}(r^2 B_r B_\theta)}{\mathrm{d}r} + r^2 \frac{\mathrm{d}(B_z B_\theta)}{\mathrm{d}z}\right]. \tag{13.209}$$

Using Equation (13.202) and integrating Equation (13.209) from $z = -h$ to $z = h$ with $h \ll r$ we obtain

$$\dot{M}\frac{\mathrm{d}r v_\theta}{\mathrm{d}r} = r^2 B_z B_\theta. \tag{13.210}$$

If the thickness δ of the transition region in which the magnetic field exerts a torque on the accretion disk satisfies the condition $\delta \ll r$ then Equation (13.210) can be approximated as

$$\dot{M}\left(\frac{r}{\delta}\right)v_\theta \simeq r^2 B_z B_\theta. \tag{13.211}$$

We assume that azimuthal gas motions with the accretion disk become dominated by x-ray pulsar magnetic fields when the conditions $v_\theta \simeq v_f = (2GM/r)^{1/2}$ and $B_\theta \simeq B_z$ are satisfied. Substituting these conditions into Equation (13.211) and assuming that the transition radius r_0 occurs when the Kepler orbital velocity $v_K = \sqrt{GM/r}$ equals the Alfvén velocity we have

$$\delta \simeq \frac{4v_K^2}{v_A^2} h \sim 4h. \tag{13.212}$$

We note that the presence of a magnetic field increases the thickness of an accretion disk.

If I and Ω are the moment of inertia and angular velocity respectively then the time rate of change of neutron-star angular momentum is

$$\frac{d(I\Omega)}{dt} = \dot{M}l_0 + N \tag{13.213}$$

where $\dot{M}$ is the mass accretion rate, $l_0 = (GMr_0)^{1/2}$ is the specific angular momentum at the transition radius r_0 and N is the torque exerted from outside r_0. We assume that N in Equation (13.213) is negligible. The time rate of change of I and Ω can be written

$$\frac{dI}{dt} = \frac{dI}{dM} \dot{M} \tag{13.214}$$

and

$$\frac{d\Omega}{dt} = -\Omega \frac{\dot{P}}{P}. \tag{13.215}$$

Substituting Equations (13.214) and (13.215) into Equation (13.213) and neglecting N we obtain

$$\frac{\dot{P}}{P} = \frac{\dot{M}}{M} \left[\frac{M}{I} \frac{dI}{dM} - \frac{l_0}{l_\star} \right] \simeq \frac{\dot{M}}{M} \left[1 - \frac{l_0}{l_\star} \right] \tag{13.216}$$

with $l_\star = I\Omega/M$. If l_0 satisfies the condition $l_0 \gg l_\star$ and we assume $r_0 \simeq r_A$ with r_A given in Equation (13.137) then Equation (13.216) becomes

$$-\frac{\dot{P}}{P} = \frac{\dot{M}}{M} \frac{(GM)^{1/2}r_A^{1/2}}{(I\Omega/M)}. \tag{13.217}$$

It is convenient to rewrite Equation (13.137) as

$$r_A \simeq 3 \times 10^8 \dot{M}_{17}^{-2/7} \mu_{30}^{4/7} \left(\frac{M}{M_\odot} \right)^{-1/7} \text{cm} \tag{13.218}$$

where the mass accretion rate $\dot{M}_{17}$ is in units of $10^{17}\,\text{g s}^{-1}$ and the neutron star magnetic moment $\mu_{30} = B_p^2 r^3/2$ is in units of 10^{30}. Substituting Equation (13.218) into Equation (13.217) with the accretion luminosity equal to

$$L = \dot{M} \frac{GM}{R} \tag{13.219}$$

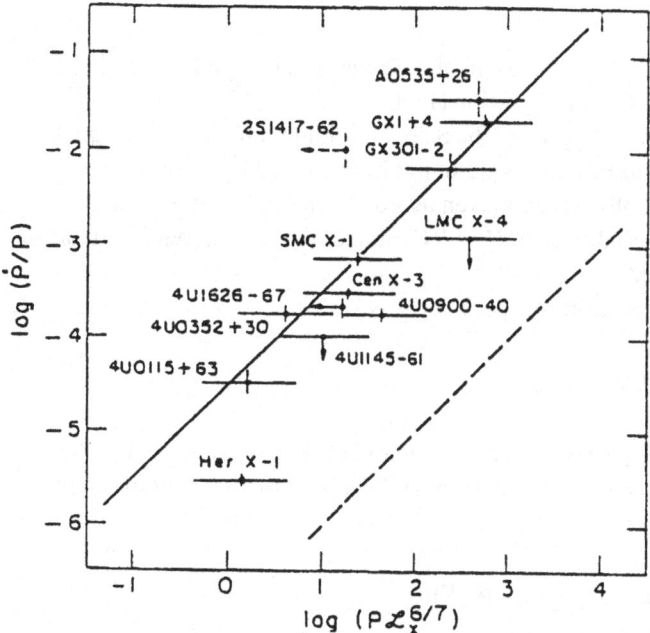

Figure 13.14. The relation discussed in text between the fractional change of pulse period, $\dot P/P$ and the parameter (pulse period × luminosity$^{6/7}$) for a number of x-ray pulsars. The units of $\dot P/P$, P and L are yr^{-1}, s and 10^{37} erg s^{-1} respectively. The solid line is the expected relation, assuming x-ray pulsars are neutron stars with predicted physical properties. The dashed line would apply if x-ray pulsars were $\simeq 1M_\odot$ white dwarfs. From Joss and Rappaport (1984).

we find

$$\frac{-\dot P}{P} \simeq \text{const}\, M^{-3/7}\mu_{30}^{-3/7}R^{6/7}PL^{6/7}. \tag{13.220}$$

Equation (13.220) implies that the rate of x-ray pulsar spin-up is proportional to $PL^{6/7}$. The latter proportionality is found to be approximately satisfied for x-ray pulsars.

For sufficiently high angular velocities or low-mass accretion rates the stellar angular velocity is greater than the Keplerian angular velocity $\omega_K = \sqrt{GM/r_0^3}$ with r_0 equal to the boundary radius between the magnetosphere of the x-ray pulsar and accretion disk (i.e. transition radius). Under such physical conditions corotation between the neutron star magnetosphere and accretion disk cannot exist because the force on the disk is outwards. It follows that mass accretion can cause x-ray pulsars to spin-down rather than spin-up, which is their more generally observed behavior.

Observational evidence for the relevance of Equation (13.220) is shown in Figure 13.14.

Problems

1. Derive Equations (13.5–13.7) by calculating the G_{00}, G_{11} and G_{22} components of Equation (12.66) and using Equation (13.4).

2. Evaluate R_{00}, R_{11} and R_{22} using the Schwarzschild metric given in Equation (12.88) and then from Equations (12.63), (12.69) and (12.84) derive the Oppenheimer–Volkoff equation given as Equation (13.17) (and (13.31)).

3. a. Show that the enthalpy $H = U + PV$ satisfies the thermodynamic relation
$$dH = TdS + VdP.$$
   b. The equation of motion of a perfect gas is

$$\frac{\partial v}{\partial t} + (v \cdot \nabla)v = -\frac{\nabla P}{\rho}.$$

Assuming isentropic motion (and therefore no viscosity) show that the vorticity $\nabla \times v$) is independent of time. *Hint*: Use the result of part a and the vector identity

$$\frac{1}{2}\nabla v^2 = v \times (\nabla \times v) + (v \cdot \nabla)v.$$

   c. Explain why $\nabla \times u_s = 0$ in a superfluid (Equation (13.56)). *Hint*: Superfluid particles are in a single quantum state and $u_s(r)$ is the velocity field of particles in this single quantum state. The wave function is $\psi(r) = a(r)\exp(i\gamma(r))$ and we have $mu_s\psi = -i\hbar\nabla\psi$.

4. Use the Oppenheimer–Volkoff equation and determine numerically how P, ρ and T vary in the outer layers of a neutron star (i.e. $\rho \le 10^{11}\,\mathrm{g\,cm^{-3}}$). Assume $M_r = M$, $R = 10\,\mathrm{km}$ and $P = 0$ and $T = 10^7\,\mathrm{K}$ at $r = R$. You may assume that the equation of state is that of a nondegenerate gas in the outermost layer ($\rho \lesssim 3 \times 10^4\,\mathrm{g\,cm^{-3}}$), a nonrelativistic degenerate-electron gas for $3 \times 10^4 \le \rho \le 10^6\,\mathrm{g\,cm^{-3}}$ and an ultrarelativistic degenerate gas for $10^6 \le \rho \le 10^{11}\,\mathrm{g\,cm^{-3}}$. What is the thickness of the outer layer of the neutron star?

5. Consider a spherical star of mass M and radius R where the gravitational field is sufficiently strong that the Oppenheimer–Volkoff equation must be used to describe hydrostatic equilibrium. Assume that density ρ is constant and determine how pressure P depends on radial coordinate r. The pressure P must clearly satisfy the condition $P < \infty$ everywhere throughout the interior of a star. What limit does the condition $P < \infty$ place on the ratio GM/R for a star of assumed uniform density?

6. The outer layers (crust) of a cooling neutron star crystallize. For assumed densities of 10^{11}, 10^9, 10^7 and $10^5\,\mathrm{g\,cm^{-3}}$ estimate the temperatures at which crystallization occurs.

7. Assume that the gas in the interior of a neutron star consists entirely of neutrons, protons, electrons and ^{56}Fe nuclei. Estimate the number densities of each of

these particles at $\rho = 10^{14}, 2 \times 10^{14}$ and $3 \times 10^{14}\, \mathrm{g\,cm^{-3}}$.

8. Estimate the density required for π^- mesons to be produced by neutron–neutron collisions in the interior of a neutron star. The mass of a π^- meson is 139.6 MeV.

9. Show that the sound velocity v_s of a perfect relativistic gas is given by the equation

$$v_s^2 = \left.\frac{\partial P}{\partial \rho}\right|_s.$$

Hint: Let ρ_1, P_1 and the velocity v_1 be perturbations in the rest frame of the unperturbed gas and use the equation $T^{ik}_{,k} = 0$ with $T^{ik} = (\rho c^2 + P)u^i u^k + Pg^{ik}$.

10. Find the speed of sound in a ideal Fermi gas at zero temperature.

11. Show that the speed of sound v_s derived in Problem 9 and the adiabatic index

$$\Gamma_1 = \left.\frac{n}{P}\frac{\partial P}{\partial n}\right|_s$$

are related by the equation $v_s^2 = \Gamma_1 \dfrac{Pc^2}{P + \rho c^2}$.

12. A particle orbits a black hole. Considering orbital motion as geodesic motion find $\Omega = d\phi/dt$ in Boyer–Linquist coordinates and show that Ω approaches the Keplerian orbital angular velocity $\sqrt{GM/r^3}$ as the angular momentum $a \to 0$.

13. a. The binary companion of a $1.6\,M_\odot$ neutron star expands and becomes a red giant. Estimate how rapidly the neutron star can accrete mass spherically assuming that it lies inside the envelope of the red giant. Can a $1.6\,M_\odot$ black hole accrete mass more rapidly? Explain.

  b. Assume that a $1\,M_\odot$ white dwarf is surrounded by the envelope of a $1\,M_\odot$ red-giant binary companion. The distance between the center of mass of the two stars is 10^{11} cm. Estimate the drag force on the white dwarf assuming that the density of the surrounding gas is $10^{-5}\,\mathrm{g\,cm^{-3}}$ and find the initial rate of decrease of the center-of-mass distance. The radius of the white dwarf is 7×10^8 cm.

Appendix 1 Physical and astronomical constants

A1.1 Physical constants

Gravitational constant	$G = 6.673 \times 10^{-8}\,\mathrm{dyn\,cm^2\,g^{-2}}$
Speed of light	$c = 2.9979 \times 10^{10}\,\mathrm{cm\,s^{-1}}$
Planck's constant	$h = 6.626\,19 \times 10^{-27}\,\mathrm{erg\,s}$
	$h/2\pi = \hbar = 1.054\,59 \times 10^{-27}\,\mathrm{erg\,s}$
Stefan–Boltzmann constant	$\sigma = ac/4$
	$= 5.669\,61 \times 10^{-5}\,\mathrm{erg\,cm^{-2}\,s^{-1}\,K^{-4}}$
Boltzmann's constant	$k = 1.380\,62 \times 10^{-16}\,\mathrm{erg\,K^{-1}}$
Electron volt	$1\,\mathrm{eV} = 1.601\,84 \times 10^{-12}\,\mathrm{erg}$
Electron charge	$e = 4.803\,25 \times 10^{-10}\,\mathrm{esu}$
Fine-structure constant	$\alpha = e^2/\hbar c = 1/137.036$
Electron mass	$m = 9.1095 \times 10^{-28}\,\mathrm{g}$
Proton mass	$m_\mathrm{p} = 1.672\,65 \times 10^{-24}\,\mathrm{g}$
Neutron mass	$m_\mathrm{n} = 1.674\,92 \times 10^{-24}\,\mathrm{g}$
Atomic mass unit (a.m.u.)	$m_\mathrm{u} = 1.660\,57 \times 10^{-24}\,\mathrm{g}$
(1/12 times mass of ^{12}C atom)	
Rydberg constant (for H)	$R_\mathrm{H} = 1.096\,78 \times 10^5\,\mathrm{cm^{-1}}$
Thomson cross section	$\sigma_\mathrm{T} = 0.665 \times 10^{-24}\,\mathrm{cm^2}$
Bohr radius	$a_0 = 0.529\,177 \times 10^{-8}\,\mathrm{cm}$
Avogadro's number	$N_\mathrm{A} = 6.022 \times 10^{23}\,\mathrm{mol^{-1}}$

A1.2 Astronomical constants

Solar mass	$M_\odot = 1.989 \times 10^{33}\,\mathrm{g}$
Solar luminosity	$L_\odot = 3.826 \times 10^{33}\,\mathrm{erg\,s^{-1}}$
Solar radius	$R_\odot = 6.9598 \times 10^{10}\,\mathrm{cm}$
Effective temperature of Sun	$T_\odot = 5770\,\mathrm{K}$
Parsec	$1\,\mathrm{pc} = 3.0856 \times 10^{18}\,\mathrm{cm}$
Light year	$1\,\mathrm{l.y.} = 9.4605 \times 10^{17}\,\mathrm{cm}$
	$= 6.324 \times 10^4\,\mathrm{AU}$
Astronomical unit (AU)	$1\,\mathrm{AU} = 1.495\,978 \times 10^{13}\,\mathrm{cm}$

Appendix 2 Further comments on the Dirac equation

In Section 3.2 we derived the Dirac equation. The matrices $\alpha^k (k = 1, 2, 3)$ and β whose anticommutation relations are given in Equation (3.43) are 4×4 matrices, and therefore this latter equation becomes somewhat more precise if its right-hand side is multiplied by the 4×4 unit matrix. The matrix representations of α^k and β are not uniquely defined. It is convenient to explicitly represent them as

$$\alpha^k = \begin{pmatrix} 0 & \sigma_k \\ \sigma_k & 0 \end{pmatrix} \text{ and } \beta = \begin{pmatrix} I & 0 \\ 0 & -I \end{pmatrix} \qquad (k = 1, 2, 3), \tag{A2.1}$$

where σ_k are the 2×2 Pauli matrices

$$\sigma_1 = \sigma_x = \begin{pmatrix} 0 & 1 \\ 1 & 0 \end{pmatrix}, \sigma_2 = \sigma_y = \begin{pmatrix} 0 & -i \\ i & 0 \end{pmatrix}, \sigma_3 = \sigma_z = \begin{pmatrix} 1 & 0 \\ 0 & -1 \end{pmatrix} \tag{A2.2}$$

and $I = \begin{pmatrix} 1 & 0 \\ 0 & 1 \end{pmatrix}$ is the 2×2 unit matrix.

As discussed in textbooks on relativistic quantum mechanics, the 4×4 gamma matrices $\gamma^\mu = (\gamma^0, \gamma^k)$ $[\mu = 0, 1, 2, 3; k = 1, 2, 3]$ with $\gamma^0 = \beta$ and $\gamma^k = \beta \alpha^k$ are introduced in order to express the Dirac equation in a relativistically covariant form. As can be readily shown by direct substitution gamma matrices satisfy the anticommutation relations

$$\gamma^\mu \gamma^\nu + \gamma^\nu \gamma^\mu = 2g^{\mu\nu}, \tag{A2.3}$$

with the metric tensor $g^{\mu\nu}$ defined by $g^{00} = 1$, $g^{kk} = -1$, $k = 1, 2, 3$ and $g^{\mu\nu} = 0, \mu \neq \nu$. The trace of a gamma matrix is defined as $\sum_{\sigma=0}^{3} \gamma_{\sigma\sigma}^\mu$ where σ is a spinor index and μ is a vector index. As can be readily shown, the traces of gamma matrices are all zero.

The V–A weak-interaction coupling term briefly discussed in Chapter 7 (see Equation 7.49) contains the factor $\bar{\psi}_1 (\gamma^\mu - \gamma_5 \gamma^\mu) \psi_2$, where γ_5 is defined as $\gamma_5 = i\gamma^0 \gamma^1 \gamma^2 \gamma^3$. This 4×4 matrix has the interesting property of changing sign under the parity operation. In mathematical terms this means $\beta \gamma_5 \beta = -\gamma_5$ and therefore γ_5 transforms under Lorentz transformations as a pseudo-scalar. As will be shown below, the $\bar{\psi} \gamma^\mu \psi$ term in Equation

(7.49) transforms as a vector under Lorentz transformations. It follows that the corresponding $\bar{\psi}\gamma_5\gamma^\mu\psi$ term in Equation (7.49) is an axial (pseudo-)vector because it is the product of a pseudoscalar and vector. Parity is not conserved in weak interactions because interaction-matrix elements are a sum of a vector and axial vector. However, it is known to be conserved in both electromagnetic and strong interactions.

The Dirac wave function $\psi(x)$ is a 4×1 column matrix, whereas $\psi^\dagger$ is the four-component row matrix $(\psi_a^\star, \psi_b^\star, \psi_c^\star, \psi_d^\star)$, where '$\star$' indicates 'complex conjugate'. $\bar{\psi}$ is defined as $\psi^\dagger\gamma_0$ with $\gamma_0 = g_{00}\gamma^0$ and $g_{00} = 1$. Under a Lorentz transformation, coordinates transform as $x'^\mu = a_\nu^\mu x^\nu$ and the Dirac wave functions transform as $\bar{\psi}'(x') = S(a)\psi(x)$, where $S(a)$ is a function of the Lorentz transformation a_ν^μ. Although we will not prove it here, it can be shown that a transformation matrix $S(a)$ exists with the following properties:

$$S^{-1}(a)\gamma^\mu S(a) = a_\nu^\mu\gamma^\nu, \tag{A2.4}$$
$$S^{-1}(a) = \gamma_0 S^\dagger\gamma_0. \tag{A2.5}$$

If Equations (A2.4) and (A2.5) are satisfied, then

$$\begin{aligned}
a_\nu^\mu\bar{\psi}(x)\gamma^\nu\psi(x) &= a_\nu^\mu\psi^\dagger(x)\gamma_0\gamma^\nu\psi(x) = \psi^\dagger(x)\gamma_0 S^{-1}\gamma^\mu S\psi(x) \\
&= \psi^\dagger(x)\gamma_0\gamma_0 S^\dagger\gamma_0\gamma^\mu S\psi(x) = \psi^\dagger(x)S^\dagger\gamma_0\gamma^\mu S\psi(x) \\
&= \psi'^\dagger(x')\gamma_0\gamma^\mu\psi'(x') = \bar{\psi}'(x')\gamma^\mu\psi'(x') \tag{A2.6}
\end{aligned}$$

and therefore by definition $\bar{\psi}(x)\gamma^\nu\psi(x)$ transforms as a vector. In Equation (A2.6) we have used the obvious equality $\gamma_0\gamma_0 = 1$ and the equality $\psi'^\dagger(x') = \psi^\dagger(x)S^\dagger$, which follows from the definition of $S(a)$, namely $\psi'(x') = S(a)\psi(x)$. The proof that the transformation $S(a)$ exists is given in textbooks on relativistic quantum mechanics.

Because γ^ν (and also β and α^k) are 4×4 matrices, there exist four linearly independent solutions of the Dirac equation. Two of these solutions have positive energy and correspond to the two spin polarization states of the free electron. On the other hand, the neutrino is known to have only one polarization state. The helicity,

$$\Sigma\cdot p/|p|, \left(\text{with } \Sigma = \begin{pmatrix} \sigma & 0 \\ 0 & \sigma \end{pmatrix}\right)$$

of a neutrino is always negative, which means that its spin direction is always opposite to its momentum p, whereas the helicity of the antineutrino is always positive because spin and momentum are in the same direction. Historically, negative energy states led Dirac to predict the existence of positrons; however, the Dirac equation is a single-particle equation, and quantum field theory is required to formulate a rigorous theoretical justification for particles and antiparticles. The mathematical correspondence that exists between negative-energy solutions of the Dirac equation and positive-energy solutions for positrons is quite useful in performing certain quantum electrodynamic calculations. The negative-energy solutions of the Dirac equation have important physical implications. The measured value of the Compton cross section, which is the relativistic generalization of the Thomson cross section derived in Chapter 2, depends on the presence of negative-energy solutions to the Dirac equation. At temperatures greater than 10^8 K the opacities in stellar interiors are reduced significantly because the Compton cross section is less than the Thomson cross section.

Appendix 3 Mathematical appendix

In this appendix we summarize mathematical proofs that are required to complete some derivations given in the main text.

A3.1 Plane-wave expansion (Equation (6.107))

A plane wave moving along the z-axis with wave number k can be expanded as

$$e^{ikr\cos\theta} = \sum_{l=0}^{\infty} A_l j_l(kr) P_l(\cos\theta), \tag{A3.1}$$

where $P_l(\cos\theta)$ are the Legendre polynomials and $j_l(kr)$ the spherical Bessel functions described in Chapters 4 and 6, respectively. We will prove that the constants A_l equal those given in Equation (6.107), namely $A_l = i^l(2l + 1)$.

Since the constants A_l are independent of θ we can evaluate them at any angle. At $\theta = 0$ Equation A3.1 becomes

$$e^{ikr} = \sum_{l=0}^{\infty} A_l j_l(kr). \tag{A3.2}$$

Recalling that $P_l(-\cos\theta) = (-1)^l P_l(\cos\theta)$ in Equation (A3.1), and taking the complex conjugate of Equation (A3.2), we obtain

$$A_l^{\star} = (-1)^l A_l = e^{-i\pi l} A_l. \tag{A3.3}$$

If we assume that $A_l^{\star} = e^{-i\pi l/2} a_l$ with a_l real, then Equation (A3.3) is satisfied, and therefore A_l can be expressed as $i^l a_l$. The derivatives of the spherical Bessel functions satisfy the recursion relation

$$\frac{dj_l(\rho)}{d\rho} = \frac{l}{2l+1} j_{l-1}(\rho) - \frac{l+1}{2l+1} j_{l+1}(\rho), \tag{A3.4}$$

where $\rho = kr$. The derivative with respect to ρ of the left-hand side of Equation (A3.1) equals i times the series expansion. Therefore, if we take the ρ derivative of both sides of Equation (A3.1), equate terms of equal l value on both sides of the resultant equation and then use Equation (A3.4), we find the recursion relation

$$a_l = \left(\frac{l+1}{2l+3}\right) a_{l+1} + \frac{l}{2l-1} a_{l-1}. \tag{A3.5}$$

Equation (A3.5) leads directly to the relation $a_l = 2l + 1$, and therefore Equation (6.107) is proven, since $A_l = i^l a_l = i^l(2l + 1)$.

A3.2 The gamma function

The gamma function given in Equation (6.90) satisfies the relations

$$\Gamma(z + 1) = \int_0^\infty t^z e^{-t} dt = \lim_{N \to \infty} \int_0^N t^z e^{-t} dt$$

$$= \lim_{N \to \infty} \left[-N^z e^{-N} + z \int_0^N t^{z-1} e^{-t} dt \right] = z\Gamma(z), \tag{A3.6}$$

where we have performed an integration by parts.

A3.3 The Sterling formula

The Sterling formula (approximation) given by Equation (3.56) can be proven in the following manner. The gamma function defined above can be rewritten as

$$\int_0^\infty e^{z \ln t - t} dt. \tag{A3.7}$$

It can readily be shown that the exponent in Equation (A3.7) has a maximum at $t = z$. Letting $t = z + y$, we obtain

$$\Gamma(z + 1) = e^{-z} \int_{-z}^\infty e^{z \ln(z+y) - y} dy = e^{-z} \int_{-z}^\infty e^{z \ln z + z \ln(1 + y/z) - y} dy$$

$$= z^z e^{-z} \int_{-z}^\infty e^{z \ln(1 + y/z) - y} dy. \tag{A3.8}$$

Making the Taylor expansion

$$\ln(1 + y/z) \simeq \frac{y}{z} - \frac{1}{2}\left(\frac{y}{z}\right)^2 + \frac{1}{3}\left(\frac{y}{z}\right)^3 + \cdots \tag{A3.9}$$

and change of variable $y = \sqrt{z}u$, Equation (A3.8) becomes

$$\Gamma(z + 1) \simeq z^z e^{-z} \int_{-z}^\infty e^{-y^3/2z + y^3/3z^2} dy = z^z e^{-z} \sqrt{z} \int_{-\sqrt{z}}^\infty e^{-u^2/2 + u^3/3\sqrt{z}} du. \tag{A3.10}$$

When z is very large, the second term in the exponent of Equation (A3.10) can be neglected and the lower limit of the integral approaches $-\infty$. Equation (A3.10) is true for positive integers n, and therefore it becomes

$$n! = \Gamma(n + 1) = \sqrt{2\pi n} n^n e^{-n}, \tag{A3.11}$$

which proves Equation (3.56).

The form of the Sterling approximation actually used in deriving distribution functions in Chapter 3 can be obtained rather simply. We write

$$\ln N! = \sum_{n=1}^{N} \ln(n) \simeq \int_{1}^{N} \ln(n)\, dn = N \ln N - N + 1. \tag{A3.12}$$

For large N Equation (A3.12) becomes

$$\ln N! = N(\ln N - 1). \tag{A3.13}$$

A3.4   Proof of Equation (6.99)

If z is the imaginary number contained in Equation (6.99), then this equation can be written as

$$\Gamma(1 + z)\Gamma(1 - z) = \frac{z}{\sin z}. \tag{A3.14}$$

The recursion relation given by Equation (A3.6) is valid when z is imaginary for both positive and negative z. Therefore, we have (see Equation 6.94)

$$\Gamma(1 - z) = -z\Gamma(-z). \tag{A3.15}$$

We will prove Equation (A3.14) by first solving for $\Gamma(z)$ and $\Gamma(-z)$.

For both real and imaginary values of z the function $\sin z$ can be represented as the infinite product

$$\sin z = z \prod_{n=1}^{\infty} \left(1 - \frac{z^2}{n^2\pi^2}\right) \tag{A3.16}$$

or

$$\frac{\sin z}{z} = \prod_{n=1}^{\infty} \left(1 - \frac{z^2}{n^2\pi^2}\right). \tag{A3.17}$$

The above formula for $\sin z$ is plausible because both left- and right-hand sides of Equation (A3.17) approach unity as $z \to 0$, and moreover they have the same zeros. A rigorous derivation of the infinite-product representation of $\sin z$ is given by Evgrafov (1978).

Because

$$\left(1 - \frac{t}{n}\right)^n \to e^{-t} \text{ as } n \to \infty, \tag{A3.18}$$

the gamma function becomes

$$\Gamma(z) = \int_{0}^{t} t^{z-1} e^{-t}\, dt = \int_{0}^{n} \left(1 - \frac{t}{n}\right)^n t^{z-1}\, dt \qquad (n \to \infty). \tag{A3.19}$$

Integrating Equation (A3.19) by parts implies that

$$\Gamma(z) = \lim_{n \to \infty} \frac{n! n^z}{z(z+1)\ldots(z+n)}, \tag{A3.20}$$

and therefore

$$\frac{1}{\Gamma(z)} = \lim_{n \to \infty} \frac{z(z+1)\ldots(z+n)}{n! n^z}$$

$$= \lim_{n \to \infty} \left[n^{-z} e^{z[1+(1/2)+\ldots+(1/n)]} z \prod_{m=1}^{n} \left(1 + \frac{z}{m}\right) e^{-z/m} \right]. \tag{A3.21}$$

The Euler constant γ is defined by the formula

$$\gamma = \lim_{n \to \infty} \left(1 + \frac{1}{2} + \frac{1}{3} + \ldots + \frac{1}{n} - \ln n\right) = 0.577\,2156, \tag{A3.22}$$

and therefore the first two terms on the right-hand side of Equation (A3.21) become

$$\lim_{n \to \infty} \left[n^{-z} e^{z[1+(1/2)+\ldots+(1/n)]} \right] =$$

$$\exp \left\{ \lim_{n \to \infty} \left[z\left(1 + \frac{1}{2} + \ldots + \frac{1}{n}\right) - z \ln n \right] \right\} = e^{\gamma z}. \tag{A3.23}$$

Substituting Equation (A3.23) into (A3.21), we obtain

$$\frac{1}{\Gamma(z)} = z e^{-\gamma z} \prod_{m=1}^{\infty} \left(1 + \frac{z}{m}\right) e^{-z/m}. \tag{A3.24}$$

Equation (A3.24) is valid for $\Gamma(-z)$, and therefore

$$\frac{1}{\Gamma(z)\Gamma(-z)} = -\frac{z}{\pi} \sin \pi z. \tag{A3.25}$$

Using the recursion relations given by Equations (A3.6) and (A3.15), we obtain Equation (A3.14), which for $z = in$ is the same as Equation (6.99). We note that the final equality in Equation (6.99) is true because $\sin(in) = i \sinh(n)$. This follows directly from the equality

$$\sin z = \frac{e^{iz} - e^{-iz}}{2i}. \tag{A3.26}$$

A3.5 The beta function

The beta function, which is defined as

$$B(m, n) = \int_0^1 t^{m-1}(1-t)^{n-1} \, dt, \tag{A3.27}$$

is useful for deriving the asymptotic formula given in Equation (6.89). We first prove the equality

$$B(m, n) = \frac{\Gamma(m)\Gamma(n)}{\Gamma(m + n)}. \tag{A3.28}$$

If we make the change of variable $t = \sin^2 \theta$, it follows almost immediately that Equation (A3.28) can be rewritten as

$$B(m, n) = 2 \int_0^{\pi/2} \sin^{2m-1} \theta \cos^{2n-1} \theta d\theta. \tag{A3.29}$$

Substituting $z = t^2$ into Equation (A3.6), we find that

$$\Gamma(m) = \int_0^\infty z^{m-1} e^{-z} dz = 2 \int_0^\infty t^{2m-1} e^{-t^2} dt, \tag{A3.30}$$

and a similar equation results for $\Gamma(n)$. It follows that

$$\Gamma(m)\Gamma(n) = 4 \left(\int_0^\infty t^{2m-1} e^{-t^2} dt \right) \left(\int_0^\infty s^{2n-1} e^{-s^2} ds \right)$$

$$= 4 \int_0^\infty \int_0^\infty t^{2m-1} s^{2n-1} e^{-(t^2+s^2)} ds dt. \tag{A3.31}$$

Introducing the polar coordinates $t = r \cos \theta$ and $s = r \sin \theta$ in Equation (A3.31) and using Equation (A3.29), we obtain

$$\Gamma(m)\Gamma(n) = 4 \left[\int_{r=0}^\infty r^{2(m+n)-1} e^{-r^2} dr \right] \left[\int_{\theta=0}^{\pi/2} \cos^{2m-1} \theta \sin^{2n-1} \theta d\theta \right]$$

$$= 2\Gamma(m + n)B(n, m) = 2\Gamma(m + n)B(m, n). \tag{A3.32}$$

The second equality in (A3.32) follows from Equation (A3.29), whereas the symmetry relation $B(n, m) = B(m, n)$ used in the final equality is a consequence of the definition of $B(m, n)$. If the substitution $t = 1 - s$ is made in Equation (A3.27), then the above symmetry relation follows directly.

A3.6   Confluent hypergeometric functions

The standard form for the confluent hypergeometric equation given by Equation (6.87) is

$$zy'' + [c - z]y' - ay = 0. \tag{A3.33}$$

The Pochhammer symbols are defined as

$$(a_0) = 1, \ldots, (a)_n = a(a + 1), \ldots, (a + n - 1) \qquad (n = 1, 2, 3). \tag{A3.34}$$

From their definition it follows that

$$(a)_n = \frac{\Gamma(a + n)}{\Gamma(a)} \qquad (n = 0, 1, 2), \tag{A3.35}$$

and therefore Equation (6.98) is the same as

$$_1F_1(a, c; z) = \sum_{n=0}^{\infty} \frac{(a)_n}{(c)_n} z^n. \tag{A3.36}$$

From Equations (A3.35), (A3.28) and (A3.27) we find the identity

$$\frac{(a)_n}{(c)_n} = \frac{\Gamma(c)}{\Gamma(a)\Gamma(c-a)} \int_0^1 t^{a+n-1}(1-t)^{c-a-1} dt, \tag{A3.37}$$

where $c > a > 0$. Substituting Equation (A3.37) into (A3.36), we obtain

$$_1F_1(a, c; z) = \frac{\Gamma(c)}{\Gamma(a)\Gamma(c-a)} \int_0^1 t^{a-1}(1-t)^{c-a-1} \left(\sum_{n=0}^{\infty} \frac{(zt)^n}{n!} \right) dt. \tag{A3.38}$$

The infinite sum in Equation (A3.38) is an exponential and therefore it becomes

$$_1F_1(a, c; z) = \frac{\Gamma(c)}{\Gamma(a)\Gamma(c-a)} \int_0^1 e^{zt} t^{a-1}(1-t)^{c-a-1} dt. \tag{A3.39}$$

If we substitute $t = 1 - u$ in Equation (A3.39), we find the equality

$$_1F_1(a, c; z) = e^z {}_1F_1(c-a, c; -z). \tag{A3.40}$$

The variable transformation used to derive Equation (A3.40) is known as Kummer's transformation.

Making the change of variable $y = z^{1-c}v$ in Equation (A3.33), we obtain the equation

$$zv'' + (2 - c - z)v' - (1 + a - c)v = 0. \tag{A3.41a}$$

Comparing (A3.33) and (A3.41a), we infer that

$$v = {}_1F_1(1 + a - c, 2 - c; z) \qquad (c \neq 2, 3, 4) \tag{A3.41b}$$

is a solution of (A3.41a), which implies that a second solution of the confluent hyper-geometric equation given by Equation (A3.33) is

$$y_2 = z^{1-c} {}_1F_1(1 + a - c, 2 - c; z) \qquad (c \neq 2, 3, 4). \tag{A3.42}$$

If c is not an integer greater than one, then y_2 is linearly independent of $_1F_1(a, c; z)$ and the general solution of Equation (A3.33) becomes

$$y = C_1 F_1(a, c; z) + C_2 z^{1-c} {}_1F_1(1 + a - c, 2 - c; z), \tag{A3.43}$$

where C_1 and C_2 are two constants which depend on the outgoing wave boundary conditions.

Our next task is to find an asymptotic (i.e. $z \to \infty$) formula for $_1F_1(a, c; z)$. From Equations (A3.39) and (A3.40) we have

$$_1F_1(a, c; z) = \frac{\Gamma(c)}{\Gamma(a)\Gamma(c-a)} e^z \int_0^1 e^{-zt} t^{c-a-1}(1-t)^{a-1} dt. \tag{A3.44}$$

The integral in Equation (A3.44) can be written as

$$\int_0^1 e^{-zt}t^{c-a-1}(1-t)^{a-1}\,dt = \int_0^\infty e^{-zt}t^{c-a-1}(1-t)^{a-1}\,dt$$

$$- \int_1^\infty e^{-zt}t^{c-a-1}(1-t)^{a-1}\,dt. \tag{A3.45}$$

If we substitute $s = zt$ in the first integral on the right-hand side of Equation (A3.45) and $v = z(t-1)$ in the second integral, the right-hand side of Equation (A3.45) becomes

$$z^{a-c}\int_0^\infty e^{-s}s^{c-a-1}\left(1-\frac{s}{z}\right)^{a-1}\,ds - e^{-z}\int_0^\infty e^{-v}\left(1+\frac{v}{z}\right)^{c-a-1}\left(\frac{-v}{z}\right)^{a-1}\,dv.$$

$$\tag{A3.46}$$

For $z \gg s$ and $z \gg v$ we can set

$$\left(1-\frac{s}{z}\right)^{a-1} \simeq 1 \text{ and } \left(\frac{v}{z}\right)^{a-1} \simeq 0. \tag{A3.47}$$

It follows that Expression (A3.46) becomes

$$\sim z^{a-c}\int_0^\infty e^{-s}s^{c-a-1}\,ds = z^{a-c}\Gamma(c-a), \tag{A3.48}$$

and therefore for very large z, Equation (A3.44) is

$$_1F_1(a,c,z) \sim \frac{\Gamma(c)}{\Gamma(a)}z^{a-c}e^z, \tag{A3.49}$$

which is the second term in Equation (6.89).

From Equation (A3.42) the other linearly independent asymptotic solution of the confluent hypergeometric equation is

$$y_2 \sim z^{1-c}{}_1F_1(1+a-c,2-c;z). \tag{A3.50}$$

Using Equations (A3.40) and (A3.49) and then setting $c = 1$, we see that Equation (A3.50) becomes

$$y_2 \sim z^{1-c}e^z\frac{\Gamma(2-c)}{\Gamma(1-a)}e^{-z}(-z)^{1-a-2+c} = \frac{(-z)^{-a}}{\Gamma(1-a)}. \tag{A3.51}$$

The expression for y_2 given by Equation (A3.51) is the same as the first term in Equation (6.49) for $c = 1$ and differs from it by only a constant for other allowed values of c. Equation (6.95), which is the correct expression for the Rutherford scattering cross section, verifies that in Equation (6.89) we have used the correct summation of the two linearly independent asymptotic solutions (A3.49) and (A3.51).

A3.7 The delta function

The one-dimensional delta function $\delta(x - x_0)$ is defined by the equation

$$\int \delta(x - x_0) f(x) \, dx = f(x_0),\tag{A3.52}$$

where it has been assumed that x_0 is within the range of integration. The delta function may be differentiated. Using integration by parts, we have

$$\int_{-\infty}^{+\infty} dx \, \delta'(x) f(x) = -f'(0).\tag{A3.53}$$

Expressing $f(x)$ in terms of its Fourier transform $F(k)$ and then taking the Fourier transform of $F(k)$, we obtain

$$f(x) = \int_{-\infty}^{+\infty} \frac{dk}{2\pi} e^{-ikx} \int_{-\infty}^{+\infty} dx' \, e^{ikx'} f(x').\tag{A3.54}$$

Exchanging the order of integration gives

$$f(x) = \int_{-\infty}^{+\infty} dx' f(x') \int_{-\infty}^{+\infty} \frac{dk}{2\pi} e^{ik(x'-x)} = \int_{-\infty}^{+\infty} dx' f(x') \delta(x' - x),\tag{A3.55}$$

and therefore

$$\delta(x' - x) = \int_{-\infty}^{+\infty} \frac{dk}{2\pi} e^{ik(x'-x)}.\tag{A3.56}$$

Other important properties of the delta function are:

$$x\delta(x) = 0,\tag{A3.57a}$$

$$\delta(x) = \delta(-x),\tag{A3.57b}$$

$$\delta(ax) = \frac{\delta(x)}{|a|}.\tag{A3.57c}$$

Equation (A3.57c) is true because

$$\int \delta(ax) f(x) \, dx = \frac{1}{|a|} \int \delta(y) f\left(\frac{y}{|a|}\right) dy = \frac{1}{|a|} f(0).\tag{A3.58}$$

To prove Equation (7.84) we divide the range of integration into small intervals i and use Equation (A3.58) to obtain

$$\int \delta[f(x)] \, dx = \sum_i \int_i \delta[f(x)] \, dx = \sum_i \int_i \delta[f(x_i) + f'(x_i)(x_i - x'_i)] \, dx'_i$$

$$= \sum_j \frac{\delta(x_j - x'_j)}{|f'(x_j)|} \quad (x_j \text{ are simple zeros of } f(x)),\tag{A3.59}$$

and therefore Equation (7.84) is proven. A special case of (A3.59) is

$$\delta(x^2 - a^2) = \frac{1}{2a} [\delta(x - a) + (\delta(x + a)] \quad (a > 0).\tag{A3.60}$$

Some important representations of the delta function are

$$\delta(x) = \lim_{\varepsilon \to 0^+} \frac{\sin(x/\varepsilon)}{\pi x} \tag{A3.61a}$$

$$= \lim_{\varepsilon \to 0^+} \frac{\varepsilon}{\pi(x^2 + \varepsilon^2)} \tag{A3.61b}$$

$$= \lim_{\varepsilon \to 0^+} \frac{\exp(-x^2/\varepsilon)}{(\pi\varepsilon)^{1/2}} \tag{A3.61c}$$

$$= \lim_{\varepsilon \to 0} \frac{\theta(x + \varepsilon) - \theta(x)}{\varepsilon}, \tag{A3.61d}$$

where the step function $\theta(x)$ in the final equation equals 1 for $x > 0$ and 0 for $x < 0$.

In three dimensions we have

$$\int f(\mathbf{r}')\delta^3(\mathbf{r} - \mathbf{r}')\,d^3r' = f(\mathbf{r}), \tag{A3.62a}$$

$$\delta^3(\mathbf{r}) = \frac{1}{(2\pi)^3} \int_{-\infty}^{+\infty} e^{i\mathbf{k}\cdot\mathbf{r}}\,d^3k, \tag{A3.62b}$$

$$\nabla^2\left(\frac{1}{r}\right) = -4\pi\delta(\mathbf{r}). \tag{A3.62c}$$

The gravitational potential $\Phi(\mathbf{r})$ caused by a mass–density distribution satisfies the equation

$$\Phi(\mathbf{r}) = -\int d^3r'\,\frac{\rho(\mathbf{r}')}{|\mathbf{r} - \mathbf{r}'|}. \tag{A3.63}$$

Using Equation (A3.62c), we obtain

$$\nabla^2\Phi(\mathbf{r}) = -\int d^3r'\left(\nabla^2\frac{1}{|\mathbf{r} - \mathbf{r}'|}\right)\rho(\mathbf{r}') = 4\pi\rho(\mathbf{r}). \tag{A3.64}$$

Equation (A3.64) shows that the gravitational potential satisfies Poisson's equation. This was already proven in Section 2.2 without the use of the delta function.

Appendix 4 Polytropes and the isothermal gas sphere

Analytic solutions of the Lane–Emden equation, given by Equation (2.146), can be found for polytropic indices $n = 0, 1$ and 5. For $n = 0$ the Lane–Emden equation becomes

$$\frac{1}{\xi^2} \frac{d}{d\xi}\left(\xi^2 \frac{d\theta}{d\xi}\right) = -1. \tag{A4.1}$$

Integration of Equation (A4.1) gives

$$\xi^2 \frac{d\theta}{d\xi} = -\frac{1}{3}\xi^3 - C, \tag{A4.2}$$

with C equal to a constant. A second integration yields

$$\theta = D + \frac{C}{\xi} - \frac{1}{6}\xi^2. \tag{A4.3}$$

Since $\theta \to \infty$ as $\xi \to 0$, we choose $C = 0$, and therefore Equation (A4.3) reduces to

$$\theta = D - \frac{1}{6}\xi^2. \tag{A4.4}$$

The further condition $\theta = 1$ at $\xi = 0$ implies that $D = 1$, and the analytic solution to Equation (A4.1) becomes

$$\theta = 1 - \frac{1}{6}\xi^2. \tag{A4.5}$$

Substituting $\theta = \chi/\xi$ and $n = 1$ in Equation (2.146) transforms the Lane–Emden equation into

$$\frac{d^2\chi}{d\xi^2} = -\chi, \tag{A4.6}$$

which has the obvious solution

$$\chi = C\sin(\xi - \delta), \tag{A4.7}$$

where C and δ are constants. Equation (A4.7) is equivalent to the equation

$$\theta = C\frac{\sin(\xi - \delta)}{\xi}. \tag{A4.8}$$

The boundary condition $\theta \to 1$ as $\xi \to 0$ reduces Equation (A4.8) to

$$\theta = \frac{\sin \xi}{\xi} \tag{A4.9}$$

whose first zero occurs at $\xi = \pi$. We remark that the Lane–Emden equation (2.146) is a linear equation for $n = 0$ and $n = 1$.

For $n = 5$ the Lane–Emden equation becomes

$$\frac{d^2\theta}{d\xi^2} + \frac{2}{\xi}\frac{d\theta}{d\xi} + \theta^5 = 0. \tag{A4.10}$$

Making the substitutions $\xi = e^{-t}$ and $\theta = \frac{1}{\sqrt{2}}e^{t/z}\chi(t)$ into Equation (A4.10), we obtain

$$4\frac{d^2\chi}{dt^2} - \chi + \chi^5 = 0. \tag{A4.11}$$

Defining $\psi = \frac{d\chi}{dt}$, we find the equality

$$\frac{d\psi}{dt} = \frac{d\left(\frac{d\chi}{dt}\right)}{dt} = \frac{d\left(\frac{d\chi}{dt}\right)}{d\chi}\frac{d\chi}{dt} = \psi\frac{d\psi}{d\chi}, \tag{A4.12}$$

which transforms Equation (A4.11) into

$$4\psi\frac{d\psi}{d\chi} - \chi + \chi^5 = 0. \tag{A4.13}$$

Integrating Equation (A4.13) leads to

$$2\psi^2 - \frac{1}{2}\theta^2 + \frac{1}{6}\theta^6 = C, \tag{A4.14}$$

with C equal to a constant. The boundary conditions $\theta = 1$ and $d\theta/d\xi = 0$ at $\xi = 0$ are equivalent to $\chi = 0$ and $d\chi/dt = \psi = 0$ as $t \to \infty$. Therefore the constant C equals zero, and Equation (A4.14) becomes

$$\psi^2 = \frac{1}{4}\theta^2 - \frac{1}{12}\theta^6, \tag{A4.15}$$

which is the same as

$$\psi = \frac{d\chi}{dt} = \frac{1}{2}\chi\left(1 - \frac{1}{3}\chi^4\right)^{1/2}. \tag{A4.16}$$

Integrating Equation (A4.16), we obtain

$$2 \int \frac{d\chi}{\chi \left(1 - \frac{1}{3}\chi^4\right)^{1/2}} = \int dt = t + D, \tag{A4.17}$$

where D is a constant. Substituting $\chi^4 = 3\cos^2\theta$ in Equation (A4.17) gives

$$t + D = -\int \sec\theta \, d\theta = -\ln(\sec\theta + \tan\theta). \tag{A4.18}$$

Taking the exponent of both sides of Equation (A4.18) implies that

$$e^{-D}e^{-t} = \sec\theta + \tan\theta. \tag{A4.19}$$

Recalling that $\xi = e^{-t}$ and $\chi^2 = \sqrt{3}\cos\theta$ enables us to transform Equation (A4.19) into the equation

$$e^{-D}\xi = \frac{\sqrt{3}}{\chi^2} + \left(\frac{3}{\chi^4} - 1\right)^{1/2}. \tag{A4.20}$$

Using the definitions $\xi = e^{-t}$ and $\theta = \frac{1}{\sqrt{2}}e^{t/2}\chi(t)$ in Equation (A4.20), we find that

$$\theta = \left(\frac{\sqrt{3}\,e^D}{1 + e^{2D}\xi^2}\right)^{1/2}. \tag{A4.21}$$

The boundary condition $\theta = 1$ at $\xi = 0$ implies that $e^D = \frac{1}{\sqrt{3}}$, and therefore Equation (A4.21) becomes

$$\theta = 1 \Big/ \left(1 + \frac{1}{3}\xi^2\right)^{1/2}. \tag{A4.22}$$

As shown in Section 1.4, the equation for an isothermal gas sphere can be expressed as

$$\frac{1}{\xi}\frac{d}{d\xi}\left[\xi^2 \frac{d\psi}{d\xi}\right] = e^{-\psi}. \tag{A4.23}$$

To obtain a solution of Equation (A4.23) we make the power-law expansion

$$\psi = \sum_{n=0}^{\infty} a_n \xi^n. \tag{A4.24}$$

It can readily be shown by substituting Equation (A4.24) into Equation (A4.23) that $a_0 = a_1 = 0$, and only even powers of ξ survive. Therefore, we have

$$\frac{d\psi}{d\xi} = 2a_2\xi + 4a_4\xi^3 + 6a_6\xi^5 + \dots, \tag{A4.25}$$

$$\xi^2 \frac{d\psi}{d\xi} = 2a_2\xi^3 + 4a_4\xi^5 + 6a_6\xi^7 + \dots, \tag{A4.26}$$

$$\frac{1}{\xi^2} \frac{d\left(\xi^2 \frac{d\psi}{d\xi}\right)}{d\xi} = 6a_2 + 20a_4\xi^2 + 42a_6\xi^4. \tag{A4.27}$$

At $\xi = 0$ the right-hand side of Equation (A4.23) is equal to zero, and therefore $6a_2 = 1$ or $a_2 = 1/6$. It follows from Equations (A4.27) and (A4.23) that

$$1 + 20a_2\xi^2 + 42a_6\xi^4 + \ldots = e^{-\psi}. \tag{A4.28}$$

Taking the derivative of Equation (A4.28) leads to

$$40a_4\xi + 168a_6\xi^3 + \ldots = e^{-\psi}(2a_2\xi + 4a_4\xi^3 + 6a_6\xi^5 + \ldots). \tag{A4.29}$$

Factoring out ξ in Equation (A4.29) and setting $\xi = 0$ implies that

$$40a_4 = (-1)(2a_2) \text{ or } a_4 = -\frac{1}{120}. \tag{A4.30}$$

Taking the derivative of Equation (A4.29) with respect to ξ and equating terms of the first order in ξ at $\xi = 0$ leads to

$$336a_6 = 4a_2^2 - 8a_4. \tag{A4.31}$$

Substituting $a_2 = \frac{1}{6}$ and $a_4 = -\frac{1}{120}$ in Equation (A4.31) gives $a_6 = \frac{1}{1890}$. We have thus obtained the solution for Problem 1.4.

It can readily be shown by direct substitution that Equation (1.63) has the solution $\rho \propto 1/r^2$, which is valid for $r \neq 0$.

Appendix 5 Solutions to selected problems

A5.1 Chapter 1

1. a. An orbit is a hyperbola for $\varepsilon > 1(E > 0)$, a parabola for $\varepsilon = 1(E = 0)$, and an ellipse for $\varepsilon < 1$ $(E < 0)$.

    b. Since $r = \hat{x}a\cos\theta + \hat{y}b\sin\theta$ and $dr = (-\hat{x}a\sin\theta + \hat{y}b\cos\theta)d\theta$, we have

    $$\frac{1}{2}\int r \times dr = \frac{ab\hat{z}}{2}\int_0^{2\pi}(\cos^2\theta + \sin^2\theta)d\theta = \pi ab\hat{z}. \tag{A5.1}$$

4. The solution to problem 4 is given in Appendix 4.

8. $Z = Z(t) = M_{\mathrm{h}}/M_{\mathrm{g}}$ where M_{g} and M_{h} are equal to the total interstellar mass and heavy-element mass respectively. Let $\delta M_{\mathrm{s}} = $ the mass of stellar remnants produced and $p\delta M_{\mathrm{s}} = $ the yield of heavy elements. We have

    $$\delta M_{\mathrm{h}} = p\delta M_{\mathrm{s}} - Z\delta M_{\mathrm{h}}. \tag{A5.2}$$

    Since $\delta M_{\mathrm{s}} = -\delta M_{\mathrm{g}}$, equation (A5.2) becomes

    $$\delta M_{\mathrm{h}} = (-p + Z)\delta M_{\mathrm{g}}. \tag{A5.3}$$

    From the definition of $Z(t)$ given above we obtain

    $$\delta Z = \delta\left(\frac{M_{\mathrm{h}}}{M_{\mathrm{g}}}\right) = \frac{1}{M_{\mathrm{g}}}(\delta M_{\mathrm{h}} - Z\delta M_{\mathrm{g}}). \tag{A5.4}$$

    Therefore,

    $$\delta Z = \frac{1}{M_{\mathrm{g}}}(-p + Z - Z)\delta M_{\mathrm{g}} = -p\frac{\delta M_{\mathrm{g}}}{M_{\mathrm{g}}}, \tag{A5.5}$$

    and we conclude that

    $$Z(t) = -p\ln\frac{M_{\mathrm{g}}(t)}{M_{\mathrm{g}}(0)}. \tag{A5.6}$$

A5.2   **Chapter 2**

1.   Consider a thin spherical mass shell whose inner and outer radii are r_1 and r_2 respectively. This mass shell is sufficiently thin that the density $\rho(r')$ within it can be assumed constant. Let r' equal the radius of the mass shell, r the distance of some element of the mass shell to an external point, and R the distance between the center of the mass shell to the same external point.

The gravitational potential $\Phi(R)$ becomes

$$\Phi(R) = -G\int\frac{\rho(r')}{r}2\pi r'^2\sin\theta d\theta dr'$$

$$= -2\pi\rho G\int_{r_1}^{r_2}r'^2 dr'\int_0^\pi\frac{\sin\theta}{r}d\theta. \tag{A5.7}$$

From the law of cosines we have

$$r^2 = r'^2 + R^2 - 2r'R\cos\theta, \tag{A5.8}$$

and therefore

$$\frac{dr}{r'R} = \frac{\sin\theta}{r}d\theta. \tag{A5.9}$$

Substituting Equation (A5.9) into Equation (A5.7) and assuming that $R > r_2$, we obtain

$$\Phi(R) = -\frac{2\pi\rho G}{R}\int_{r_1}^{r_2}r'dr'\int_{R-r'}^{R+r'}dr$$

$$= -\frac{4\pi\rho G}{R}\int_{r_1}^{r_2}r'^2 dr' = -\frac{GM}{R}, \tag{A5.10}$$

where $M = \dfrac{4\pi}{3}\rho(r_2^3 - r_1^3)$.

For R inside the mass shell (i.e. $R < r_1$) we have

$$\Phi(R) = -\frac{2\pi\rho G}{R}\int_{r_1}^{r_2}r'dr'\int_{r'-R}^{r'+R}dr$$

$$= -4\pi\rho G\int_{r_1}^{r_2}r'dr' = -2\pi\rho G(r_2^2 - r_1^2), \tag{A5.11}$$

which is a constant. Equations (A5.10) and (A5.11) imply that the gravitational force from a thin mass shell is $-GM/R^2$ outside the mass shell and zero inside. Since gravitational potentials from different mass shells are additive, we conclude that for spherical stars the gravitational acceleration at distance R from the stellar center is $-GM(R)/R^2$, where $M(R)$ is the mass interior to radial distance R.

In Chapter 12 we shall discuss general relativity and show that the Schwarzschild metric holds for a spherical mass distribution. Since $M(R)$ must be zero inside a spherical mass shell (otherwise $M(R)/R$ would diverge as $R \to 0$), space is flat there, and consequently gravitational forces must vanish, because in general relativity, gravitational forces are associated with the curvature of space.

3. a. From the first law of thermodynamics we have $TdS = dQ = dU + PdV$ for reversible processes. We can express dU and dS as

$$dU = \left(\frac{\partial U}{\partial V}\right)_T dV + \left(\frac{\partial U}{\partial T}\right)_V dT \qquad (A5.12)$$

and

$$dS = \left(\frac{\partial S}{\partial V}\right)_T dV + \left(\frac{\partial S}{\partial T}\right)_V dT \qquad (A5.13)$$

respectively. The above equations imply that

$$dS = \frac{1}{T}\left[\left(\frac{\partial U}{\partial V}\right)_T + P\right]dV + \frac{1}{T}\left(\frac{\partial U}{\partial T}\right)_V dT. \qquad (A5.14)$$

Since dS is a perfect differential, the integrability condition implies that

$$\frac{\partial}{\partial T}\frac{\partial S}{\partial V} = \frac{\partial}{\partial V}\frac{\partial S}{\partial T}. \qquad (A5.15)$$

Equations (A5.14) and (A5.15) lead to the equation

$$\frac{\partial}{\partial T}\left[\frac{1}{T}\left(\frac{\partial U}{\partial V}\right)_T + \frac{P}{T}\right] = \frac{\partial}{\partial V}\left[\frac{1}{T}\left(\frac{\partial U}{\partial T}\right)_V\right]. \qquad (A5.16)$$

From Equation (A5.16) we have

$$\left(\frac{\partial U}{\partial V}\right)_T = T\left(\frac{\partial P}{\partial T}\right)_V - P. \qquad (A5.17)$$

b. Treating a photon gas as a particle gas with $p = h\nu/c$, $E = h\nu$ and velocity $v = c$, we can readily calculate pressure and internal density as we did for an ideal gas. It follows that the radiation pressure is one-third of the internal energy density.

4. For blackbody radiation the internal energy U can be written $U = VU_r(T)$, with V the volume. Therefore, we obtain

$$\left(\frac{\partial U}{\partial V}\right)_T = U_r(T) \text{ and } \left(\frac{\partial U}{\partial T}\right)_V = V\frac{dU_r}{dT}. \qquad (A5.18)$$

Because $P = \frac{1}{3}U_r(T)$, the integrability condition used in solving Problem 2.3 becomes

$$\frac{\partial}{\partial T}\left[\frac{1}{T}\left(U_r + \frac{U_r}{3}\right)\right] = \frac{1}{T}\frac{dU_r}{dT}, \qquad (A5.19)$$

with $U_r = U_r(T)$. Equation (A5.19) implies that

$$\frac{dU_r}{U_r} = 4\frac{dT}{T},$$

(A5.20)

and therefore $U_r = aT^4$.

6.   The gravitational potential energy of a polytrope of index $n = 3$ is

$$\Omega = \frac{-3}{(5-n)}\frac{GM^2}{R}.$$

(A5.21)

The virial theorem implies that binding energy is one-half the gravitational energy for a nonrelativistic gas. Therefore, the binding energy of a one-solar-mass main-sequence star is approximately

$$\frac{1}{2}|\Omega| \sim \frac{3}{(5-3)}\frac{GM_\odot^2}{R_\odot} \sim 3 \times 10^{48}\,\text{erg}.$$

(A5.22)

It follows that its pre-main-sequence lifetime τ is

$$\tau \sim \frac{3 \times 10^{48}}{L_\odot} \sim 3 \times 10^7 \text{ years}.$$

(A5.23)

7.   The energy release per synthesized α particle is $4m_pc^2 - m_{He}c^2$. It follows that the nuclear energy release per initial gram of hydrogen is $\simeq 6 \times 10^{18}\,\text{erg g}^{-1}$ and the main-sequence lifetime of a $1\,M_\odot$ star $\simeq 10^{10}$ years. For $\mu = 0.7$ Equation (2.135) implies that the temperature at the center of a $1\,M_\odot$ main-sequence star is

$$T \simeq \frac{Gm_p\mu M_\odot}{kR_\odot} \simeq 16 \times 10^6\,K.$$

(A5.24)

9.   The equations of radiative and hydrostatic equilibrium are

$$\frac{L}{4\pi r^2} = \frac{-ac}{3\kappa\rho}\frac{dT^4}{dr}$$

(A5.25)

and

$$\frac{dP}{dr} = -\rho\frac{GM}{r^2}.$$

(A5.26)

If we multiply Equation (A5.25) by ρGM, it becomes

$$\rho\frac{GM}{r^2} = -\frac{4\pi ac}{3\kappa}\frac{GM}{L}\frac{dT^4}{dr}.$$

(A5.27)

From Equations (A5.26) and (A5.27) we obtain

$$dP = \frac{4\pi ac}{3\kappa}\frac{GM}{L}dT^4 \equiv KdT^4$$

(A5.28)

Integrating Equation (A5.28) gives

$$P = KT^4,$$ (A5.29)

where $P = 0$ at $r = R$ and we have assumed that T is small at $r = R$. Equations (A5.25) and (A5.29) imply that

$$\frac{L\,dr}{4\pi r^2} = \frac{-4ack\,dT}{\mu m_p 3\kappa K}.$$ (A5.30)

Integrating Equation (A5.30) from the surface to radial distance r ($r < R$), we obtain

$$T = \frac{L\mu m_p 3\kappa K}{16\pi ack}\left(\frac{1}{r} - \frac{1}{R}\right).$$ (A5.31)

10. The pressure and temperature must be continuous across the discontinuity, and therefore from the ideal gas law we obtain

$$\frac{\rho_H kT}{\mu_H m_p} = \frac{\rho_{He} kT}{\mu_{He} m_p}.$$ (A5.32)

Since $\mu_{He} = \frac{4}{3}$ and $\mu_H = \frac{1}{2}$, Equation (A5.32) gives

$$\frac{\rho_{He}}{\rho_H} = \frac{4/3}{1/2} = \frac{8}{3}.$$ (A5.33)

A5.3 Chapter 3

1. a. The average speed of a particle is

$$\bar{v} = 4\pi \left(\frac{m}{2\pi kT}\right)^{3/2} \int_0^\infty v^3 \exp(-mv^2/2kT)\,dv$$

$$= 4\pi \left(\frac{m}{2\pi kT}\right)^{3/2} \left(\frac{2kT}{m}\right)^2 \int_0^\infty x^3 \exp(-x^2)\,dx$$

$$= 2\left(\frac{2kT}{\pi m}\right)^{1/2} \int_0^\infty z e^{-z}\,dz = \left(\frac{8kT}{\pi m}\right)^{1/2}.$$ (A5.34)

We note that the most probable speed v_p is obtained by requiring that the derivative of $v^2 \exp(-mv^2/2kT)$ with respect to v equal zero. This gives $v_p = (2kT/m)^{1/2}$.

b. The mean square velocity

$$\overline{v^2} = \left(\frac{8kT}{\pi m}\right)\int_0^\infty z^{3/2} e^{-z}\,dz = \frac{3}{2}\left(\frac{2kT}{m}\right),$$ (A5.35)

and therefore $\frac{1}{2}m\overline{v^2} = \frac{3}{2}kT$.

c. The mean square fluctuation

$$\langle \Delta v^2 \rangle = \langle (v - \bar{v})^2 \rangle = \bar{v^2} - 2\bar{v}^2 + \bar{v}^2$$

$$= \bar{v^2} - \bar{v}^2 = \left(\frac{3\pi - 8}{\pi}\right)\left(\frac{kT}{m}\right). \tag{A5.36}$$

4.

a. $E_n = \dfrac{(n\hbar)^2}{2m}\dfrac{1}{r^2} - \dfrac{e^2}{r}$ (n is a positive integer). Requiring that $\partial E_n / \partial r = 0$ gives

$$r_n = n^2 a_0 = \frac{\hbar^2 n^2}{me^2}$$

and

$$E_n = \frac{-me^4}{2\hbar^2 n^2}. \tag{A5.37}$$

b. In Equation (3.30) the two lowest energy levels are

$$E_{111} = \frac{h^2}{8mL^2}(1 + 1 + 1) = \frac{3h^2}{8mL^2} \tag{A5.38}$$

and

$$E_{211} = E_{121} = E_{112} = \frac{h^2}{8mL^2}(4 + 1 + 1) = \frac{3h^2}{4mL^2}. \tag{A5.39}$$

A zero-point energy exists because if we choose $n_x = n_y = n_z = 0$, then the wave function is equal to zero everywhere. We remark that the factor of 2 for electron degeneracy given in Equation (3.32) could also be accounted for by choosing $n_x, n_y, n_z = \pm 1, \pm 2, \pm 3 \ldots$

It is also well known (see Chapter 6) that the energy levels of a harmonic oscillator are $E_n = h\nu(n + \frac{1}{2})$, with $\frac{1}{2}h\nu$ equal to the zero-point energy. According to classical physics, a particle can be at rest with zero energy and momentum. The existence of a zero-point energy follows directly from the uncertainty relation $\Delta x^2 \Delta p^2 \geq \hbar/4$.

7. In a Fermi gas the quantum state of lowest energy has nonzero energy. On the other hand, in a Bose gas each particle occupies a zero-energy quantum state as $T \to 0$. From our discussion of boson statistics, the number density n of a non-relativistic Bose gas becomes

$$n = \frac{gm^{3/2}}{\sqrt{2\pi^2\hbar^3}}\int_0^\infty \frac{\sqrt{\varepsilon}\,d\varepsilon}{e^{(\varepsilon - \mu)/kT} - 1}, \tag{A5.40}$$

where g equals the degeneracy factor $2S + 1$, with $S = 0, 1, 2, \ldots$ In Equation (A5.40) the chemical potential μ must be negative because n must be positive, and as $T \to 0$, μ must $\to 0$. For an assumed number density n, the critical temperature

$T = T_c$ at which $\mu \to 0$ can be obtained from the equation

$$n = \frac{g(mkT_c)^{3/2}}{\sqrt{2\pi^2\hbar^3}} \int_0^\infty \frac{\sqrt{x}\,dx}{e^x - 1}, \tag{A5.41}$$

where $x = \varepsilon/kT$ and the integral has the numerical value of 2.31. From Equation (A5.41) the critical temperature for Bose condensation becomes

$$T_c = \frac{h^2}{2.31mk}\left(\frac{n}{4\pi\sqrt{2}}\right)^{2/3}. \tag{A5.42}$$

When μ becomes equal to zero, Equation (A5.42) no longer correctly describes the particle number density because we have taken it proportional to $\sqrt{\varepsilon}$, which is obviously not true at $\varepsilon = 0$. Since particles that condense to the ground state have zero energy, the energy per unit volume for $T < T_c$ is

$$\frac{E}{V} = \frac{gm^{3/2}(kT)^{5/2}}{\sqrt{2\pi^2\hbar^3}} \int_0^\infty \frac{x^{3/2}\,dx}{e^x - 1}. \tag{A5.43}$$

The number density of particles with $\varepsilon > 0$ is

$$n(\varepsilon > 0) = \frac{g(mkT)^{3/2}}{\sqrt{2\pi^2\hbar^3}} \int_0^\infty \frac{\sqrt{x}\,dx}{e^x - 1} = n\left(\frac{T}{T_c}\right)^{3/2} \tag{A5.44}$$

and therefore the number density of zero-energy particles becomes

$$n(\varepsilon = 0) = n\left[1 - \left(\frac{T}{T_c}\right)^{3/2}\right] \tag{A5.45}$$

8.   a. Since $F = U - TS$, we have

$$dF = -S\,dT - P\,dV. \tag{A5.46}$$

We can also express dF as

$$dF = \left(\frac{\partial F}{\partial T}\right)_V dT + \left(\frac{\partial F}{\partial V}\right)_T dV. \tag{A5.47}$$

A comparison between Equations (A5.46) and (A5.47) shows that

$$P = -\left(\frac{\partial F}{\partial V}\right)_T \quad \text{and} \quad S = -\left(\frac{\partial F}{\partial T}\right)_V$$

and

$$U = -T^2\frac{\partial(F/T)}{\partial T} = -T\frac{\partial F}{\partial T} + F = TS + F,$$

which is correct from the definition of F.

    b. Since $dF = dU - S\,dT - T\,dS$, it follows that under conditions of constant V

and T the free energy F must decrease because by the second law of thermodynamics the entropy S must increase in irreversible processes.

c. If $P_n = g_n \exp(-E_n/kT)/Z$ is the probability that an energy level is occupied,
then we can define the entropy as

$$S = k \sum_n P_n \ln P_n \tag{A5.48}$$

and therefore

$$TS = -kT \sum_n P_n \ln P_n = -kT \sum_n P_n(-\ln Z - E_n/kT)$$

$$= kT \ln Z + \sum_n P_n E_n = kT \ln Z + U. \tag{A5.49}$$

Comparing the above equation with the definition of F gives

$$F = -kT \ln Z. \tag{A5.50}$$

9.   From Equation (3.107) we have

$$1 \gg e^{\mu/kT} = \frac{n}{2(2\pi)^{3/2}} \left(\frac{h^2}{mkT}\right)^{3/2}, \tag{A5.51}$$

and therefore

$$n \left(\frac{h^2}{mkT}\right)^{3/2} \ll 1, \tag{A5.52}$$

when Boltzmann statistics apply.

A5.4 Chapter 4

4.   If we neglect proton spin, the degeneracy of the energy level E_n of the hydrogen
atom is

$$2 \sum_{l=0}^{n-1} (2l + 1) = \frac{4n(n-1)}{2} + 2n = 2n^2. \tag{A5.53}$$

The above equality follows almost directly because the summation

$$\sum_{l=1}^{n-1} l = \frac{n(n-1)}{2}. \tag{A5.54}$$

5.   The Heisenberg equation of motion of a dynamical variable that does not depend
explicitly on time is

$$i\hbar \frac{dF}{dt} = (FH - HF) \equiv [F, H], \tag{A5.55}$$

where H is the Hamiltonian. If we let $F = r \cdot p$, and note that its time average must equal zero for a periodic system, then the Heisenberg equation of motion becomes

$$0 = i\hbar \frac{d}{dt} \langle r \cdot p \rangle = \langle [(r \cdot p, H] \rangle. \tag{A5.56}$$

The commutator

$$[r \cdot p, H] = \left[xp_x + yp_y + zp_z, \frac{p_x^2 + p_y^2 + p_z^2}{2m} + V(x, y, z) \right]$$

$$= \frac{p_x^2 + p_y^2 + p_z^2}{m} - \left(x\frac{\partial V}{\partial x} + y\frac{\partial V}{\partial y} + z\frac{\partial V}{\partial z} \right)$$

$$= 2T - r \cdot \nabla V, \tag{A5.57}$$

with T equal to the kinetic energy. Equation (A5.57) is the virial theorem in quantum mechanics and implies that

$$2\langle T \rangle = \langle r \cdot \nabla V \rangle \tag{A5.58}$$

The Coulomb potential is spherically symmetric and inversely proportional to the radii. Therefore, Equation (A5.58) becomes

$$2\langle T \rangle = |\langle V \rangle|, \tag{A5.59}$$

which is similar to the virial theorem for nonrelativistic masses in Newtonian gravitation. This proves the stated relationship between binding energy and potential energy. We leave it to the reader to perform the calculation directly with hydrogenic wave functions.

6. Let $\sigma_n(Z, \varepsilon)$ equal the cross section for photoionization from the energy level with principal quantum number n of a hydrogenic atom with nuclear charge Z. The frequency of the ionizing photon and excess kinetic energy of the electron are related by the equation

$$h\nu = Z^2 R_y \left(\frac{1}{n^2} + \varepsilon \right), \tag{A5.60}$$

where $R_y = 13.6\,\text{eV}$. Photoionization electrons will eventually recombine and produce recombination lines,

$$p + e^- \rightarrow H(n, l) + h\nu. \tag{A5.61}$$

The principle of detailed balance implies that in thermodynamic equilibrium the rate of recombination to level n must equal the rate of photoionization from level n. It follows that the recombination coefficient $\alpha_n(T)$ in units of $\text{cm}^{-3}\,\text{s}^{-1}$ is related to σ_n by the equation

$$n_p n_e \alpha_n(T) = n_H \int_{R_y}^{\infty} U_\nu c\sigma(\nu)(1 - e^{-h\nu/kT})\,d\nu, \tag{A5.62}$$

where U_ν is the density of blackbody radiation per unit frequency, $n_p n_e / n_H$ satisfies the Saha equation and the factor $(1 - \exp(-h\nu/kT))$ corrects for stimulated emission.

A5.5 Chapter 5

2.   The time average energy density of a monochromatic electromagnetic wave with amplitude $E_0 = H_0$ is

$$\frac{\langle E_0^2 \sin^2 \omega t + H_0^2 \cos^2 \omega t \rangle}{8\pi} = \frac{E_0^2}{8\pi}. \tag{A5.63}$$

The corresponding intensity of radiation is $I(\theta, \phi) = I_0 \delta(\theta - \theta_0) \delta(\phi - \phi_0)$. Since the radiation energy density is I_0/c, we conclude that

$$I_0 = \frac{cE_0^2}{8\pi}. \tag{A5.64}$$

3.   Consider some point P at a distance r from a star of radius R. If a ray drawn from point P intersects the star, then the intensity in that direction is B. The intensity in other directions is zero. It follows that the flux is

$$F = \int I \cos \theta \, d\Omega = B \int_0^{2\pi} \int_0^{\theta_c} \sin \theta \cos \theta \, d\theta d\phi, \tag{A5.65}$$

where $\theta_c = \sin^{-1}(R/r)$. Performing the integral in Equation (A5.65) gives

$$F = \pi B(1 - \cos^2 \theta_c) = \pi B \left(\frac{R}{r}\right)^2. \tag{A5.66}$$

6.   Since the probability that an atom emits a collisionally broadened line during the time interval $t_0 - (t_0 + dt_0)$ is

$$\exp(-t_0/\tau_c) \frac{dt_0}{\tau_c},$$

where τ_c is the mean collisional lifetime, the intensity of emitted radiation becomes

$$\phi_\nu = \text{constant} \int_0^\infty \left[\frac{\sin \pi (\nu - \nu_0) t_0}{\pi (\nu - \nu_0)}\right]^2 \exp\left(\frac{-t_0}{\tau_c}\right) \frac{dt_0}{\tau_c}$$

$$= \frac{1}{2\pi^2 \tau_c} \frac{1}{(\nu - \nu_0)^2 + (1/2\pi\tau_c)^2}, \tag{A5.67}$$

where the above constant has been determined by the requirement

$$\int_0^\infty \phi_\nu \, d\nu = 1. \tag{A5.68}$$

8. When the radiative force per unit mass $\kappa L/4\pi cr^2$ exceeds the gravitational force per unit mass GM/r^2, we can define an effective gravitational potential

$$\frac{-G_{eff}M}{r} = \frac{1}{r}\left(GM - \frac{\kappa L}{4\pi c}\right),$$
(A5.69)

with r equal to the radius of the star. Since the kinetic energy per unit mass at a large distance from the star must equal the effective gravitational potential at the stellar surface, we conclude that

$$v^2 = \frac{2}{r}\left(\frac{\kappa L}{4\pi c} - GM\right).$$
(A5.70)

15. a. Hydrostatic equilibrium implies that

$$dP = -\rho(z)g(z)\,dz,$$
(A5.71)

where $\rho(z) = \Sigma_i n_i m_i = n(z)m$, with m the average mass of the atmospheric constituents. The ideal gas law is

$$P = nkT.$$
(A5.72)

Therefore, for an isothermal atmosphere with g independent of z we have

$$\frac{dP}{P} = \frac{dn}{n} = \frac{-dz}{H},$$
(A5.73)

where $H = kT/mg$ is the scale height. Integrating Equation (A5.73) gives

$$P(z) = P(z_0)e^{-mg(z-z_0)/kT},$$
(A5.74)

with z_0 the elevation of the surface. Equation (A5.74) predicts that $P(z) \to 0$ as $z \to \infty$.

b. If we substitute $g(z) = g(z_0)(z_0/z)^2$ into Equation (A5.71) and integrate from elevation z_0 to $z > z_0$, we find that

$$P(z) = P(z_0)\exp[-(GMm/kT)(1/z_0 - 1/z)].$$
(A5.75)

Equation (A5.75), which correctly accounts for the z dependence of g, implies that $P(z) \to$ constant as $z \to \infty$.

A5.6 Chapter 6

2. From Equations (6.23) and (6.26) it follows that the reaction rate r satisfies the proportionality

$$r \propto n_1 n_2 \int_0^\infty v(E)\sigma(E)\exp(-E/kT)E^{1/2}\,dE,$$
(A5.76)

where (as discussed in the text above Equation (6.44)) $\sigma(E) = P_l(E)\sigma_n(E)$, with $P_l(E)$

the penetration factor and σ_n the nuclear cross section. In the weak-screening approximation, which is usually relevant in stellar nuclear-burning regions, the Coulomb potential is modified by the small constant negative factor $U_0 = -Z_1 Z_2 e^2/\lambda_0$ given in Equation (6.45). Because U_0 is negative, particles of energy E are as effective in barrier penetration and nuclear interaction as those of energy $E - U_0 > E$. This implies that relation (A5.76) becomes

$$r \propto n_1 n_2 \int_0^\infty E^{1/2} \exp(-E/kT) P_l(E - U_0) \sigma_n(E - U_0)\, dE$$

$$= n_1 n_2 \int_{-U_0}^\infty (E' + U_0)^{1/2} \exp[-(E' + U_0)/kT] P(E') \sigma_n(E')\, dE', \quad \text{(A5.77)}$$

where in the second relation in (A5.77) we have made the variable change $E' = E - U_0$. Because U_0 is small, the lower limit of the integral in Equation (A5.77) can be set equal to zero, and only the change in the exponential factor is appreciable. Therefore, in the weak-screening approximation, we have

$$r \propto n_1 n_2 \exp(Z_1 Z_2 e^2/\lambda_0) \int E^{1/2} \exp(-E/kT) P(E) \sigma_n(E)\, dE, \quad \text{(A5.78)}$$

and Equation (6.46) is proven.

3. An ion sphere of radius R consists of Z uniformly distributed electrons surrounding a nucleus of charge Z. The average number of electrons within a radius $r < R$ of the nucleus is

$$q(r) = \left(\frac{r}{R}\right)^3 Ze. \quad \text{(A5.79)}$$

Therefore, the electrostatic energy caused by the electron–electron interaction becomes

$$U_{ee} = \int_0^R \frac{q(r)}{r}\, dq(r) = \frac{3(Ze)^2}{R^6} \int_0^R r^4\, dr = \frac{3}{5} \frac{(Ze)^2}{R}. \quad \text{(A5.80)}$$

The corresponding electron–nucleus interaction energy is

$$U_{en} = -Ze \int_0^R \frac{dq(r)}{r} = \frac{-3(Ze)^2}{2R}. \quad \text{(A5.81)}$$

Therefore, the total electrostatic energy becomes

$$U_{ee} + U_{en} = \frac{-9}{10} \frac{(Ze)^2}{R} = 17.6 Z^{5/3} \left(\frac{\rho}{\mu_e}\right)^{1/3} \text{eV}, \quad \text{(A5.82)}$$

where ρ and μ_e are the density and electron molecular weight respectively.

Consider two interacting nuclei with charges Z_1 and Z_2. In the strong screening approximation the penetration enhancement factor U_0 discussed in Problem 6.2 is the difference in electrostatic energy between two well-separated ion spheres and ion spheres of charge Z_1 and Z_2 with a common origin. Therefore from Equation (A5.82) we have

$$U_0 = -17.6\left(\frac{\rho}{\mu_e}\right)^{1/3}[(Z_1 + Z_2)^{5/3} - Z_1^{5/3} - Z_2^{5/3}]\,\text{eV}. \tag{A5.83}$$

The strong screening approximation is only applicable at very high densities (see Problem 6.4).

8. a. If the energy width of a compound nuclear state is Γ, then because the line has a Lorentzian shape, its mean lifetime τ is given by the equation

$$\Gamma\tau = \hbar. \tag{A5.84}$$

Equation (A5.84) is similar to the Heisenberg uncertainty principle, which relates uncertainties in position and momentum. In this part of the problem we assume that the spin and angular momentum of the interacting nuclei are zero. Suppose that two colliding nuclei are initially in state 1 and therefore the amplitude of this state at $t = 0$ (see the discussion of time-dependent perturbation theory given in Chapter 4) is $a_1(0) = 1$. The initial (i.e. $t = 0$) amplitude of the unstable compound nuclear state is $a_2(0) = 0$. Using time-dependent perturbation theory and the fact that the mean lifetime of the compound nuclear state is $\tau = \hbar/\Gamma$, we find that

$$i\hbar\dot{a}_2 = -(\Gamma/2)a_2 + H_{21}e^{-i(E_1 - E_2)t/\hbar}. \tag{A5.85}$$

If H_{21} is assumed zero and $a_2(0)$ taken nonzero, then Equation (A5.85) implies that

$$|a_2(t)|^2 = |a_2(0)|^2 e^{-t/\tau}, \tag{A5.86}$$

which implies that τ is the mean lifetime. Equation (A5.85) can be written as

$$i\hbar\frac{d(a_2 e^{\Gamma t/2\hbar})}{dt} = H_{21}e^{-i(E_1 - E_2 + i\Gamma/2)t/\hbar}, \tag{A5.87}$$

and therefore

$$i\hbar a_2 e^{\Gamma t/2\hbar} = \int_0^t H_{21}e^{-i(E_1 - E_2 + i\Gamma/2)t'/\hbar}\,dt'. \tag{A5.88}$$

From Equation (A5.88)

$$|a_2(t)|^2 = \frac{|H_{21}|^2}{(E_1 - E_2)^2 + \Gamma^2/4}. \tag{A5.89}$$

Suppose that the density of initial states of the interacting nuclei is $n_1(E_1)$. From time-dependent perturbation theory it follows that a compound nucleus, once excited, can decay back into initial states with a mean lifetime τ_1 given by

$$\frac{1}{\tau_1} = \frac{\Gamma_1}{\hbar} = \frac{2\pi}{\hbar}|H_{12}|^2 n_1(E_1) = \frac{2\pi}{\hbar}|H_{21}|^2 n_1(E_1). \tag{A5.90}$$

The second equality in Equation (A5.90) follows because $|H_{21}|^2 = |H_{12}|^2$. From Equations (A5.89) and (A5.90) the decay rate back into the initial channel 1 is

$$\frac{|a_2(t)|^2}{\tau_1} = \frac{|H_{21}|^2}{(E_1 - E_2)^2 + \Gamma^2/4} \frac{\Gamma_1}{\hbar}$$

$$= \frac{1}{2\pi\hbar n_1(E_1)} \frac{\Gamma_1(E_1)^2}{(E_1 - E_2)^2 + \Gamma^2/4}. \tag{A5.91}$$

The density of initial states is

$$n_1(E)\,dE = \frac{V}{(2\pi)^3} 4\pi k^2 \frac{dk}{dE}\,dE, \tag{A5.92}$$

where $dE/dk = \hbar^2 k/m = \hbar v$ and m is the reduced mass of the colliding particles. The flux of initial particles times $\sigma(1 \to 2 \to 1)$ must equal the decay rate into channel 1, and therefore Equations (A5.91) and (A5.92) imply that

$$\sigma(1 \to 2 \to 1) = \frac{\pi}{k_1^2} \frac{\Gamma_1^2}{(E_1 - E_2)^2 + \Gamma^2/4}. \tag{A5.93}$$

b. If $l = 0$, then $2J + 1 = (2S_1 + 1)(2S_2 + 1)$, and the expression given in part c of this problem reduces to the result in part a.

c. Suppose the interacting nuclei have spins S_1 and S_2. In addition, if the compound nucleus has total angular momentum J, any of its $2J + 1$ substates can be formed. Therefore, Equation (A5.93) is multiplied by a factor

$$\frac{(2J + 1)}{(2S_1 + 1)(2S_2 + 1)}. \tag{A5.94}$$

A5.9 Chapter 9

3.   The alternating symbol ε_{ijk} is defined as

$$\varepsilon_{ijk} = \begin{cases} +1 & \text{if } i,j,k \text{ are a cyclic permutation of } 1, 2, 3. \\ -1 & \text{if } i,j,k \text{ are an anticyclic permutation of } 1, 2, 3. \\ 0 & \text{if any two (or three) indices are equal.} \end{cases}$$

Summation over similarly labeled indices is understood. The k component of the velocity is $v_k = (\omega \times r)_k = \varepsilon_{klm}\omega_l r_m$, and therefore

$$(\nabla \times \mathbf{v})_i = \varepsilon_{ijk} \frac{\partial}{\partial x_j} v_k = \varepsilon_{ijk} \varepsilon_{klm} \frac{\partial \omega_l r_m}{\partial x_j}$$

$$= \varepsilon_{kij} \varepsilon_{klm} \frac{\partial \omega_l r_m}{\partial x_j} = (\delta_{il}\delta_{jm} - \delta_{im}\delta_{jl}) \frac{\partial \omega_l r_m}{\partial x_j}$$

$$= (\delta_{il}\delta_{jj} - \delta_{ij}\delta_{ji})\omega_l = (3-1)\omega_i = 2\omega_i, \tag{A5.95}$$

where $\delta_{jj} = 3$, because the Kronecker delta function is 1 if its two indices are the same and zero otherwise.

9.   Bessel's differential equation is

$$x^2 \frac{d^2 y}{dx^2} + x \frac{dy}{dx} + (x^2 - n^2)y = 0. \tag{A5.96}$$

The general solution of the above equation is

$$y = C_1 J_n(x) + C_2 Y_n(x). \tag{A5.97}$$

The function $J_n(x)$ has a finite limit as $x \to 0$ and is called a Bessel function of the first kind.

     Substituting $\Phi_{\pm} = \exp(\pm kz) J_0(kR)$ into Laplace's equation, given in the problem, we find that

$$\frac{d^2 J_0(kR)}{dR^2} + \frac{dJ_0(kR)}{R\,dR} + k^2 J_0(kR) = 0. \tag{A5.98}$$

Dividing the above equation by k^2 shows it to be equivalent to Bessel's equation with $n = 0$.

A5.10 Chapter 10

4.   From vector calculus we have

$$\frac{1}{2}\nabla \times (\mathbf{B} \times \mathbf{r}) = \frac{\mathbf{B}}{2}\nabla \cdot \mathbf{r} - (\mathbf{B} \cdot \nabla)\frac{\mathbf{r}}{2} = \mathbf{B} = \nabla \times \mathbf{A}, \tag{A5.99}$$

because $\mathbf{B}$ is uniform and $\nabla \cdot \mathbf{r} = 3$.

5.   The magnetic vector potential A and current density J are related by the equation

$$A(x_1, y_1, z_1) = \frac{1}{c} \int \frac{J(x_2, y_2, z_2)}{|r_1 - r_2|} dV_2. \tag{A5.100}$$

Assume that the current is confined to a thin wire loop. The volume element dV_2 can then be chosen as a short section of wire of length dl_2. Since the current density J is I/S, with I the loop current and S its cross section, we have $J dV_2 = I dl_2$. Therefore, the magnetic vector potential of a short section of wire is

$$dA = \frac{I}{c} \frac{dl_2}{|r_1 - r_2|}. \tag{A5.101}$$

For simplicity we assume that A is measured at some point $(0, y_1, z_1)$ and the current loop is confined to the $z = 0$ plane. Therefore

$$A = \frac{I}{c} \oint \frac{dl_2}{|r_1 - r_2|}, \tag{A5.102}$$

where integration is around the loop. Taking the point $(0, y_1, z_1)$ very distant from the current loop implies that first-order variations in $|r_1 - r_2|$ depend only on the coordinate y_2 and not on x_2. Therefore,

$$|r_1 - r_2| \simeq r_1 - y_2 \sin\theta, \tag{A5.103}$$

where r_1 is the distance from the center of the loop at the origin of the coordinate system to the point $(0, y_1, z_1)$ at which A is measured, θ the angle between r_1 and the z-axis and y_2 the y-coordinate of some point on the current loop. Because we have chosen the point of measurement to be $(0, y_1, z_1)$, contributions to A from integrating dy_2 around the loop cancel, and from Equation (A5.102) A has only the $\hat{x}$ component:

$$A(0, y_1, z_1) = \frac{\hat{x}}{c} \frac{I}{r_1} \oint \left(1 + \frac{y_2 \sin\theta}{r_1}\right) dx_2. \tag{A5.104}$$

The integrations $\oint dx_2 = 0$ and $\oint y_2 \, dx_2 = S$. Therefore, Equation (A5.104) becomes

$$A(0, y_1, z_1) = \frac{IS \sin\theta}{cr_1^2}. \tag{A5.105}$$

Since the magnetic moment is I/c times the area S, we conclude that

$$A = \frac{m \times r}{r^3}. \tag{A5.106}$$

6.   We have

$$\frac{d^2 m_\perp}{dt^2} = m \sin\theta \, \omega^2, \tag{A5.107}$$

with θ the angle between the rotation axis and magnetic moment vector m. Equating the rotational-energy loss rate to the magnetic-dipole radiation gives

$$\frac{d\left(\frac{1}{2} I \omega^2\right)}{dt} = \frac{dE}{dt} = -\frac{2m^2 \sin^2\theta \, \omega^4}{3c^3}. \tag{A5.108}$$

If we substitute $z = \omega_0/\omega$ and

$$K = \frac{2m^2 \sin^2\theta \, \omega_0^2}{3c^3 I} \tag{A5.109}$$

into Equation (A5.108), we obtain

$$\frac{dz^2}{dt} = 2K.$$

The latter equation has the solution

$$t = \frac{1}{2}\frac{z^2}{K} + \text{constant}, \tag{A5.110}$$

with $\omega = \omega_0$ at $t = 0$. The age of the pulsar is measured from the time its spin rate was high. If we set $t = 0$ at the present time, then its age becomes

$$t_{age} = -\frac{1}{2K}. \tag{A5.111}$$

A5.12 Chapter 12

1.   a. We can always express T^{ik} as

$$T^{ik} = \frac{1}{2}(T^{ik} + T^{ki}) + \frac{1}{2}(T^{ik} - T^{ki}), \tag{A5.112}$$

which is the sum of a symmetric and antisymmetric tensor.

   b. Because A_{ik} is an antisymmetric tensor and S^{ik} a symmetric tensor, we have

$$A_{ik}S^{ik} = - A_{ki}S^{ik}. \tag{A5.113}$$

The above indices i and k are dummy indices and therefore can be interchanged to give

$$A_{ik}S^{ik} = - A_{ki}S^{ik} = - A_{ik}S^{ik}, \tag{A5.114}$$

which must be zero because it is the negative of itself. It can readily be shown that if a tensor is symmetric (antisymmetric) in one coordinate system, it is symmetric (antisymmetric) in all coordinate systems.

   c. This follows because T^i_i transforms as

$$T^i_i = \frac{\partial \bar{x}^j}{\partial x^i}\frac{\partial x^i}{\partial \bar{x}^j}\bar{T}^j_j = \bar{T}^j_j. \tag{A5.115}$$

Similarly, it follows that T_{ii} and T^{ii} are not invariant under coordinate transformations.

2.   a. Since the contravariant metric tensor g^{ji} is reciprocal to the covariant tensor g_{kj}, we have

$$g_{kj}g^{ji} = \delta^i_k, \tag{A5.116}$$

and therefore the Kronecker delta is obviously a mixed tensor.

   b. In local Cartesian coordinates, $\Gamma^i_{jk} = 0$. It follows that Γ^i_{jk} is not a tensor because if a tensor is zero in one coordinate system it is zero in any coordinate system.

8.   In terms of the gravitational potential Φ and universal time t, the Newtonian equation of motion is

$$\frac{d^2x^\alpha}{dt^2} = -\frac{\partial\Phi}{\partial x^\alpha} \qquad (\alpha = 1, 2, 3). \tag{A5.117}$$

The four-dimensional form of the above equation is

$$\frac{d^2x^\alpha}{d\lambda^2} + \frac{\partial\Phi}{\partial x^\alpha}\left(\frac{dt}{d\lambda}\right)^2 = 0, \tag{A5.118}$$

$$\frac{d^2t}{d\lambda^2} = 0. \tag{A5.119}$$

From the geodesic equation the only nonzero components of the affine connection are

$$\Gamma^\alpha_{00} = \frac{\partial\Phi}{\partial x^\alpha} \qquad (\alpha = 1, 2, 3), \tag{A5.120}$$

and the only nonzero components of the Riemann tensor become

$$R^\alpha_{0\beta 0} = -R^\alpha_{00\beta} = \frac{\partial\Phi}{\partial x^\alpha\partial x^\beta} \qquad (\alpha = 1, 2, 3). \tag{A5.121}$$

A5.13 Chapter 13

3.   a. From the definition of the enthalpy H and the first law of thermodynamics we obtain

$$dH = dU + P\,dV + V\,dP = T\,dS + V\,dP. \tag{A5.122}$$

b. From the given vector identity the equation of motion becomes

$$\frac{\partial \boldsymbol{v}}{\partial t} + \frac{1}{2}\nabla v^2 + (\nabla \times \boldsymbol{v}) \times \boldsymbol{v} = -\frac{\nabla P}{\rho}. \tag{A5.123}$$

Taking the curl of the above equation gives

$$\frac{\partial \boldsymbol{\omega}}{\partial t} + \nabla \times (\boldsymbol{\omega} \times \boldsymbol{v}) = \frac{1}{\rho^2}\nabla\rho \times \nabla P, \tag{A5.124}$$

with $\nabla \times \boldsymbol{v} = \boldsymbol{\omega}$. For isentropic flow the pressure equals $P(\rho, s) = P(\rho)$, and consequently the right-hand side of Equation (A5.124) vanishes and the vorticity $\boldsymbol{\omega}$ satisfies the equation

$$\frac{\partial \boldsymbol{\omega}}{\partial t} + \nabla \times (\boldsymbol{\omega} \times \boldsymbol{v}) = 0. \tag{A5.125}$$

Using Stokes' theorem, the circulation $\Gamma = \oint \boldsymbol{v} \cdot d\boldsymbol{l}$ round a closed loop becomes

$$\Gamma = \oint \boldsymbol{v} \cdot d\boldsymbol{l} = \int_A \omega \cdot \hat{\boldsymbol{n}} \, dA. \tag{A5.126}$$

As in our discussion of magnetic flux constancy, given in Section 10.1, the time derivative of Γ has an explicit time dependence that depends on the area of the loop. Therefore, we have

$$\frac{d\Gamma}{dt} = \int_A \frac{\partial \omega}{\partial t} \cdot \hat{\boldsymbol{n}} \, dA + \oint \omega \cdot (\boldsymbol{v} \times d\boldsymbol{l}). \tag{A5.127}$$

Using

$$\omega \cdot (\boldsymbol{v} \times d\boldsymbol{l}) = (\omega \times \boldsymbol{v}) \cdot d\boldsymbol{l} \tag{A5.128}$$

and Stokes' theorem, Equation (A5.126) becomes

$$\frac{d\Gamma}{dt} = \int_A \left[\frac{\partial \omega}{\partial t} + \nabla \times (\omega \times \boldsymbol{v}) \right] \cdot \hat{\boldsymbol{n}} \, dA, \tag{A5.129}$$

which by Equation (A5.125) is

$$\frac{d\Gamma}{dt} = 0. \tag{A5.130}$$

The above relation is known as Kelvin's circulation theorem and is mathematically similar to the conservation-of-magnetic-flux theorem.

c. From quantum mechanics the particle current density is

$$J = \frac{1}{2}(\psi p \psi^\dagger + \text{complex conjugate}), \tag{A5.131}$$

with the momentum operator $p = -i\hbar\nabla$. If we assume that the particle wave function $\psi = a \exp(i\gamma(r))$, with a equal to a constant, then the particle current density becomes

$$J = -i\hbar a^2 \nabla \gamma(r). \tag{A5.132}$$

Since each particle contributes a current density

$$J = ma^2 \boldsymbol{u}_s(r), \tag{A5.133}$$

with $\boldsymbol{u}_s(r)$ the velocity field in a single quantum state, Equations (A5.132) and (A5.133) imply that

$$\boldsymbol{u}_s = \frac{-\hbar}{m} \nabla \gamma(r). \tag{A5.134}$$

The curl of a gradient is zero, and consequently

$$\nabla \times \boldsymbol{u}_\text{s} = 0. \tag{A5.135}$$

Integrating around a closed path gives

$$\oint \boldsymbol{u}_\text{s} \cdot \text{d}\boldsymbol{l} = \frac{-\hbar}{m} \oint \nabla\gamma \cdot \text{d}\boldsymbol{l} = \frac{-\hbar}{m} 2\pi n, \tag{A5.136}$$

with $n = 0, \pm 1, \pm 2, \ldots$ The final equality follows because the wave function must be single valued, which means that phase at any point can change by any integral multiply value of 2π.

5.   If we assume that a star has constant density ρ, then the Oppenheimer–Volkoff equation given by Equation (13.17) becomes

$$-\frac{\text{d}P}{\text{d}r} = 4\pi Gr \frac{(\rho c^2 + P(r))(\rho c^2/3 + P(r))}{(1 - 8\pi G\rho c^2 r^2/3)}. \tag{A5.137}$$

Integrating the above equation from the surface where $r = R$ and $P = 0$, we obtain

$$\frac{P(r) + \rho c^2}{3P(r) + \rho c^2} = \left(\frac{1 - 8\pi G\rho R^2/3}{1 - 8\pi G\rho r^2/3}\right)^{1/2}, \tag{A5.138}$$

where $\rho = 3M/4\pi R^3$ for $r < R$.

Substituting for ρ in Equation (A5.138) gives

$$\frac{P(r)}{\rho c^2} = \frac{[1 - (2GM/R)]^{1/2} - [1 - (2GMr^2/R^3)]^{1/2}}{[1 - (2GMr^2/R^3)]^{1/2} - 3[1 - (2GM/R)]^{1/2}}. \tag{A5.139}$$

The pressure given by Equation (A5.139) becomes infinite when its denominator vanishes. This occurs at

$$\Gamma_\infty = 9R^2 - \frac{4R^3}{GM}, \tag{A5.140}$$

and GM/R must satisfy the inequality

$$\frac{GM}{R} < \frac{4}{9}. \tag{A5.141}$$

It should be emphasized that stars with uniform density cannot exist.

10.   The energy density of a degenerate Fermi gas is

$$\rho c^2 = \frac{2}{h^3} \int_0^{p_\text{F}} [(pc^2)^2 + m^2 c^4]^{1/2} 4\pi p^2 \, \text{d}p, \tag{A5.142}$$

and the pressure is equal to

$$P = \frac{2}{h^3} \int_0^{p_\text{F}} \frac{1}{3} p^2 c^2 [(pc)^2 + m^2 c^4]^{-1/2} 4\pi p^2 \, \text{d}p. \tag{A5.143}$$

Therefore, for a relativistic, ideal, degenerate Fermi gas we have

$$v_s^2 = \frac{dP}{d\rho} = \frac{dP/dp_F}{d\rho/dp_F} = \frac{1}{3} \frac{p_F^2 c^4}{[(p_F c)^2 + m^2 c^4]} \simeq \frac{c^2}{3}\left[1 - \frac{m^2 c^4}{E_F^2}\right], \tag{A5.144}$$

and as $\rho \to \infty$, $v_s \to c/\sqrt{3}$.

11. Since $\rho c^2/n$ is the energy density per particle, the first law of thermodynamics for a relativistic gas becomes

$$d(\rho c^2/n) = -P\,d\left(\frac{1}{n}\right) + T\,dS, \tag{A5.145}$$

where n is the particle number density and s the entropy per particle. Equation (A5.145) implies that

$$c^2\,d\rho = (\rho c^2 + P)\frac{dn}{n} + nT\,dS, \tag{A5.146}$$

and therefore for variations under conditions of constant entropy we have

$$\frac{d\ln\rho}{d\ln n} = \frac{\rho c^2 + P}{\rho c^2}. \tag{A5.147}$$

Equation (A5.147) is equivalent to

$$\frac{dn}{c^2\,d\rho} = \frac{n}{\rho c^2 + P}. \tag{A5.148}$$

Since by definition $\Gamma_1 = (n/P)(dP/dn)$ and $v_s^2 = dP/d\rho$, we obtain

$$\frac{v_s^2}{\Gamma_1} = \frac{dP/d\rho}{(n/p)(dP/d\rho)} = \frac{P}{n}\frac{dn}{d\rho} = \frac{Pc^2}{\rho c^2 + P}. \tag{A5.149}$$

References

Chapter 1

Aller, L. H. (1991), *Atoms, Stars and Nebulae*, third edition (Cambridge University Press: Cambridge).

Bessell, M. S. and Stringfellow, G. S. (1993), *Annu. Rev. Astron. Astrophys.*, **31**, 433.

Bethe, H. (1939), cited in Chandrasekhar (1939).

Bethe, H. and Critchfield, C. H. (1938), *Phys. Rev.*, **54**, 248.

Boesgaard, A. M. and Steigman, G. (1985), *Annu. Rev. Astron. Astrophys.*, **23**, 319.

Boss, A. P. (1996), *Physics Today*, Vol. 49, No. 9, p. 32.

Boyer, R. H. and Lindquist, R. W. (1967), *J. Math. Phys.* **8**, 265.

Burbidge, E. M., Burbidge, G. R., Fowler, W. A. and Hoyle, F. (1957), *Rev. Mod. Phys.*, **29**, 547.

Burton, B. (1988), in *Galactic and Extragalactic Radio Astronomy*, 2nd edn, edited by G. L. Verschuur and K. I. Kellermann (Springer: New York).

Cameron, A. G. W. (1955), *Astrophys. J.*, **121**, 144.

Cameron, A. G. W. (1957), *Chalk River Rpt.*, CRL-41.

Chandrasekhar, S. (1931a), *Phil. Mag.*, **11**, 592.

Chandrasekhar, S. (1931b), *Astrophys. J.*, **74**, 81.

Chandrasekhar, S. (1939), *An Introduction to the Study of Stellar Structure* (University of Chicago Press: Chicago).

Clayton, D. D. (1968), *Principles of Stellar Evolution and Nucleosynthesis* (McGraw-Hill: New York).

Colgate, S. A. and White, R. H. (1966), *Astrophys. J.*, **143**, 626.

Eddington, A. S. (1926), *The Internal Constitution of the Stars* (Cambridge University Press: Cambridge).

Fowler, R. H. (1926), *Mon. Not. Roy. Astron. Soc.*, **87**, 114.

Giacconi, R., Gursky, H., Paolini, F. R. and Rossi, B. B. (1962), *Phys. Rev. Lett.*, **9**, 439.

Gilmore, G., King, I., van der Kruit, I. (1990), *The Milky Way as a Galaxy* (University Science Books: Mill Valley).

Gleise, W. (1978), in *The HR Diagram* (IAU Symp. 80), edited by A. G. Davis Philip and P. S. Hayes (Reidel).

Goldstein, H. (1950), *Classical Mechanics* (Addison-Wesley: Cambridge).

Gray, D. F. (1997), *Nature*, **385**, 795.

Hewish, A., Bell, S. J., Pilkington, J. D. H., Scott, P. F. and Collins, R. A. (1968), *Nature*, **217**, 709.

Hulse, R. A. and Taylor, J. H. (1975), *Astrophys. J. Lett.*, **195**, L51.

Jaschek, C. and Jaschek, M. (1987), *The Classification of Stars* (Cambridge University Press: Cambridge).

Johnson, H. L. and Morgan, W. W. (1953), *Astrophys. J.*, **117**, 313.

Kerr, R. P. (1963), *Phys. Rev. Lett.*, **11**, 237.

Laplace, P. S. (1795), *Le Systéme du Monde*, Vol. II (Paris). English edition: *The System of the World* (W. Flint: London, 1809).

Matthews, T. A. and Sandage, A. R. (1963), *Astrophys. J.*, **138**, 30.

Minkowski, R. (1941), *Pub. A. S. P.* **53**, 224.

Motz, L. (1970), *Astrophysics and Stellar Structure* (Ginn: Waltham).

Oppenheimer, J. R. and Snyder, H. (1939), *Phys.*

Rev., **56**, 455.

Oppenheimer, J. R. and Volkoff, G. M. (1939), *Phys. Rev.*, **55**, 374.

Petschek, A. G., editor (1990), *Supernovae* (Springer-Verlag: New York).

Popper, D. M. (1980), *Annu. Rev. Astron. Astrophys.* **18**, 115.

Renzini, A. and Pecci, F. F. (1988), *Annu. Rev. Astron Astrophys.*, **26**, 199.

Rose, W. K. (1973), *Astrophysics* (Holt, Rinehart and Winston: New York).

Rose, W. K. (1989), *Stars, Galaxies and Cosmology* (Kendall/Hunt: Dubuque).

Rowan-Robinson, M. (1985), *The Cosmological Distance Ladder* (Freeman and Co.: New York).

Sandage, A. R. (1957), *Astrophys. J.*, **125**, 435.

Sandage, A. R. (1986), *Annu. Rev. Astron. Astrophys.*, **24**, 421.

Sargent, A. I. and Welch, W. J. (1993), *Annu. Rev. Astron. Astrophys.*, **31**, 297.

Schwarzschild, K. (1916), *Sitzungsber. Dtsch. Akad. Wiss. Berlin, Kl. Math. Phys. Tech.*, p. 189.

Schwarzschild, M. (1957), *Structure and Evolution of the Stars* (Princeton University Press: Princeton).

Shapiro, S. L. and Teukolsky, S. A. (1983), *Black Holes, White Dwarfs and Neutron Stars* (John Wiley: New York).

Unsöld, A. (1969), *The New Cosmos* (Springer-Verlag: New York).

von Weizsäcker, C. F. (1938), *Phys. Zs.*, **39**, 633.

Woolf, N., Schwarzschild, M. and Rose, W. K. (1964), *Astrophys. J.*, **140**, 833.

Yang, J., Turner, M. S., Steigman, G., Schramm, D. N. and Olive, K. A. (1984), *Astrophys. J.*, **281**, 493.

Zinn, R. J. (1985), *Astrophys. J.*, **125**, 435.

Chapter 2

Chandrasekhar, S. (1939), *An Introduction to the Study of Stellar Structure* (University of Chicago Press: Chicago).

Clayton, D. D. (1968), *Principles of Stellar Evolution and Nucleosynthesis* (McGraw-Hill): New York).

Rose, W. K. (1973), *Astrophysics* (Holt, Rinehart and Winston: New York).

Schwarzschild, M. (1957), *Structure and Evolution of the Stars* (Princeton University Press: Princeton).

Chapter 3

Boyd, T. J. M. and Sanderson, J. J. (1969), *Plasma Dynamics* (Barnes and Noble: New York).

Clayton, D. D. (1968), *Principles of Stellar Evolution and Nucleosynthesis* (McGraw-Hill: New York).

Field, G. B., Somerville, W. B. and Dressler, K. (1966), *Annu. Rev. Astron. Astrophys.*, **4**, 210.

Landau, L. D. and Lifschitz, E. M. (1958), *Statistical Physics* (Pergamon Press: London).

Landau, L. D. and Lifschitz, E. M. (1960), *Mechanics* (Pergamon Press: Oxford).

Mayer, J. E. and Mayer, M. G. (1940), *Statistical Mechanics* (John Wiley: New York).

Schiff, L. I. (1955), *Quantum Mechanics*, second edition (McGraw-Hill: New York).

Tolman, R. C. (1938), *The Principles of Statistical Mechanics* (Oxford University Press: Oxford).

Chapter 4

Bransden, B. H. and Joachain, C. J. (1983), *Physics of Atoms and Molecules* (Longman Scientific and Technical: England).

Clayton, D. D. (1968), *Principles of Stellar Evolution and Nucleosynthesis* (McGraw-Hill: New York).

Cox, A. N. and Tabor, T. E. (1976), *Astrophys. J.*, **31**, 271.

Iglesias, C. A., Rogers, F. J. and Wilson, B. G. (1990), *Astrophys. J.*, **360**, 221.

Schiff, L. I. (1955), *Quantum Mechanics*, second edition (McGraw-Hill: New York).

Weissbluth, M. (1978), *Atoms and Molecules* (Academic Press: San Diego).

Chapter 5

Athay, G. (1986), in *Physics of the Sun*, Vol. 2, edited by P. A. Sturrock *et al.* (D. Reidel: Dordrecht).

Jaschek, C. and Jaschek, M. T. (1987), *The Classification of Stars* (Cambridge University Press: Cambridge).

Mihalas, D. (1978), *Stellar Atmospheres*, second edition (Freeman: San Francisco).

Rose, W. K. (1973), *Astrophysics* (Holt, Rinehart and Winston: New York).

Rose, W. K. (1984), *Astrophys. J.*, **285**, 237.

Chapter 6

Arnett, W. D. and Truran, J. W., editors (1985), *Nucleosynthesis* (University of Chicago Press: Chicago).

Arnett, W. D. (1996), *Supernovae and Nucleosynthesis* (Princeton University Press: Princeton).

Casten, R. F. (1990), *Nuclear Structure from a Simple Perspective* (Oxford University Press: Oxford).

Clayton, D. D. (1968), *Principles of Stellar Evolution and Nucleosynthesis* (McGraw-Hill: New York).

Cowan, J. J. and Rose, W. K. (1977), *Astrophys. J.*, **212**, 149.

Cowan, J. J., Thielemann, F.-K. and Truran, J. W. (1991), *Annu. Rev. Astron. Astrophys.*, **29**, 447.

Cowley, C. R. (1995), *An Introduction to Cosmochemistry* (Cambridge University Press: Cambridge).

Jackson, D. F. (1970), *Nuclear Reactions* (Chapman and Hall: London).

Oberhummer, H., editor (1991), *Nuclei in the Cosmos* (University of Chicago Press: Chicago).

Rolfs, L. E. and Rodney, W. S. (1988), *Cauldrons in the Cosmos* (University of Chicago Press: Chicago).

Siemens, P. J. and Jensen, A. S. (1987), *Elements of Nuclei* (Addison-Wesley: Redwood City).

Wong, S. S. M. (1990), *Introductory Nuclear Physics* (Prentice-Hall: Englewood Cliffs).

Chapter 7

Bahcall, J. N. (1989), *Neutrino Astrophysics* (Cambridge University Press: Cambridge).

Cottingham, W. N. and Greenwood, D. A. (1986), *An Introduction to Nuclear Physics* (Cambridge University Press: Cambridge).

Festa, G. C. and Ruderman, M. A. (1969), *Phys. Rev.*, **180**, 1227.

Greiner, W. and Müller, B. (1993), *Gauge Theory of Weak Interactions* (Springer-Verlag: Berlin).

Haxton, W. C. (1995), *Annu. Rev. Astrom. Astrophys.*, **33**, 459.

Kim, C. W. and Pevsner, A. (1993), *Neutrinos in Physics and Astrophysics* (Harwood Academic: Langhorne).

Perkins, D. H. (1987), *Introduction to High Energy Physics*, third edition (Addison-Wesley: Menlo Park).

Renton, P. (1990), *Electroweak Interactions* (Cambridge University Press: Cambridge).

Renton, P. (1990), *Electroweak Interactions* (Cambridge University Press: Cambridge).

Shapiro, S. L. and Teukolsky, S. A. (1983), *Black Holes, White Dwarfs and Neutron Stars* (John Wiley: New York).

Chapter 8

Christy, R. F. (1966), *Astrophys. J.*, **144**, 108.

Gough, D. and Toomre, J. (1991), *Annu. Rev. Astron. Astrophys.*, **29**, 627.

Harper, R. and Rose, W. K. (1970), *Astrophys. J.*, **162**, 963.

Libbrecht, K. G. and Woodward, M. F. (1990). *Nature*, **345**, 779.

Richtmyer, R. D. and Morton, K. W. (1967), *Difference Methods for Initial Value Problems*, second edition (Interscience: New York).

Rose, W. K. (1967), *Astrophys. J.*, **150**, 193.

Rose, W. K. (1973), *Astrophysics* (Holt, Rinehart and Winston: New York).

Sandage, A. R. (1972), *Q.J.R. Astron. Soc.*, **13**, 202.

Schwarzschild, M. and Härm, R. (1965), *Astrophys. J.*, **142**, 855.

Smith, R. L. and Rose, W. K. (1972), *Astrophys. J.*, **176**, 395.

Weigert, A. (1966), *Zs. f. Astrophys.*, **64**, 395.

Winget, D. E. (1988), in *IAU Symp. 123: Advances in Helio- and Astroseismology*, edited by J. Christensen-Dalsgaard and S. Frandsen (Reidel: Dordrecht).

Winget, D. E., Nather, R. E., Clemens, J. C. *et al.* (1994), *Astrophys. J.*, **430**, 839.

Chapter 9

Rose, W. K. (1992), in *Astrophysical Disks*, edited by S. F. Dermott, J. H. Hunter and R. E. Wilson, *Ann. N. Y. Acad. Sci.*, Vol. 675.

Schwarzschild, M. (1957), *Structure and Evolution of the Stars* (Princeton University Press: Princeton).

Shakura, N. I. and Sunyaev, R. A. (1976), *Mon. Not. Roy. Astron. Soc.*, **175**, 613.

Shapiro, S. L. and Teukolsky, S. A. (1983), *Black Holes, White Dwarfs and Neutron Stars* (John Wiley: New York).

Spitzer, L. (1978), *Physical Processes in the Interstellar Medium* (John Wiley: New York).

Symon, K. R. (1960), *Mechanics*, second edition (Addison-Wesley: Reading).

Chapter 10

Boyd, T. J. M. and Sanderson, J. J. (1969), *Plasma Dynamics* (Barnes and Noble: New York).

Ferraro, V. C. A. and Plumpton, D. (1966). *An Introduction to Magneto-Fluid Mechanics*, second edition (Oxford University Press: Oxford).

Hardee, P. and Rose, W. K. (1976), *Astrophys. J.*, **210**, 533.

Krause, F. and Rädler, K.-H. (1980), *Mean-Field Magnetohydrodynamics and Dynamo Theory* (Pergamon: Oxford).

Ladbury, R. (1996), *Physics Today*, Vol. 49, No. 9, p. 17.

Levy, E. H. and Rose, W. K. (1974), *Astrophys. J.*, **193**, 419.

Parker, E. N. (1963), *interplanetary Dynamical Processes* (Interscience: New York).

Parker, E. N. (1979), *Cosmic Magnetic Fields* (Oxford University Press: Oxford).

Rose, W. K. (1987), *Astrophys. J.*, **313**, 146.

Rose, W. K. (1989), *Astrophys. J.*, **337**, 91.

Sturrock, P. A. (1994), *Plasma Physics* (Cambridge University Press: Cambridge).

Taylor, J. H. and Stinebring, D. R. (1986), *Annu. Rev. Astron. Astrophys.*, **24**, 285.

Thompson, D. J. (1994), in *The Second Compton Symposium*, edited by C. E. Fichtel, N. Gehrels and J. P. Norris (AIP: New York).

Weber, E. J. and Davis, L. (1967), *Astrophys. J.*, **148**, 217.

Chapter 11

Arnett, W. D., Bahcall, J. N., Kirshner, R. P. and Woosley, S. E. (1989), *Annu. Rev. Astron. Astrophys.* **27**, 629.

Arnett, W. D. (1996), *Supernovae and Nucleosynthesis* (Princeton University Press: Princeton).

Bahcall, J. N. (1989), *Neutrino Astrophysics* (Cambridge University Press: Cambridge).

Bode, M. F. and Evans, A., editors (1989), *Classical Novae* (John Wiley: New York).

Branch, D. and Tammann, G. A. (1992), *Annu. Rev. Astron. Astrophys.* **30**, 359.

D'Antona, F. and Mazzitelli, I. (1990), *Annu. Rev. Astron. Astrophys.* **28**, 139.

Gehrels, N., Leventhal, M. and MacCallum C. J. (1988), in *Nuclear Spectroscopy of Astrophysical Sources*, AIP Conf. Proc. No. 170, edited by N. Gehrels, U. G. Share (AIP: New York).

Marschall, X. (1994), *Astronomy* **February**, 41.

Oberhummer, H., editor (1991), *Nuclei in the Cosmos* (Springer-Verlag: Berlin).

Petschek, A. G., editor (1990), *Supernovae* (Springer-Verlag: New York).

Rose, W. K. (1968), *Astrophys. J.*, **152**, 245.

Rose, W. K. (1969), *Astrophys. J.*, **155**, 491.

Rose, W. K. and Smith, R. L. (1972), *Astrophys. J.*, **176**, 395.

Shapiro, S. L. and Teukolsky, S. A. (1983), *Black Holes, White Dwarfs and Neutron Stars* (John Wiley: New York).

Shu, F. H. (1992), *The Physics of Astrophysics.*, Vol. II. *Gas Dynamics* (University Science Books: Mill Valley).

van den Bergh (1975), in *Stars and Stellar Systems*, Vol. 5. *Galaxies and the Universe*, edited by A. R. Sandage *et al.*, (University of Chicago Press: Chicago).

Volker, W. (1990), *Annu. Rev. Astron. Astrophys.* **28**, 103.

Weiler, K. W. and Sramek, R. A. (1988), *Annu. Rev. Astron. Astrophys.* **26**, 295.

Witteborn, F. C. *et al.*, (1989). *Astrophys. J. Lett*, **338**, L9.

Chapter 12

Adler, R., Bazin, M. and Schiffer, M. (1965), *Introduction to General Relativity* (McGraw-Hill: New York).

Backer, D. C. and Hellings, R. W. (1986), *Annu. Rev. Astron. Astrophys.* **24**, 537.

Hulse, R. A. and Taylor, J. H. (1975), *Astrophys. J.*, **195**, L51.

Landau, L. D. and Lifschitz, E. M. (1962), *Classical Theory of Fields* (Pergamon Press: Oxford).

Lightman, A. P., Press, W. H., Price, R. H. and Teukolsky, S. A. (1975), *Problem Book in Relativity and Gravitation* (Princeton University Press: Princeton).

Misner, C. W., Thorne, K. S. and Wheeler, J. A. (1973), *Gravitation* (Freeman: San Francisco).

Robertson, H. P. and Noonan, T. W. (1968), *Relativity and Cosmology* (Saunders: Philadelphia).

Schultz, B. F. (1985), *A First Course in General Relativity* (Cambridge University Press: Cambridge).

Shapiro, S. L. and Teukolsky, S. A. (1983), *Black Holes, White Dwarfs and Neutron Stars* (John Wiley: New York).

Straumann, N. (1984), *General Relativity and Relativistic Astrophysics* (Springer-Verlag: Berlin).

Taylor, J. H. and Weissberg, J. M. (1989), *Astrophys. J.*, **345**, 434.

Weinberg, S. (1972), *Gravitation and Cosmology* (John Wiley: New York).

Chapter 13

Harrison, B. K., Thorne, K. S., Wakano, M. and Wheeler, J. A. (1965), *Gravitation Theory and Gravitational Collapse* (University of Chicago Press: Chicago).

Ghosh, P. and Lamb, F. K. (1979), *Astrophys. J.*, **234**, 296.

Gursky, H. and Ruffini, R. (1975), *Neutron Stars, Black Holes and Binary X-ray Stars* (Reidel: Dordrecht).

Joss, P. C. and Rappaport, S. A. (1984), *Annu. Rev. Astron. Astrophys.* **22**, 471.

Lawden, D. F. (1982), *An Introduction to Tensor Calculus, Relativity and Cosmology*, third edition (John Wiley: New York).

Lewin, W. H. G. and Joss, P. C. (1977), *Nature*, **270**, 211.

Lewin, W. H. G., van Paradijs, J. and van den Heuvel, E. P. J., editors (1995), *X-ray Binaries* (Cambridge University Press: Cambridge).

Pringle, J. E. (1981), *Annu. Rev. Astron. Astrophys.* **19**, 137.

Rose, W. K. (1973), *Astrophysics* (Holt, Rinehart and Winston: New York).

Rose, W. K. (1992), in *Astrophysical Disks*, edited by Dermott, S. F., Hunter, J. H. and Wilson, R. E., *Ann. N Y. Acad. Sci.*, Vol. 675.

Rose, W. K. (1995), *Mon. Not. Roy. Astron. Soc.*, **276**, 1191.

Rybicki, G. B. and Lightman, A. P. (1979), *Radiative Processes in Astrophysics* (John Wiley: New York).

Shapiro, S. L. and Teukolsky, S. A. (1983), *Black Holes, White Dwarfs and Neutron Stars* (John Wiley: New York).

Straumann, N. (1984), *General Relativity and Relativistic Astrophysics* (Springer-Verlag: Berlin).

Sunyaev, R. and Titarchuk, L. (1980), *Astron.*

Astrophys. **86**, 121.

van der Klis, M. (1989), *Annu. Rev. Astron. Astrophys.* **27**, 517.

White, N. E., Swank, J. H. and Holt, S. S. (1983), *Astrophys. J.*, **270**, 711.

Appendix 2

Greiner, W. (1990), *Relativistic Quantum Mechanics* (Springer-Verlag: Berlin).

Messiah, A. (1962), *Quantum Mechanics*, Vol. II (North Holland: Amsterdam).

Appendix 3

Andrews, L. C. (1985), *Special Functions for Engineers and Applied Mathematicians* (MacMillan: New York).

Evgrafov, M. A. (1978), *Analytic Functions* (Dover: New York).

Spiegel, M. R. (1974), *Fourier Analysis* (McGraw-Hill: New York).

Thompson, W. J. (1994), *Angular Momentum* (Wiley: New York).

Appendix 4

Chandrasekhar, S. (1957), *Stellar Structure* (Dover: New York).

Stephenson, G. and Radmore, P. M. (1990), *Advanced Mathematical Methods for Engineering and Science Students* (Cambridge University Press: Cambridge).

Index

Printed in the United States
By Bookmasters